Springer Tracts in Modern Physics
Volume 226

Managing Editor: G. Höhler, Karlsruhe

Editors: A. Fujimori, Chiba
J. Kühn, Karlsruhe
Th. Müller, Karlsruhe
F. Steiner, Ulm
J. Trümper, Garching
C. Varma, California
P. Wölfle, Karlsruhe

Available online at
SpringerLink.com

Starting with Volume 165, Springer Tracts in Modern Physics is part of the [SpringerLink] service. For all customers with standing orders for Springer Tracts in Modern Physics we offer the full text in electronic form via [SpringerLink] free of charge. Please contact your librarian who can receive a password for free access to the full articles by registration at:

springerlink.com

If you do not have a standing order you can nevertheless browse online through the table of contents of the volumes and the abstracts of each article and perform a full text search.

There you will also find more information about the series.

Springer Tracts in Modern Physics

Springer Tracts in Modern Physics provides comprehensive and critical reviews of topics of current interest in physics. The following fields are emphasized: elementary particle physics, solid-state physics, complex systems, and fundamental astrophysics.
Suitable reviews of other fields can also be accepted. The editors encourage prospective authors to correspond with them in advance of submitting an article. For reviews of topics belonging to the above mentioned fields, they should address the responsible editor, otherwise the managing editor.
See also springer.com

Managing Editor

Gerhard Höhler

Institut für Theoretische Teilchenphysik
Universität Karlsruhe
Postfach 69 80
76128 Karlsruhe, Germany
Phone: +49 (7 21) 6 08 33 75
Fax: +49 (7 21) 37 07 26
Email: gerhard.hoehler@physik.uni-karlsruhe.de
www-ttp.physik.uni-karlsruhe.de/

Elementary Particle Physics, Editors

Johann H. Kühn

Institut für Theoretische Teilchenphysik
Universität Karlsruhe
Postfach 69 80
76128 Karlsruhe, Germany
Phone: +49 (7 21) 6 08 33 72
Fax: +49 (7 21) 37 07 26
Email: johann.kuehn@physik.uni-karlsruhe.de
www-ttp.physik.uni-karlsruhe.de/~jk

Thomas Müller

Institut für Experimentelle Kernphysik
Fakultät für Physik
Universität Karlsruhe
Postfach 69 80
76128 Karlsruhe, Germany
Phone: +49 (7 21) 6 08 35 24
Fax: +49 (7 21) 6 07 26 21
Email: thomas.muller@physik.uni-karlsruhe.de
www-ekp.physik.uni-karlsruhe.de

Fundamental Astrophysics, Editor

Joachim Trümper

Max-Planck-Institut für Extraterrestrische Physik
Postfach 13 12
85741 Garching, Germany
Phone: +49 (89) 30 00 35 59
Fax: +49 (89) 30 00 33 15
Email: jtrumper@mpe.mpg.de
www.mpe-garching.mpg.de/index.html

Solid-State Physics, Editors

Atsushi Fujimori
Editor for The Pacific Rim

Department of Physics
University of Tokyo
7-3-1 Hongo, Bunkyo-ku
Tokyo 113-0033, Japan
Email: fujimori@wyvern.phys.s.u-tokyo.ac.jp
http://wyvern.phys.s.u-tokyo.ac.jp/welcome_en.html

C. Varma
Editor for The Americas

Department of Physics
University of California
Riverside, CA 92521
Phone: +1 (951) 827-5331
Fax: +1 (951) 827-4529
Email: chandra.varma@ucr.edu
www.physics.ucr.edu

Peter Wölfle

Institut für Theorie der Kondensierten Materie
Universität Karlsruhe
Postfach 69 80
76128 Karlsruhe, Germany
Phone: +49 (7 21) 6 08 35 90
Fax: +49 (7 21) 6 08 77 79
Email: woelfle@tkm.physik.uni-karlsruhe.de
www-tkm.physik.uni-karlsruhe.de

Complex Systems, Editor

Frank Steiner

Institut für Theoretische Physik
Universität Ulm
Albert-Einstein-Allee 11
89069 Ulm, Germany
Phone: +49 (7 31) 5 02 29 10
Fax: +49 (7 31) 5 02 29 24
Email: frank.steiner@uni-ulm.de
www.physik.uni-ulm.de/theo/qc/group.html

F. Jegerlehner

The Anomalous Magnetic Moment of the Muon

Springer

Friedrich Jegerlehner
Humboldt-Universität zu Berlin
Institut für Physik
Theorie der Elementarteilchen
Newtonstr. 15
12489 Berlin, Germany
fjeger@physik.hu-berlin.de

F. Jegerlehner, *The Anomalous Magnetic Moment of the Muon*, STMP 226 (Springer, Berlin Heidelberg 2007), DOI 10.1007/ 978-3-540-72634-0

Library of Congress Control Number: 2007929738

Physics and Astronomy Classification Scheme (PACS):
MUOSN PROPERTIES   14.60.EF,
RADIATIVE CORRECTIONS ELECTROMAGNETIC   13.40.KS,
... ELECTROWEAK   12.15.LK,
PHOTONS INTERACTIONS WITH HADRONS   13.60.−R

ISSN print edition: 0081-3869
ISSN electronic edition: 1615-0430
e-ISBN 978-3-540-72634-0

This work is subject to copyright. All rights are reserved, whether the whole or part of the material is concerned, specifically the rights of translation, reprinting, reuse of illustrations, recitation, broadcasting, reproduction on microfilm or in any other way, and storage in data banks. Duplication of this publication or parts thereof is permitted only under the provisions of the German Copyright Law of September 9, 1965, in its current version, and permission for use must always be obtained from Springer. Violations are liable for prosecution under the German Copyright Law.

Springer is a part of Springer Science+Business Media
springer.com
© Springer-Verlag Berlin Heidelberg 2008

The use of general descriptive names, registered names, trademarks, etc. in this publication does not imply, even in the absence of a specific statement, that such names are exempt from the relevant protective laws and regulations and therefore free for general use.

Cover production: eStudio Calamar S.L., F. Steinen-Broo, Pau/Girona, Spain

The closer you look the more there is to see

Preface

It seems to be a strange enterprise to attempt write a physics book about a single number. It was not my idea to do so, but why not. In mathematics, maybe, one would write a book about π. Certainly, the muon's anomalous magnetic moment is a very special number and today reflects almost the full spectrum of effects incorporated in today's Standard Model (SM) of fundamental interactions, including the electromagnetic, the weak and the strong forces. The muon $g - 2$, how it is also called, is a truly fascinating theme both from an experimental and from a theoretical point of view and it has played a crucial role in the development of QED which finally developed into the SM by successive inclusion of the weak and the strong interactions. The topic has fascinated a large number of particle physicists, last but not least it was always a benchmark for theory as a monitor for effects beyond what was known at the time. As an example, nobody could believe that a muon is just a heavy version of an electron; why should nature repeat itself, it hardly can make sense. The first precise muon $g - 2$ experiment at CERN answered that question: yes the muon is just a heavier replica of the electron! Today we know we have a threefold replica world, there exist three families of leptons, neutrinos, up-quarks and down-quarks, and we know we need them to get in a way for free a tiny breaking at the per mill level of the fundamental symmetry of time-reversal invariance, by a phase in the family-mixing matrix. At least three families must be there to allow for this possibility. This symmetry breaking also know as CP–violation is mandatory for the existence of all normal matter in our universe which clustered into galaxies, stars, planets, and after all allowed life to develop. Actually, this observed matter–antimatter asymmetry, to our present knowledge, cries for additional CP-violating interactions, beyond what is exhibited in the SM. And maybe it is a_μ which already gives us a hint how such a basic problem could find its solution. The muon was the first replica particle found. At the time, the existence of the muon surprised physicists so much that the Nobel laureate Isidor I. Rabi exclaimed, "Who ordered that?". But the muon is special in many other respects and its unique properties allow us to play experiment and theory to the extreme in precision.

One of the key points of the anomalous magnetic moment is its simplicity as an observable. It has a classical static meaning while at the same time it is a highly non-trivial quantity reflecting the quantum structure of nature in many facets. This simplicity goes along with an unambiguous definition and a well-understood quasi-classical behavior in a static perfectly homogeneous magnetic field. At the same time the anomalous magnetic moment is tricky to calculate in particular if one wants to know it precisely. To start with, the problem is the same as for the electron, and how tricky it was one may anticipate if one considers the 20 years it took for the most clever people of the time to go form Dirac's prediction of the gyromagnetic ratio $g = 2$ to the anomalous $g - 2 = \alpha/\pi$ of Schwinger.

Today the single number $a_\mu = (g_\mu - 2)/2$ in fact is an overlay of truly many numbers, in a sense hundreds or thousands (as many as there are Feynman diagrams contributing), of different signs and sizes and only if each of these numbers is calculated with sufficient accuracy the correct answer can be obtained; if one single significant contribution fails to be correct also our single number ceases to have any meaning beyond that wrong digit. So high accuracy is the requirement and challenge.

For the unstable short-lived muon which decays after about 2 micro seconds, for a long time nobody knew how one could measure its anomalous magnetic moment. Only when parity violation was discovered by end of the 1950s one immediately realized how to polarize muons, how to study the motion of the spin in a magnetic field and how to measure the Larmor precession frequency which allows to extract a_μ. The muon $g - 2$ is very special, it is in many respects much more interesting than the electron $g - 2$, and the $g - 2$ of the τ; for example, we are not even able to confirm that $g_\tau \sim 2$ because the τ is by far too short-lived to allow for a measurement of its anomaly with presently available technology. So the muon is a real lucky case as a probe for investigating physics at the frontier of our knowledge. By now, with the advent of the recent muon $g - 2$ experiment, performed at Brookhaven National Laboratory with an unprecedented precision of 0.54 parts per million, the anomalous magnetic moment of the muon is not only one of the most precisely measured quantities in particle physics, but theory and experiment lie apart by three standard deviations, the biggest "discrepancy" among all well measured and understood precision observables at present.

This promises nearby new physics, which future accelerator experiments are certainly going to disentangle. It may indicate that we are at the beginning of a new understanding of fundamental physics beyond or behind the SM. Note, however, that this is a small deviation and usually a five-standard deviation is required to be accepted as a real deviation, i.e. there is a small chance that the gap is a statistical fluctuation only.

One would expect that it is very easy to invent new particles and/or interactions to account for the missing contribution from the theory side. Surprisingly other experimental constraints, in particular the absence of any other real deviation from the SM, make it hard to find a simple explanation.

Most remarkably, in spite of these tensions between different experiments, the minimal supersymmetric extension of the standard model, which promised new physics to be "around the corner", is precisely what could fit. So the presently observed deviation in $g-2$ of the muon feeds hopes that the end of the SM is in sight.

About the book: in view of the fact that there now exist a number of excellent more or less extended reviews, rather than adding another topical report, I tried to write a self-contained book not only about the status of the present knowledge on the anomalous magnetic moment of the muon, but also reminding the reader about its basic context and the role it played in developing the basic theoretical framework of particle theory. After all, the triumph this scientific achievement marks, for both theory and experiment, has its feedback on its roots as it ever had in the past. I hope it makes the book more accessible for non-experts and it is the goal to reach a broader community to learn about this interesting topic without compromising with resect to provide a basic understanding of what it means.

So the books is addressed to graduate students and experimenters interested in deepening some theoretical background and to learn in some detail how it really works. Thus, the book is not primarily addressed to the experts, but nevertheless gives an up-to-date status report on the topic. Knowledge of special relativity and quantum mechanics and a previous encounter with QED are expected.

While the structural background of theory is indispensable for putting into perspective its fundamental aspects, it is in the nature of the theme that numbers and the comparison with the experiment play a key role in this book.

The book is organized as follows: Part I presents a brief history of the subject followed in Chap. 2 by an outline of the concepts of quantum field theory and an introduction into QED, including one-loop renormalization and a calculation of the leading lepton anomaly as well as some tools like the renormalization group, scalar QED for pions and a sketch of QCD. Chapter 3 first discusses the motion of leptons in an external field in the classical limit and then overviews the profile of the physics which comes into play and what is the status for the electron and the muon $g-2$'s. The basic concept and tools for calculating higher-order effects are outlined.

In Part II the contributions to the muon $g-2$ are discussed in detail. Chapter 4 reviews the QED calculations. Chapter 5 is devoted to the hadronic contributions, in particular to the problems of evaluating the leading vacuum polarization contributions from electron–positron annihilation data. Also hadronic light-by-light scattering is critically reviewed. Chapter 6 describes the principle of the experiment in some detail as well as some other background relevant for determining $g_\mu - 2$. The final Chap. 7 gives a detailed comparison of theory with the experiment and discusses possible impact for physics beyond the standard theory and future perspectives.

Acknowledgments and Thanks

It is a pleasure to thank all my friends and colleagues for the many stimulating discussions which contributed to the existence of this book. I am specially grateful to my colleagues at the Humboldt University at Berlin for the kind hospitality and support.

For careful and critical reading of various chapters of the manuscript my special thanks go to Oliver Baer, Beat Jegerlehner, Dominik Stöckinger, Oleg Tarasov and Graziano Venanzoni. I am particularly grateful to Wolfgang Kluge for his invaluable help in preparing the manuscript, for his careful reading of the book, his continuous interest, critical remarks, for many interesting discussions and advice.

I have particularly profited from numerous enlightening discussions with Simon Eidelman, Andreas Nyffeler, Heiri Leutwyler, Jürg Gasser, Gilberto Colangelo, Klaus Jungmann, Klaus Mönig, Achim Stahl, Mikhail Kalmykov, Rainer Sommer and Oleg Tarasov.

Thanks to B. Lee Roberts and members of the E821 collaboration for many helpful discussions over the years and for providing me some of the illustrations. I am grateful also to Keith Olive, Dominik Stöckinger and Sven Heinemeyer for preparing updated plots which are important additions to the book. Many thanks also to Günter Werth and his collaborators who provided the pictures concerning the electron and ion traps and for critical comments.

I have received much stimulation and motivation from my visits to Frascati and I gratefully acknowledge the kind hospitality extended to me by Frascati National Laboratory and the KLOE group.

Much pleasure came with the opportunities of European Commission's Training and Research Networks EURODAFNE and EURIDICE under the guidance of Giulia Pancheri and the TARI Project lead by Wolfgang Kluge which kept me in steady contact with a network of colleagues and young researchers which have been very active in the field and contributed substantially to the progress. In fact results from the Marseille, Lund/Valencia, Bern, Vienna, Karlsruhe/Katowice, Warsaw and Frascati nodes were indispensable for preparing this book. The work was supported in part by EC-Contracts HPRN-CT-2002-00311 (EURIDICE) and RII3-CT-2004-506078 (TARI).

Ultimately, my greatest thanks go to my wife Marianne whose constant encouragement, patience and understanding was essential for the completion of this book.

Wildau/Berlin,   *Friedrich Jegerlehner*
March 2007

Contents

Part I Basic Concepts, Introduction to QED, g − 2 in a Nutshell, General Properties and Tools

1 Introduction 3
  References 17

2 Quantum Field Theory and Quantum Electrodynamics 23
  2.1 Quantum Field Theory Background 23
      2.1.1 Concepts, Conventions and Notation 23
      2.1.2 C, P, T and CPT 30
  2.2 The Origin of Spin 34
  2.3 Quantum Electrodynamics 44
      2.3.1 Perturbation Expansion, Feynman Rules 46
      2.3.2 Transition Matrix–Elements, Particle–Antiparticle Crossing 51
      2.3.3 Cross Sections and Decay Rates 53
  2.4 Regularization and Renormalization 55
      2.4.1 The Structure of the Renormalization Procedure .. 55
      2.4.2 Dimensional Regularization 58
  2.5 Tools for the Evaluation of Feynman Integrals 65
      2.5.1 $\epsilon = 4 - d$ Expansion, $\epsilon \to +0$ 65
      2.5.2 Bogolubov-Schwinger Parametrization 66
      2.5.3 Feynman Parametric Representation 66
      2.5.4 Euclidean Region, Wick–Rotations 67
      2.5.5 The Origin of Analyticity 69
      2.5.6 Scalar One–Loop Integrals 72
      2.5.7 Tensor Integrals 74
  2.6 One–Loop Renormalization 76
      2.6.1 The Photon Propagator and the Photon Self–Energy ... 76
      2.6.2 The Electron Self–Energy 86
      2.6.3 Charge Renormalization 92

             2.6.4 Dyson– and Weinberg–Power-Counting Theorems 100
             2.6.5 The Running Charge and the Renormalization Group .. 102
             2.6.6 Bremsstrahlung and the Bloch-Nordsieck Prescription .. 112
      2.7 Pions in Scalar QED and Vacuum Polarization by Vector
           Mesons 121
      2.8 Note on QCD: The Feynman Rules and the Renormalization
           Group 125
      References 131

3   **Lepton Magnetic Moments: Basics** 135
      3.1 Equation of Motion for a Lepton in an External Field 135
      3.2 Magnetic Moments and Electromagnetic Form Factors 140
             3.2.1 Main Features: An Overview 140
             3.2.2 The Anomalous Magnetic Moment of the Electron 162
             3.2.3 The Anomalous Magnetic Moment of the Muon 166
      3.3 Structure of the Electromagnetic Vertex in the SM 168
      3.4 Dipole Moments in the Non–Relativistic Limit 172
      3.5 Projection Technique 173
      3.6 Properties of the Form Factors 179
      3.7 Dispersion Relations 181
             3.7.1 Dispersion Relations and the Vacuum Polarization 182
      3.8 Dispersive Calculation of Feynman Diagrams 190
      References 198

Part II A Detailed Account of the Theory, Outline of Concepts of the Experiment, Status and Perspectives

4   **Electromagnetic and Weak Radiative Corrections** 207
      4.1 $g-2$ in Quantum Electrodynamics 207
             4.1.1 One–Loop QED Contribution 209
             4.1.2 Two–Loop QED Contribution 209
             4.1.3 Three–Loop QED Contribution 213
             4.1.4 Four–Loop QED Contribution 218
             4.1.5 Five–Loop QED Contribution 222
      4.2 Weak Contributions 224
             4.2.1 Weak One–Loop Effects 228
             4.2.2 Weak Two–Loop Effects 229
      References 259

5   **Hadronic Effects** 263
      5.1 Vacuum Polarization Effects and e^+e^- Data 264
             5.1.1 Integrating the Experimental Data and Estimating
                    the Error 275
             5.1.2 The Cross–Section $e^+e^- \to$ Hadrons 277

| | 5.1.3 $R(s)$ in Perturbative QCD................................282
| | 5.1.4 Non–Perturbative Effects, Operator Product Expansion 286
| 5.2 | Leading Hadronic Contribution to $(g-2)$ of the Muon 289
| | 5.2.1 Addendum I: The Hadronic Contribution
| | to the Running Fine Structure Constant 295
| | 5.2.2 Addendum II: τ Spectral Functions
| | vs. e^+e^- Annihilation Data........................296
| | 5.2.3 Digression: Exercises on the Low Energy Contribution . 298
| 5.3 | Higher Order Contributions................................304
| 5.4 | Hadronic Light–by–Light Scattering 310
| | 5.4.1 Calculating the Hadronic LbL Contribution...........314
| | 5.4.2 Sketch on Hadronic Models 316
| | 5.4.3 Pion–pole Contribution 323
| | 5.4.4 The $\pi^0\gamma\gamma$ Transition Form Factor 325
| | 5.4.5 A Summary of Results 337
| References 340

6 The $g-2$ Experiments 347
- 6.1 Overview on the Principle of the Experiment................347
- 6.2 Particle Dynamics 352
- 6.3 Magnetic Precession for Moving Particles...................355
  - 6.3.1 $g-2$ Experiment and Magic Momentum 358
- 6.4 Theory: Production and Decay of Muons 362
- 6.5 Muon $g-2$ Results..365
- 6.6 Ground State Hyperfine Structure of Muonium...............367
- 6.7 Single Electron Dynamics and the Electron $g-2$ 369
- References 373

7 Comparison Between Theory and Experiment and Future Perspectives 375
- 7.1 Experimental Results Confront Standard Theory 375
- 7.2 New Physics in $g-2$......................................381
  - 7.2.1 Anomalous Couplings 392
  - 7.2.2 Supersymmetry 393
- 7.3 Perspectives for the Future 406
- References 411

Index 421

Part I

Basic Concepts, Introduction to QED, g – 2 in a Nutshell, General Properties and Tools

1

Introduction

The book gives an introduction to the basics of the anomalous magnetic moments of leptons and reviews the current state of our knowledge of the anomalous magnetic moment $(g-2)$ of the muon and related topics. The muon usually is denoted by μ. Recent $g-2$ experiments at Brookhaven National Laboratory (BNL) in the USA have reached the impressive precision of 0.54 parts per million [1]. The anomalous magnetic moment of the muon is now one of the most precisely measured quantities in particle physics and allows us to test relativistic local *Quantum Field Theory* (QFT) in its depth, with unprecedented accuracy. It puts severe limits on deviations from the standard theory of elementary particles and at the same time opens a window to new physics. The book describes the fascinating story of uncovering the fundamental laws of nature to the deepest by an increasingly precise investigation of a single observable. The anomalous magnetic moment of the muon not only encodes all the known but also the as of yet unknown non-Standard-Model physics[1]. The latter, however, is still hidden and is waiting to be discovered on the way to higher precision which allows us to see smaller and smaller effects.

In order to understand what is so special about the muon anomalous magnetic moment we have to look at leptons in general. The muon (μ^-), like the much lighter electron (e^-) or the much heavier tau (τ^-) particle, is one of the 3 known charged leptons: elementary spin 1/2 fermions of electric charge -1 in units of the positron charge e, as free relativistic one particle states described by the Dirac equation. Each of the leptons has its positively charged

---

[1] As a matter of principle, an experimentally determined quantity always includes all effects, known and unknown, existing in the real world. This includes electromagnetic, strong, weak and gravitational interactions, plus whatever effects we might discover in future.

antiparticle, the positron e^+, the μ^+ and the τ^+, respectively, as required by any local relativistic quantum field theory [2][2].

Of course the charged leptons are never really free, they interact electromagnetically with the photon and weakly via the heavy gauge bosons W and Z, as well as very much weaker also with the Higgs. Puzzling enough, the three leptons have identical properties, except for the masses which are given by $m_e = 0.511$ MeV, $m_\mu = 105.658$ MeV and $m_\tau = 1776.99$ MeV, respectively. In reality, the lepton masses differ by orders of magnitude and actually lead to a very different behavior of these particles. As mass and energy are equivalent according to Einstein's relation $E = mc^2$, heavier particles in general decay into lighter particles plus kinetic energy. An immediate consequence of the very different masses are the very different lifetimes of the leptons. Within the *Standard Model* (SM) of elementary particle interactions the electron is stable on time scales of the age of the universe, while the μ has a short lifetime of $\tau_\mu = 2.197 \times 10^{-6}$ seconds and the τ is even more unstable with a lifetime $\tau_\tau = 2.906 \times 10^{-13}$ seconds only. Also, the decay patterns are very different: the μ decays very close to 100% into electrons plus two neutrinos ($e\bar{\nu}_e\nu_\mu$), however, the τ decays to about 65% into hadronic states $\pi^-\nu_\tau, \pi^-\pi^0\nu_\tau, \cdots$ while the main leptonic decay modes only account for 17.36% $\mu^-\bar{\nu}_\mu\nu_\tau$ and 17.85% $e^-\bar{\nu}_e\nu_\tau$, respectively. This has a dramatic impact on the possibility to study these particles experimentally and to measure various properties precisely. The most precisely studied lepton is the electron, but the muon can also be explored with extreme precision. Since the muon, the much heavier partner of the electron, turns out to be much more sensitive to hypothetical physics beyond the SM than the electron itself, the muon is much more suitable as a "crystal ball" which could give us hints about not yet uncovered physics. The reason is that some effects scale with powers of m_ℓ^2, as we will see below. Unfortunately, the τ is so short lived, that corresponding experiments are not possible with present technology.

A direct consequence of the pronounced mass hierarchy is the fundamentally different role the different leptons play in nature. While the stable electrons, besides protons and neutrons, are everywhere in ordinary matter, in atoms, molecules, gases, liquids, metals, other condensed matter states etc., muons seem to be very rare and their role in our world is far from obvious. Nevertheless, even though we may not be aware of it, muons as cosmic ray particles are also part of our everyday life. They are continuously created when highly energetic particles from deep space, mostly protons, collide with atoms from the Earth's upper atmosphere. The initial collisions create pions which then decay into muons. The highly energetic muons travel at nearly the speed of light down through the atmosphere and arrive at ground level at

---

[2]Dirac's theory of electrons, positrons and photons was an early version of what later developed into *Quantum Electrodynamics* (QED), as it is known since around 1950.

a rate of about 1 muon per cm² and minute. The relativistic *time dilatation* thereby is responsible that the muons have time enough to reach the ground. As we will see later the basic mechanisms observed here are the ones made use of in the muon $g - 2$ experiments. Also remember that the muon was discovered in cosmic rays by Anderson & Neddermeyer in 1936 [3], a few years after Anderson [4] had discovered antimatter in form of the positron, a "positively charged electron" as predicted by Dirac, in cosmic rays in 1932.

Besides charge, spin, mass and lifetime, leptons have other very interesting static (classical) electromagnetic and weak properties like the magnetic and electric *dipole moments*. Classically the dipole moments can arise from either electrical *charges* or *currents*. A well known example is the circulating current, due to an orbiting particle with electric charge e and mass m, which exhibits a magnetic dipole moment $\boldsymbol{\mu}_L = \frac{1}{2c} e\, \boldsymbol{r} \times \boldsymbol{v}$ given by

$$\boldsymbol{\mu}_L = \frac{e}{2mc} \boldsymbol{L} \tag{1.1}$$

where $\boldsymbol{L} = m\, \boldsymbol{r} \times \boldsymbol{v}$ is the orbital angular momentum ($\boldsymbol{r}$ position, $\boldsymbol{v}$ velocity). An electrical dipole moment can exist due to relative displacements of the centers of positive and negative electrical charge distributions. Thus both electrical and magnetic properties have their origin in the electrical *charges and their currents*. Magnetic charges are not necessary to obtain magnetic moments. This aspect carries over from the basic asymmetry between electric and magnetic phenomena in Maxwell's equations. While electric charges play the fundamental role of the sources of the electromagnetic fields, elementary magnetic charges, usually called magnetic monopoles, are absent. A long time ago, Dirac [5] observed that the existence of magnetic charges would allow us to naturally explain the quantization of both the electric charge e and the magnetic charge m. They would be related by

$$em = \frac{1}{2} n\hbar c \;, \quad \text{where } n \text{ is an integer.}$$

Apparently, nature does not make use of this possibility and the question of the existence of magnetic monopoles remains a challenge for the future in particle physics.

Whatever the origin of magnetic and electric moments are, they contribute to the electromagnetic interaction Hamiltonian (interaction energy) of the particle with magnetic and electric fields

$$\mathcal{H} = -\boldsymbol{\mu}_m \cdot \boldsymbol{B} - \boldsymbol{d}_e \cdot \boldsymbol{E} \;, \tag{1.2}$$

where $\boldsymbol{B}$ and $\boldsymbol{E}$ are the magnetic and electric field strengths and $\boldsymbol{\mu}_m$ and $\boldsymbol{d}_e$ the magnetic and electric dipole moment operators. Usually, we measure magnetic moments in units of the *Bohr magneton*

$$\mu_0 = e\hbar/2mc \tag{1.3}$$

and the *spin operator*
$$S = \frac{\hbar \boldsymbol{\sigma}}{2} \tag{1.4}$$

is replacing the angular momentum operator $\boldsymbol{L}$. Thus, generalizing the classical form (1.1) of the orbital magnetic moment, one writes (see Sect. 3.1)

$$\boldsymbol{\mu}_m = g\, Q\, \mu_0\, \frac{\boldsymbol{\sigma}}{2}\ , \quad \boldsymbol{d}_e = \eta\, Q\, \mu_0\, \frac{\boldsymbol{\sigma}}{2}\ , \tag{1.5}$$

where σ_i ($i = 1, 2, 3$) are the Pauli spin matrices, Q is the electrical charge in units of e, $Q = -1$ for the leptons $Q = +1$ for the antileptons. The equations are defining the gyromagnetic ratio g (g-factor) and its electric pendant η, respectively, quantities exhibiting important dynamical information about the leptons as we will see later.

The magnetic interaction term gives rise to the well known *Zeeman effect*: atomic spectra show a level splitting

$$\Delta E = \frac{e}{2mc}\,(\boldsymbol{L} + g\boldsymbol{S}) \cdot \boldsymbol{B} = g_J\, \mu_0\, m_j\, B\ .$$

The second form gives the result evaluated in terms of the relevant quantum numbers. m_j is the 3rd component of the total angular momentum $\boldsymbol{J} = \boldsymbol{L} + \boldsymbol{S}$ in units of $\hbar$ and takes values $m_j = -j, -j+1, \cdots, j$ with $j = l \pm \frac{1}{2}$. g_J is Landé's g-factor[3]. If spin is involved one calls it *anomalous Zeeman effect*. The latter obviously is suitable to study the magnetic moment of the electron by investigating atomic spectra in magnetic fields.

The anomalous magnetic moment is an *observable*[4] which can be relatively easily studied experimentally from the motion of the lepton in an external magnetic field. The story started in 1925 soon after Goudsmit and Uhlenbeck [6] had postulated that an electron had an intrinsic angular momentum of $\frac{1}{2}\hbar$, and that associated with this spin angular momentum there is a magnetic dipole moment equal to $e\hbar/2mc$, which is the Bohr magneton μ_0. The

---

[3] The Landé g_J may be calculated based on the *"vector model"* of angular momentum composition:

$$(\boldsymbol{L} + g\boldsymbol{S}) \cdot \boldsymbol{B} = \frac{(\boldsymbol{L} + g\boldsymbol{S}) \cdot \boldsymbol{J}}{J}\, \frac{\boldsymbol{J} \cdot \boldsymbol{B}}{J} = \frac{(\boldsymbol{L} + g\boldsymbol{S}) \cdot (\boldsymbol{L} + \boldsymbol{S})}{J^2}\, J_z B$$
$$= \frac{L^2 + gS^2 + (g+1)\,\boldsymbol{L}\cdot\boldsymbol{S}}{J^2}\, m_j \hbar B = \frac{(g+1)\,J^2 - (g-1)\,L^2 + (g-1)\,S^2}{2J^2}\, m_j \hbar B$$

where we have eliminated $\boldsymbol{L} \cdot \boldsymbol{S}$ using $J^2 = L^2 + S^2 + 2\boldsymbol{L} \cdot \boldsymbol{S}$. Using $J = j(j+1)\,\hbar$ etc. we find

$$g_J = 1 + (g-1)\,\frac{j(j+1) - l(l+1) + s(s+1)}{2j(j+1)}\ .$$

With the Dirac value $g = 2$ we find the usual textbook expression.

[4] A quantity which is more or less directly accessible in an experiment. In general small corrections based on well understood and established theory are necessary for the interpretation of the experimental data.

important question "is $(\mu_m)_e$ precisely equal to μ_0", or "is $g = 1$" in our language, was addressed by Back and Landé in 1925 [7]. Their conclusion, based on a study of numerous experimental investigations on the Zeeman effect, was that the magnetic moment of the electron $(\mu_m)_e$ was consistent with the Goudsmit and Uhlenbeck postulate. In fact, the analysis was not conclusive, as we know, since they did not really determine g. Soon after Pauli had formulated the quantum mechanical treatment of the electron spin in 1927 [8], where g remains a free parameter, Dirac presented his relativistic theory in 1928 [9].

The Dirac theory predicted, unexpectedly, $g = 2$ for a free electron [9], twice the value $g = 1$ known to be associated with orbital angular momentum. After first experimental confirmations of Dirac's prediction $g_e = 2$ for the electron (Kinster and Houston 1934) [10], which strongly supported the Dirac theory, yet within relatively large experimental errors at that time, it took about 20 more years of experimental efforts to establish that the electrons magnetic moment actually exceeds 2 by about 0.12%, the first clear indication of the existence of an "anomalous"[5] contribution

$$a_\ell \equiv \frac{g_\ell - 2}{2}, \quad (\ell = e, \mu, \tau) \tag{1.6}$$

to the magnetic moment [11]. By end of the 1940's the breakthrough in understanding and handling renormalization of QED (Tomonaga, Schwinger, Feynman, and others around 1948 [12]) had made unambiguous predictions of higher order effects possible, and in particular of the leading (one–loop diagram) contribution to the anomalous magnetic moment

$$a_\ell^{\text{QED}(1)} = \frac{\alpha}{2\pi}, \quad (\ell = e, \mu, \tau) \tag{1.7}$$

by Schwinger in 1948 [13] (see Sect. 2.6.3 and Chap. 3). This contribution is due to quantum fluctuations via virtual electron photon interactions and in QED is universal for all leptons. The history of the early period of enthusiasm and worries in the development and first major tests of QED as a renormalizable covariant local quantum field theory is elaborated in great detail in the fascinating book by Schweber [14] (concerning $g-2$ see Chap. 5, in particular).

In 1947 Nafe, Nelson and Rabi [15] reported an anomalous value by about 0.26% in the hyperfine splitting of hydrogen and deuterium, which was quickly confirmed by Nagle et al. [16], and Breit [17] suggested a possible anomaly $g \neq 2$ of the magnetic moment of the electron. Soon after, Kusch and Foley [18], by a study of the hyperfine–structure of atomic spectra in a constant magnetic field, presented the first precision determination of the magnetic moment of the electron $g_e = 2.00238(10)$ in 1948, just before the theoretical result had

---

[5]The anomalous magnetic moment is called anomalous for historic reasons, as a deviation from the classical result. In QED or any QFT higher order effects, so called radiative corrections, are the normal case, which does not make such phenomena less interesting.

been settled. Together with Schwinger's result $a_e^{(2)} = \alpha/(2\pi) \simeq 0.00116$ (which accounts for 99% of the anomaly) this provided one of the first tests of the virtual quantum corrections, usually called radiative corrections, predicted by a relativistic quantum field theory. The discovery of the fine structure of the hydrogen spectrum (Lamb–shift) by Lamb and Retherford [19] and the corresponding calculations by Bethe, Kroll & Lamb and Weisskopf & French [20] was the other triumph of testing the new level of theoretical understanding with precision experiments. These successes had a dramatic impact in establishing quantum field theory as a general framework for the theory of elementary particles and for our understanding of the fundamental interactions. It stimulated the development of QED[6] in particular and the concepts of quantum field theory in general. With the advent of non–Abelian gauge theories, proposed by Yang and Mills (YM) [22] in 1954, and after 't Hooft and Veltman [23] found the missing clues to understanding and handling them on the quantum level, many years later in 1971, the SM [24] (Glashow, Weinberg, Salam 1981/1987) finally emerged as a comprehensive theory of weak, electromagnetic and strong interactions. The strong interactions had emerged as Quantum Chromodynamics (QCD) [25] (Fritzsch, Gell-Mann, Leutwyler 1973), exhibiting the property of Asymptotic Freedom (AF) [26] (Gross, Politzer and Wilczek 1973). All this structure today is crucial for obtaining sufficiently precise predictions for the anomalous magnetic moment of the muon as we will see.

The most important condition for the anomalous magnetic moment to be a useful monitor for testing a theory is its unambiguous predictability within that theory. The predictability crucially depends on the following properties of the theory:

1. it must be a local relativistic quantum field theory and
2. it must be renormalizable.

As a consequence $g - 2$ vanishes at tree level. This means that g cannot be an independently adjustable parameter in any renormalizable QFT, which in turn implies that $g-2$ is a calculable quantity and the predicted value can be confronted with experiments. As we will see $g-2$ can in fact be both predicted as well as experimentally measured with very high accuracy. By confronting precise theoretical predictions with precisely measured experimental data it is possible to subject the theory to very stringent tests and to find its possible limitation.

The particle–antiparticle duality [2], also called crossing or charge conjugation property, which is a basic consequence of any relativistic local QFT, implies in the first place that particles and antiparticles have identical masses and spins. In fact, charge conjugation turned out not to be a universal symmetry of the world of elementary particles. Since, in some sense, an antiparticle

---

[6]Today we understand QED as an Abelian gauge theory. This important structural property was discovered by Weyl [21] in 1929.

is like a particle propagating backwards in time, *charge conjugation* C has to be considered together with *time-reversal* T (time-reflection), which in a relativistic theory has to go together with *parity* P (space-reflection). Besides C, T and P are the two other basic discrete transformation laws in particle physics. A well known fundamental prediction which relates C, P and T is the CPT theorem: the product of the three discrete transformations, taken in any order, is a symmetry of any relativistic QFT. Actually, in contrast to the individual transformations C, P and T, which are symmetries of the electromagnetic– and strong–interactions only, CPT is a universal symmetry and it is this symmetry which guarantees that particles and antiparticles have identical masses as well as equal lifetimes[7]. But also the dipole moments are very interesting quantities for the study of the discrete symmetries mentioned.

To learn about the properties of the dipole moments under such transformations we have to look at the interaction Hamiltonian (1.2). In particular the behavior under parity and time-reversal is of interest. Naively, one would expect that electromagnetic (QED) and strong interactions (QCD) are giving the dominant contributions to the dipole moments. However, both preserve P and T and thus the corresponding contributions to (1.2) must conserve these symmetries as well. A glimpse at (1.5) tells us that both the magnetic and the electric dipole moment are proportional to the spin vector $\boldsymbol{\sigma}$ which transforms as an axial vector. Thus, on the one hand, both $\boldsymbol{\mu}_m$ and $\boldsymbol{d}_e$ are axial vectors. On the other hand, the electromagnetic fields $\boldsymbol{E}$ and $\boldsymbol{B}$ transform as a vector (polar vector) and an axial vector, respectively. An axial vector changes sign under T but not under P, while a vector changes sign under P but not under T. We observe that to the extent that P and/or T are conserved only the magnetic term $-\boldsymbol{\mu}_m \cdot \boldsymbol{B}$ is allowed while an electric dipole term $-\boldsymbol{d}_e \cdot \boldsymbol{E}$ is forbidden and hence we must have $\eta = 0$ in (1.5). Since the weak interactions violate parity maximally, weak contributions cannot be excluded by the parity argument. However, T (by the CPT–theorem equivalent to CP) is also violated by the weak interactions, but only via fermion family mixing in the Yukawa sector of the SM (see below). It turns out that, at least for light particles like the known leptons, effects are much smaller. So electric dipole

---

[7]In some cases particle and antiparticle although of different flavor may have the same conserved quantum numbers and mix. Examples of such mixing phenomena are $K^0 - \bar{K}^0$–oscillations or $B^0 - \bar{B}^0$–oscillations. The time evolution of the neutral kaon system, for example, is described by

$$i\frac{d}{dt}\begin{pmatrix} K^0 \\ \bar{K}^0 \end{pmatrix} = H \begin{pmatrix} K^0 \\ \bar{K}^0 \end{pmatrix}, \quad H \equiv M - \frac{i}{2}\Gamma$$

where M and Γ are Hermitian 2×2 matrices, the mass and the decay matrices. The corresponding eigenvalues are $\lambda_{L,S} = m_{L,S} - \frac{i}{2}\gamma_{L,S}$. CPT invariance in this case requires the diagonal elements of $\mathcal{M}$ to be equal. In fact $|m_{K^0} - m_{\bar{K}^0}| < 4.4 \times 10^{-19}$ GeV (90%CL) provides the best test of CPT, while the mass eigenstates K_L and K_S exhibit a mass difference $\Delta m = m_{K_L} - m_{K_S} = 3.483 \pm 0.006 \times 10^{-12}$ MeV.

moments are suppressed by approximate T invariance at the level of second order weak interactions (for a theoretical review see [27]). In fact experimental bounds tell us that they are very tiny [28][8]

$$|d_e| < 1.6 \times 10^{-27} \, e \cdot \text{cm at } 90\% \text{ C.L.} \tag{1.8}$$

This will also play an important role in the interpretation of the $g-2$ experiments as we will see later. A new dedicated experiment for measuring the muon electric dipole moment in a storage ring is under discussion [29].

As already mentioned, the anomalous magnetic moment of a lepton is a dimensionless quantity, a pure number, which may be computed order by order as a perturbative expansion in the fine structure constant α in QED, and beyond QED, in the SM of elementary particles or extensions of it. As an effective interaction term an anomalous magnetic moment is induced by the interaction of the lepton with photons or other particles. It corresponds to a dimension 5 operator and since a renormalizable theory is constrained to exhibit terms of dimension 4 or less only, such a term must be absent for any fermion in any renormalizable theory at tree level. It is the absence of such a possible *Pauli term* that leads to the prediction $g = 2 + O(\alpha)$. On a formal level it is the requirement of renormalizability which forbids the presence of a Pauli term in the Lagrangian defining the theory (see Sect. 2.4.2).

In 1956 a_e was already well measured by Crane et al. [30] and Berestetskii et al. [31] pointed out that the sensitivity of a_ℓ to short distance physics scales like

$$\frac{\delta a_\ell}{a_\ell} \sim \frac{m_\ell^2}{\Lambda^2} \tag{1.9}$$

where Λ is an UV cut–off characterizing the scale of new physics. It was therefore clear that the anomalous magnetic moment of the muon would be a much better probe for possible deviations from QED. However, parity violation of weak interaction was not yet known at that time and nobody had an idea how to measure a_μ.

As already discussed at the beginning of this introduction, the origin of the vastly different behavior of the three charged leptons is due to the very different masses μ_ℓ, implying completely different lifetimes $\tau_e = \infty$, $\tau_\ell = 1/\Gamma_\ell \propto 1/G_F^2 m_\ell^5$ ($\ell = \mu, \tau$) and vastly different decay patterns. G_F is the Fermi constant, known from weak radioactive decays. In contrast to muons, electrons exist in atoms which opens the possibility to investigate a_e directly via the spectroscopy of atoms in magnetic fields. This possibility does not exist for muons[9]. However, Crane et al. [30] already used a different method to measure a_e. They produced polarized electrons by shooting high–energy electrons on a gold foil. The part of the electron bunch which is scattered at right angles, is partially polarized and trapped in a magnetic field, where

---

[8]The unit $e \cdot$ cm is the dipole moment of an e^+e^-–pair separated by 1 cm. Since $d = \frac{\eta}{2} \frac{e\hbar c}{2mc^2}$, the conversion factor needed is $\hbar c = 1.9733 \cdot 10^{-11}$ MeV cm and $e = 1$.
[9]We discard here the possibility to form and investigate muonic atoms.

spin precession takes place for some time. The bunch is then released from the trap and allowed to strike a second gold foil, which allows to analyze the polarization and to determine a_e. Although this technique is in principle very similar to the one later developed to measure a_μ, it is obvious that in practice handling the muons in a similar way is not possible. One of the main questions was: how is it possible to polarize such short lived particles like muons?

After the proposal of parity violation in weak transitions by Lee and Yang [32] in 1957, it immediately was realized that muons produced in weak decays of the pion ($\pi^+ \to \mu^+ +$ neutrino) should be longitudinally polarized. In addition, the decay positron of the muon ($\mu^+ \to e^+ + 2$ neutrinos) could indicate the muon spin direction. This was confirmed by Garwin, Lederman and Weinrich [33] and Friedman and Telegdi [34][10]. The first of the two papers for the first time determined $g_\mu = 2.00$ within 10% by applying the muon spin precession principle (see Chap. 6). Now the road was free to seriously think about the experimental investigation of a_μ.

It should be mentioned that at that time the nature of the muon was quite a mystery. While today we know that there are three lepton–quark families with identical basic properties except for differences in masses, decay times and decay patterns, at these times it was hard to believe that the muon is just a heavier version of the electron ($\mu - e$–puzzle). For instance, it was expected that the μ exhibited some unknown kind of interaction, not shared by the electron, which was responsible for the much higher mass. So there was plenty of motivation for experimental initiatives to explore a_μ.

The big interest in the muon anomalous magnetic moment was motivated by Berestetskii's argument of dramatically enhanced short distance sensitivity. As we will see later, one of the main features of the anomalous magnetic moment of leptons is that it mediates helicity flip transitions. The *helicity* is the projection of the spin vector onto the momentum vector which defines the direction of motion and the velocity. If the spin is parallel to the direction of motion the particle is right–handed, if it is antiparallel it is called left–handed[11]. For massless particles the helicities would be conserved by the SM interactions and helicity flips would be forbidden. For massive particles helicity flips are allowed and their transition amplitude is proportional to the mass of the particle. Since the transition probability goes with the modulus square of the amplitude, for the lepton's anomalous magnetic moment this implies, generalizing (1.9), that quantum fluctuations due to heavier particles or contributions from higher energy scales are proportional to

---

[10] The latter reference for the first time points out that P and C are violated simultaneously, in fact P is maximally violated while CP is to very good approximation conserved in this decay.

[11] Handedness is used here in a naive sense of the "right–hand rule". Naive because the handedness defined in this way for a massive particle is frame dependent. The proper definition of handedness in a relativistic QFT is in terms of the chirality (see Sect. 2.2). Only for massless particles the two different definitions of handedness coincide.

$$\frac{\delta a_\ell}{a_\ell} \propto \frac{m_\ell^2}{M^2} \qquad (M \gg m_\ell)\,, \tag{1.10}$$

where M may be

- the mass of a heavier SM particle, or
- the mass of a hypothetical heavy state beyond the SM, or
- an energy scale or an ultraviolet cut–off where the SM ceases to be valid.

On the one hand, this means that the heavier the new state or scale the harder it is to see (it decouples as $M \to \infty$). Typically the best sensitivity we have for nearby new physics, which has not yet been discovered by other experiments. On the other hand, the sensitivity to "new physics" grows quadratically with the mass of the lepton, which means that the interesting effects are magnified in a_μ relative to a_e by a factor $(m_\mu/m_e)^2 \sim 4 \times 10^4$. This is what makes the anomalous magnetic moment of the muon a_μ the predestinated "monitor for new physics". By far the best sensitivity we have for a_τ the measurement of which however is beyond present experimental possibilities, because of the very short lifetime of the τ.

The first measurement of the anomalous magnetic moment of the muon was performed at Columbia in 1960 [35] with a result $a_\mu = 0.00122(8)$ at a precision of about 5%. Soon later in 1961, at the CERN cyclotron (1958–1962) the first precision determination became available [36, 37]. Surprisingly, nothing special was observed within the 0.4% level of accuracy of the experiment. It was the first real evidence that the muon was just a heavy electron. In particular this meant that the muon was point–like and no extra short distance effects could be seen. This latter point of course is a matter of accuracy and the challenge to go further was evident.

The idea of a muon storage rings was put forward next. A first one was successfully realized at CERN (1962–1968) [38, 39, 40]. It allowed to measure a_μ for both μ^+ and μ^- at the same machine. Results agreed well within errors and provided a precise verification of the CPT theorem for muons. An accuracy of 270 ppm was reached and an insignificant 1.7 σ (1 σ = 1 Standard Deviation (SD)) deviation from theory was found. Nevertheless the latter triggered a reconsideration of theory. It turned out that in the estimate of the three–loop $O(\alpha^3)$ QED contribution the leptonic light–by–light scattering part (dominated by the electron loop) was missing. Aldins et al. [41] then calculated this and after including it, perfect agreement between theory and experiment was obtained.

One also should keep in mind that the first theoretical successes of QED predictions and the growing precision of the a_e experiments challenged theoreticians to tackle the much more difficult higher order calculations for a_e as well as for a_μ. Soon after Schwinger's result Karplus and Kroll 1949 [42] calculated the two–loop term for a_e. In 1957, shortly after the discovery of parity violation and a first feasibility proof in [33], dedicated experiments to explore a_μ were discussed. This also renewed the interest in the two–loop calculation which was reconsidered, corrected and extended to the muon by Sommerfield [43] and Petermann [44], in the same year. Vacuum polarization insertions with fermion loops with leptons different from the external one were calculated in [45, 46]. About 10 years later with the new generation of $g-2$ experiments at the first muon storage ring at CERN $O(\alpha^3)$ calculations were started by Kinoshita [47], Lautrup and De Rafael [48] and Mignaco and Remiddi [49]. It then took about 30 years until Laporta and Remiddi [50] found a final analytic result in 1996. Many of these calculations would not have been possible without the pioneering *computer algebra* programs, like ASHMEDAI[51], SCHOONSHIP [52, 53] and REDUCE [54]. More recently Vermaseren's FORM [55] package evolved into a standard tool for large scale calculations. Commercial software packages like MACSYMA or the more up-to-date ones MATEMATICA and MAPLE, too, play an important role as advanced tools to solve difficult problems by means of computers. Of course, the dramatic increase of computer performance and the use of more efficient computing algorithms have been crucial for the progress achieved. In particular calculations like the ones needed for $g-2$ had a direct impact on the development of these computer algebra systems.

In an attempt to overcome the systematic difficulties of the first a second *muon storage ring* was built (1969–1976) [56, 57]. The precision of 7 ppm reached was an extraordinary achievement at that time. For the first time the m_μ^2/m_e^2-enhanced hadronic contribution came into play. Again no deviations were found. With the achieved precision the muon $g-2$ remained a benchmark for beyond the SM theory builders ever since. Only 20 years later the BNL experiment E821, again a muon storage ring experiment, was able to set new standards in precision. Now, at the present level of accuracy the complete SM is needed in order to be able to make predictions at the appropriate level of precision. As already mentioned, at present further progress is hampered somehow by difficulties to include properly the non–perturbative strong interaction part. At a certain level of precision *hadronic effects* become important and we are confronted with the question of how to evaluate them reliably. At low energies QCD gets strongly interacting and a perturbative calculation is not possible. Fortunately, analyticity and unitarity allow us to express the leading hadronic vacuum polarization contributions via a dispersion relation (analyticity) in terms of experimental data [58]. The key relation here is the optical theorem (unitarity) which determines the imaginary part of the vacuum polarization amplitude through the total cross section for electron–positron annihilation into hadrons. First estimations were performed in [59, 60, 61] after the discovery of the ρ- and the ω–resonances[12], and in [64], after first e^+e^- cross–section measurements were performed at

---

[12]The ρ is a $\pi\pi$ resonance which was discovered in pion nucleon scattering $\pi^- + p \to \pi^-\pi^0 p$ and $\pi^- + p \to \pi^-\pi^+ n$ [62] in 1961. The neutral ρ^0 is a tall resonance in the $\pi^+\pi^-$ channel which may be directly produced in e^+e^-–annihilation and plays a key role in the evaluation of the hadronic contributions to a_μ^{had}. The ρ contributes about 70% to a_μ^{had} which clearly demonstrates the non-perturbative nature of the

the colliding beam machines VEPP-2 and ACO in Novosibirsk [65] and Orsay [66], respectively. One drawback of this method is that now the precision of the theoretical prediction of a_μ is limited by the accuracy of experimental data. We will say more on this later on.

The success of the CERN muon anomaly experiment and the progress in the consolidation of the SM, together with given possibilities for experimental improvements, were a good motivation for Vernon Hughes and other interested colleagues to push for a new experiment at Brookhaven. There the intense proton beam of the Alternating Gradient Synchrotron (AGS) was available which would allow to increase the statistical accuracy substantially [67]. The main interest was a precise test of the electroweak contribution due to virtual W and Z exchange, which had been calculated immediately after the renormalizability of the SM had been settled in 1972 [68]. An increase in precision by a factor 20 was required for this goal. On the theory side the ongoing discussion motivated, in the early 1980's already, Kinoshita and his collaborators to start the formidable task to calculate the $O(\alpha^4)$ contribution with more than one thousand four–loop diagrams. The direct numerical evaluation was the only promising method to get results within a reasonable time. Early results [69, 70] could be improved continuously [71] and this work is still in progress. Increasing computing power was and still is a crucial factor in this project. Here only a small subset of diagrams are known analytically (see Sect. 4.1 for many more details and a more complete list of references). The size of this contribution is about 6 σ's in terms of the present experimental accuracy and thus mandatory for the interpretation of the experimental result.

The other new aspect, which came into play with the perspectives of a substantially more accurate experiment, concerned the hadronic contributions, which in the early 1980's were known with rather limited accuracy only. Much more accurate e^+e^-–data from experiments at the electron positron storage ring VEPP-2M at Novosibirsk allowed a big step forward in the evaluation of the leading hadronic vacuum polarization effects [70, 72, 73] (see also [74]). A more detailed analysis based on a complete up–to–date collection of data followed about 10 years later [75]. Further improvements were possible thanks to new hadronic cross section measurements by BES II [76] (BEPC ring) at Beijing and by CMD-2 [77] at Novosibirsk. More recently, cross section measurements via the radiative return mechanism by KLOE [78] (DAΦNE ring) at Frascati became available. The new results are in fair agreement with the new CMD-2 and SND data [79, 80]. Attempts to include τ spectral functions via isospin relations will be discussed in Sect. 5.2.2. A radiative return experiment is in progress at BABAR [81] and a new energy scan experiment by CLEO-c [82].

---

hadronic effects. The ω–resonances was discovered as a $\pi^+\pi^0\pi^-$ peak shortly after the ρ in proton–antiproton annihilation $p\bar{p} \to \pi^+\pi^+\pi^0\pi^-\pi^-$ [63].

The physics of the anomalous magnetic moments of leptons has challenged the particle physics community for more than 50 years now and experiments as well as theory in the meantime look rather intricate. For a long time a_e and a_μ provided the most precise tests of QED in particular and of relativistic local QFT as a common framework for elementary particle theory in general.

Of course it was the hunting for deviations from theory and the theorists speculations about "new physics around the corner" which challenged new experiments again and again. The reader may find more details about historical aspects and the experimental developments in the interesting recent review: "The 47 years of muon g-2" by Farley and Semertzidis [83].

Until about 1975 searching for "new physics" via a_μ in fact essentially meant looking for physics beyond QED. As we will see later, also standard model hadronic and weak interaction effect carry the enhancement factor $(m_\mu/m_e)^2$, and this is good news and bad news at the same time. Good news because of the enhanced sensitivity to many details of SM physics like the weak gauge boson contributions, bad news because of the enhanced sensitivity to the hadronic contributions which are very difficult to control and in fact limit our ability to make predictions at the desired precision. This is the reason why quite some fraction of the book will have to deal with these hadronic effects (see Chap. 5).

The pattern of lepton anomalous magnetic moment physics which emerges is the following: a_e is a quantity which is dominated by QED effects up to very high precision, presently at the .66 parts per billion (ppb) level! The sensitivity to hadronic and weak effects as well as the sensitivity to physics beyond the SM is very small. This allows for a very solid and model independent (essentially pure QED) high precision prediction of a_e. The very precise experimental value and the very good control of the theory part in fact allows us to determine the fine structure constant α with the highest accuracy in comparison with other methods (see Sect. 3.2.2). A very precise value for α of course is needed as an input to be able to make precise predictions for other observables like a_μ, for example. While a_e, theory wise, does not attract too much attention, although it requires to push QED calculation to high orders, a_μ is a much more interesting and theoretically challenging object, sensitive to all kinds of effects and thus probing the SM to much deeper level (see Chap. 4). Note that in spite of the fact that a_e has been measured about 829 times more precisely than a_μ the sensitivity of the latter to "new physics" is still about 52 times larger. The experimental accuracy achieved in the past few years at BNL is at the level of 0.54 parts per million (ppm) and better than the accuracy of the theoretical predictions which are still obscured by hadronic uncertainties. A small discrepancy at the 2 to 3 σ level persisted [84, 85, 86] since the first new measurement in 2000 up to the one in 2004 (four independent measurements during this time), the last for the time being (see Chap. 7). Again, the "disagreement" between theory and experiment, suggested by the first BLN measurement, rejuvenated the interest in the subject and entailed a reconsideration of the theory predictions. The most prominent error found this time in previous calculations concerned the problematic hadronic light–by–light scattering contribution which turned out to be in error by a sign [87]. The change improved the agreement between theory and experiment by about 1 σ. Problems with the hadronic e^+e^-–annihilation data used to evaluate the hadronic vacuum polarization contribution led to a similar shift in opposite direction, such that a small discrepancy persists.

Speculations about what kind of effects could be responsible for the deviation will be presented in Sect. 7.2. No real measurement yet exists for a_τ. Bounds are in agreement with SM expectations[13] [88]. Advances in experimental techniques one day could promote a_τ to a new "telescope" which would provide new perspectives in exploring the short distance tail of the unknown real world, we are continuously hunting for. The point is that the relative weights of the different contributions are quite different for the τ in comparison to the μ.

In the meantime activities are expected to go on to improve the impressive level of precision reached by the muon $g - 2$ experiment E821 at BNL. Since the error was still dominated by statistical errors rather than by systematic ones, further progress is possible in any case. But also new ideas to improve on sources of systematic errors play an important role for future projects. Plans for an upgrade of the Brookhaven experiment in USA or a similar project J-PARC in Japan are expected to be able to improve the accuracy by a factor 5 or 10, respectively [89]. For the theory such improvement factors are a real big challenge and require much progress in our understanding of non–perturbative strong interaction effects. In addition, challenging higher order computations have to be pushed further within the SM and beyond. Another important aspect: the large hadron collider LHC at CERN will go into operation soon and will certainly provide important hints about how the SM has to be completed by new physics. Progress in the theory of a_μ will come certainly in conjunction with projects [in the state of realization] to measure hadronic electron–positron annihilation cross–sections with substantially improved accuracy (see Sect. 7.3). These cross sections are an important input for reducing the hadronic vacuum polarization uncertainties which yield the dominating source of error at present. In any case there is good reason to expect also in future interesting promises of physics beyond the SM from this "crystal ball" of particle physicists.

Besides providing a summary of the status of the physics of the anomalous magnetic moment of the muon, the aim of this book is an introduction to the theory of the magnetic moments of leptons also emphasizing the fundamental principles behind our present understanding of elementary particle theory. Many of the basic concepts are discussed in details such that physicists with only some basic knowledge of quantum field theory and particle physics should get the main ideas and learn about the techniques applied to get theoretical

---

[13] Theory predicts $(g_\tau - 2)/2 = 117721(5) \times 10^{-8}$; the experimental limit from the LEP experiments OPAL and L3 is $-0.052 < a_\tau < 0.013$ at 95% CL.

predictions of such high accuracy, and why it is possible to measure anomalous magnetic moments so precisely.

Once thought as a QED test, today the precision measurement of the anomalous magnetic moment of the muon is a test of most aspects of the SM with the electromagnetic, the strong and the weak interaction effects and beyond, maybe supersymmetry is responsible for the observed deviation.

There are many excellent and inspiring introductions and reviews on the subject [90, 91, 92, 93, 94, 95, 96, 97, 98, 99, 100, 101, 102, 103, 104, 106, 107], which were very helpful in writing this book. For a recent rewiew see also [108].

After completion of this work a longer review article appeared [109], which especially reviews the experimental aspects in much more depth than this book. For a recent reanalysis of the light–by–light contribution I refer the reader to [110], which presents the new estimate $a_\mu^{\mathrm{LbL}} = (110 \pm 40) \times 10^{-11}$. Another update is comparing electron, muon and tau anomalous magnetic moments [111].

A last minute update was necessary to include the new result [112] (June 2007) on the universal $O(\alpha^4)$ term, which implies a 7 σ shift in α. Note that with α defined via a_e the change in the universal part of $g-2$ only modifies the bookkeeping but does not affect the final result as $a_e^{\mathrm{uni}} = a_\mu^{\mathrm{uni}}$ and the non-universal part of a_e only accounts for $(a_e^{\mathrm{exp}} - a_e^{\mathrm{uni}})/a_e^{\mathrm{exp}} = 3.8$ parts per billion.

References

1. G. W. Bennett et al. [Muon (g-2) Collaboration], Phys. Rev. Lett. **92** (2004) 161802
2. P. A. M. Dirac, Proc. Roy. Soc. A **126** (1930) 360
3. C. D. Anderson, S. H. Neddermeyer, Phys. Rev. **50** (1936) 263; S. H. Neddermeyer, C. D. Anderson, Phys. Rev. **51** (1937) 884
4. C. D. Anderson, Phys. Rev. **43** (1933) 491.
5. P. A. M. Dirac, Proc. Roy. Soc. A **133** (1931) 60; Phys. Rev. **74** (1948) 817
6. G. E. Uhlenbeck, S. Goudsmit, Naturwissenschaften **13** (1925) 953; Nature **117** (1926) 264
7. E. Back, A. Landé, *Zeemaneffekt und Multiplettstruktur der Spektrallinien*, 1st edn (J. Springer, Berlin 1925), pp. 213; see also: H. A. Bethe, E. E. Salpeter, *Quantum Mechanics of One- and Two-Electron Systems*, Handbuch der Physik, Band XXXV, Atoms I (Springer Verlag, Berlin, 1957)
8. W. Pauli, Zeits. Phys. **43** (1927) 601
9. P. A. M. Dirac, Proc. Roy. Soc. A **117** (1928) 610; A **118** (1928) 351
10. L. E. Kinster, W. V. Houston, Phys. Rev. **45** (1934) 104
11. P. Kusch, Science **123** (1956) 207; Physics Today **19** (1966) 23
12. S. Tomonaga, Riken Iho, Progr. Theor. Phys. **1** (1946) 27; J. Schwinger, Phys. Rev. **74** (1948) 1439; R. P. Feynman, Phys. Rev. **76** (1949) 749; F. Dyson, Phys. Rev. **75** (1949) 486, ibid. 1736
13. J. S. Schwinger, Phys. Rev. **73** (1948) 416

14. S. S. Schweber, *QED and the Men Who Made It: Dyson, Feynman, Schwinger, and Tomonaga*, 1st edn (Princeton University Press, Princeton 1994) pp. 732
15. J. E. Nafe, E. B. Nelson, I. I. Rabi, Phys. Rev. **71** (1947) 914
16. D. E. Nagle, R. S. Julian, J. R. Zacharias, Phys. Rev. **72** (1947) 971
17. G. Breit, Phys. Rev. **72** (1947) 984
18. P. Kusch, H. M. Foley, Phys. Rev. **73** (1948) 421; Phys. Rev. **74** (1948) 250
19. W. E. Lamb Jr, R. C. Retherford, Phys. Rev. **72** (1947) 241
20. H. A. Bethe, Phys. Rev. **72** (1947) 339; N. M. Kroll, W. E. Lamb Jr, Phys. Rev. **75** (1949) 388; V. Weisskopf, J. B. French, Phys. Rev. **75** (1949) 1240
21. H. Weyl, I. Zeits. Phys. **56** (1929) 330
22. C. N. Yang, R. L. Mills, Phys. Rev. **96** (1954) 191
23. G. 't Hooft, Nucl. Phys. B **33** (1971) 173; **35** (1971) 167; G. 't Hooft, M. Veltman, Nucl. Phys. B **50** (1972) 318
24. S. L. Glashow, Nucl. Phys. B **22** (1961) 579; S. Weinberg, Phys. Rev. Lett. **19** (1967) 1264; A. Salam, Weak and electromagnetic interactions. In: *Elementary Particle Theory*, ed by N. Svartholm, (Amquist and Wiksells, Stockholm 1969) pp. 367–377
25. H. Fritzsch, M. Gell-Mann, H. Leutwyler, Phys. Lett. **47B** (1973) 365
26. H. D. Politzer, Phys. Rev. Lett. **30** (1973) 1346; D. Gross, F. Wilczek, Phys. Rev. Lett. **30** (1973) 1343
27. W. Bernreuther, M. Suzuki, Rev. Mod. Phys. **63** (1991) 313 [Erratum-ibid. **64** (1992) 633]
28. B. C. Regan, E. D. Commins, C. J. Schmidt, D. DeMille, Phys. Rev. Lett. **88** (2002) 071805
29. F. J. M. Farley et al., Phys. Rev. Lett. **93** (2004) 052001; M. Aoki et al. [J-PARC Letter of Intent]: *Search for a Permanent Muon Electric Dipole Moment at the $\times 10^{-24} e\cdot$ cm Level*, http://www-ps.kek.jp/jhf-np/LOIlist/pdf/L22.pdf
30. W. H. Luisell, R. W. Pidd, H. R. Crane, Phys. Rev. **91** (1953) 475; ibid. **94** (1954) 7; A. A. Schupp, R. W. Pidd, H. R. Crane, Phys. Rev. **121** (1961) 1; H. R. Crane, Sci. American **218** (1968) 72
31. V. B. Berestetskii, O. N. Krokhin, A. X. Klebnikov, Zh. Eksp. Teor. Fiz. **30** (1956) 788 [Sov. Phys. JETP **3** (1956) 761]; W. S. Cowland, Nucl. Phys. B **8** (1958) 397
32. T. D. Lee, C. N. Yang, Phys. Rev. **104** (1956) 254
33. R. L. Garwin, L. Lederman, M. Weinrich, Phys. Rev. **105** (1957) 1415
34. J. I. Friedman, V .L. Telegdi, Phys. Rev. **105** (1957) 1681
35. R. L. Garwin, D. P. Hutchinson, S. Penman, G. Shapiro, Phys. Rev. **118** (1960) 271
36. G. Charpak, F. J. M. Farley, R. L. Garwin, T. Muller, J. C. Sens, V. L. Telegdi, A. Zichichi, Phys. Rev. Lett. **6** (1961) 128; G. Charpak, F. J. M. Farley, R. L. Garwin, T. Muller, J. C. Sens, A. Zichichi, Nuovo Cimento **22** (1961) 1043; Phys. Lett. **1B** (1962) 16
37. G. Charpak, F. J. M. Farley, R. L. Garwin, T. Muller, J. C. Sens, A. Zichichi, Nuovo Cimento **37** (1965) 1241
38. F. J. M. Farley, J. Bailey, R. C. A. Brown, M. Giesch, H. Jöstlein, S. van der Meer, E. Picasso, M. Tannenbaum, Nuovo Cimento **45** (1966) 281
39. J. Bailey et al., Phys. Lett. B **28** (1968) 287

40. J. Bailey, W. Bartl, G. von Bochmann, R. C. A. Brown, F. J. M. Farley, M. Giesch, H. Jöstlein, S. van der Meer, E. Picasso, R. W. Williams, Nuovo Cimento A **9** (1972) 369
41. J. Aldins, T. Kinoshita, S. J. Brodsky, A. J. Dufner, Phys. Rev. Lett. **23** (1969) 441; Phys. Rev. D **1** (1970) 2378
42. R. Karplus, N. M. Kroll, Phys. Rep. C **77** (1950) 536
43. C. M. Sommerfield, Phys. Rev. **107** (1957) 328; Ann. Phys. (N.Y.) **5** (1958) 26
44. A. Petermann, Helv. Phys. Acta **30** (1957) 407; Nucl. Phys. **5** (1958) 677
45. H. Suura, E. Wichmann, Phys. Rev. **105** (1957) 1930; A. Petermann, Phys. Rev. **105** (1957) 1931
46. H. H. Elend, Phys. Lett. **20** (1966) 682; Erratum-ibid. **21** (1966) 720
47. T. Kinoshita, Nuovo Cim. B **51** (1967) 140
48. B. E. Lautrup, E. De Rafael, Phys. Rev. **174** (1968) 1835
49. J. A. Mignaco, E. Remiddi, Nuovo Cim. A **60** (1969) 519
50. S. Laporta, E. Remiddi, Phys. Lett. B **379** (1996) 283
51. M. I. Levine, *ASHMEDAI*. Comput. Phys. I (1967) 454; R. C. Perisho, *ASHMEDAI User's Guide*, U.S. AEC Report No. COO-3066-44, 1975
52. M. J. G. Veltman, *SchoonShip*, CERN Report 1967
53. H. Strubbe, Comput. Phys. Commun. **8** (1974) 1; Comput. Phys. Commun. **18** (1979) 1
54. A. C. Hearn, *REDUCE User's Manual*, Stanford University, Report No. ITP-292, 1967, rev. 1968; A. C. Hearn, *REDUCE 2 User's Manual*. Stanford Artificial Intelligence Project Memo AIM-133, 1970 (unpublished)
55. J. A. M. Vermaseren, *Symbolic manipulation with FORM*, Amsterdam, Computer Algebra Nederland, 1991.
56. J. Bailey et al. [CERN Muon Storage Ring Collaboration], Phys. Lett. B **67** (1977) 225 [Phys. Lett. B **68** (1977) 191]
57. J. Bailey et al. [CERN-Mainz-Daresbury Collaboration], Nucl. Phys. B **150** (1979) 1
58. N. Cabbibo, R. Gatto, Phys. Rev. Lett. **4** (1960) 313, Phys. Rev. **124** (1961) 1577
59. C. Bouchiat, L. Michel, J. Phys. Radium **22** (1961) 121
60. L. Durand, III., Phys. Rev. **128** (1962) 441; Erratum-ibid. **129** (1963) 2835
61. T. Kinoshita, R. J. Oakes, Phys. Lett. **25**B (1967) 143
62. A. R. Erwin, R. March, W. D. Walker, E. West, Phys. Rev. Lett. **6** (1961) 628; D. Stonehill et al., Phys. Rev. Lett. **6** (1961) 624; E. Pickup, D. K. Robinson, E. O. Salant, Phys. Rev. Lett. **7** (1961) 192; D. McLeod, S. Richert, A. Silverman, Phys. Rev. Lett. **7** (1961) 383; D. D. Carmony, R. T. Van de Walle, Phys. Rev. Lett. **8** (1962) 73; J. Button, G. R. Kalbfleisch, G. R. Lynch, B. C. Maglic, A. H. Rosenfeld, M. L. Stevenson, Phys. Rev. **126** (1962) 1858
63. B. C. Maglic, L. W. Alvarez, A. H. Rosenfeld, M. L. Stevenson, Phys. Rev. Lett. **7** (1961) 178; N. H. Xuong, G. R. Lynch, Phys. Rev. Lett. **7** (1961) 327; A. Pevsner et al., Phys. Rev. Lett. **7** (1961) 421; M. L. Stevenson, L. W. Alvarez, B. C. Maglic, A. H. Rosenfeld, Phys. Rev. **125** (1962) 687
64. M. Gourdin, E. De Rafael, Nucl. Phys. B **10** (1969) 667
65. V. L. Auslander, G. I. Budker, Ju. N. Pestov, A. V. Sidorov, A. N. Skrinsky. A. G. Kbabakhpashev, Phys. Lett. B **25** (1967) 433
66. J. E. Augustin et al., Phys. Lett. B **28** (1969), 503, 508, 513, 517

67. C. Heisey et al., *A new precision measurement of the muon g-2 value at the level of 6.35 ppm*, Brookhaven AGS Proposal 821 (1985), revised (1986). Design Report for AGS 821 (1989)
68. R. Jackiw, S. Weinberg, Phys. Rev. D **5** (1972) 2396; I. Bars, M. Yoshimura, Phys. Rev. D **6** (1972) 374; G. Altarelli, N. Cabibbo, L. Maiani, Phys. Lett. B **40** (1972) 415; W. A. Bardeen, R. Gastmans, B. Lautrup, Nucl. Phys. B **46** (1972) 319; K. Fujikawa, B. W. Lee, A. I. Sanda, Phys. Rev. D **6** (1972) 2923
69. T. Kinoshita, W. B. Lindquist, Phys. Rev. Lett. **47** (1981) 1573
70. T. Kinoshita, B. Nizic, Y. Okamoto, Phys. Rev. Lett. **52** (1984) 717; Phys. Rev. D **31** (1985) 2108
71. T. Kinoshita, M. Nio, Phys. Rev. Lett. **90** (2003) 021803; Phys. Rev. D **70** (2004) 113001
72. L. M. Barkov et al., Nucl. Phys. B **256** (1985) 365
73. J. A. Casas, C. Lopez, F. J. Ynduráin, Phys. Rev. D **32** (1985) 736
74. F. Jegerlehner, Z. Phys. C **32** (1986) 195
75. S. Eidelman, F. Jegerlehner, Z. Phys. C **67** (1995) 585
76. J. Z. Bai et al. [BES Collaboration], Phys. Rev. Lett. **84** (2000) 594; Phys. Rev. Lett. **88** (2002) 101802
77. R. R. Akhmetshin et al. [CMD-2 Collaboration], Phys. Lett. B **578** (2004) 285
78. A. Aloisio et al. [KLOE Collaboration], Phys. Lett. B **606** (2005) 12
79. V. M. Aulchenko et al. [CMD-2 Collaboration], JETP Lett. **82** (2005) 743 [Pisma Zh. Eksp. Teor. Fiz. **82** (2005) 841]; R. R. Akhmetshin et al., JETP Lett. **84** (2006) 413 [Pisma Zh. Eksp. Teor. Fiz. **84** (2006) 491]; hep-ex/0610021
80. M. N. Achasov et al. [SND Collaboration], J. Exp. Theor. Phys. **103** (2006) 380 [Zh. Eksp. Teor. Fiz. **130** (2006) 437]
81. B. Aubert et al. [BABAR Collaboration], Phys. Rev. D **70** (2004) 072004; **71** (2005) 052001; **73** (2006) 012005; **73** (2006) 052003
82. S. Dytman, Nucl. Phys. B (Proc. Suppl.) **131** (2004) 213
83. F. J. M. Farley, Y. K. Semertzidis, Prog. Part. Nucl. Phys. **52** (2004) 1
84. H. N. Brown et al. [Muon (g-2) Collaboration], Phys. Rev. D **62** (2000) 091101
85. H. N. Brown et al. [Muon (g-2) Collaboration], Phys. Rev. Lett. **86** (2001) 2227
86. G. W. Bennett et al. [Muon (g-2) Collaboration], Phys. Rev. Lett. **89** (2002) 101804 [Erratum-ibid. **89** (2002) 129903]
87. M. Knecht, A. Nyffeler, Phys. Rev. D **65** (2002) 073034; A. Nyffeler, hep-ph/0210347 (and references therein)
88. K. Ackerstaff et al. [OPAL Collab.], Phys. Lett. B **431** (1998) 188; M. Acciarri et al. [L3 Collab.], Phys. Lett. B **434** (1998) 169; W. Lohmann, Nucl. Phys. B (Proc. Suppl.) **144** (2005) 122
89. B. L. Roberts Nucl. Phys. B (Proc. Suppl.) **131** (2004) 157; R. M. Carey et al., Proposal of the BNL Experiment E969, 2004 (www.bnl.gov/henp/docs/pac0904/P969.pdf); J-PARC Letter of Intent L17, B. L. Roberts contact person
90. J. Bailey, E. Picasso, Progr. Nucl. Phys. **12** (1970) 43
91. B. E. Lautrup, A. Peterman, E. de Rafael, Phys. Reports 3C (1972) 193
92. F. Combley, E. Picasso, Phys. Reports **14C** (1974) 1
93. F. J. M. Farley, Contemp. Phys. **16** (1975) 413
94. T. Kinoshita, *Quantum Electrodynamics*, 1st edn (World Scientific, Singapore 1990) pp. 997, and contributions therein

95. F. J. M. Farley, E. Picasso, In: *Quantum Electrodynamics*, ed T. Kinoshita (World Scientific, Singapore 1990) pp. 479–559
96. T. Kinoshita, W. J. Marciano, In: *Quantum Electrodynamics*, ed. T. Kinoshita (World Scientific, Singapore 1990) pp. 419–478
97. V. W. Hughes, T. Kinoshita, Rev. Mod. Phys. **71** (1999) S133
98. A. Czarnecki, W. J. Marciano, Nucl. Phys. B (Proc. Suppl.) **76** (1999) 245
99. A. Czarnecki, W. J. Marciano, Phys. Rev. D **64** (2001) 013014
100. V. W. Hughes, *The anomalous magnetic moment of the muon*. In: Intern. School of Subnuclear Physics: 39th Course: New Fields and Strings in Subnuclear Physics, Erice, Italy, 29 Aug - 7 Sep 2001; Int. J. Mod. Phys. A **18S1** (2003) 215
101. E. de Rafael, *The muon g-2 revisited*. In: XVI Les Rencontres de Physique de la Vallee d'Aoste: Results and Perspectives in Particle Physics, La Thuile, Aosta Valley, Italy, 3–9 Mar 2002; hep-ph/0208251
102. A. Nyffeler, Acta Phys. Polon. B **34** (2003) 5197; Nucl. Phys. B (Proc. Suppl.) **131** (2004) 162; hep-ph/0305135
103. M. Knecht, *The anomalous magnetic moment of the muon: A theoretical introduction*, In: 41st International University School of Theoretical Physics: Flavor Physics (IUTP 41), Schladming, Styria, Austria, 22–28 Feb 2003; hep-ph/0307239
104. M. Passera, J. Phys. G **31** (2005) R75
105. K. P. Jungmann, *Precision Measurements at the Frontiers of Standard Theory: The Magnetic Anomaly of Leptons*, DPG Frühjahrstagung, Berlin, 4–9 March 2005
106. P. J. Mohr, B. N. Taylor, Rev. Mod. Phys. **72** (2000) 351; Rev. Mod. Phys. **77** (2005) 1
107. T. Kinoshita, Nucl. Phys. B (Proc. Suppl.) **144** (2005) 206
108. K. Melnikov, A. Vainshtein, *Theory of the muon anomalous magnetic moment*, (Springer, Berlin, 2006) 176 p
109. J. P. Miller, E. de Rafael, B. L. Roberts, Rept. Prog. Phys. **70** (2007) 795
110. J. Bijnens, J. Prades, Mod. Phys. Lett. A **22** (2007) 767
111. M. Passera, Nucl. Phys. Proc. Suppl. **169** (2007) 213
112. T. Aoyama, M. Hayakawa, T. Kinoshita, M. Nio, arXiv:0706.3496 [hep-ph]

2

Quantum Field Theory and Quantum Electrodynamics

One of the main reasons why quantities like the anomalous magnetic moment of the muon attract so much attention is their prominent role in basic tests of QFT in general and of Quantum Electrodynamics (QED) and the Standard Model (SM) in particular. QED and the SM provide a truly basic framework for the properties of elementary particles and allow to make unambiguous theoretical predictions which may be confronted with clean experiments which allow to control systematic errors with amazing precision. In order to set up notation we first summarize some basic concepts. The reader familiar with QED, its renormalization and leading order radiative corrections may skip this introductory section, which is a modernized version of material covered by classical textbooks [1, 2]. Since magnetic moments of elementary particles are intimately related to the spin the latter plays a key role for this book. In a second section, therefore, we will have a closer look at how the concept of spin comes into play in quantum field theory.

2.1 Quantum Field Theory Background

2.1.1 Concepts, Conventions and Notation

We briefly sketch some basic concepts and fix the notation. A relativistic quantum field theory (QFT), which combines special relativity with quantum mechanics [3], is defined on the configuration space of space–time events described by points (**contravariant** vector)

$$x^\mu = (x^0, x^1, x^2, x^3) = (x^0, \boldsymbol{x}) \,;\, \mathrm{x}^0 = \mathrm{t}\,(= \mathrm{time})$$

in Minkowski space with metric

$$g_{\mu\nu} = g^{\mu\nu} = \begin{pmatrix} 1 & 0 & 0 & 0 \\ 0 & -1 & 0 & 0 \\ 0 & 0 & -1 & 0 \\ 0 & 0 & 0 & -1 \end{pmatrix}.$$

F. Jegerlehner: *Quantum Field Theory and Quantum Electrodynamics*, STMP 226, 23–133 (2008)
DOI 10.1007/978-3-540-72634-0_2 © Springer-Verlag Berlin Heidelberg 2008

The metric defines a scalar product[1]

$$x \cdot y = x^0 y^0 - \boldsymbol{x} \cdot \boldsymbol{y} = g_{\mu\nu} x^\mu y^\nu = x^\mu x_\mu$$

invariant under Lorentz transformations, which include

1. rotations
2. special Lorentz transformations (boosts)

The set of linear transformations (Λ, a)

$$x^\mu \to x^{\mu'} = \Lambda^\mu{}_\nu x^\nu + a^\mu \tag{2.1}$$

which leave invariant the **distance**

$$(x - y)^2 = g_{\mu\nu}(x^\mu - y^\mu)(x^\nu - y^\nu) \tag{2.2}$$

between two events x and y form the **Poincaré group** $\mathcal{P}$. $\mathcal{P}$ includes the Lorentz transformations and the translations in time and space.

Besides the Poincaré invariance, also space reflections (called parity) P and time reversal T, defined by

$$Px = P(x^0, \boldsymbol{x}) = (x^0, -\boldsymbol{x}), \quad Tx = T(x^0, \boldsymbol{x}) = (-x^0, \boldsymbol{x}), \tag{2.3}$$

play an important role. They are symmetries of the electromagnetic (QED) and the strong interactions (QCD) but are violated by weak interactions. The proper orthochronous transformations $\mathcal{P}_+^\uparrow$ do not include P and T, which requires the constraints $\det \Lambda = 1$ and $\Lambda^0{}_0 \geq 0$.

Finally, we will need the totally antisymmetric pseudo–tensor

$$\varepsilon^{\mu\nu\rho\sigma} = \begin{cases} +1 & (\mu\nu\rho\sigma) \text{ even permutation of } (0123) \\ -1 & (\mu\nu\rho\sigma) \text{ odd permutation of } (0123) \\ 0 & \text{otherwise} \end{cases},$$

which besides $g^{\mu\nu}$ is the second numerically Lorentz–invariant (L–invariant) tensor.

In QFT relativistic particles are described by quantum mechanical states[2], like $|\ell^-(\boldsymbol{p}, r)\rangle$ for a lepton ℓ^- of momentum $\boldsymbol{p}$ and 3rd component of spin r [4]

---

[1] As usual we adopt the summation convention: Repeated indices are summed over unless stated otherwise. For Lorentz indices $\mu, \cdots = 0, 1, 2, 3$ summation only makes sense (i.e. respects L–invariance) between upper (contravariant) and lower (covariant) indices and is called **contraction**.

[2] A relativistic quantum mechanical system is described by a state vector $|\psi\rangle \in \mathcal{H}$ in Hilbert space, which transforms in a specific way under $\mathcal{P}_+^\uparrow$. We denote by $|\psi'\rangle$ the state transformed by $(\Lambda, a) \in \mathcal{P}_+^\uparrow$. Since the system is required to be invariant, transition probabilities must be conserved

$$|\langle \phi' | \psi' \rangle|^2 = |\langle \phi | \psi \rangle|^2. \tag{2.4}$$

(Wigner states). Spin will be considered in more detail in the next section. These states carry L–invariant mass $p^2 = m^2$ and spin s, and may be obtained by applying corresponding *creation operators* $a^+(\boldsymbol{p}, r)$ to the ground state $|0\rangle$, called vacuum:

$$|\boldsymbol{p}, r\rangle = a^+(\boldsymbol{p}, r) |0\rangle \,. \tag{2.7}$$

The energy of the particle is $p^0 = \omega_p = \sqrt{\boldsymbol{p}^2 + m^2}$. The hermitian adjoints of the creation operators, the *annihilation operators* $a(\boldsymbol{p}, r) \doteq (a^+(\boldsymbol{p}, r))^+$, annihilate a state of momentum $\boldsymbol{p}$ and 3rd component of spin r,

$$a(\boldsymbol{p}, r)|\boldsymbol{p}', r'\rangle = (2\pi)^3 \, 2\omega_p \, \delta^{(3)}(\boldsymbol{p} - \boldsymbol{p}') \, \delta_{rr'} \, |0\rangle$$

and since the vacuum is empty, in particular, they annihilate the vacuum

$$a(\boldsymbol{p}, r) |0\rangle = 0 \,. \tag{2.8}$$

The creation and annihilation operators for leptons (spin 1/2 fermions), a and a^+, and the corresponding operators b and b^+ for the antileptons, satisfy the canonical *anticommutation relations* (Fermi statistics)

Therefore, there must exist a unitary operator $U(\Lambda, a)$ such that

$$|\psi\rangle \rightarrow |\psi'\rangle = U(\Lambda, a) |\psi\rangle \in \mathcal{H}$$

and $U(\Lambda, a)$ must satisfy the group law:

$$U(\Lambda_2, a_2) \, U(\Lambda_1, a_1) = \omega U(\Lambda_2 \Lambda_1, \Lambda_2 a_1 + a_2) \,.$$

This means that $U(\Lambda, a)$ is a **representation up to a phase** ω (ray representation) of $\mathcal{P}_+^\uparrow$. Without loss of generality one can choose $\omega = \pm 1$ (Wigner 1939).

The generators of $\mathcal{P}_+^\uparrow$ are the relativistic energy–momentum operator P_μ

$$U(a) \equiv U(1, a) = e^{i \, P_\mu a^\mu} = 1 + i \, P_\mu a^\mu + \ldots \tag{2.5}$$

and the relativistic angular momentum operator $M_{\mu\nu}$

$$U(\Lambda) \equiv U(\Lambda, 0) = e^{\frac{i}{2} \omega^{\mu\nu} M_{\mu\nu}} = 1 + \frac{i}{2} \omega^{\mu\nu} M_{\mu\nu} + \ldots \tag{2.6}$$

Since for infinitesimal transformations we have

$$\Lambda^\mu{}_\nu = \delta^\mu{}_\nu + \omega^\mu{}_\nu \quad \text{with} \quad \omega_{\mu\nu} = -\omega_{\nu\mu},$$

the generators $M_{\mu\nu}$ are antisymmetric:

$$M_{\mu\nu} = -M_{\nu\mu} \,.$$

By unitarity of $U(\Lambda, a)$, P_μ and $M_{\mu\nu}$ are Hermitian operators on the Hilbert space. The generator of the time translations P_0 represents the Hamiltonian H of the system ($H \equiv P_0$) and determines the **time evolution**. If $|\psi\rangle = |\psi\rangle_H$ is a Heisenberg state, which coincides with the Schrödinger state $|\psi(0)\rangle_S$ at $t = 0$, then $|\psi(t)\rangle_S = e^{-iHt} |\psi(0)\rangle_S$ represents the state of the system at time t.

$$\{a(\boldsymbol{p},r), a^+(\boldsymbol{p}',r')\} = \{b(\boldsymbol{p},r), b^+(\boldsymbol{p}',r')\} = (2\pi)^3\, 2\omega_p\, \delta^{(3)}(\boldsymbol{p}-\boldsymbol{p}')\, \delta_{rr'} \quad (2.9)$$

with all other anticommutators vanishing. Note, the powers of 2π appearing at various places are convention dependent. Corresponding creation and annihilation operators for photons (spin 1 bosons) satisfy the *commutation relations* (Bose statistics)

$$[c(\boldsymbol{p},\lambda), c^+(\boldsymbol{p}',\lambda')] = (2\pi)^3\, 2\omega_p\, \delta^{(3)}(\boldsymbol{p}-\boldsymbol{p}')\, \delta_{\lambda\lambda'} \,. \quad (2.10)$$

In configuration space particles have associated fields [5, 6, 7]. The leptons are represented by Dirac fields $\psi_\alpha(x)$, which are four–component spinors $\alpha = 1, 2, 3, 4$, and the photon by the real vector potential field $A^\mu(x)$ from which derives the electromagnetic field strength tensor $F^{\mu\nu} = \partial^\mu A^\nu - \partial^\nu A^\mu$. The free fields are represented in terms of the creation and annihilation operators

$$\psi_\alpha(x) = \sum_{r=\pm 1/2} \int \mathrm{d}\mu(p) \left\{ u_\alpha(\boldsymbol{p},r)\, a(\boldsymbol{p},r)\, \mathrm{e}^{-ipx} + v_\alpha(\boldsymbol{p},r)\, b^+(\boldsymbol{p},r)\, \mathrm{e}^{ipx} \right\}$$

(2.11)

for the fermion, and

$$A_\mu(x) = \sum_{\lambda=\pm} \int \mathrm{d}\mu(p) \left\{ \varepsilon_\mu(p,\lambda)\, c(\boldsymbol{p},\lambda)\, \mathrm{e}^{-ipx} + \mathrm{h.c.} \right\} \quad (2.12)$$

for the photon (h.c. = hermitian conjugation). The Fourier transformation has to respect that the physical state is on the mass–shell and has positive energy (*spectral condition*: $p^2 = m^2$, $p^0 \geq m$, $m \geq 0$), thus $p^0 = \omega_p = \sqrt{m^2 + \boldsymbol{p}^2}$ and

$$\int \mathrm{d}\mu(p) \cdots \equiv \int \frac{\mathrm{d}^3 p}{2\omega_p (2\pi)^3} \cdots = \int \frac{\mathrm{d}^4 p}{(2\pi)^3} \Theta(p^0) \delta(p^2 - m^2) \cdots$$

Note that Fourier amplitudes $\mathrm{e}^{\mp ipx}$ in (2.11) and (2.12), because of the on–shell condition $p^0 = \omega_p$, are plane wave (free field) solutions of the *Klein-Gordon equation*: $(\Box_x + m^2)\, \mathrm{e}^{\mp ipx} = 0$ or the *d' Alembert equation* $(\Box_x)\, \mathrm{e}^{\mp ipx} = 0$ for the photon where $m_\gamma = 0$. Therefore, the fields themselves satisfy the Klein-Gordon or the d' Alembert equation, respectively. The "amplitudes" u, v and ε_μ, appearing in (2.11) and (2.12) respectively, are classical one–particle wave functions (plane wave solutions) satisfying the free field equations in momentum space[3]. Thus u the lepton wavefunction and v the

---

[3] Our convention for the four–dimensional Fourier transformation for general (off-shell) fields, reads (all integrations from $-\infty$ to $+\infty$)

$$\tilde{\psi}(p) = \int \mathrm{d}^4 x\, \mathrm{e}^{ipx} \psi(x)\,, \quad \tilde{A}^\mu(p) = \int \mathrm{d}^4 x\, \mathrm{e}^{ipx} A^\mu(x)\,. \quad (2.13)$$

The inverse transforms then take the form

$$\psi(x) = \int \frac{\mathrm{d}^4 p}{(2\pi)^4}\, \mathrm{e}^{-ipx} \tilde{\psi}(p)\,, \quad A^\mu(x) = \int \frac{\mathrm{d}^4 p}{(2\pi)^4}\, \mathrm{e}^{-ipx} \tilde{A}^\mu(p)\,, \quad \delta^{(4)}(x) = \int \frac{\mathrm{d}^4 p}{(2\pi)^4}\, \mathrm{e}^{-ipx}$$

antilepton wavefunction are four–spinors, c–number solutions of the Dirac equations,

$$(\slashed{p} - m)\, u_\alpha(\boldsymbol{p}, r) = 0 \,, \quad \text{for the lepton}$$
$$(\slashed{p} + m)\, v_\alpha(\boldsymbol{p}, r) = 0 \,, \quad \text{for the antilepton.} \tag{2.14}$$

As usual, we use the short notation $\slashed{p} \doteq \gamma^\mu p_\mu = \gamma^0 p^0 - \boldsymbol{\gamma p}$ (repeated indices summed over). Note that the relations (2.14) directly infer that the Dirac field is a solution of the Dirac equation $(i\gamma^\mu \partial_\mu - m)\, \psi(x) = 0$.

The γ–*matrices* are 4×4 matrices which satisfy the **Dirac algebra:**[4]

$$\{\gamma^\mu, \gamma^\nu\} = \gamma^\mu \gamma^\nu + \gamma^\nu \gamma^\mu = 2 g^{\mu\nu} \tag{2.15}$$

The L–invariant parity odd matrix γ_5 (under parity $\gamma^0 \to \gamma^0$, $\gamma^i \to -\gamma^i$ $i = 1, 2, 3$)

$$\gamma_5 = i\gamma^0 \gamma^1 \gamma^2 \gamma^3 \,;\quad \gamma_5^2 = 1 \,;\quad \gamma_5 = \gamma_5^+ \tag{2.16}$$

satisfies the anticommutation relation

$$\{\gamma_5, \gamma^\mu\} = \gamma_5 \gamma^\mu + \gamma^\mu \gamma_5 = 0 \tag{2.17}$$

and is required for the formulation of parity violating theories like the weak interaction part of the Standard Model (SM) and for the projection of Dirac fields to left–handed (L) and right–handed (R) chiral fields

---

and hence the derivative with respect to x^μ turns into multiplication by the *four–momentum* $-ip_\mu$: $\partial_\mu \psi(x) \to -ip_\mu \tilde{\psi}(p)$ etc.

[4] Dirac's γ–matrices are composed from Pauli matrices. In quantum mechanics spacial rotations are described by the group of unitary, unimodular ($\det U = 1$) complex 2×2 matrix transformations $SU(2)$ rather than by classical $O(3)$ rotations. The structure constants are given by ϵ_{ikl} ($i, k, l = 1, 2, 3$) the fully antisymmetric permutation tensor. The generators of $SU(2)$ are given by $T_i = \frac{\sigma_i}{2}$; σ_i ($i = 1, 2, 3$) in terms of the 3 hermitian and traceless *Pauli matrices*

$$\sigma_1 = \begin{pmatrix} 0 & 1 \\ 1 & 0 \end{pmatrix}, \quad \sigma_2 = \begin{pmatrix} 0 & -i \\ i & 0 \end{pmatrix}, \quad \sigma_3 = \begin{pmatrix} 1 & 0 \\ 0 & -1 \end{pmatrix}$$

one of which (σ_3) is diagonal. The properties of the Pauli matrices are

$$[\sigma_i, \sigma_k] = 2i\epsilon_{ikl} \sigma_l \,,\quad \{\sigma_i, \sigma_k\} = 2\delta_{ik}$$
$$\sigma_i^+ = \sigma_i \,,\quad \sigma_i^2 = 1 \,,\quad \text{Tr}\, \sigma_i = 0$$
$$\sigma_i \sigma_k = \frac{1}{2} \{\sigma_i, \sigma_k\} + \frac{1}{2} [\sigma_i, \sigma_k] = \delta_{ik} + i\epsilon_{ikl} \sigma_l$$

As usual we denote by $[A, B] = AB - BA$ the commutator, by $\{A, B\} = AB + BA$ the anticommutator. Dirac's γ–matrices in standard representation (as an alternative to the helicity representation, considered below) are

$$\gamma^0 = \begin{pmatrix} 1 & 0 \\ 0 & -1 \end{pmatrix}, \quad \gamma^i = \begin{pmatrix} 0 & \sigma_i \\ -\sigma_i & 0 \end{pmatrix}, \quad \gamma_5 = \begin{pmatrix} 0 & 1 \\ 1 & 0 \end{pmatrix}$$

$$\psi_R = \Pi_+\psi \ ; \ \ \psi_L = \Pi_-\psi \tag{2.18}$$

where

$$\Pi_\pm = \frac{1}{2}(1 \pm \gamma_5) \tag{2.19}$$

are hermitian chiral projection matrices[5]

$$\Pi_+ + \Pi_- = 1 \ , \ \Pi_+\Pi_- = \Pi_-\Pi_+ = 0 \ , \ \Pi_-^2 = \Pi_- \ \text{and} \ \Pi_+^2 = \Pi_+ \ .$$

Note that $\psi^+\psi$ or u^+u, which might look like the natural analog of $|\psi|^2 = \psi^*\psi$ of the lepton wave function in quantum mechanics, are not scalars (invariants) under Lorentz transformations. In order to obtain an invariant we have to sandwich the matrix A which implements hermitian conjugation of the Dirac matrices $A\gamma_\mu A^{-1} = \gamma_\mu^+$. One easily checks that we may identify $A = \gamma^0$. Thus defining the *adjoint spinor* by $\bar{\psi} \doteq \psi^+\gamma^0$ we may write $\psi^+ A\psi = \bar{\psi}\psi$ etc.

The standard basis of 4×4 matrices in four–spinor space is given by the 16 elements

$$\Gamma_i = 1 \ , \ \gamma_5 \ , \ \gamma^\mu \ , \ \gamma^\mu\gamma_5 \ \text{and} \ \sigma^{\mu\nu} = \frac{i}{2}[\gamma^\mu,\gamma^\nu] \ . \tag{2.21}$$

The corresponding products $\bar{\psi}\Gamma_i\psi$ are scalars in spinor space and transform as ordinary scalar (S), pseudo–scalar (P), vector (V), axial–vector (A) and tensor (T), respectively, under Lorentz transformations.

The Dirac spinors satisfy the normalization conditions

$$\begin{array}{lll}
\bar{u}(p,r)\gamma^\mu u(p,r') = 2\,p^\mu \delta_{rr'} \ , & \bar{v}(p,r)\gamma^\mu v(p,r') = 2\,p^\mu \delta_{rr'} \\
\bar{u}(p,r)v(p,r') = 0 \ \ \ \ , & \bar{u}(p,r)u(p,r) = 2m\,\delta_{rr'} \\
\bar{v}(p,r)u(p,r') = 0 \ \ \ \ , & \bar{v}(p,r)v(p,r) = -2m\,\delta_{rr'}
\end{array} \tag{2.22}$$

---

[5] Usually, the quantization of a massive particle with spin is defined relative to the z-axis as a standard frame. In general, the direction of polarization $\boldsymbol{\xi}$, $\boldsymbol{\xi}^2 = 1$ in the rest frame may be chosen arbitrary. For a massive fermion of momentum p

$$\Pi_\pm = \frac{1}{2}\left(1 \pm \gamma_5 \slashed{n}\right)$$

define the general from of covariant spin projection operators, where n is a space like unit vector orthogonal to p

$$n^2 = -1 \ ; \ n \cdot p = 0 \ .$$

The general form of n is obtained by applying Lorentz–boost L_p to the polarization vector in the rest frame

$$n = L_p\left(0, \boldsymbol{\xi}\right) = \left(\frac{\boldsymbol{p} \cdot \boldsymbol{\xi}}{m}, \boldsymbol{\xi} + \frac{\boldsymbol{p} \cdot \boldsymbol{\xi}}{m(p^0+m)}\boldsymbol{p}\right) \ . \tag{2.20}$$

When studying polarization phenomena the polarization vectors n enter as independent additional vectors in covariant decompositions of amplitudes, besides the momentum vectors.

and completeness relations

$$\sum_r u(p,r)\bar{u}(p,r) = \slashed{p} + m \ , \ \sum_r v(p,r)\bar{v}(p,r) = \slashed{p} - m \ . \tag{2.23}$$

For the photon the *polarization vector* $\varepsilon_\mu(p,\lambda)$ satisfies the normalization

$$\varepsilon_\mu(p,\lambda)\varepsilon^{\mu*}(p,\lambda') = -\delta_{\lambda\lambda'} \ , \tag{2.24}$$

the completeness relation

$$\sum_{\lambda=\pm} \varepsilon_\mu(p,\lambda)\varepsilon_\nu^*(p,\lambda) = -g_{\mu\nu} + p_\mu f_\nu + p_\nu f_\mu \ , \tag{2.25}$$

and the absence of a scalar mode requires

$$p_\mu \varepsilon^\mu(p,\lambda) = 0 \ . \tag{2.26}$$

The "four-vectors" f in the completeness relation are arbitrary gauge dependent quantities, which must drop out from physical quantities. Gauge invariance, i.e. invariance under *Abelian gauge transformations* $A_\mu \to A_\mu - \partial_\mu \alpha(x)$, $\alpha(x)$ an arbitrary scalar function, amounts to the invariance under the substitutions

$$\varepsilon_\mu \to \varepsilon_\mu + \lambda\, p_\mu \ ; \ \lambda \ \text{an arbitrary constant} \tag{2.27}$$

of the polarization vectors. One can prove that the polarization "vectors" for massless spin 1 fields can not be covariant. The non–covariant terms are always proportional to p_μ, however.

Besides a definite relativistic transformation property, like

$$U(\Lambda,a)\psi_\alpha(x)U^{-1}(\Lambda,a) = D_{\alpha\beta}(\Lambda^{-1})\psi_\beta(\Lambda x + a) \ ,$$

for a Dirac field, where $D(\Lambda)$ is a four–dimensional (non–unitary) representation of the group $SL(2,C)$ which, in contrast to $L_+^\uparrow$ itself, exhibits true *spinor representations* (see Sect. 2.2). The fields are required to satisfy Einstein causality: "no physical signal may travel faster than light", which means that commutators for bosons and anticommutators for fermions must vanish outside the light cone

$$[A_\mu(x), A_\nu(x')] = 0 \ , \ \{\psi_\alpha(x), \bar{\psi}_\beta(x')\} = 0 \ \text{for} \ (x-x')^2 < 0 \ .$$

This is only possible if all fields exhibit two terms, a creation and an annihilation part, and for charged particles this means that to each particle an antiparticle of the same mass and spin but of opposite charge must exist [8]. In addition, and equally important, causality requires spin $1/2$, $3/2$, $\cdots$ particles to be fermions quantized with anticommutation rules and hence necessarily have to fulfill the *Pauli exclusion principle* [9], while spin 0, 1, $\cdots$ must be bosons to be quantized by normal commutation relations [10]. Note that neutral particles only, like the photon, may be their own antiparticle, the field then has to be real.

2.1.2 C, P, T and CPT

In QED as well as in QCD, not however in weak interactions, interchanging particles with antiparticles defines a symmetry, *charge conjugation C*. It is mapping particle into antiparticle creation and annihilation operators and vice versa:
$$a(\boldsymbol{p},r) \stackrel{C}{\leftrightarrow} b(\boldsymbol{p},r) \;,\quad a^+(\boldsymbol{p},r) \stackrel{C}{\leftrightarrow} b^+(\boldsymbol{p},r) \;,$$
up to a phase. For the Dirac field charge conjugation reads (see 2.35)
$$\psi_\alpha(x) \stackrel{C}{\to} C_{\alpha\beta}\bar{\psi}_\beta^T(x) \tag{2.28}$$
with (X^T = transposition of the matrix or vector X)
$$C = \mathrm{i}\left(\gamma^2\gamma^0\right) = -\mathrm{i}\begin{pmatrix} 0 & \sigma_2 \\ \sigma_2 & 0 \end{pmatrix} \;. \tag{2.29}$$

Properties of C are:
$$C^T = -C \;,\quad C\gamma^\mu C^{-1} = -\left(\gamma^\mu\right)^T \;,$$
and for the spinors charge conjugation takes the form
$$(Cu)^T = \bar{v} \quad\text{and}\quad (Cv)^T = \bar{u} \;, \tag{2.30}$$
which may be verified by direct calculation.

As under charge conjugation the charge changes sign, also the electromagnetic current must change sign
$$U(C)\,j^\mu_{\mathrm{em}}(x)\,U^{-1}(C) = -j^\mu_{\mathrm{em}}(x) \;. \tag{2.31}$$
Notice that for any contravariant four–vector j^μ we may write the parity transformed vector $(j^0, -\boldsymbol{j}) \equiv j_\mu$ as a covariant vector. We will use this notation in the following.

Since the electromagnetic interaction $\mathcal{L}^{\mathrm{QED}}_{\mathrm{int}} = e j^\mu_{\mathrm{em}}(x)A_\mu(x)$ respects C–, P– and T–invariance[6] separately, we immediately get the following transformation properties for the photon field:

---

[6] Any transformation which involves time-reversal T must be implemented as a anti–unitary transformation $\bar{U}(T)$, because the Hamiltonian cannot be allowed to change sign by the requirement of positivity of the energy (Wigner 1939). **Anti–unitarity** is defined by the properties
$$\bar{U}(\alpha|\psi\rangle + \beta|\phi\rangle) = \alpha^* \bar{U}|\psi\rangle + \beta^* \bar{U}|\phi\rangle = \alpha^*|\psi'\rangle + \beta^*|\phi'\rangle \tag{2.32}$$
and
$$\langle\psi'|\phi'\rangle = \langle\psi|\phi\rangle^* \;. \tag{2.33}$$
The complex conjugation of matrix elements is admitted by the fact that it also preserves the probability $|\langle\psi|\phi\rangle|^2$. Because of the complex conjugation of matrix elements an anti–unitary transformation implies a **Hermitian transposition** of states and operators.

2.1 Quantum Field Theory Background

$$\begin{aligned}
U(C)\, A^\mu(x)\, U^{-1}(C) &= -A^\mu(x) \\
U(P)\, A^\mu(x)\, U^{-1}(P) &= (PA)^\mu(Px) = A_\mu(Px) \\
\bar{U}(T)\, A^\mu(x)\, \bar{U}^{-1}(T) &= -(TA)^\mu(Tx) = A_\mu(Tx) \ .
\end{aligned} \quad (2.34)$$

Notice that the charge parity for the photon is $\eta_C^\gamma = -1$.

For the Dirac fields C, P and T take the form

$$\begin{aligned}
U(C)\, \psi_\alpha(x)\, U^{-1}(C) &= \mathrm{i}\left(\gamma^2\gamma^0\right)_{\alpha\beta} \bar{\psi}_\beta^T(x) \\
U(P)\, \psi_\alpha(x)\, U^{-1}(P) &= \left(\gamma^0\right)_{\alpha\beta} \psi_\beta(Px) \\
\bar{U}(T)\, \psi_\alpha(x)\, \bar{U}^{-1}(T) &= \mathrm{i}\left(\gamma^2\gamma_5\right)_{\alpha\beta} \bar{\psi}_\beta^T(Tx)
\end{aligned} \quad (2.35)$$

where the phases have been chosen conveniently. We observe that, in contrast to the boson fields, the transformation properties of the Dirac fields are by no means obvious; they follow from applying C, P and T to the Dirac equation.

A very important consequence of *relativistic local quantum field theory* is the validity of the CPT–**theorem:** Any Poincaré ($\mathcal{P}_+^\uparrow$) [special Lorentz transformations, rotations plus translations] invariant field theory with normal commutation relations [bosons satisfying commutation relations, fermions anticommutation relations] is CPT invariant.

Let $\Theta = CPT$ where C, P and T may be applied in any order. There exists an anti–unitary operator $\bar{U}(\Theta)$ which (with an appropriate choice of the phases) is transforming scalar, Dirac and vector fields according to

$$\begin{aligned}
\bar{U}(\Theta)\, \phi(x)\, \bar{U}^{-1}(\Theta) &= \phi^*(-x) \\
\bar{U}(\Theta)\, \psi(x)\, \bar{U}^{-1}(\Theta) &= \mathrm{i}\gamma_5 \psi(-x) \\
\bar{U}(\Theta)\, A_\mu(x)\, \bar{U}^{-1}(\Theta) &= -A_\mu(-x) \ ,
\end{aligned} \quad (2.36)$$

and which leaves the vacuum invariant: $\bar{U}(\Theta)|0\rangle = |0\rangle$ up to a phase. The CPT-theorem asserts that the transformation $\bar{U}(\Theta)$ under very general conditions is a symmetry of the theory (Lüders 1954, Pauli 1955, Jost 1957) [11].

The basic reason for the validity of the CPT-theorem is the following: If we consider a Lorentz transformation $\Lambda \in \mathrm{L}_+^\uparrow$ represented by a unitary operator $U(\boldsymbol{\chi}, \boldsymbol{\omega} = \boldsymbol{n}\theta)$ ($\boldsymbol{\chi}$ parametrizing a Lorentz–boost, $\boldsymbol{\omega}$ parametrizing a rotation), then the operator $U(\boldsymbol{\chi}, \boldsymbol{n}(\theta + 2\pi)) = -U(\boldsymbol{\chi}, \boldsymbol{n}\theta)$ is representing the same L–transformation. In a local quantum field theory the mapping $\Lambda \to -\Lambda$ for $\Lambda \in \mathrm{L}_+^\uparrow$, which is equivalent to the requirement that $\Theta : x \to -x$ must be a symmetry: the invariance under four–dimensional reflections.

Consequences of CPT are that modulus of the charges, masses, g–factors and lifetimes of particles and antiparticles must be equal. Consider a one particle state $|\psi\rangle = |e, \boldsymbol{p}, \boldsymbol{s}\rangle$ where e is the charge, $\boldsymbol{p}$ the momentum and $\boldsymbol{s}$ the spin. The CPT conjugate state is given by $|\tilde{\psi}\rangle = |-e, \boldsymbol{p}, -\boldsymbol{s}\rangle$. The state $|\psi\rangle$ is an eigenstate of the Hamiltonian which is describing the time evolution of the free particle:

$$\mathcal{H}|\psi\rangle = E|\psi\rangle \quad (2.37)$$

and the CPT conjugate relation reads $\tilde{\mathcal{H}}|\tilde{\psi}\rangle = E|\tilde{\psi}\rangle$. Since $\tilde{\mathcal{H}} = \mathcal{H}$ by the CPT theorem, we thus have

$$\mathcal{H}|\tilde{\psi}\rangle = E|\tilde{\psi}\rangle \ . \tag{2.38}$$

At $\boldsymbol{p} = 0$ the eigenvalue E reduces to the mass and therefore the two eigenvalue equations say that the mass of particle and antiparticle must be the same:

$$\bar{m} = m \ . \tag{2.39}$$

The equality of the g–factors may be shown in the same way, but with a Hamiltonian which describes the interaction of the particle with a magnetic field $\boldsymbol{B}$. Then (2.37) holds with eigenvalue

$$E = m - g\left(\frac{e\hbar}{2mc}\right)\boldsymbol{s}\cdot\boldsymbol{B} \ . \tag{2.40}$$

The CPT conjugate state ($e \to -e$, $\boldsymbol{s} \to -\boldsymbol{s}$, $m \to \bar{m}$, $g \to \bar{g}$, $\boldsymbol{B} \to \boldsymbol{B}$) according to (2.38) will have the same eigenvalue

$$E = \bar{m} - \bar{g}\left(\frac{e\hbar}{2\bar{m}c}\right)\boldsymbol{s}\cdot\boldsymbol{B} \ . \tag{2.41}$$

and since $\bar{m} = m$ we must have

$$\bar{g} = g \tag{2.42}$$

For the proof of the equality of the lifetimes

$$\bar{\tau} = \tau \tag{2.43}$$

we refer to the textbook [12]. Some examples of experimental tests of CPT, relevant in our context, are (see [13])

$\|q_{e^+} + q_{e^-}\|/e$	$< 4 \times 10^{-8}$
$(m_{e^+} - m_{e^-})/m_{\mathrm{average}}$	$< 8 \times 10^{-9}$ 90% CL
$(g_{e^+} - g_{e^-})/g_{\mathrm{average}}$	$(-0.5 \pm 2.1) \times 10^{-12}$
$(g_{\mu^+} - g_{\mu^-})/g_{\mathrm{average}}$	$(-2.6 \pm 1.6) \times 10^{-8}$
$(\tau_{\mu^+} - \tau_{\mu^-})/\tau_{\mathrm{average}}$	$(2 \pm 8) \times 10^{-5}$.

The best test of CPT comes from the neural kaon mass difference

$$\left|\frac{m_{\overline{K}^0} - m_{K^0}}{m_{K^0}}\right| \leq 10^{-18} \ .$$

The existence of a possible electric dipole moment we have discussed earlier on p. 10 of the Introduction. An electric dipole moment requires a T violating theory and the CPT theorem implies that equivalently CP must be violated. In fact, CP invariance alone (independently of CPT and T) gives important predictions relating decay properties of particles and antiparticles. We are interested here particularly in μ–decay, which plays a crucial role in the muon $g - 2$ experiment. Consider a matrix element for a particle a with spin $\boldsymbol{s}_a$

at rest decaying into a bunch of particles b, c, $\cdots$ with spins s_b, s_c, $\cdots$ and momenta p_b, p_c, $\cdots$:

$$\mathcal{M} = \langle p_b, s_b; p_c, s_c; \cdots |\mathcal{H}_{\text{int}}|0, s_a\rangle \,. \tag{2.44}$$

Under CP we have to substitute $s_a \to s_{\bar{a}}$, $p_a \to -p_{\bar{a}}$, etc. such that, provided $\mathcal{H}_{\text{int}}$ is CP symmetric we obtain

$$\bar{\mathcal{M}} = \langle -p_{\bar{b}}, s_{\bar{b}}; -p_{\bar{c}}, s_{\bar{c}}; \cdots |\mathcal{H}_{\text{int}}|0, s_{\bar{a}}\rangle \equiv \mathcal{M} \,. \tag{2.45}$$

The modulus square of these matrix–elements gives the transition probability for the respective decays, and (2.45) tells us that the decay rate of a particle into a particular configuration of final particles is identical to the decay rate of the antiparticle into the same configuration of antiparticles with all momenta reversed.

For the muon decay $\mu^- \to e^- \bar{\nu}_e \nu_\mu$, after integrating out the unobserved neutrino variables, the decay electron distribution is of the form

$$\frac{\mathrm{d} N_{e^-}}{\mathrm{d} x \, \mathrm{d} \cos\theta} = A(x) + B(x)\, \hat{s}_\mu \cdot \hat{p}_{e^-} \,, \tag{2.46}$$

where $x = 2 p_{e^-}/m_\mu$ with p_{e^-} the electron momentum in the muon rest frame and $\cos\theta = \hat{s}_\mu \cdot \hat{p}_{e^-}$, $\hat{s}_\mu$ and $\hat{p}_{e^-}$ the unit vectors in direction of s_μ and p_{e^-}.

The corresponding expression for the antiparticle decay $\mu^+ \to e^+ \nu_e \bar{\nu}_\mu$ reads

$$\frac{\mathrm{d} N_{e^+}}{\mathrm{d} x \, \mathrm{d} \cos\theta} = \bar{A}(x) + \bar{B}(x)\, \hat{s}_\mu \cdot \hat{p}_{e^+} \,, \tag{2.47}$$

and therefore for all angles and all electron momenta

$$A(x) + B(x) \cos\theta = \bar{A}(x) - \bar{B}(x) \cos\theta$$

or

$$A(x) = \bar{A}(x) \,, \quad B(x) = -\bar{B}(x) \,. \tag{2.48}$$

It means that the decay asymmetry is equal in magnitude but opposite in sign for μ^- and μ^+. This follows directly from CP and independent of the type of interaction (V-A,V+A,S,P or T) and whether P is violated or not. In spite of the fact that the SM exhibits CP violation (see the Introduction to Sect. 4.2), as implied by a CP violating phase in the quark family mixing matrix in the charged weak current, in μ–decay CP violation is a very small higher order effect and by far too small to have any detectable trace in the decay distributions, i.e. CP symmetry is perfectly realized in this case. The strong correlation between the muon polarization and charge on the one side (see Chap. 6) and the decay electron/positron momentum is a key element of tracing spin polarization information in the muon $g-2$ experiments.

CP violation, and the associated T violation plays an important role in determining the electric dipole moment of electrons and muons. In principle

it is possible to test T invariance in μ–decay by searching for T odd matrix elements like
$$s_e \cdot (s_\mu \times p_e) \,. \tag{2.49}$$
This is very difficult and has not been performed. A method which works is the study of the effect of an electric dipole moment on the spin precession in the muon $g-2$ experiment. This will be studied in Sect. 6.3.1 on p. 361.

The best limit for the electron (1.8) comes from investigating T violation in Thallium (^{205}Tl) where the EDM is enhanced by the ratio $R = d_\mathrm{atom}/d_e$, which in the atomic Thallium ground state studied is $R = -585$. Investigated are $\boldsymbol{v} \times \boldsymbol{E}$ terms in high electrical fields $\boldsymbol{E}$ in an atomic beam magnetic–resonance device [14].

2.2 The Origin of Spin

As promised at the beginning of the chapter the intimate relation of the anomalous magnetic moment to spin is a good reason to have a closer look at how spin comes into play in particle physics. The spin and the magnetic moment of the electron became evident from the deflection of atoms in an inhomogeneous magnetic field and the observation of the fine structure by optical spectroscopy [15, 16]. Spin is the intrinsic "self–angular momentum" of a point–particle and when it was observed by Goudsmit and Uhlenbeck it was completely unexpected. The question about the origin of spin is interesting because it is not obvious how a point–like object can possess its own angular momentum. A first theoretical formulation of spin in quantum mechanics was given by Pauli in 1927 [17], where spin was introduced as a new degree of freedom saying that there are two kinds of electrons in a doublet.

In modern relativistic terms, in the SM, particles and in particular leptons and quarks are considered to be massless originally, as required by chiral symmetry. All particles acquire their mass due to symmetry breaking via the Higgs mechanism: a scalar neutral Higgs field H develops a non–vanishing vacuum expectation value v and particles moving in the corresponding Bose condensate develop an effective mass. In the SM, in the physical *unitary gauge* a *Yukawa interaction* term upon a shift $H \to H + v$

$$\mathcal{L}_\mathrm{Yukawa} = \sum_f \frac{G_f}{\sqrt{2}} \bar\psi_f \psi_f H \to \sum_f \left(m_f \bar\psi_f \psi_f + \frac{m_f}{v} \bar\psi_f \psi_f H \right) \tag{2.50}$$

induces a fermion mass term with mass $m_f = \frac{G_f}{\sqrt{2}} v$ where G_f is the *Yukawa coupling*.

In the massless state there are actually two independent electrons characterized by positive and negative helicities (chiralities) corresponding to right–handed (R) and left–handed (L) electrons, respectively, which do not "talk" to each other. Helicity h is defined as the projection of the spin vector onto the direction of the momentum vector

$$h \doteq S \frac{p}{|p|} \quad (2.51)$$

as illustrated in Fig. 2.1 and transform into each other by space-reflections P (parity). Only after a fermion has acquired a mass, helicity flip transitions as effectively mediated by an anomalous magnetic moment (see below) are possible. In a renormalizable QFT an anomalous magnetic moment term is not allowed in the Lagrangian. It can only be a term induced by radiative corrections and in order not to vanish requires chiral symmetry to be broken by a corresponding mass term.

Angular momentum has to do with rotations, which form the rotation group $O(3)$. Ordinary 3–space rotations are described by orthogonal 3×3 matrices R ($RR^T = R^TR = I$ where I is the unit matrix and R^T denotes the transposed matrix) acting as $x' = Rx$ on vectors x of three–dimensional Euclidean position space $\mathbf{R}^3$. Rotations are preserving scalar products between vectors and hence the length of vectors as well as the angles between them. Multiplication of the rotation matrices is the group operation and of course the successive multiplication of two rotations is non–commutative $[R_1, R_2] \neq 0$ in general. The rotation group is characterized by the Lie algebra $[\mathcal{J}_i, \mathcal{J}_j] = \varepsilon_{ijk}\mathcal{J}_k$, where the $\mathcal{J}_i$'s are normalized skew symmetric 3×3 matrices which generate the infinitesimal rotations around the x, y and z axes, labeled by $i, j, k = 1, 2, 3$. By ε_{ijk} we denoted the totally antisymmetric Levi-Civita tensor. The Lie algebra may be written in the form of the angular momentum algebra

$$[J_i, J_j] = \mathrm{i}\varepsilon_{ijk} J_k \quad (2.52)$$

by setting $\mathcal{J}_i = -\mathrm{i}J_i$, with Hermitian generators $J_i = J_i^+$. The latter form is well known from quantum mechanics (QM). In quantum mechanics rotations have to be implemented by unitary representations $U(R)$ ($UU^+ = U^+U = I$ and U^+ is the Hermitian conjugate of U) which implement transformations of the state vectors in physical Hilbert space $|\psi\rangle' = U(R)|\psi\rangle$ for systems rotated relative to each other. Let J_i be the generators of the infinitesimal transformations of the group $O(3)$, the angular momentum operators, such that a finite rotation of magnitude $|\omega| = \theta$ about the direction of $n = \omega/\theta$ may be represented by $U(R(\omega)) = \exp{-\mathrm{i}\omega J}$ (ω_i, $i = 1, 2, 3$ a real rotation vector). While for ordinary rotations the J_k's are again 3×3 matrices, in fact the lowest dimensional matrices which satisfy (2.52) in a non–trivial manner are 2×2 matrices. The corresponding Lie algebra is the one of the group $SU(2)$ of unitary 2×2 matrices U with determinant unity: $\det U = 1$. It is a simply connected group and in fact it is the universal covering group of $O(3)$, the latter being doubly connected. Going to $SU(2)$ makes rotations a single

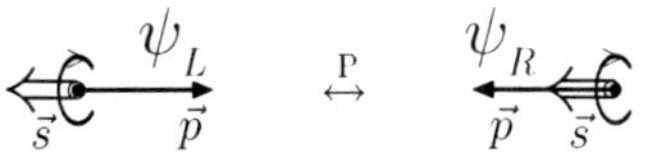

Fig. 2.1. Massless "electrons" have fixed helicities

valued mapping in parameter space which is crucial to get the right phases in the context of QM. Thus $SU(2)$ is lifting the two–fold degeneracy of $O(3)$. As a basic fact in quantum mechanics rotations are implemented as unitary representations of $SU(2)$ and not by $O(3)$ in spite of the fact that the two groups share the same abstract Lie algebra, characterized by the structure constants ε_{ijk}. Like $O(3)$, the group $SU(2)$ is of order $r = 3$ (number of generators) and rank $l = 1$ (number of diagonal generators). The generators of a unitary group are hermitian and the special unitary transformations of determinant unity requires the generators to be traceless. The canonical choice is $J_i = \frac{\sigma_i}{2}$; σ_i the Pauli matrices

$$\sigma_1 = \begin{pmatrix} 0 & 1 \\ 1 & 0 \end{pmatrix}, \quad \sigma_2 = \begin{pmatrix} 0 & -i \\ i & 0 \end{pmatrix}, \quad \sigma_3 = \begin{pmatrix} 1 & 0 \\ 0 & -1 \end{pmatrix} \quad (2.53)$$

There is one diagonal operator $S_3 = \frac{\sigma_3}{2}$ the 3rd component of spin. The eigenvectors of S_3 are

$$U(r = \tfrac{1}{2}, -\tfrac{1}{2}) = \begin{pmatrix} 1 \\ 0 \end{pmatrix}, \begin{pmatrix} 0 \\ 1 \end{pmatrix}. \quad (2.54)$$

characterized by the eigenvalues of $\frac{1}{2}, -\frac{1}{2}$ of S_3 called spin up [↑] and spin down [↓], respectively. The eigenvectors represent the possible independent states of the system: two in our case. They thus span a two–dimensional space of complex vectors which are called two–spinors. Thus $SU(2)$ is acting on the space of spinors, like $O(3)$ is acting on ordinary configuration space vectors. From the two non–diagonal matrices we may form the two ladder operators: $S_{\pm 1} = \frac{1}{2}(\sigma_1 \pm i\sigma_2)$

$$S_{+1} = \begin{pmatrix} 0 & 1 \\ 0 & 0 \end{pmatrix}, \quad S_{-1} = \begin{pmatrix} 0 & 0 \\ 1 & 0 \end{pmatrix}$$

which map the eigenvectors into each other and hence change spin by one unit. The following figure shows the simplest case of a so called *root diagram*: the full dots represent the two states labeled by the eigenvalues $S_3 = \pm\frac{1}{2}$ of the diagonal operator. The arrows, labeled with $S_{\pm 1}$ denote the transitions between the different states, as implied by the Lie algebra:

$$\begin{array}{c} S_{+1} \\ \bullet \longrightarrow \bullet \quad S_3 \\ -\tfrac{1}{2} \quad S_{-1} \quad +\tfrac{1}{2} \end{array}$$

The simplest non–trivial representation of $SU(2)$ is the so called fundamental representation, the one which defines $SU(2)$ itself and hence has dimension two. It is the one we just have been looking at. There is only one fundamental representation for $SU(2)$, because the complex conjugate U^* of a representation U which is also a representation, and generally a new one, is equivalent to the original one. The fundamental representation describes intrinsic angular momentum $\frac{1}{2}$ with two possible states characterized by the eigenvalues of the diagonal generator $\pm\frac{1}{2}$. The *fundamental representations* are basic because all others may be constructed by taking tensor products of fundamental representations. In the simplest case of a product of two spin $\frac{1}{2}$ vectors, which are called (two component) spinors $u_i v_k$ may describe a spin zero (anti–parallel spins [↑↓]) or a spin 1 (parallel spins [↑↑]).

In a relativistic theory, described in more detail in the previous section, one has to consider the Lorentz group $L_+^\uparrow$ of proper (preserving orientation of space–time [+]) orthochronous (preserving the direction of time [↑]) Lorentz transformations Λ, in place of the rotation group. They include besides the rotations $R(\boldsymbol{\omega})$ the Lorentz boosts (special Lorentz transformations) $L(\boldsymbol{\chi})^7$ by velocity $\boldsymbol{\chi}$. Now rotations do not play any independent role as they are not a Lorentz invariant concept. Correspondingly, purely spatial 3–vectors like the spin vector $\boldsymbol{S} = \frac{\boldsymbol{\sigma}}{2}$ do not have an invariant meaning. However, the three–vector of Pauli matrices $\boldsymbol{\sigma}$ may be promoted to a four–vector of 2×2 matrices:

$$\sigma_\mu \doteq (1, \boldsymbol{\sigma}) \text{ and } \hat{\sigma}_\mu \doteq (1, -\boldsymbol{\sigma}) \qquad (2.56)$$

which will play a key role in what follows. Again, the L–transformations $\Lambda \in L_+^\uparrow$ on the classical level in (relativistic) quantum mechanics have to be replaced by the simply connected universal covering group with identical Lie algebra, which is $SL(2,C)$, the group of unimodular ($\det U = 1$) complex 2×2 matrix transformations U, with matrix multiplication as the group operation. The group $SL(2,C)$ is related to $L_+^\uparrow$ much in the same way as $SU(2)$ to $O(3)$, namely, the mapping $U_\Lambda \in SL(2,C) \to \Lambda \in L_+^\uparrow$ is two–to–one and the two–fold degeneracy of elements in $L_+^\uparrow$ is lifted in $SL(2,C)$.

The key mapping establishing a linear one–to–one correspondence between real four–vectors and Hermitian 2×2 matrices is the following: with any real four–vector x^μ in Minkowski space we may associate a Hermitian 2×2 matrix

$$x^\mu \to X = x^\mu \sigma_\mu = \begin{pmatrix} x^0 + x^3 & x^1 - ix^2 \\ x^1 + ix^2 & x^0 - x^3 \end{pmatrix} \qquad (2.57)$$

with

$$\det X = x^2 = x^\mu x_\mu \;, \qquad (2.58)$$

---

[7]The special L–transformation $L(p)$ which transforms from a state in the rest frame $(m, \boldsymbol{0})$ to a state of momentum p^μ may be written as

$$\begin{aligned} L^i{}_j &= \delta^i{}_j + \hat{p}_i \hat{p}_j (\cosh \beta - 1) \\ L^i{}_0 &= L^0{}_i = \hat{p}_i \sinh \beta \\ L^0{}_0 &= \cosh \beta \end{aligned} \qquad (2.55)$$

with $\hat{\boldsymbol{p}} = \boldsymbol{p}/|\boldsymbol{p}|$, $\cosh \beta = \omega_p/m$, $\sinh \beta = |\boldsymbol{p}|/m$ and $\tanh \beta = |\boldsymbol{p}|/\omega_p = v$ the velocity of the state.

while every Hermitian 2×2 matrix X determines a real four vector by

$$X \to x^\mu = \frac{1}{2}\mathrm{Tr}\,(X\sigma^\mu)\;. \tag{2.59}$$

An element $U \in SL(2,C)$ provides a mapping

$$X \to X' = UXU^+ \quad \text{i.e.} \quad x'^\mu \sigma_\mu = x^\nu U \sigma_\nu U^+ \tag{2.60}$$

between Hermitian matrices, which preserves the determinant

$$\det X' = \det U \det X \det U^+ = \det X\;, \tag{2.61}$$

and corresponds to the real linear transformation

$$x^\mu \to x'^\mu = \Lambda^\mu{}_\nu\, x^\nu \tag{2.62}$$

which satisfies $x'^\mu x'_\mu = x^\mu x_\mu$ and therefore is a Lorentz transformation.

The Lie algebra of $SL(2,C)$ is the one of $L_+^\uparrow$ and thus given by 6 generators: $\boldsymbol{J}$ for the **rotations** and $\boldsymbol{K}$ for the Lorentz **boosts**, satisfying

$$[J_i, J_k] = \mathrm{i}\epsilon_{ikl} J_l\;,\; [J_i, K_k] = \mathrm{i}\epsilon_{ikl} K_l\;,\; [K_i, K_k] = -\mathrm{i}\epsilon_{ikl} J_l \tag{2.63}$$

as a coupled algebra of the J_i's and K_i's. Since these generators are Hermitian $\boldsymbol{J} = \boldsymbol{J}^+$ and $\boldsymbol{K} = \boldsymbol{K}^+$ the group elements $\mathrm{e}^{-\mathrm{i}\boldsymbol{\omega}\boldsymbol{J}}$ and $\mathrm{e}^{\mathrm{i}\boldsymbol{\chi}\boldsymbol{K}}$ are **unitary**[8]. This algebra can be decoupled by the linear transformation

$$\boldsymbol{A} = \frac{1}{2}\,(\boldsymbol{J} + \mathrm{i}\boldsymbol{K})\;,\quad \boldsymbol{B} = \frac{1}{2}\,(\boldsymbol{J} - \mathrm{i}\boldsymbol{K}) \tag{2.64}$$

under which the Lie algebra takes the form

$$\boldsymbol{A} \times \boldsymbol{A} = \mathrm{i}\boldsymbol{A}\;,\quad \boldsymbol{B} \times \boldsymbol{B} = \mathrm{i}\boldsymbol{B}\;,\quad [A_i, B_j] = 0 \tag{2.65}$$

of two decoupled angular momentum algebras. Since $\boldsymbol{A}^+ = \boldsymbol{B}$ and $\boldsymbol{B}^+ = \boldsymbol{A}$, the new generators are not Hermitian any more and hence give rise to **non–unitary** irreducible representations. These are **finite dimensional** and evidently characterized by a pair (A, B), with $2A$ and $2B$ integers. The dimension of the representation (A, B) is $(2A+1) \cdot (2B+1)$. The angular momentum of the representation (A, B) decomposes into $J = A + B, A + B - 1, \cdots |A - B|$. Massive particle states are constructed starting from the rest frame where J is the spin and the state corresponds to a multiplet of $2J+1$ degrees of freedom.

---

[8]In $SL(2,C)$ the Lie algebra obviously has the 2×2 matrix representation $J_i = \sigma_i/2$ $K_i = \pm \mathrm{i}\,\sigma_i/2$ in terms of the Pauli matrices, however, $\boldsymbol{K}^+ = -\boldsymbol{K}$ is non–Hermitian and the corresponding finite dimensional representation non–unitary. Unitary representations of the Lorentz group, required to implement relativistic covariance on the Hilbert space of physical states, are necessarily infinite dimensional. Actually, the two possible signs of K_i indicated exhibits that there are two different inequivalent representations.

The crucial point is that in relativistic QM besides the mass of a state also the spin has an invariant (reference–frame independent) meaning. There exist exactly two **Casimir operators**, invariant operators commuting with all generators (2.5) and (2.6) of the Poincaré group $\mathcal{P}_+^\uparrow$. One is the mass operator

$$M^2 = P^2 = g_{\mu\nu} P^\mu P^\nu \qquad (2.66)$$

the other is

$$L^2 = g_{\mu\nu} L^\mu L^\nu \; ; \quad L^\mu \doteq \frac{1}{2} \varepsilon^{\mu\nu\rho\sigma} P_\nu M_{\rho\sigma} \; , \qquad (2.67)$$

where L^μ is the Pauli-Lubansky operator. These operators characterize mass m and spin j of the states in an invariant way: $M^2|p,j,j_3;\alpha\rangle = p^2|p,j,j_3;\alpha\rangle$ and $L^2|p,j,j_3;\alpha\rangle = -m^2 j(j+1)|p,j,j_3;\alpha\rangle$.

The classification by (A,B) together with (2.64) shows that for $SL(2,C)$ we have two inequivalent fundamental two–dimensional representations: $(\frac{1}{2},0)$ and $(0,\frac{1}{2})$. The transformations may be written as a unitary rotation times a hermitian boost as follows[9]:

$$\begin{aligned} U_\Lambda &= U(\boldsymbol{\chi},\boldsymbol{\omega}) = D^{(\frac{1}{2})}(\Lambda) = e^{\boldsymbol{\chi}\frac{\boldsymbol{\sigma}}{2}} e^{-i\boldsymbol{\omega}\frac{\boldsymbol{\sigma}}{2}} & \text{for} & (\frac{1}{2},0) \\ \bar{U}_\Lambda &= U^+_{\Lambda^{-1}} = \bar{D}^{(\frac{1}{2})}(\Lambda) = e^{-\boldsymbol{\chi}\frac{\boldsymbol{\sigma}}{2}} e^{-i\boldsymbol{\omega}\frac{\boldsymbol{\sigma}}{2}} & \text{for} & (0,\frac{1}{2}) \end{aligned} \qquad (2.68)$$

While σ_μ (2.56) is a covariant vector

$$U_\Lambda \sigma_\mu U_\Lambda^+ = \Lambda^\nu{}_\mu \sigma_\nu \qquad (2.69)$$

with respect to the representation $U_\Lambda = D^{(\frac{1}{2})}(\Lambda)$, the vector $\hat{\sigma}_\mu$ (2.56) is covariant with respect to $\bar{U}_\Lambda = \bar{D}^{(\frac{1}{2})}(\Lambda)$

$$\bar{U}_\Lambda \hat{\sigma}_\mu \bar{U}_\Lambda^+ = \Lambda^\nu{}_\mu \hat{\sigma}_\nu \; . \qquad (2.70)$$

Note that

$$U(\boldsymbol{\chi},\boldsymbol{n}\theta) \quad \text{and} \quad U(\boldsymbol{\chi},\boldsymbol{n}(\theta+2\pi)) = -U(\boldsymbol{\chi},\boldsymbol{n}\theta) \qquad (2.71)$$

represent the same Lorentz transformation. U_Λ is therefore a double–valued representation of $L_+^\uparrow$.

An important theorem [18] says that a massless particle of helicity λ may be only in the representations satisfying

$$(A,B) = (A, A-\lambda)$$

where $2A$ and $2(A-\lambda)$ are non–negative integer numbers. Thus the simplest representations for massless fields are the spin 1/2 states

---

[9] Again, these finite dimensional representations U_Λ, U_P (below), etc. should not be confused with the corresponding infinite dimensional unitary representations $U(\Lambda)$, $U(P)$, etc acting on the Hilbert space of physical states considered in the preceding section.

$$\lambda = +\tfrac{1}{2} : (\tfrac{1}{2},0) \quad \text{right-handed } (R)$$
$$-\tfrac{1}{2} : (0,\tfrac{1}{2}) \quad \text{left-handed } (L) \tag{2.72}$$

of helicity $+\tfrac{1}{2}$ and $-\tfrac{1}{2}$, respectively.

The finite dimensional irreducible representations of $SL(2,C)$ to mass 0 and spin j are one-dimensional and characterized by the helicity $\lambda = \pm j$. To a given spin $j > 0$ there exist exactly two helicity states. Each of the two possible states is invariant by itself under $L_+^\uparrow$, however, the two states get interchanged under parity transformations:

$$U_P \, h \, U_P^{-1} = -h \; . \tag{2.73}$$

Besides the crucial fact of the validity of the spin–statistics theorem (valid in any relativistic QFT), here we notice an other important difference between spin in non–relativistic QM and spin in QFT. In QM spin 1/2 is a system of two degrees of freedom as introduced by Pauli, while in QFT where we may consider the massless case we have two independent singlet states. Parity P, as we know, acts on four–vectors like $Px = (x^0, -\boldsymbol{x})$ and satisfies $P^2 = 1$[10]. With respect to the rotation group O_3, P^2 is just a rotation by the angle 2π and thus in the context of the rotation group P has no special meaning. This is different for the Lorentz group. While

$$U_P \boldsymbol{J} = \boldsymbol{J} U_P \tag{2.74}$$

commutes

$$U_P \boldsymbol{K} = -\boldsymbol{K} U_P \tag{2.75}$$

does not. As a consequence, we learn that

$$U_P U(\boldsymbol{\chi}, \boldsymbol{n}\,\theta) = U(-\boldsymbol{\chi}, \boldsymbol{n}\,\theta) U_P \tag{2.76}$$

and hence

$$U_P U_\Lambda = \bar{U}_\Lambda U_P \; . \tag{2.77}$$

Thus under parity a left–handed massless fermion is transformed into a right–handed one and vice versa, which of course is also evident from Fig. 2.1, if we take into account that a change of frame by a Lorentz transformation (velocity $v \leq c$) cannot flip the spin of a massless particle.

The necessity to work with $SL(2,C)$ becomes obvious once we deal with spinors. On a classical level, two–spinors or Weyl spinors w are elements of a vector space V of two complex entries, which transform under $SL(2,C)$ by matrix multiplication: $w' = Uw$, $w \in V$, $U \in SL(2,C)$

$$w = \begin{pmatrix} a \\ b \end{pmatrix} \; ; \; a,b \in C \; . \tag{2.78}$$

---

[10] Note that while $P^2 = 1$ the phase η_P of its unitary representation U_P is constrained by $U_P^2 = \pm 1$ only, i.e. $\eta_P = \pm 1$ or $\pm i$.

2.2 The Origin of Spin

Corresponding to the two representations there exist two local Weyl spinor fields (see (2.11))

$$\varphi_a(x) = \sum_{r=\pm 1/2} \int d\mu(p) \left\{ u_a(p,r)\, a(\boldsymbol{p},r)\, e^{-ipx} + v_a(p,r)\, b^+(\boldsymbol{p},r)\, e^{ipx} \right\}$$

$$\chi_a(x) = \sum_{r=\pm 1/2} \int d\mu(p) \left\{ \hat{u}_a(p,r)\, a(\boldsymbol{p},r)\, e^{-ipx} + \hat{v}_a(p,r)\, b^+(\boldsymbol{p},r)\, e^{ipx} \right\},$$

(2.79)

with two components $a = 1,2$, which satisfy the Weyl equations

$$i\,(\hat{\sigma}^\mu \partial_\mu)_{ab}\, \varphi_b(x) = m\chi_a(x)$$
$$i\,(\sigma^\mu \partial_\mu)_{ab}\, \chi_b(x) = m\varphi_a(x) \ .$$

(2.80)

The appropriate one–particle wave functions $u(p,r)$ etc. may be easily constructed as follows: for a massive particle states are constructed by starting in the rest frame where rotations act as ($\omega = |\boldsymbol{\omega}|$, $\hat{\boldsymbol{\omega}} = \boldsymbol{\omega}/\omega$)

$$D^{(\frac{1}{2})}(R(\boldsymbol{\omega})) = \bar{D}^{(\frac{1}{2})}(R(\boldsymbol{\omega})) = e^{-i\boldsymbol{\omega}\cdot\frac{\boldsymbol{\sigma}}{2}} = \mathbf{1}\cos\frac{\omega}{2} - i\boldsymbol{\sigma}\cdot\hat{\boldsymbol{\omega}}\sin\frac{\omega}{2} \ . \quad (2.81)$$

Notice that this $SU(2)$ rotation is a rotation by half of the angle, only, of the corresponding classical O_3 rotation. Here the non–relativistic construction of the states applies and the spinors at rest are given by (2.54). The propagating particles carrying momentum $\boldsymbol{p}$ are then obtained by performing a Lorentz–boost to the states at rest. A boost $L(\boldsymbol{p})$ (2.55) of momentum $\boldsymbol{p}$ is given by $D^{(\frac{1}{2})}(L(\boldsymbol{p})) = e^{\boldsymbol{\chi}\cdot\frac{\boldsymbol{\sigma}}{2}} = N^{-1}\,(p^\mu\sigma_\mu + m)$ and $\bar{D}^{(\frac{1}{2})}(L(\boldsymbol{p})) = e^{-\boldsymbol{\chi}\cdot\frac{\boldsymbol{\sigma}}{2}} = N^{-1}\,(p^\mu\hat{\sigma}_\mu + m)$, respectively, in the two basic representations. $N = (2m\,(p^0 + m))^{-\frac{1}{2}}$ is the normalization factor. The one–particle wave functions (two-spinors) of a Weyl particle and its antiparticle are thus given by

$$u(p,r) = N^{-1}\,(p^\mu\sigma_\mu + m)\, U(r) \quad \text{and} \quad v(p,r) = N^{-1}\,(p^\mu\sigma_\mu + m)\, V(r) \ ,$$

respectively, where $U(r)$ and $V(r) = -i\sigma_2 U(r)$ are the rest frame spinors (2.54). The last relation one has to require for implementing the charge conjugation property for the spinors (2.30) in terms of the matrix (2.29). For the adjoint representation, similarly,

$$\hat{u}(p,r) = N^{-1}\,(p^\mu\hat{\sigma}_\mu + m)\, U(r) \quad \text{and} \quad \hat{v}(p,r) = -N^{-1}\,(p^\mu\hat{\sigma}_\mu + m)\, V(r) \ .$$

The $-$ sign in the last equation, $(-1)^{2j}$ for spin j, is similar to the $-i\sigma_2$ in the relation between U and V, both are required to make the fields local and with proper transformation properties. We can easily derive (2.80) now. We may write $\hat{\sigma}_\mu p^\mu = \omega_p \mathbf{1} - \boldsymbol{\sigma}\cdot\boldsymbol{p} = 2|\boldsymbol{p}|(\frac{\omega_p}{2|\boldsymbol{p}|}\mathbf{1} - h)$ where $h \equiv \frac{\boldsymbol{\sigma}}{2}\cdot\frac{\boldsymbol{p}}{|\boldsymbol{p}|}$ is the helicity operator, and for massless states, where $\omega_p = |\boldsymbol{p}|$, we have $\hat{\sigma}_\mu p^\mu = 2|\boldsymbol{p}|(\frac{1}{2} - h)$ a projection operator on states with helicity $-\frac{1}{2}$, while $\sigma_\mu p^\mu = 2|\boldsymbol{p}|(\frac{1}{2} + h)$ a projection operator on states with helicity $+\frac{1}{2}$. Furthermore, we observe that $p^\mu p^\nu \hat{\sigma}_\mu \sigma_\nu = p^\mu p^\nu \sigma_\mu \hat{\sigma}_\nu = p^2 = m^2$ and one easily verifies the Weyl equations using the given representations of the wave functions.

In the massless limit $m \to 0$: $p^0 = \omega_p = |\boldsymbol{p}|$ we obtain two decoupled equations

$$\mathrm{i}\,(\hat{\sigma}^\mu \partial_\mu)_{ab}\,\varphi_b(x) = 0$$
$$\mathrm{i}\,(\sigma^\mu \partial_\mu)_{ab}\,\chi_b(x) = 0 \ .$$

In momentum space the fields are just multiplied by the helicity projector and the equations say that the massless fields have fixed helicities:

$$(\frac{1}{2},0): \quad \varphi \sim \psi_R \qquad (0,\frac{1}{2}): \quad \chi \sim \psi_L \qquad (2.82)$$

which suggests to rewrite the transformations as

$$\psi_{a\,L,R}(x) \to \psi'_{a\,L,R}(x') = (\Lambda_{L,R})_{ab}\,\psi_{b\,L,R}(\Lambda x) \qquad (2.83)$$

with

$$(\Lambda_{L,R})_{ab} = \left(\mathrm{e}^{\pm\chi\frac{\sigma}{2}}\mathrm{e}^{-\mathrm{i}\omega\frac{\sigma}{2}}\right)_{ab} \quad (\Lambda_R^+ = \Lambda_L^{-1}) \ . \qquad (2.84)$$

Using $\sigma_2\sigma_i\sigma_2 = -\sigma_i^*$ one can show that $\sigma_2\Lambda_L\sigma_2 = \Lambda_R^*$. Thus, $\psi_L^c \equiv \sigma_2\psi_L^*$ (up to arbitrary phase) is defining a charge conjugate spinor which transforms as $\psi_L^c \sim \psi_R$. Indeed $\Lambda_R\psi_L^c = \Lambda_R\sigma_2\psi_L^* = \sigma_2\Lambda_L^*\psi_L^* = \sigma_2\psi_L^{*'} = \psi_L^{c'}$ and thus $\psi_L^c \equiv \sigma_2\psi_L^* \equiv \varphi \sim \psi_R$. Similarly, $\psi_R^c \equiv \sigma_2\psi_R^* \equiv \chi \sim \psi_L$. We thus learn, that for massless fields, counting particles and antiparticles separately, we may consider all fields to be left-handed. The second term in the field, the antiparticle creation part, in each case automatically includes the right-handed partners.

The Dirac field is the bispinor field obtained by combining the irreducible fields $\varphi_a(x)$ and $\chi_a(x)$ into one reducible field $(\frac{1}{2},0) \oplus (0,\frac{1}{2})$. It is the natural field to be used to describe fermions participating parity conserving interactions like QED and QCD. Explicitly, the Dirac field is given by

$$\psi_\alpha(x) = \begin{pmatrix} \varphi_a \\ \chi_a \end{pmatrix}(x) = \sum_r \int d\mu(p) \left\{ u_\alpha(p,r)\,a(\boldsymbol{p},r)\,\mathrm{e}^{-\mathrm{i}px} + v_\alpha(p,r)\,b^+(\boldsymbol{p},r)\,\mathrm{e}^{\mathrm{i}px} \right\}$$

where

$$u_\alpha = \begin{pmatrix} u_a \\ \hat{u}_a \end{pmatrix} \ ; \quad v_\alpha = \begin{pmatrix} v_a \\ \hat{v}_a \end{pmatrix} \ . \qquad (2.85)$$

$\psi_\alpha(x)$ satisfies the **Dirac equation**:

$$(\mathrm{i}\gamma^\mu \partial_\mu - m)_{\alpha\beta}\,\psi_\beta(x) = 0$$

where

$$\gamma^\mu \doteq \begin{pmatrix} 0 & \sigma^\mu \\ \hat{\sigma}^\mu & 0 \end{pmatrix} \qquad (2.86)$$

are the Dirac matrices in the helicity representation (Weyl basis).

The Dirac equation is nothing but the Weyl equations written in terms of the bispinor ψ. Note that a Dirac spinor combines a right–handed Weyl spinor of a particle with a right–handed Weyl spinor of its antiparticle. For $m = 0$, the Dirac operator $i\gamma^\mu \partial_\mu$ in momentum space is $\not{p} = \gamma^\mu p_\mu$. Thus the Dirac equation just is the helicity eigenvalue equation:

$$\gamma^\mu p_\mu \tilde\psi(p) \doteq \begin{pmatrix} 0 & \sigma^\mu p_\mu \\ \hat\sigma^\mu p_\mu & 0 \end{pmatrix} \begin{pmatrix} \tilde\varphi \\ \tilde\chi \end{pmatrix}(p) = 2|\boldsymbol{p}| \begin{pmatrix} 0 & (\tfrac{1}{2}+h) \\ (\tfrac{1}{2}-h) & 0 \end{pmatrix} \begin{pmatrix} \tilde\varphi \\ \tilde\chi \end{pmatrix}(p) = 0 \,. \tag{2.87}$$

Under parity $\psi_\alpha(x)$ transforms into itself

$$\psi_\alpha(x) \to \eta_P (\gamma^0)_{\alpha\beta} \psi_\beta(Px)$$

where γ^0 just interchanges $\varphi \leftrightarrow \chi$ and hence takes the form

$$\gamma^0 \doteq \begin{pmatrix} 0 & 1 \\ 1 & 0 \end{pmatrix}\,.$$

The irreducible components φ and χ are eigenvectors of the matrix

$$\gamma_5 \doteq \begin{pmatrix} 1 & 0 \\ 0 & -1 \end{pmatrix}$$

and the projection operators (2.19) projecting back to the Weyl fields according to (2.18)[11].

The kinetic term of the Dirac Lagrangian decomposes into a L and a R part $\mathcal{L}_{\text{Dirac}} = \bar\psi \gamma^\mu \partial_\mu \psi = \bar\psi_R \gamma^\mu \partial_\mu \psi_R + \bar\psi_L \gamma^\mu \partial_\mu \psi_L$ (4 degrees of freedom). A Dirac mass term $m\bar\psi\psi = m\,(\bar\psi_L\psi_R + \bar\psi_R\psi_L)$ breaks chiral symmetry as it is non–diagonal in the Weyl fields and induces helicity flip transitions as required by the anomalous magnetic moment in a renormalizable QFT. A remark concerning hadrons. It might look somewhat surprising that hadrons, which are composite particles made of *colored quarks* and *gluons*, in many respects look like "elementary particles" which are well described as Wigner particles (if one switches off the electromagnetic interaction which cause a

---

[11] The standard representation of the Dirac field/algebra, described in Sect. 2.1.1, is adapted to a simple interpretation in the rest frame (requires $m \neq 0$). It may be obtained from the ones in the Weyl basis ("helicity" representation) by a similarity transformation $\mathcal{S}$

$$\psi(x) = \mathcal{S}\,\psi^{\text{helicity}}(x)\,,\quad \gamma_\mu = \mathcal{S}\,\gamma_\mu^{\text{helicity}}\mathcal{S}^{-1}\,,\quad \mathcal{S} = \mathcal{S}^{-1} = \frac{1}{\sqrt{2}} \begin{pmatrix} 1 & 1 \\ 1 & -1 \end{pmatrix}$$

such that

$$u(0, r) = \sqrt{2m}\begin{pmatrix} U(r) \\ 0 \end{pmatrix}\,,\quad v(0,r) = \sqrt{2m}\begin{pmatrix} 0 \\ V(r) \end{pmatrix}$$

in the standard basis.

serious IR problem which spoils the naive Wigner state picture as we will describe below), particles of definite mass and spin and charge quantized in units of e and have associated electromagnetic form factors and in particular a definite magnetic moment. However, the gyromagnetic ratio g_P from the relation $\boldsymbol{\mu}_P = g_P\, e\hbar/(2m_P\, c)\, \boldsymbol{s}$ turns out to be $g_P \sim 2.8$ or $a_P = (g_P - 2)/2 \sim 0.4$ showing that the proton is not really a Dirac particle and its anomalous magnetic moment indicates that the proton is not a point particle but has internal structure. This was first shown long time ago by atomic beam magnetic deflection experiments [19], before the nature of the muon was clarified. For the latter it was the first measurement at CERN which yielded $a_\mu = 0.00119(10)$ [20] and revealed the muon to be just a heavy electron. Within errors at that time the muon turned out to have the same value of the anomalous magnetic moment as the electron, which is known to be due to virtual radiative corrections.

The analysis of the spin structure on a formal level, discussing the quantum mechanical implementation of relativistic symmetry principles, fits very naturally with the observed spin phenomena. In particular the existence of the fundamental spin $\frac{1}{2}$ particles which must satisfy Pauli's exclusion principle has dramatic consequences for real life. Without the existence of spin as an extra fundamental quantum number in general and the spin $\frac{1}{2}$ fermions in particular, stability of nuclei against Coulomb collapse and of stars against gravitational collapse would be missing and the universe would not be ours.

2.3 Quantum Electrodynamics

The lepton–photon interaction is described by QED, which is structured by local $U(1)$ gauge invariance[12]

$$\begin{aligned}\psi(x) &\to \mathrm{e}^{-\mathrm{i}e\alpha(x)}\psi(x) \\ A_\mu(x) &\to A_\mu(x) - \partial_\mu\alpha(x)\ ,\end{aligned} \qquad (2.88)$$

with an arbitrary scalar function $\alpha(x)$, implying lepton–photon interaction according to *minimal coupling*, which means that we have to perform the substitution $\partial_\mu \to D_\mu = \partial_\mu - \mathrm{i}eA_\mu(x)$ in the Dirac equation $(\mathrm{i}\gamma^\mu\partial_\mu - m)\psi(x) = 0$ of a free lepton[13]. This implies that the electromagnetic interaction is described by the bare Lagrangian

---

[12] The known elementary particle interactions, the strong, electromagnetic and weak forces, all derive from a local gauge symmetry principle. This was first observed by Weyl [21] for the Abelian QED and later extended to non–Abelian gauge theories by Yang and Mills [22]. The gauge symmetry group governing the Standard Model of particle physics is $SU(3)_c \otimes SU(2)_L \otimes U(1)_Y$.

[13] The modified derivative $D_\mu = \partial_\mu - \mathrm{i}eA_\mu(x)$ is called *covariant derivative*. e is the *gauge coupling*. The minimal substitution promotes the global gauge symmetry of the free Dirac Lagrangian to a local gauge symmetry of the electron–photon system, i.e. the interacting system has more symmetry than the free electron.

$$\mathcal{L}^{\text{QED}} = -\frac{1}{4}F_{\mu\nu}F^{\mu\nu} - \frac{1}{2}\xi^{-1}(\partial_\mu A^\mu)^2 + \bar{\psi}(i\gamma^\mu D_\mu - m)\psi$$
$$= \mathcal{L}_{0A}^{\xi} + \mathcal{L}_{0\psi} + e j_{\text{em}}^\mu(x) A_\mu(x) \ , \tag{2.89}$$

and the corresponding field equations read[14]

$$(i\gamma^\mu \partial_\mu - m)\psi(x) = -e : A_\mu(x)\gamma^\mu \psi(x) :$$
$$(\Box g^{\mu\nu} - (1-\xi^{-1})\partial^\mu \partial^\nu) A_\nu(x) = -e : \bar{\psi}(x)\gamma^\mu \psi(x) : \ . \tag{2.90}$$

The interaction part of the Lagrangian is

$$\mathcal{L}_{\text{int}} = e j_{\text{em}}^\mu(x) A_\mu(x) \ , \tag{2.91}$$

while the bilinear free field parts $\mathcal{L}_{0A}^{\xi}$ and $\mathcal{L}_{0\psi}$ define the propagators of the photon and the leptons, respectively (given below). As in classical electrodynamics the gauge potential A^μ is an auxiliary field which exhibits unphysical degrees of freedom, and is not uniquely determined by Maxwell's equations. In order to get a well defined photon propagator a gauge fixing condition is required. We adopt the linear covariant Lorentz gauge : $\partial_\mu A^\mu = 0$, which is implemented via the Lagrange multiplier method, with Lagrange multiplier $\lambda = 1/\xi$, ξ is called gauge parameter[15]. The gauge invariance of physical quantities infers that they do not depend on the gauge parameter.

Above we have denoted by e the charge of the electron, which by convention is taken to be negative. In the following we will explicitly account for the sign of the charge and use e to denote the positive value of the charge of the positron. The charge of a fermion f is then given by $Q_f e$, with Q_f the charge of a fermion in units of the positron charge e. A collection of charged fermions f enters the electromagnetic current as

$$j_{\text{em}}^\mu = \sum_f Q_f \bar{\psi}_f \gamma^\mu \psi_f \ , \tag{2.92}$$

for the leptons alone $j_{\text{em}}^{\mu\,\text{lep}} = -\sum_\ell \bar{\psi}_\ell \gamma^\mu \psi_\ell$ ($\ell = e, \mu, \tau$). If not specified otherwise $\psi(x)$ in the following will denote a lepton field carrying negative charge $-e$.

One important object we need for our purpose is the *unitary* scattering matrix S which encodes the perturbative lepton–photon interaction processes and is given by

$$S = T\left(e^{i\int d^4 x\, \mathcal{L}_{\text{int}}^{(1)}(x)}\right)\bigg|_{\otimes} \ . \tag{2.93}$$

---

[14]The prescription : $\cdots$: means *Wick ordering* of products of fields: write the fields in terms of creation and annihilation operators and order them such that all annihilation operators are to the right of all creation operators, assuming the operators to commute (bosons) or to anticommute (fermions). This makes the vacuum expectation value of the field product vanish.

[15]The parametrization of the gauge dependence by the inverse of the Lagrange multiplier $\xi = 1/\lambda$ is just a commonly accepted convention.

The prescription ⊗ says that all graphs (see below) which include vacuum diagrams (disconnected subdiagrams with no external legs) as factors have to be omitted. This corresponds to the proper normalization of the S–operator. Unitarity requires

$$SS^+ = S^+S = 1 \quad \Leftrightarrow \quad S^+ = S^{-1} \tag{2.94}$$

and infers the conservation of quantum mechanical transition probabilities. The prescription T means time ordering of all operators, like

$$T\{\phi(x)\phi(y)\} = \Theta(x^0 - y^0)\phi(x)\phi(y) \pm \Theta(y^0 - x^0)\phi(y)\phi(x) \tag{2.95}$$

where the + sign holds for boson fields and the − sign for fermion fields. Under the T prescription all fields are commuting (bosons) or anticommuting (fermions). All fields in (2.93) may be taken to be free fields. With the help of S we may calculate the basic objects of a QFT, the *Green functions*. These are the vacuum expectation values of *time ordered* or *chronological* products of fields like the electromagnetic correlator

$$G_{\mu,\alpha\beta}(x,y,\bar{y}) \doteq \langle 0|T\{A_\mu(x)\psi_\alpha(y)\bar{\psi}_\beta(\bar{y})\}|0\rangle. \tag{2.96}$$

2.3.1 Perturbation Expansion, Feynman Rules

The full Green functions of the interacting fields like $A^\mu(x)$, $\psi(x)$, etc. can be expressed completely in terms of corresponding free fields via the *Gell-Mann Low formula* [23] (interaction picture)

$$\langle 0|T\{A_\mu(x)\psi_\alpha(y)\bar{\psi}_\beta(\bar{y})\}|0\rangle = $$
$$\langle 0|T\left\{A_\mu^{(0)}(x)\psi_\alpha^{(0)}(y)\bar{\psi}_\beta^{(0)}(\bar{y})\,\mathrm{e}^{\mathrm{i}\int \mathrm{d}^4x'\,\mathcal{L}_{\mathrm{int}}^{(0)}(x')}\right\}|0\rangle_\otimes = \sum_{n=0}^{N} \frac{\mathrm{i}^n}{n!}\int \mathrm{d}^4z_1\cdots \mathrm{d}^4z_n$$
$$\langle 0|T\left\{A_\mu^{(0)}(x)\psi_\alpha^{(0)}(y)\bar{\psi}_\beta^{(0)}(\bar{y})\,\mathcal{L}_{\mathrm{int}}^{(0)}(z_1)\cdots\mathcal{L}_{\mathrm{int}}^{(0)}(z_n)\right\}|0\rangle_\otimes + O(\mathrm{e}^{N+1})$$
$$\tag{2.97}$$

with $\mathcal{L}_{\mathrm{int}}^{(0)}(x)$ the interaction part of the Lagrangian. On the right hand side all fields are free fields and the vacuum expectation values can be computed by applying the known properties of free fields. Expanding the exponential as done in (2.97) yields the perturbation expansion. The evaluation of the formal perturbation series is not well defined and requires regularization and renormalization, which we will discuss briefly below. In a way the evaluation is simple: one writes all free fields in terms of the creation and annihilation operators and applies the canonical anticommutation (fermions) and the canonical commutation (bosons) relations to bring all annihilation operators to the right, where they annihilate the vacuum $\cdots a(\boldsymbol{p},r)|0\rangle = 0$ and the creation operators to the left where again they annihilate the vacuum $0 = \langle 0|b^+(\boldsymbol{p},r)\cdots$, until no operator is left over (Wick ordering) [24]. The only non–vanishing contribution comes from the complete contraction of all fields in pairs, where a pairing corresponds to a propagator as a factor. The rules for the evaluation of all possible contributions are known as the Feynman Rules.

The Feynman Rules

1) draw all vertices as points in a plane: external ones with the corresponding external fields $\psi(y_i)$, $\bar\psi(\bar y_j)$ or $A^\mu(x_k)$ attached to the point, and the internal interaction vertices $-ie\bar\psi\gamma_\mu\psi A^\mu(z_n)$ with three fields attached to the point z_n.
2) contract all fields in pairs represented by a line connecting the two vertices, thereby fields of different particles are to be characterized by different types of lines.

As a result one obtains a Feynman diagram.

The field pairings define the free propagators

$$\overline{\psi(y) \cdots \bar\psi(\bar y)} \Leftrightarrow iS_\mathrm{F}(y-\bar y) \text{ and } \overline{A^\mu(x_1) \cdots A^\nu(x_2)} \Leftrightarrow iD^{\mu\nu}(x_1-x_2)$$

given by the vacuum expectation values of the pair of time–ordered free fields,

$$iS_{\mathrm{F}\,\alpha\beta}(y-\bar y) \doteq \langle 0|T\{\psi(y)_\alpha \bar\psi(\bar y)_\beta\}|0\rangle$$
$$iD^{\mu\nu}(x_1-x_2) \doteq \langle 0|T\{A^\mu(x_1) A^\nu(x_2)\}|0\rangle \ .$$

The latter may easily be calculated using the free field properties.

Feynman diagrams translate into Feynman integrals via the famous Feynman rules given by Fig. 2.2 in momentum space.

In configuration space all interaction vertices in (2.97) are integrated over. The result thus is a Feynman integral. In fact the perturbation expansion is not yet well defined. In order to have a well defined starting point, the theory has to be *regularized* [25] and parameter and fields have to be *renormalized* in order to obtain a well defined set of renormalized Green functions. The problems arise because propagators are singular functions (so called distributions) the products of them are not defined at coinciding space–time arguments (short–distance [coordinate space] or ultra–violet [momentum space] singularities). An example of such an ill–defined product is the Fermion loop contribution to the photon propagator:

$$iS_\mathrm{F}(x-y)_{\alpha\beta}\,(-ie\gamma_\mu)_{\beta\gamma}\,iS_\mathrm{F}(y-x)_{\gamma\delta}\,(-ie\gamma_\nu)_{\delta\alpha} \ .$$

The ambiguity in general can be shown to be a local distribution, which for a renormalizable theory is of the form [27]

$$a\delta(x-y) + b^\mu\partial_\mu\delta(x-y) + c\,\Box\,\delta(x-y) + d^{\mu\nu}\partial_\mu\partial_\nu\delta(x-y)$$

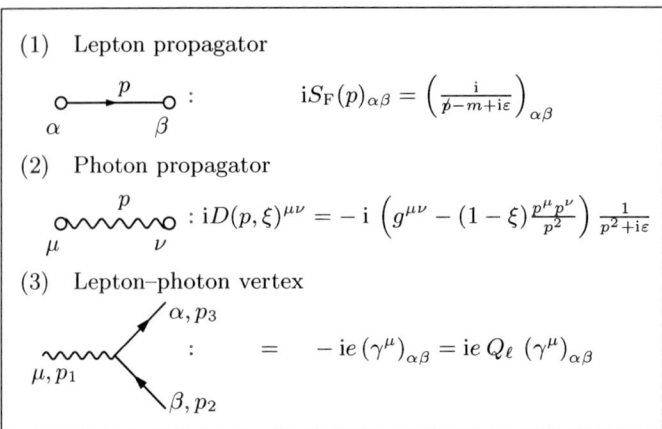

Fig. 2.2. Feynman rules for QED (I)

with derivatives up to second order at most, which, in momentum space, is a second order polynomial in the momenta[16]. The regularization we will adopt is dimensional regularization [32], where the space–time dimension is taken to be d arbitrary to start with (see below).

In momentum space each line has associated a d–momentum p_i and at each vertex momentum conservation holds. Because of the momentum conservation δ–functions many d–momentum integrations become trivial. Each loop, however, has associated an independent momentum (the loop–momentum) l_i which has to be integrated over

$$\frac{1}{(2\pi)^d} \int d^d l_i \cdots \qquad (2.98)$$

---

[16] The mathematical problems with the point–like structure of elementary particles and with covariant quantization of the photons hindered the development of QFT for a long time until the break through at the end of the 1940s [26]. In 1965 Tomonaga, Schwinger and Feynman were honored with the Nobel Prize "for their fundamental work in quantum electrodynamics, with deep–ploughing consequences for the physics of elementary particles". For non–Abelian gauge theories like the modern strong interaction theory *Quantum Chromodynamics* (QCD) [29, 30] and the electroweak *Standard Model* [31], the proper quantization, regularization and renormalization was another obstacle which was solved only at the beginning of the 1970s by 't Hooft and Veltman. They were awarded the Nobel Prize in 1999 "for elucidating the quantum structure of electroweak interactions in physics". They have placed particle physics theory on a firmer mathematical foundation. They have in particular shown how the theory, beyond QED, may be used for precise calculations of physical quantities. Needless to say that these developments were crucial for putting precision physics, like the one with the anomalous magnetic moments, on a fundamental basis.

in d space–time dimensions. For each closed fermion loop a factor -1 has to be applied because of Fermi statistics. There is an overall d–momentum conservation factor $(2\pi)^d \, \delta^{(d)}(\sum p_{i\,\text{external}})$. Note that the lepton propagators as well as the vertex insertion $ie\gamma_\mu$ are matrices in spinor space, at each vertex the vertex insertion is sandwiched between the two adjacent propagators:

$$\cdots \mathrm{i} S_F(p)_{\alpha\gamma} \, (-\mathrm{i} e \gamma_\mu)_{\gamma\delta} \, \mathrm{i} S_F(p')_{\delta\beta} \cdots$$

Since any renormalizable theory exhibits fermion fields not more than bilinear, as a conjugate pair $\bar\psi \cdots \psi$, fermion lines form open strings

$$[\Pi^n_{i=1}(S\gamma)_i] \; S = \text{\scriptsize(diagram)} \qquad (2.99)$$

of matrices in spinor space

$$[S_F(p_1)\,\gamma_{\mu_1}\,S_F(p_2)\,\gamma_{\mu_2}\cdots\gamma_{\mu_n}\,S_F(p_{n+1})]_{\alpha\beta}$$

or closed strings (fermion loops),

$$\mathrm{Tr}\,[\Pi^n_{i=1}(S\gamma)_i] = \text{\scriptsize(diagram)} \qquad (2.100)$$

which correspond to a trace of a product of matrices in spinor space:

$$\mathrm{Tr}\,[S_F(p_1)\,\gamma_{\mu_1}\,S_F(p_2)\,\gamma_{\mu_2}\cdots S_F(p_n)\,\gamma_{\mu_n}]\;.$$

Closed *fermion loops* actually contribute with two different orientations. If the number of vertices is odd the two orientations yield traces in spinor space of opposite sign such that they cancel provided the two contributions have equal weight. If the number of vertices is even the corresponding traces in spinor space contribute with equal sign, i.e. it just makes a factor of two in the equal weight case. In QED in fact the two orientations have equal weight due to the charge conjugation invariance of QED and is called *Furry's theorem* [33]. As already mentioned, each Fermion loop carries a factor -1 due the Fermi statistics. All this is easy to check using the known properties of the Dirac fields[17].

For a given set of external vertices and a given order n of perturbation theory (n internal vertices) one obtains a sum over all possible complete contractions, where each one may be represented by a Feynman diagram Γ. The Fourier transform (FT) thus, for each connected component of a diagram, is given by expressions of the form

---

[17] Note that in QCD the corresponding closed quark loops with quark–gluon vertices behave differently because of the color matrices at each vertex. The trace of the product of color matrices in general has an even as well as an odd part.

$$\mathrm{FT}\,\langle 0|T\left\{A_\mu(x_1)\cdots\psi_\alpha(y_1)\cdots\bar\psi_\beta(\bar y_1)\cdots\right\}|0\rangle_{\mathrm{connected}} =$$

$$= (-\mathrm{i})^F\,(2\pi)^d\delta^{(d)}(\sum p_{\mathrm{ext}})\left(\Pi_{i=1}^N\int\frac{\mathrm{d}^d l_i}{(2\pi)^d}\right)$$

$$\times\sum_\Gamma \Pi_{i\in L_\ell,i\notin \bar L_f}\,\mathrm{i}S_{\mathrm{F}}(p_i)\,(-\mathrm{i}e\gamma_{\mu_i})\left[\Pi_{f\in\bar L_f}\mathrm{i}S_{\mathrm{F}}(p_f)\right]\Pi_{j\in L_\gamma}\mathrm{i}D^{\mu_j\nu_j}(q_j)\;,$$

where L_ℓ is the set of lepton lines, L_γ the set of photon lines and $\bar L_f$ the set of lines starting with an external $\bar\psi$ field, N the number of independent closed loops and F the number of closed fermion loops. Of course, spinor indices and Lorentz indices must contract appropriately, and momentum conservation must be respected at each vertex and over all. The basic object of our interest is the Green function associated with the electromagnetic vertex dressed by external propagators:

$$G_{\mu,\alpha\beta}(x,y,z)\doteq\langle 0|T\left\{A_\mu(x)\psi_\alpha(y)\bar\psi_\beta(z)\right\}|0\rangle =$$
$$\int \mathrm{d}x'\mathrm{d}y'\mathrm{d}z'\,\mathrm{i}D'_{\mu\nu}(x'-x)\,\mathrm{i}S'_{\mathrm{F}\alpha\alpha'}(y'-y)\left(\mathrm{i}\Gamma^\nu_{\alpha'\beta'}(x',y',z')\right)\mathrm{i}S'_{\mathrm{F}\beta'\beta}(z'-z)$$

which graphically may be represented as follows

with *one particle irreducible*[18] (1PI) dressed vertex

where $\mathrm{i}D'_{\mu\nu}(x'-x)$ is a full photon propagator, a photon line dressed with all radiative corrections:

---

[18] Diagrams which cannot be cut into two disconnected diagrams by cutting a single line. 1PI diagrams are the building blocks from which any diagram may be obtained as a tree of 1PI "blobs".

$\mathrm{i} D'_{\mu\nu}(x'-x) =$ [diagram] $=$ [diagrams] $+ \cdots$

and $\mathrm{i} S'_{F\alpha\alpha'}(y'-y)$ is the full lepton propagator, a lepton line dressed by all possible radiative corrections

$\mathrm{i} S'_{F\alpha\alpha'}(y'-y) =$ [diagrams] $+ \cdots$

The tools and techniques of calculating these objects as a perturbation series in lowest non–trivial order will be developed in the next section.

2.3.2 Transition Matrix–Elements, Particle–Antiparticle Crossing

The Green functions from the point of view of a QFT are building blocks of the theory. However, they are not directly observable objects. The physics is described by quantum mechanical transition matrix elements, which for scattering processes are encoded in the scattering matrix. For QED the latter is given formally by (2.93). The existence of a S–matrix requires that for very early and for very late times ($t \to \mp\infty$) particles behave as free scattering states. For massless QED, the electromagnetic interaction does not have finite range (Coulomb's law) and the scattering matrix does not exist in the naive sense. In an order by order perturbative approach the problems manifest themselves as an infrared (IR) problem. As we will see below, nevertheless a suitable redefinition of the transition amplitudes is possible, which allows a perturbative treatment under appropriate conditions. Usually, one is not directly interested in the S–matrix as the latter includes the identity operator I which describes through–going particles which do not get scattered at all. It is customary to split off the identity from the S–matrix and to define the T–**matrix** by

$$S = I + \mathrm{i}\,(2\pi)^4\,\delta^{(4)}(P_f - P_i)\,T \;, \tag{2.101}$$

with the overall four–momentum conservation factored out. In spite of the fact, that Green functions are not observables they are very useful to understand important properties of the theory. One of the outstanding features of a QFT is the *particle–antiparticle crossing* property which states that in a scattering amplitude an incoming particle [antiparticle] is equivalent to an outgoing antiparticle [particle] and vice versa. It means that the same function, namely an appropriate time–ordered Green function, at the same time describes several processes. For example, muon pair production in electron positron annihilation $e^+e^- \to \mu^+\mu^-$ is described by amplitudes which at the same time describe electron–muon scattering $e^-\mu^- \to e^-\mu^-$ or whatever process we can obtain by bringing particles from one side of the reaction balance to the other side as an antiparticle etc. Another example is muon decay $\mu^+ \to e^+ \nu_e \bar{\nu}_\mu$ and neutrino scattering $\nu_\mu e^- \to \mu^- \nu_e$. For the electromagnetic vertex it relates properties of the electrons [leptons, quarks] to properties of the positron [antileptons, antiquarks].

Since each external free field on the right hand side of (2.97) exhibits an annihilation part and a creation part, each external field has two interpretations, either as an incoming particle or as an outgoing antiparticle. For the adjoint field incoming and outgoing get interchanged. This becomes most obvious if we invert the field decomposition (2.11) for the Dirac field which yields the corresponding creation/annihilation operators

$$a(\bm{p},r) = \bar{u}(\bm{p},r)\gamma^0 \int d^3x\, e^{ipx}\, \psi(x) \ , \quad b^+(\bm{p},r) = \bar{v}(\bm{p},r)\gamma^0 \int d^3x\, e^{-ipx}\, \psi(x) \ .$$

Similarly, inverting (2.12) yields

$$c(\bm{p},\lambda) = -\varepsilon^{\mu*}(\bm{p},\lambda)\, i \int d^3x\, e^{ipx}\, \overleftrightarrow{\partial}_0 A_\mu(x)$$

and its hermitian conjugate for the photon, with $f(x) \overleftrightarrow{\partial}_\mu g(x) \equiv f(x)\partial_\mu g(x) - (\partial_\mu f(x)) g(x)$. Since these operators create or annihilate scattering states, the above relations provide the bridge between the Green functions, the vacuum expectation values of time–ordered fields, and the scattering matrix elements. This is how the crossing property between different physical matrix elements comes about. The S–matrix elements are obtained from the Green functions by the Lehmann, Symanzik, Zimmermann [34] (LSZ) reduction formula: the external full propagators of the Green functions are omitted (multiplication by the inverse full propagator, i.e. no radiative corrections on external amputated legs) and replaced by an external classical one particle wave function and the external momentum is put on the mass shell. Note that the on–shell limit only exists after the amputation of the external one particle poles. Graphically, at lowest order, the transition from a Green function to a T matrix–element for a lepton line translates into

$$\lim_{\slashed{q} \to m} -i(\slashed{p} - m) \quad \raisebox{-2pt}{\includegraphics{}} \quad \to \quad \raisebox{-2pt}{\includegraphics{}} \quad = \quad u(p,r) \cdots$$

and a corresponding operation has to be done for all the external lines of the Green function.

The set of relations for QED processes is given in Table 2.1.

We are mainly interested in the electromagnetic vertex here, where the crossing relations are particularly simple, but not less important. From the 1PI vertex function $\Gamma^\mu(p_1,p_2)$ we obtain
the electron form factor for $e^-(p_1) + \gamma(q) \to e^-(p_2)$

$$T = \bar{u}(p_2,r_2)\Gamma^\mu(p_1,p_2)u(p_1,r_1) \ ,$$

the positron form factor for $e^+(-p_2) + \gamma(q) \to e^+(-p_1)$
$$T' = \bar{v}(p_2, r_2)\Gamma^\mu(-p_2, -p_1)v(p_1, r_1) \,,$$
and the e^+e^--annihilation amplitude of $e^-(p_1) + e^+(-p_2) \to \gamma(-q)$
$$T'' = \bar{v}(p_2, r_2)\Gamma^\mu(p_1, p_2)u(p_1, r_1) \,.$$

Given the T matrix–elements, the bridge to the experimental numbers is given by the cross–sections and decay rates, which we present for completeness here.

2.3.3 Cross Sections and Decay Rates

The **differential cross section** for a two particle collision
$$A(p_1) + B(p_2) \to C(p'_1) + D(p'_2) \cdots$$
is given by
$$d\sigma = \frac{(2\pi)^4 \delta^{(4)}(P_f - P_i)}{2\sqrt{\lambda(s, m_1^2, m_2^2)}} \, |T_{fi}|^2 \, d\mu(p'_1)d\mu(p'_2) \cdots$$

$s = (p_1 + p_2)^2$ is the square of the total CM energy and $\lambda(x,y,z) = x^2 + y^2 + z^2 - 2xy - 2xz - 2yz$ is a two body phase–space function. In the CM frame (see the figure):

Table 2.1. Rules for the treatment of external legs in the evaluation of T–matrix elements

Scattering state	Graphical representation	Wave function
Dirac particles: incoming particle		$u(p, r)$
incoming antiparticle		$\bar{v}(p, r)$
outgoing particle		$\bar{u}(p, r)$
outgoing antiparticle		$v(p, r)$
Photon: incoming photon		$\varepsilon^\mu(p, r)$
outgoing photon		$\varepsilon^{\mu *}(p, r)$

$$\sqrt{\lambda} = \sqrt{\lambda(s, m_1^2, m_2^2)} = 2\,|\,\boldsymbol{p}\,|\sqrt{s} \qquad (2.102)$$

where $\boldsymbol{p} = \boldsymbol{p}_i$ is the three–momentum of the initial state particle A.

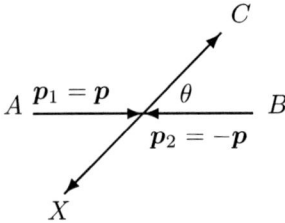

The total cross–section follows by integration over all phase space

$$\sigma = \int d\sigma \,.$$

Finally, we consider the decay of unstable particles. The differential **decay rate** for $A \to B + C + \cdots$ is given by

$$d\Gamma = \frac{(2\pi)^4 \delta^{(4)}(P_f - P_i)}{2 m_1} \,|\,T_{fi}\,|^2 \, d\mu(p_1') d\mu(p_2') \cdots$$

By "summing" over all possible decay channels we find the total width

$$\Gamma = \oint d\Gamma = \frac{1}{\tau}, \qquad (2.103)$$

where τ is the **lifetime** of the particle, which decays via the exponential decay law

$$N(t) = N_0 \, \mathrm{e}^{-t/\tau}\,. \qquad (2.104)$$

Cross sections are measured typically by colliding beams of stable particles and their antiparticles like electrons (e^-), positrons (e^+), protons (p) or antiprotons ($\bar{p}$). The beam strength of an accelerator or storage ring required for accelerating and collimating the beam particles is determined by the particle flux or *luminosity* L, the number of particles per cm^2 and *seconds*. The energy of the machine determines the resolution

$$\lambda = \frac{hc}{E_{\mathrm{c.m.}}} \simeq \frac{1.2\,\mathrm{GeV}}{E_{\mathrm{c.m.}}(\mathrm{GeV})} \times 10^{-15}\,\mathrm{m}\,,$$

while the luminosity determines the collision rate

$$\frac{\Delta N}{\Delta t} = L \cdot \sigma\,,$$

and the cross–section σ is thus given by dividing the observed event rate by the luminosity

$$\sigma = \frac{1}{L}\,\frac{\Delta N}{\Delta t}\,. \qquad (2.105)$$

2.4 Regularization and Renormalization

The vertex and self–energy functions, as well as all other Green functions, on the level of the bare theory are well defined order by order in perturbation theory only after smoothing the short distance or ultraviolet (UV) divergences by appropriate regularization. Here we assume QED or the SM to be regularized by dimensional regularization [32]. By going to lower dimensional space–times the features of the theory, in particular the symmetries, remain the same, however, the convergence of the Feynman integrals gets improved. For a renormalizable theory, in principle, one can always choose the dimension low enough, $d < 2$, such that the integrals converge. By one or two partial integrations one can analytically continue the integrals in steps from d to $d+1$, such that the perturbation expansion is well defined for $d = 4 - \epsilon$ with ϵ a small positive number. For $\epsilon \to 0$ ($d \to 4$) the perturbative series in the *fine structure constant* $\alpha = e^2/4\pi$ exhibits poles in ϵ:

$$A = \sum_{n=0}^{N} \alpha^n \sum_{m=0}^{n} a_{nm}(1/\epsilon)^{n-m}$$

and the limit $d \to 4$ to the real physical space–time does not exist, at first. The problems turn out to be related to the fact that the bare objects are not physical ones, they are not directly accessible to observation and require some adjustments. This in particular is the case for the bare parameters, the bare fine structure constant (electric charge) which is modified by vacuum polarization (quantum fluctuations), and the bare masses. Also the bare fields are not the ones which interpolate suitably to the physical states they are assumed to describe. The appropriate entities are in fact obtained by a simple reparametrization in terms of new parameters and fields, which is called *renormalization*.

2.4.1 The Structure of the Renormalization Procedure

Renormalization may be performed in three steps:

(i) Shift of the mass parameters or *mass renormalization*: replace the bare mass parameters of the bare Lagrangian by renormalized ones

$$\begin{aligned} m_{f0} &= m_{f\mathrm{ren}} + \delta m_f \quad \text{for fermions} \\ M_{b0}^2 &= M_{b\mathrm{ren}}^2 + \delta M_b^2 \quad \text{for bosons} \end{aligned} \quad (2.106)$$

(ii) Multiplicative renormalization of the bare fields or *wave function renormalization*: replace the bare fields in the bare Lagrangian by renormalized ones

$$\psi_{f0} = \sqrt{Z_f}\,\psi_{f\mathrm{ren}}\,, \quad A_0^\mu = \sqrt{Z_\gamma}\,A_{\mathrm{ren}}^\mu \quad (2.107)$$

and correspondingly for the other fields of the SM. To leading order $Z_i = 1$ and hence

$$Z_i = 1 + \delta Z_i \; , \quad \sqrt{Z_i} = 1 + \frac{1}{2}\delta Z_i + \cdots \quad (2.108)$$

(iii) Vertex renormalization or *coupling constant renormalization*: substitute the bare coupling constant by the renormalized one

$$e_0 = e_{\mathrm{ren}} + \delta e \; . \quad (2.109)$$

The *renormalization theorem* states that order by order in the perturbation expansion all UV divergences showing up in physical quantities (S–matrix elements) get eliminated by an appropriate choice of the *counter terms* $\delta m_f, \delta M_b^2, \delta e$ and $\delta Z_i = Z_i - 1$. In other words, suitably normalized physical amplitudes expressed in terms of measurable physical parameters are finite in the limit $\epsilon \to 0$, i.e. they allow us to take away the regularization (cut–off $\Lambda \to \infty$ if a UV cut–off was used to regularize the bare theory). Note that for Green functions, which are not gauge invariant in general, also the fictitious gauge parameter has to be renormalized in order to obtain finite Green functions.

The reparametrization of the bare Lagrangian (2.89) in terms of renormalized quantities reads

$$\begin{aligned}
\mathcal{L}^{\mathrm{QED}} &= -\frac{1}{4} F_{\mu\nu\,0}(x) F_0^{\mu\nu}(x) - \frac{1}{2}\xi_0^{-1}\left(\partial_\mu A_0^\mu(x)\right)^2 + \bar{\psi}_0(x)\,(i\gamma^\mu\partial_\mu - m_0)\,\psi_0(x) \\
&\quad - e_0 \bar{\psi}_0(x)\,\gamma^\mu \psi_0(x)\, A_{\mu\,0}(x) \\
&= \mathcal{L}^{\mathrm{QED}}_{(0)} + \mathcal{L}^{\mathrm{QED}}_{\mathrm{int}}
\end{aligned}$$

$$\begin{aligned}
\mathcal{L}^{\mathrm{QED}}_{(0)} &= -\frac{1}{4} F_{\mu\nu\,\mathrm{ren}}(x) F_{\mathrm{ren}}^{\mu\nu}(x) - \frac{1}{2}\xi_{\mathrm{ren}}^{-1}\left(\partial_\mu A_{\mathrm{ren}}^\mu(x)\right)^2 \\
&\quad + \bar{\psi}_{\mathrm{ren}}(x)\,(i\gamma^\mu\partial_\mu - m_{\mathrm{ren}})\,\psi_{\mathrm{ren}}(x)
\end{aligned}$$

$$\begin{aligned}
\mathcal{L}^{\mathrm{QED}}_{\mathrm{int}} &= -e_{\mathrm{ren}}\,\bar{\psi}_{\mathrm{ren}}(x)\,\gamma^\mu \psi_{\mathrm{ren}}(x)\,A_{\mu\,\mathrm{ren}}(x) \\
&\quad -\frac{1}{4}(Z_\gamma - 1)\,F_{\mu\nu\,\mathrm{ren}}(x) F_{\mathrm{ren}}^{\mu\nu}(x) + (Z_e - 1)\,\bar{\psi}_{\mathrm{ren}}(x)\,i\gamma^\mu\partial_\mu\psi_{\mathrm{ren}} \\
&\quad -(m_0 Z_e - m_{\mathrm{ren}})\,\bar{\psi}_{\mathrm{ren}}\psi_{\mathrm{ren}}(x) \\
&\quad -(e_0\sqrt{Z_\gamma}Z_e - e_{\mathrm{ren}})\,\bar{\psi}_{\mathrm{ren}}(x)\,\gamma^\mu\psi_{\mathrm{ren}}(x)\,A_{\mu\,\mathrm{ren}}(x) \quad (2.110)
\end{aligned}$$

with $\xi_{\mathrm{ren}} = Z_\gamma \xi_0$ the gauge fixing term remains unrenormalized (no corresponding counter term). The counter terms now are showing up in $\mathcal{L}^{\mathrm{QED}}_{\mathrm{int}}$ and may be written in terms of $\delta Z_\gamma = Z_\gamma - 1$, $\delta Z_e = Z_e - 1$, $\delta m = m_0 Z_e - m_{\mathrm{ren}}$ and $\delta e = e_0\sqrt{Z_\gamma}Z_e - e_{\mathrm{ren}}$. They are of next higher order in e^2, either $O(e^2)$ for propagator insertions or $O(e^3)$ for the vertex insertion, in leading order. The counter terms have to be adjusted order by order in perturbation theory by the renormalization conditions which define the precise physical meaning of the parameters (see below).

The Feynman rules Fig. 2.2 have to be supplemented by the rules of including the counter terms shown in Fig 2.3 in momentum space.

Obviously the propagators (two–point functions) of the photon and of the electron get renormalized according to

$$D_0 = Z_\gamma D_{\text{ren}}$$
$$S_{F\,0} = Z_e S_{F\,\text{ren}} \ . \tag{2.111}$$

The renormalized electromagnetic vertex function may be obtained according to the above rules as

$$G^\mu_{\text{ren}} = \frac{1}{\sqrt{Z_\gamma}} \frac{1}{Z_e} G^\mu_0 \tag{2.112}$$

$$= D_{\text{ren}} S_{F\,\text{ren}} \Gamma^\mu_{\text{ren}} S_{F\,\text{ren}} = \frac{1}{\sqrt{Z_\gamma}} \frac{1}{Z_e} D_0 S_{F\,0} \Gamma^\mu_0 S_{F\,0}$$

$$= \frac{1}{\sqrt{Z_\gamma}} \frac{1}{Z_e} Z_\gamma Z_e^2 \, D_{\text{ren}} S_{F\,\text{ren}} \Gamma^\mu_0 S_{F\,\text{ren}}$$

and consequently

$$\Gamma^\mu_{\text{ren}} = \sqrt{Z_\gamma} Z_e \, \Gamma^\mu_0 = \sqrt{Z_\gamma} Z_e \left\{ e_0 \gamma^\mu + \Gamma'^\mu_0 \right\} \Big|_{e_0 \to e + \delta e,\ m_0 \to m + \delta m,\ \ldots}$$

$$= \sqrt{1 + \delta Z_\gamma} \, (1 + \delta Z_e) \left\{ e \, (1 + \frac{\delta e}{e}) \gamma^\mu + \Gamma'^\mu_0 \right\}$$

$$= \left(1 + \frac{1}{2} \delta Z_\gamma + \delta Z_e + \frac{\delta e}{e} \right) e \gamma^\mu + \Gamma'^\mu_0 + \cdots \tag{2.113}$$

(1) Lepton propagator insertions

$$\underset{\alpha}{\xrightarrow{p}} \otimes \underset{\beta}{\longrightarrow} \ : \ \mathrm{i} \left(\delta Z_e \, (\slashed{p} - m) - \delta m \right)_{\alpha\beta}$$

(2) Photon propagator insertion

$$\underset{\mu}{\sim\!\!\sim\!\!\sim} \otimes \underset{\nu}{\sim\!\!\sim\!\!\sim} \ : \ -\mathrm{i} \delta Z_\gamma \left(p^2 g^{\mu\nu} - p^\mu p^\nu \right)$$

(3) Lepton–photon vertex insertion

$$\quad : \ = \ -\mathrm{i} \delta e \, (\gamma^\mu)_{\alpha\beta}$$

with α, p_3; μ, p_1; β, p_2.

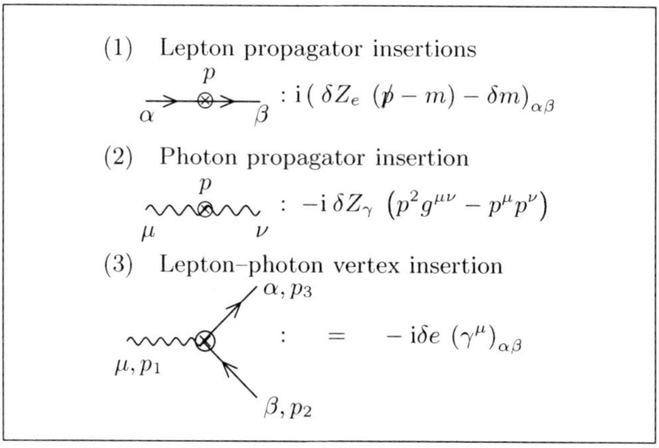

Fig. 2.3. Feynman rules for QED (II): the counter terms

where now the bare parameters have to be considered as functions of the renormalized ones:

$$e_0 = e_0(e, m), \quad m_0 = m_0(m, e) \quad \text{etc.} \tag{2.114}$$

and e, m etc. denote the renormalized parameters. The last line of (2.113) gives the perturbatively expanded form suitable for one–loop renormalization. It may also be considered as the leading n–th order renormalization if $\Gamma_0'^{\mu}$ has been renormalized to $n-1$-st order for all sub–divergences. More precisely, if we expand the exact relation of (2.113) (second last line) and include all counter terms, including the ones which follow from (2.114), up to order $n-1$ in $\Gamma_0'^{\mu}$, such that all sub–divergences of $\Gamma_0'^{\mu}$ are renormalized away, only the overall divergence of order n will be there. After including the wavefunction renormalization factors of order n as well (by calculating the corresponding propagators) the remaining overall divergence gets renormalized away by fixing $\delta e^{(n)}$, according to the last line of (2.113), by the charge renormalization condition:

$$\bar{u}(p_2, r_2) \Gamma_{\text{ren}}^{\mu}(p_1, p_2) u(p_1, r_1) = e_{\text{ren}} \bar{u}(p_2, r_2) \gamma^{\mu} u(p_1, r_1)$$

at zero photon momentum $q = p_2 - p_1 = 0$ (classical limit, *Thomson limit*).

2.4.2 Dimensional Regularization

Starting with the Feynman rules of the classical quantized Lagrangian, called bare Lagrangian, the formal perturbation expansion is given in terms of ultraviolet (UV) divergent Feynman integrals if we try to do that in $d = 4$ dimensions without UV cut–off. As an example consider the scalar one–loop self–energy diagram and the corresponding Feynman integral

$$= \frac{1}{(2\pi)^d} \int d^d k \frac{1}{k^2 - m^2 + i\varepsilon} \frac{1}{(k+p)^2 - m^2 + i\varepsilon} \underset{|k| \gg |p|, m}{\sim} \int \frac{d^d k}{k^4}$$

which is logarithmically divergent for the physical space–time dimension $d = 4$ because the integral does not fall–off sufficiently fast at large k. In order to get a well–defined perturbation expansion the theory must be ***regularized***[19].

---

[19] Often one simply chooses a cut–off (upper integration limit in momentum space) to make the integrals converge by "brute force". A cut–off may be considered to parametrize our ignorance about physics at very high momentum or energy. If the cut–off Λ is large with respect to the energy scale E of a phenomenon considered, $E \ll \Lambda$, the cut–off dependence may be removed by considering only relations between low–energy quantities (renormalization). Alternatively, a cut–off may be interpreted as the scale where one expects new physics to enter and it may serve to investigate how a quantity (or the theory) behaves under changes of the cut–off (renormalization group). In most cases simple cut–off regularization violates symmetries badly and it becomes a difficult task to make sure that one obtains the right theory when the cut–off is removed by taking the limit $\Lambda \to \infty$ after renormalization.

The regularization should respect as much as possible the symmetries of the initial bare form of the Lagrangian and of the related *Ward-Takahashi* (WT) *identities* of the "classical theory". For gauge theories like QED, QCD or the SM dimensional regularization [32] (DR) is the most suitable regularization scheme as a starting point for the perturbative approach, because it respects as much as possible the classical symmetries of a Lagrangian[20]. The idea behind DR is the following:

i) Feynman rules formally look the same in different space–time dimensions $d = n$(integer)
ii) In the UV region Feynman integrals converge the better the lower d is.

The example given above demonstrates this, in $d = 4 - \epsilon$ ($\epsilon > 0$) dimensions (just below $d = 4$) the integral is convergent. Before we specify the rules of DR in more detail, let us have a look at convergence properties of Feynman integrals.

Dyson Power Counting

The action
$$S = i \int d^d x \, \mathcal{L}_{\text{eff}} \tag{2.115}$$
measured in units of $\hbar = 1$ is dimensionless and therefore dim $\mathcal{L}_{\text{eff}} = d$ in mass units. The inspection of the individual terms yields the following dimensions for the fields:

$$\begin{array}{ll}
\bar{\psi}\gamma^\mu \partial_\mu \psi & : \text{dim } \psi = \frac{d-1}{2} \\
(\partial_\mu A_\nu - \cdots)^2 & : \text{dim } A_\mu = \frac{d-2}{2} \\
\bar{e}_0 \bar{\psi}\gamma^\mu \psi A_\mu & : \text{dim } \bar{e}_0 = \frac{4-d}{2} \Rightarrow \bar{e}_0 = e_0 \mu^{\epsilon/2}
\end{array} \tag{2.116}$$

where $\epsilon = 4 - d$, e_0 denotes the dimensionless bare coupling constant (dim $e_0 = 0$) and μ is an arbitrary mass scale. The dimension of time ordered Green functions in momentum space is then given by (the Fourier transformation $\int d^d q \, e^{-iqx} \cdots$ gives $-d$ for each field):

$$\dim G^{(n_B, 2n_F)} = n_B \frac{d-2}{2} + 2n_F \frac{d-1}{2} - (n_B + 2n_F)d$$

where
$$\begin{array}{l} n_B : \#\text{of boson fields} : G_{i\mu}, \cdots \\ 2n_F : \#\text{of Dirac fields (in pairs)} : \psi \cdots \bar{\psi} \, . \end{array}$$

It is convenient to split off factors which correspond to external propagators (see p. 50) and four–momentum conservation and to work with 1PI amplitudes, which are the objects relevant for calculating T matrix elements. The corresponding proper amputated vertex functions are of dimension

---

[20] An inconsistency problem, concerning the definition of γ_5 for $d \neq 4$, implies that the chiral WT identities associated with the parity violating weak fermion currents in the SM are violated in general (see e.g. [35]).

$$\dim \hat{G}^{\mathrm{amp}} = d - n_B \frac{d-2}{2} - 2n_F \frac{d-1}{2} \ . \qquad (2.117)$$

A generic Feynman diagram represents a Feynman integral

$$\Longleftrightarrow I_\Gamma(p) = \int \frac{\mathrm{d}^d k_1}{(2\pi)^d} \cdots \frac{\mathrm{d}^d k_m}{(2\pi)^d} J_\Gamma(p,k) \ .$$

The convergence of the integral can be inspected by looking at the behavior of the integrand for large momenta: For $k_i = \lambda \hat{k}_i$ and $\lambda \to \infty$ we find

$$\Pi_i \mathrm{d}^d k_i \, J_\Gamma(p,k) \to \lambda^{d(\Gamma)}$$

where

$$d(\Gamma) = d - n_B \frac{d-2}{2} - 2n_F \frac{d-1}{2} + \sum_{i=1}^{n}(d_i - d)$$

is called the *superficial divergence* of the 1PI diagram Γ. The sum extends over all (n) vertices of the diagram and d_i denotes the dimension of the vertex i. The $-d$ at each vertex accounts for d–momentum conservation. For a vertex exhibiting $n_{i,b}$ Bose fields, $n_{i,f}$ Fermi fields and l_i derivatives of fields we have

$$d_i = n_{i,b} \frac{d-2}{2} + n_{i,f} \frac{d-1}{2} + l_i \qquad (2.118)$$

Here it is important to mention one of the most important conditions for a QFT to develop its full predictive power: *renormalizability*. In order that $d(\Gamma)$ in (2.118) is bounded in physical space–time $d = 4$ all interaction vertices must have dimension not more than $d_i \leq 4$. An anomalous magnetic moment effective interaction term (Pauli term)

$$\delta \mathcal{L}^{\mathrm{AMM}}_{\mathrm{eff}} = \frac{ieg}{4m} \bar{\psi}(x) \sigma_{\mu\nu} \psi(x) F^{\mu\nu}(x) \ , \qquad (2.119)$$

has dimension 5 (in $d = 4$) and thus would spoil the renormalizability of the theory[21]. Such a term is thus forbidden in any renormalizable QFT. In contrast, in any renormalizable QTF the anomalous magnetic moment of a fermion is a quantity unambiguously predicted by the theory.

The relation (2.118) may be written in the alternative form

$$d(\Gamma) = 4 - n_B - 2n_F \frac{3}{2} + L(d-4) \ .$$

The result can be easily understood: the loop expansion of an amplitude has the form

---

[21] The dimension of $F^{\mu\nu}$ is 2, 1 for the photon field plus 1 for the derivative.

$$A^{(L)} = A^{(0)} \left[1 + a_1 \alpha + a_2 \alpha^2 + \cdots + a_L \alpha^L + \cdots \right] \quad (2.120)$$

where $\alpha = e^2/4\pi$ is the conventional expansion parameter. $A^{(0)}$ is the tree level amplitude which coincides with the result in $d = 4$.

We now are ready to formulate the convergence criterion which reads:

I_Γ convergent $\bowtie$ $d(\gamma) < 0$ $\forall$ 1PI sub–diagrams $\gamma \subseteq \Gamma$
I_Γ divergent $\bowtie$ $\exists\ \gamma \subseteq \Gamma$ with $d(\gamma) \geq 0$.

In $d \leq 4$ dimensions, a renormalizable theory has the following types of primitively divergent diagrams (i.e. diagrams with $d(\Gamma) \geq 0$ which may have divergent sub–integrals)[22]:

$\qquad d-2\ [2] \qquad d-3\ [1] \qquad d-4\ [0]$

$+(L_\Gamma - 1)(d-4)$ for a diagram with $L_\Gamma (\geq 1)$ loops. The list shows the non–trivial leading one–loop $d(\Gamma)$ to which per additional loop a contribution $(d-4)$ has to be added (see (2.120)), in square brackets the values for $d = 4$. Thus the dimensional analysis tells us that convergence improves for $d < 4$. For a renormalizable theory we have

- $d(\Gamma) \leq 2$ for $d = 4$.

In lower dimensions

- $d(\Gamma) < 2$ for $d < 4$

a renormalizable theory becomes super–renormalizable, while in higher dimensions

- $d(\Gamma)$ unbounded! $d > 4$

and the theory is non–renormalizable.

---

[22] According to (2.120) there are two more potentially divergent structures

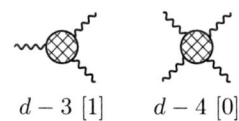

$\qquad d-3\ [1] \qquad d-4\ [0]$

with superficial degree of divergence as indicated. However, the triple photon vertex is identically zero by Furry's theorem, C odd amplitudes are zero in the C preserving QED. The four photon light–by–light scattering amplitude, due the transversality of the external physical photons, has an effective dimension $d(\Gamma)_{\text{eff}} = -4$, instead of 0, and is thus very well convergent. For the same reason, transversality of the photon self–energy, actually the photon propagator has $d(\Gamma)_{\text{eff}} = 0$ instead of 2. In both cases it is the Abelian gauge symmetry which makes integrals better convergent than they look like by naive power counting.

Dimensional Regularization

Dimensional regularization of theories with spin is defined in three steps.

1. Start with Feynman rules formally derived in $d = 4$.
2. Generalize to $d = 2n > 4$. This intermediate step is necessary in order to treat the vector and spinor indices appropriately. Of course it means that the UV behavior of Feynman integrals at first gets worse.

1) For fermions we need the $d = 2n$–dimensional Dirac algebra:

$$\{\gamma^\mu, \gamma^\nu\} = 2g^{\mu\nu}\mathbf{1} \; ; \; \{\gamma^\mu, \gamma_5\} = 0 \qquad (2.121)$$

where γ_5 must satisfy $\gamma_5^2 = 1$ and $\gamma_5^+ = \gamma_5$ such that $\frac{1}{2}(1 \pm \gamma_5)$ are the chiral projection matrices. The metric has dimension d

$$g^{\mu\nu}g_{\mu\nu} = g_\mu^\mu = d \; ; \; g_{\mu\nu} = \begin{pmatrix} 1 & 0 & \cdots \\ 0 & -1 & \\ \vdots & & \ddots \\ & & & -1 \end{pmatrix} .$$

By $\mathbf{1}$ we denote the unit matrix in spinor space. In order to have the usual relation for the adjoint spinors we furthermore require

$$\gamma^{\mu+} = \gamma^0 \gamma^\mu \gamma^0 . \qquad (2.122)$$

Simple consequences of this d–dimensional algebra are:

$$\begin{aligned}
\gamma_\alpha \gamma^\alpha &= d\,\mathbf{1} \\
\gamma_\alpha \gamma^\mu \gamma^\alpha &= (2-d)\,\gamma^\mu \\
\gamma_\alpha \gamma^\mu \gamma^\nu \gamma^\alpha &= 4g^{\mu\nu}\,\mathbf{1} + (d-4)\,\gamma^\mu \gamma^\nu \\
\gamma_\alpha \gamma^\mu \gamma^\nu \gamma^\rho \gamma^\alpha &= -2\gamma^\rho \gamma^\nu \gamma^\mu + (4-d)\gamma^\mu \gamma^\nu \gamma^\rho \text{ etc.}
\end{aligned} \qquad (2.123)$$

Traces of strings of γ–matrices are very similar to the ones in 4–dimensions. In $d = 2n$ dimensions one can easily write down $2^{d/2}$–dimensional representations of the Dirac algebra [36]. Then

$$\begin{aligned}
\text{Tr}\,\mathbf{1} &= f(d) = 2^{d/2} \\
\text{Tr}\,\prod_{i=1}^{2n-1} \gamma^{\mu_i}(\gamma^5) &= 0 \\
\text{Tr}\,\gamma^\mu \gamma^\nu &= f(d)\,g^{\mu\nu} \\
\text{Tr}\,\gamma^\mu \gamma^\nu \gamma^\rho \gamma^\sigma &= f(d)\,(g^{\mu\nu}g^{\rho\sigma} - g^{\mu\rho}g^{\nu\sigma} + g^{\mu\sigma}g^{\nu\rho}) \text{ etc.}
\end{aligned} \qquad (2.124)$$

One can show that for *renormalized quantities* the only relevant property of $f(d)$ is $f(d) \to 4$ for $d \to 4$. Very often the **convention** $f(d) = 4$ (for

2.4 Regularization and Renormalization

any d) is adopted. Bare quantities and the related *minimally subtracted* MS or modified minimally subtracted $\overline{\text{MS}}$ quantities (see below for the precise definition) **depend** upon this convention (by terms proportional to $\ln 2$).

In **anomaly free** theories we can assume γ_5 to be fully anticommuting! But then

$$\text{Tr}\,\gamma^\mu\gamma^\nu\gamma^\rho\gamma^\sigma\gamma_5 = 0 \quad \text{for all} \quad d \neq 4! \tag{2.125}$$

The 4–dimensional object

$$4\mathrm{i}\varepsilon^{\mu\nu\rho\sigma} = \text{Tr}\,\gamma^\mu\gamma^\nu\gamma^\rho\gamma^\sigma\gamma_5 \quad \text{for} \quad d = 4$$

cannot be obtained by dimensional continuation if we use an anticommuting γ_5 [36].

Since fermions do not have self interactions they only appear as closed fermion loops, which yield a trace of γ–matrices, or as a fermion string connecting an external $\psi\cdots\bar\psi$ pair of fermion fields. In a transition amplitude $|T|^2 = \text{Tr}\,(\cdots)$ we again get a trace. Consequently, in principle, we have eliminated all γ's! Commonly one writes a covariant tensor decomposition into invariant amplitudes, like, for example,

$$\cdots = \mathrm{i}\Gamma^\mu = -\mathrm{i}e\left\{\gamma^\mu A_1 + \mathrm{i}\sigma^{\mu\nu}\frac{q_\nu}{2m}A_2 + \gamma^\mu\gamma_5 A_3 + \cdots\right\}$$

where μ is an external index, q^μ the photon momentum and $A_i(q^2)$ are scalar form factors.

2) **External momenta** (and external indices) **must** be taken $d = 4$ dimensional, because the number of independent "form factors" in covariant decompositions depends on the dimension, with a fewer number of independent functions in lower dimensions. Since four functions cannot be analytic continuation of three etc. we have to keep the external structure of the theory in $d = 4$. The reason for possible problems here is the non–trivial spin structure of the theory of interest. The following rules apply:

External momenta : $p^\mu = (p^0, p^1, p^2, p^3, 0, \cdots, 0)$ 4 – dimensional
Loop momenta : $k^\mu = (k^0, \cdots k^{d-1})$ d – dimensional
$k^2 = (k^0)^2 - (k^1)^2 - \cdots - (k^{d-1})^2$
$pk = p^0 k^0 - \boldsymbol{p}\cdot\boldsymbol{k}$ 4 – dimensional etc.

3. Interpolation in d to complex values and extrapolation to $d < 4$.
Loop integrals now read

$$\mu^{4-d}\int\frac{\mathrm{d}^d k}{(2\pi)^d}\cdots \tag{2.126}$$

with μ an arbitrary scale parameter. The crucial properties valid in DR independent of d are: (F.P. = finite part)

a) $\int d^d k\, k_\mu f(k^2) = 0$
b) $\int d^d k\, f(k+p) = \int d^d k\, f(k)$
   which is **not** true with UV cut – off's
c) If $f(k) = f(|k|)$:
   $\int d^d k\, f(k) = \frac{2\pi^{d/2}}{\Gamma(\frac{d}{2})} \int_0^\infty dr\, r^{d-1} f(r)$
d) For divergent integrals, by analytic subtraction:
   F.P. $\int_0^\infty dr\, r^{d-1+\alpha} \equiv 0$ for arbitrary α
   so called *minimal subtraction* (MS). Consequently

$$\text{F.P.} \int d^d k\, f(k) = \text{F.P.} \int d^d k\, f(k+p) = \text{F.P.} \int d^d(\lambda k)\, f(\lambda k)\ .$$

This implies that **dimensionally regularized integrals behave like convergent integrals** and formal manipulations are justified. Starting with d sufficiently small, by partial integration, one can always find a representation for the integral which converges for $d = 4 - \epsilon$, $\epsilon > 0$ small.

In order to elaborate in more detail how DR works in practice, let us consider a generic one–loop Feynman integral

$$I_\Gamma^{\mu_1\cdots\mu_m}(p_1,\cdots,p_n) = \int d^d k\, \frac{\prod_{j=1}^m k^{\mu_j}}{\prod_{i=1}^n ((k+p_i)^2 - m_i^2 + i\varepsilon)}$$

which has superficial degree of divergence

$$d(\Gamma) = d + m - 2n \leq d - 2$$

where the bound holds for two– or more–point functions in **renormalizable** theories and for $d \leq 4$. Since the physical tensor and spin structure has to be kept in $d = 4$, by contraction with external momenta or with the metric tensor $g_{\mu_i\mu_j}$ it is always possible to write the above integral as a sum of integrals of the form

$$I_\Gamma^{\hat\mu_1\cdots\hat\mu_{m'}}(\hat p_1,\cdots,\hat p_{n'}) = \int d^d k\, \frac{\prod_{j=1}^{m'} \hat k^{\mu_j}}{\prod_{i=1}^{n'} ((k+\hat p_i)^2 - m_i^2 + i\varepsilon)}$$

where now $\hat\mu_j$ and $\hat p_i$ are $d = 4$–dimensional objects and

$$d^d k = d^4 \hat k\, d^{d-4} \bar k = d^4 \hat k\, \omega^{d-5}\, d\omega\, d\Omega_{d-4}\ .$$

In the $d - 4$–dimensional complement the integrand depends on ω only! The angular integration over $d\Omega_{d-4}$ yields

$$\int d\Omega_{d-4} = S_{d-4} = \frac{2\pi^{\epsilon/2}}{\Gamma(\epsilon/2)}\ ;\ \epsilon = d - 4\ ,$$

which is the surface of the $d-4$–dimensional sphere. Using this result we get (discarding the four–dimensional tensor indices)

$$I_\Gamma(\{\hat{p}_i\}) = \int d^4\hat{k}\, J_\Gamma(d,\hat{p},\hat{k})$$

where

$$J_\Gamma(d,\hat{p},\hat{k}) = S_{d-4}\int_0^\infty d\omega\, \omega^{d-5} f(\hat{p},\hat{k},\omega)\ .$$

Now this integral can be analytically continued to **complex** values of d. For the ω–integration we have

$$d^\omega(\Gamma) = d - 4 - 2n$$

i.e. the ω–integral converges if

$$d < 4 + 2n\ .$$

In order to avoid infrared singularities in the ω–integration one has to analytically continue by appropriate partial integration. After p–fold partial integration we have

$$I_\Gamma(\{\hat{p}_i\}) = \frac{2\pi^{\frac{d-4}{2}}}{\Gamma(\frac{d-4}{2}+p)}\int d^4\hat{k}\int_0^\infty d\omega\, \omega^{d-5+2p}\left(-\frac{\partial}{\partial\omega^2}\right)^p f(\hat{p},\hat{k},\omega)$$

where the integral is convergent in $4-2p < \mathrm{Re}\ d < 2n-m = 4-d^{(4)}(\Gamma) \geq 2$. For a renormalizable theory at most 2 partial integrations are necessary to define the theory.

2.5 Tools for the Evaluation of Feynman Integrals

2.5.1 $\epsilon = 4-d$ Expansion, $\epsilon \to +0$

For the expansion of integrals near $d=4$ we need some asymptotic expansions of Γ–functions:

$$\Gamma(1+x) = \exp\left[-\gamma x + \sum_{n=2}^\infty \frac{(-1)^n}{n}\zeta(n)x^n\right]\quad |x|\leq 1$$

$$\psi(1+x) = \frac{d}{dx}\ln\Gamma(1+x) = \frac{\Gamma'(1+x)}{\Gamma(1+x)} \stackrel{|x|<1}{=} -\gamma + \sum_{n=2}^\infty (-1)^n\zeta(n)x^{n-1}$$

where $\zeta(n)$ denotes Riemann's Zeta function. The defining functional relation is

$$\Gamma(x) = \frac{\Gamma(x+1)}{x}\ ,$$

which for $n = 0, 1, 2, \cdots$ yields $\Gamma(n+1) = n!$ with $\Gamma(1) = \Gamma(2) = 1$. Furthermore we have

$$\Gamma(x)\,\Gamma(1-x) = \frac{\pi}{\sin \pi x}$$

$$\Gamma(\frac{1}{2}+x)\,\Gamma(\frac{1}{2}-x) = \frac{\pi}{\cos \pi x}\;.$$

Important special constants are

$$\Gamma(\frac{1}{2}) = \sqrt{\pi}$$

$$\Gamma'(1) = -\gamma\;;\;\;\gamma = 0.577215\cdots\;\text{Euler's constant}$$

$$\Gamma''(1) = \gamma^2 + \zeta(2)\;;\;\;\zeta(2) = \frac{\pi^2}{6} = 1.64493\cdots$$

As a typical result of an ϵ–expansion, which we should keep in mind for later purposes, we have

$$\Gamma(1+\frac{\epsilon}{2}) = 1 - \frac{\epsilon}{2}\gamma + (\frac{\epsilon}{2})^2 \frac{1}{2}(\gamma^2 + \zeta(2)) + \cdots$$

2.5.2 Bogolubov-Schwinger Parametrization

Suppose we choose for each propagator an independent momentum and take into account momentum conservation at the vertices by δ–functions. Then, for $d = n$ integer, we use

i)
$$\frac{\mathrm{i}}{p^2 - m^2 + \mathrm{i}\varepsilon} = \int_0^\infty \mathrm{d}\alpha\,\mathrm{e}^{-\mathrm{i}\alpha(m^2 - p^2 + \mathrm{i}\varepsilon)} \tag{2.127}$$

ii)
$$\delta^{(d)}(k) = \frac{1}{(2\pi)^d} \int_{-\infty}^{+\infty} \mathrm{d}^d x\,\mathrm{e}^{\mathrm{i}kx} \tag{2.128}$$

and find that all momentum integrations are of Gaussian type. The Gaussian integrals yield

$$\int_{-\infty}^{+\infty} \mathrm{d}^d k\, P(k)\, \mathrm{e}^{\mathrm{i}(ak^2 + 2b(k\cdot p))} = P\left(\frac{-\mathrm{i}}{2b}\frac{\partial}{\partial p}\right)\left(\frac{\pi}{\mathrm{i}a}\right)^{d/2}\mathrm{e}^{-\mathrm{i}\,b^2/a\,p^2} \tag{2.129}$$

for any polynomial P. The resulting form of the Feynman integral is the so called Bogolubov-Schwinger representation.

2.5.3 Feynman Parametric Representation

Transforming pairs of α–variables in the above Bogolubov-Schwinger parametrization according to (l is denoting the pair (i,k))

$$(\alpha_i, \alpha_k) \to (\xi_l, \alpha_l) : (\alpha_i, \alpha_k) = (\xi_l \alpha_l, (1-\xi_l)\alpha_l) \tag{2.130}$$

$$\int_0^\infty \int_0^\infty d\alpha_i d\alpha_k \cdots = \int_0^\infty d\alpha_l \, \alpha_l \int_0^1 d\xi_l \cdots \quad , \tag{2.131}$$

the integrals are successively transformed into $\int_0^1 d\xi \cdots$ integrals and at the end there remains *one* α–integration only which can be performed using

$$\int_0^\infty d\alpha \, \alpha^a \, e^{-\alpha x} = \Gamma(a+1) x^{-(a+1)} \quad . \tag{2.132}$$

The result is the Feynman parametric representation. If L is the number of lines of a diagram, the Feynman integral is $L-1$–dimensional.

2.5.4 Euclidean Region, Wick–Rotations

The basic property which allows us to perform a Wick rotation is analyticity which derives from the causality of a relativistic QFT. In momentum space the Feynman propagator

$$\frac{1}{q^2 - m^2 + i\varepsilon} = \frac{1}{q^0 - \sqrt{\boldsymbol{q}^2 + m^2 - i\varepsilon}} \frac{1}{q^0 + \sqrt{\boldsymbol{q}^2 + m^2 - i\varepsilon}}$$

$$= \frac{1}{2\omega_p} \left\{ \frac{1}{q^0 - \omega_p + i\varepsilon} - \frac{1}{q^0 + \omega_p - i\varepsilon} \right\} \tag{2.133}$$

is an analytic function in q^0 with poles at $q^0 = \pm(\omega_p - i\varepsilon)$[23] where $\omega_p = \sqrt{\boldsymbol{q}^2 + m^2}$. This allows us to rotate by $\frac{\pi}{2}$ the integration path in q^0, going from $-\infty$ to $+\infty$, without crossing any singularity. In doing so, we rotate from Minkowski space to Euclidean space

$$q^0 \to -i q^d \Rightarrow q = (q^0, q^1, \ldots, q^{d-2}, q^{d-1}) \to \underline{q} = (q^1, q^2, \ldots, q^{d-1}, q^d)$$

and thus $q^2 \to -\underline{q}^2$. This rotation to the Euclidean region is called **Wick rotation**.

More precisely: analyticity of a function $\tilde{f}(q^0, \boldsymbol{q})$ in q^0 implies that the **contour integral**

$$\oint_{C(R)} dq^0 \, \tilde{f}(q^0, \boldsymbol{q}) = 0 \tag{2.134}$$

for the closed path $C(R)$ in Fig. 2.4 vanishes. If the function $\tilde{f}(q^0, \boldsymbol{q})$ falls off sufficiently fast at infinity, then the contribution from the two "arcs" goes to zero when the radius of the contour $R \to \infty$. In this case we obtain

---

[23] Note that because of the positivity of $\boldsymbol{q}^2 + m^2$ for any non–vacuum state, we have $\omega_p - i\varepsilon = \sqrt{\boldsymbol{q}^2 + m^2 - i\varepsilon}$ in the limit $\lim_{\varepsilon \to 0}$, which is always understood. The symbolic parameter ε of the $i\varepsilon$ prescription, may be scaled by any fixed positive number.

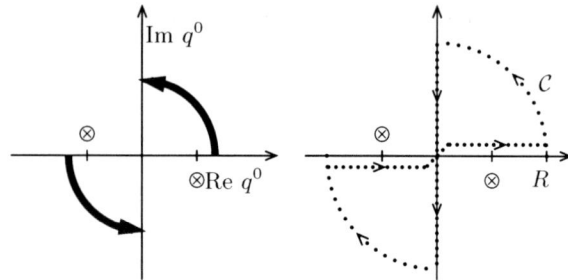

Fig. 2.4. Wick rotation in the complex q^0–plane. The poles of the Feynman propagator are indicated by ⊗'s. C is an integration contour, R is the radius of the arcs

$$\int_{-\infty}^{\infty} dq^0\, \tilde{f}(q^0, \boldsymbol{q}\,) + \int_{+i\infty}^{-i\infty} dq^0\, \tilde{f}(q^0, \boldsymbol{q}\,) = 0 \qquad (2.135)$$

or

$$\int_{-\infty}^{\infty} dq^0\, \tilde{f}(q^0, \boldsymbol{q}\,) = \int_{-i\infty}^{+i\infty} dq^0\, \tilde{f}(q^0, \boldsymbol{q}\,) = -\mathrm{i} \int_{-\infty}^{+\infty} dq^d\, \tilde{f}(-\mathrm{i}q^d, \boldsymbol{q}\,)\ , \qquad (2.136)$$

which is the Wick rotation. At least in perturbation theory, one can prove that the conditions required to allow us to perform a Wick rotation are fulfilled.

We notice that the Euclidean Feynman propagator obtained by the Wick rotation

$$\frac{1}{q^2 - m^2 + \mathrm{i}\varepsilon} \rightarrow -\frac{1}{\underline{q}^2 + m^2}$$

has no singularities (poles) and an iε–prescription is not needed any longer.

In configuration space a Wick rotation implies going to **imaginary time** $x^0 \rightarrow \mathrm{i}x^0 = x^d$ such that $qx \rightarrow -\underline{q}\underline{x}$ and hence

$$x^0 \rightarrow -\mathrm{i}x^d \Rightarrow x^2 \rightarrow -\underline{x}^2\ ,\ \ \Box_x \rightarrow -\Delta_{\underline{x}}\ ,\ \ \mathrm{i}\int d^d x \cdots \rightarrow \int d^d \underline{x} \cdots\ .$$

While in Minkowski space $x^2 = 0$ defines the light–cone $x^0 = \pm|\boldsymbol{x}|$, in the Euclidean region $\underline{x}^2 = 0$ implies $\underline{x} = 0$. Note that possible singularities on the light–cone like $1/x^2$, $\delta(x^2)$ etc. turn into singularities at the point $\underline{x} = 0$. This simplification of the singularity structure is the merit of the positive definite metric in Euclidean space.

In momentum space the Euclidean propagators are positive (discarding the overall sign) and any Feynman amplitude in Minkowski space may be obtained via

$$I_M(p) = (-\mathrm{i})^{N_\mathrm{int}}(-\mathrm{i})^{V-1} I_E(\underline{p})\big|_{p^4=\mathrm{i}p^0\,;\ m^2\to m^2-\mathrm{i}\varepsilon}$$

from its Euclidean version. Here, N_int denotes the number of internal lines (propagators) and V the number of vertices if we use the substitutions (convention dependent)

$$\frac{1}{p^2 - m^2 + \mathrm{i}\varepsilon} \to \frac{1}{\underline{p}^2 + m^2}\,;\ \mathrm{i}g_i \to \mathrm{i}\,(\mathrm{i}g_i) = -g_i\,;\ \int \mathrm{d}^d k \to \int \mathrm{d}^d \underline{k}$$

to define the Euclidean Feynman amplitudes. By g_i we denote the gauge couplings.

For the dimensionally regularized amplitudes, where potentially divergent integrals are defined via analytic continuation from regions in the complex d–plane where integrals are manifestly convergent, the terms from the arc segments can always be dropped. Also note that dimensional regularization and the power counting rules (superficial degree of divergence etc.) hold irrespective of whether we work in d–dimensional Minkowski space–time or in d–dimensional Euclidean space. The metric is obviously not important for the UV–behavior of the integrals.

The relationship between Euclidean and Minkowski quantum field theory is not only a very basic and surprising general feature of any local relativistic field theory but is a property of central practical importance for the non–perturbative approach to QFT via the Euclidean path–integral (e.g., lattice QCD). In a QFT satisfying the Wightman axioms the continuation of the vacuum–expectation values of time–ordered products of local fields (the time–ordered Green functions) from Minkowski space to four–dimensional Euclidean space is always possible [37]. Conversely, the *Osterwalder-Schrader theorem* ascertains that the Euclidean correlation functions of fields can be analytically continued to Minkowski space, provided we have a local action which satisfies the so–called reflection positivity condition [38]. Accordingly, the full Minkowski QFT including its S–matrix, if it exists, can be reconstructed from the knowledge of the Euclidean correlation functions and from a mathematical point of view the Minkowski and the Euclidean version of a QFT are completely equivalent.

2.5.5 The Origin of Analyticity

At the heart of analyticity is the causality. The time ordered Green functions which encode all information of the theory in perturbation theory are given by integrals over products of causal propagators ($z = x - y$)

$$\begin{aligned}
\mathrm{i}S_\mathrm{F}(z) &= \langle 0|T\{\psi(x)\bar{\psi}(y)\}|0\rangle \\
&= \Theta(x^0 - y^0)\langle 0|\psi(x)\bar{\psi}(y)|0\rangle - \Theta(y^0 - x^0)\langle 0|\bar{\psi}(y)\psi(x)|0\rangle \\
&= \Theta(z^0)\,\mathrm{i}S^+(z) + \Theta(-z^0)\,\mathrm{i}S^-(z)
\end{aligned} \qquad (2.137)$$

exhibiting a positive frequency part propagating forward in time and a negative frequency part propagating backward in time. The Θ function of time ordering makes the Fourier–transform to be analytic in a half–plane in momentum space. For $K(\tau = z^0) = \Theta(z^0) \mathrm{i} S^+(z)$, for example, we have

$$\tilde{K}(\omega) = \int_{-\infty}^{+\infty} \mathrm{d}\tau K(\tau)\, \mathrm{e}^{\mathrm{i}\omega\tau} = \int_0^{+\infty} \mathrm{d}\tau K(\tau)\, \mathrm{e}^{-\eta\tau} \mathrm{e}^{\mathrm{i}\xi\tau} \tag{2.138}$$

such that $\tilde{K}(\omega = \xi + \mathrm{i}\eta)$ is a regular analytic function in the **upper half** ω–**plane** $\eta > 0$. This of course only works because τ is restricted to be positive.

In a relativistically covariant world, in fact, we always need two terms (see (2.137)), a positive frequency part $\Theta(z^0 = t - t')\, S^+(z)$, corresponding to the particle propagating forward in time, and a negative frequency part $\Theta(-z^0 = t'-t) S^-(z)$, corresponding to the antiparticle propagating backward in time. The two terms correspond in momentum space to the two terms of (2.133).

Of course, for a free Dirac field we know what the Stückelberg-Feynman propagator in momentum space looks like

$$\tilde{S}_\mathrm{F}(q) = \frac{\slashed{q}+m}{q^2 - m^2 + \mathrm{i}\varepsilon}$$

and its analytic properties are manifest. It is an analytic function in q^0 with poles at $q^0 = \pm(\omega_p - \mathrm{i}\varepsilon)$ where $\omega_p = \sqrt{\boldsymbol{q}^{\,2} + m^2}$.

Analyticity is an extremely important basic property of a QFT and a powerful instrument which helps to solve seemingly purely "technical" problems as we will see. For example it allows us to perform a Wick rotation to Euclidean space and in Euclidean space a QFT looks like a classical statistical system and one can apply the methods of statistical physics to QFT [39]. In particular the numerical approach to the intrinsically non–perturbative QCD via lattice QCD is based on analyticity. The objects which manifestly exhibit the analyticity properties and are providing the bridge to the Euclidean world are the *time ordered Green functions*.

Note that by far not all objects of interest in a QFT are analytic. For example, any solution of the homogeneous (no source) Klein-Gordon equation

$$(\Box_x + m^2)\, \Delta(x-y; m^2) = 0\,,$$

like the so called positive frequency part Δ^+ or the causal commutator Δ of a free scalar field $\varphi(x)$, defined by

$$<0|\,\varphi(x),\varphi(y)\,|0> = \mathrm{i}\,\Delta^+(x-y;m^2)$$
$$[\varphi(x),\varphi(y)] = \mathrm{i}\,\Delta(x-y;m^2)\,,$$

2.5 Tools for the Evaluation of Feynman Integrals

which, given the properties of the free field, may easily be evaluated to have a representation

$$\Delta^+(z; m^2) = -\mathrm{i}\,(2\pi)^{-3} \int \mathrm{d}^4 p\, \Theta(p^0)\, \delta(p^2 - m^2)\, \mathrm{e}^{-\mathrm{i}pz}$$

$$\Delta(z; m^2) = -\mathrm{i}\,(2\pi)^{-3} \int \mathrm{d}^4 p\, \epsilon(p^0)\, \delta(p^2 - m^2)\, \mathrm{e}^{-\mathrm{i}pz}\ .$$

Thus, in momentum space, as solutions of

$$(p^2 - m^2)\, \tilde{\Delta}(p) = 0\ ,$$

only singular ones exist. For the positive frequency part and the causal commutator they read

$$\Theta(p^0)\, \delta(p^2 - m^2) \quad \text{and} \quad \epsilon(p^0)\, \delta(p^2 - m^2)\ ,$$

respectively. The Feynman propagator, in contrast, satisfies an inhomogeneous (with point source) Klein-Gordon equation

$$(\Box_x + m^2)\, \Delta_F(x - y; m^2) = -\delta^{(4)}(x - y)\ .$$

The δ function comes from differentiating the Θ function factors of the T product. Now we have

$$\langle 0|T\{\varphi(x), \varphi(y)\}|0\rangle = \mathrm{i}\,\Delta_F(x - y; m^2)$$

with

$$\Delta_F(z; m^2) = (2\pi)^{-4} \int \mathrm{d}^4 p\, \frac{1}{p^2 - m^2 + \mathrm{i}\varepsilon}\, \mathrm{e}^{-\mathrm{i}pz}$$

and in momentum space

$$(p^2 - m^2)\, \tilde{\Delta}_F(p) = 1\ ,$$

obviously has analytic solutions, a particular one being the scalar Feynman propagator

$$\frac{1}{p^2 - m^2 + \mathrm{i}\varepsilon} = \mathcal{P}\left(\frac{1}{p^2 - m^2}\right) - \mathrm{i}\pi\,\delta(p^2 - m^2)\ . \tag{2.139}$$

The $\mathrm{i}\varepsilon$ prescription used here precisely correspond to the boundary condition imposed by the time ordering prescription T in configuration space. The symbol $\mathcal{P}$ denotes the principal value; the right hand side exhibits the splitting into real and imaginary part.

Analyticity will play a crucial role later on and is the basic property from which dispersion relations derive (see Sect. 3.7).

2.5.6 Scalar One–Loop Integrals

Here we apply our tools to the simplest scalar one–loop integrals (p.i.= partial integration).

$$\frac{\mu^{4-d}}{(2\pi)^d}\int d^d\underline{k}\,\frac{1}{\underline{k}^2+m^2} = \mu^{4-d}(4\pi)^{-d/2}\int_0^\infty d\alpha\,\alpha^{-d/2}e^{-\alpha m^2}$$

convergent for $d < 2$ *** [24]

$$\stackrel{\mathrm{p.i.}}{=} -\frac{2m^2}{d-2}\mu^{4-d}(4\pi)^{-d/2}\int_0^\infty d\alpha\,\alpha^{1-d/2}e^{-\alpha m^2}$$

convergent for $d < 4$

$$= -2m^2(4\pi)^{-d/2}\frac{\Gamma(2-d/2)}{d-2}\left(\frac{m^2}{\mu^2}\right)^{d/2-2}$$

$$= -2m^2(4\pi)^{-2}\frac{2}{\epsilon}\Gamma(1+\tfrac{\epsilon}{2})\frac{1}{2-\epsilon}e^{\frac{\epsilon}{2}(\ln 4\pi - \ln\frac{m^2}{\mu^2})}$$

$$\stackrel{\epsilon\to +0}{\simeq} m^2(4\pi)^{-2}\left\{\tfrac{2}{\epsilon} - \gamma + 1 + \ln 4\pi - \ln\tfrac{m^2}{\mu^2}\right\} + O(\epsilon)$$

$$\frac{\mu^{4-d}}{(2\pi)^d}\int d^d k\,\frac{1}{\underline{k}^2+m_1^2}\,\frac{1}{(\underline{k}+\underline{p})^2+m_2^2}$$

$$= \mu^{4-d}(4\pi)^{-d/2}\int_0^\infty d\alpha_1 d\alpha_2(\alpha_1+\alpha_2)^{-d/2}e^{-(\alpha_1 m_1^2+\alpha_2 m_2^2+\frac{\alpha_1\alpha_2}{\alpha_1+\alpha_2}\underline{p}^2)}$$

$\alpha_1 = x\lambda$; $\alpha_2 = (1-x)\lambda$

$$= \mu^{4-d}(4\pi)^{-d/2}\Gamma(2-\tfrac{d}{2})\int_0^1 dx(xm_1^2+(1-x)m_2^2+x(1-x)\underline{p}^2))^{d/2-2}$$

convergent for $d < 4$

$$= (4\pi)^{-2}\tfrac{2}{\epsilon}\Gamma(1+\tfrac{\epsilon}{2})e^{\frac{\epsilon}{2}\ln 4\pi}\int_0^1 dx\,e^{-\frac{\epsilon}{2}\ln\frac{xm_1^2+(1-x)m_2^2+x(1-x)p^2}{\mu^2}}$$

$$\stackrel{\epsilon\to +0}{\simeq} (4\pi)^{-2}\left\{\tfrac{2}{\epsilon} - \gamma + \ln 4\pi - \int_0^1 dx\,\ln\frac{xm_1^2+(1-x)m_2^2+x(1-x)p^2}{\mu^2}\right\} + O(\epsilon)$$

$$\frac{\mu^{4-d}}{(2\pi)^d}\int d^d\underline{k}\,\frac{1}{\underline{k}^2+m_1^2}\,\frac{1}{(\underline{k}+\underline{p}_1)^2+m_2^2}\,\frac{1}{(\underline{k}+\underline{p}_1+\underline{p}_2)^2+m_3^2}$$

convergent for $d = 4$

$$\stackrel{\epsilon\to +0}{\simeq} (4\pi)^{-2}\int_0^\infty d\alpha_1 d\alpha_2 d\alpha_3\,\frac{1}{(\alpha_1+\alpha_2+\alpha_3)^2}e^{-(\alpha_1 m_1^2+\alpha_2 m_2^2+\alpha_3 m_3^2)}$$

$$\times e^{-\frac{\alpha_1\alpha_2\underline{p}_1^2+\alpha_2\alpha_3\underline{p}_2^2+\alpha_3\alpha_1\underline{p}_3^2}{\alpha_1+\alpha_2+\alpha_3}}$$

$\alpha_1 = xy\lambda$; $\alpha_2 = x(1-y)\lambda$; $\alpha_3 = (1-x)\lambda$; $\alpha_1+\alpha_2+\alpha_3 = \lambda$

$$= (4\pi)^{-2}\int_0^1 dy dx x\,\tfrac{1}{N}$$

$$N = x^2 y\,(1-y)\underline{p}_1^2 + x\,(1-x)(1-y)\underline{p}_2^2 + x\,(1-x)\,y\underline{p}_3^2 + xy m_1^2 + x\,(1-y)\,m_2^2 + (1-x)\,m_3^2$$

---

[24] A direct integration here yields

$$m^2(4\pi)^{-d/2}\Gamma(1-d/2)\left(\frac{m^2}{\mu^2}\right)^{d/2-2}$$

which by virtue of $\Gamma(1-d/2) = -2\Gamma(2-d/2)/(d-2)$ is the same analytic function as the one obtained via the partial integration method.

Standard Scalar One–loop Integrals ($m^2 \hat{=} m^2 - i\varepsilon$)

$$\begin{array}{c} \underset{p}{\bigcirc}{}^{m} \end{array} = \mu_0^\epsilon \int \frac{\mathrm{d}^d k}{(2\pi)^d} \frac{1}{k^2 - m^2} = -\frac{i}{16\pi^2} A_0(m),$$

defines the standard *tadpole type integral*, where

$$A_0(m) = -m^2 (\text{Reg} + 1 - \ln m^2) \tag{2.140}$$

with

$$\text{Reg} = \frac{2}{\epsilon} - \gamma + \ln 4\pi + \ln \mu_0^2 \equiv \ln \mu^2 . \tag{2.141}$$

The last identification defines the $\overline{\text{MS}}$ scheme of (modified) minimal subtraction.

$$\begin{array}{c} \underset{pm_2}{\overset{m_1}{\longrightarrow}\!\!\bigcirc\!\!\longrightarrow} \end{array} = \mu_0^\epsilon \int \frac{\mathrm{d}^d k}{(2\pi)^d} \frac{1}{(k^2 - m_1^2)((k+p)^2 - m_2^2)} = \frac{i}{16\pi^2} B_0(m_1, m_2; p^2),$$

defines the standard *propagator type integral*, where

$$B_0(m_1, m_2; s) = \text{Reg} - \int_0^1 \mathrm{d}z \, \ln(-sz(1-z) + m_1^2(1-z) + m_2^2 z - i\varepsilon) . \tag{2.142}$$

$$\begin{array}{c} \text{(triangle diagram)} \end{array} = \mu_0^\epsilon \int \frac{\mathrm{d}^d k}{(2\pi)^d} \frac{1}{(k^2 - m_1^2)((k+p_1)^2 - m_2^2)((k+p_1+p_2)^2 - m_3^2)}$$
$$= -\frac{i}{16\pi^2} C_0(m_1, m_2, m_3; p_1^2, p_2^2, p_3^2),$$

defines the standard *form factor type integral*, where

$$C_0(m_1, m_2, m_3; s_1, s_2, s_3) = \int_0^1 \mathrm{d}x \int_0^x \mathrm{d}y \frac{1}{ax^2 + by^2 + cxy + dx + ey + f} \tag{2.143}$$

with

$$\begin{aligned} a &= s_2, & d &= m_2^2 - m_3^2 - s_2, \\ b &= s_1, & e &= m_1^2 - m_2^2 + s_2 - s_3, \\ c &= s_3 - s_1 - s_2, & f &= m_3^2 - i\varepsilon . \end{aligned}$$

Remark: the regulator term Reg in (2.141) denotes the UV regulated pole term $\frac{2}{\epsilon}$ supplemented with $O(1)$ terms which always accompany the pole term and result from the ϵ–expansion of the d–dimensional integrals. While in the MS scheme just the poles $\frac{2}{\epsilon}$ are subtracted, in the modified MS scheme $\overline{\text{MS}}$

also the finite terms included in (2.141) are subtracted. The dependence on the UV cut-off $\frac{2}{\epsilon}$ in the $\overline{\text{MS}}$ scheme defined by $\text{Reg} \equiv \ln \mu^2$ is reflected in a dependence on the $\overline{\text{MS}}$ *renormalization scale* μ.

The UV–singularities (poles in ϵ at d=4) give rise to *finite extra contributions* when they are multiplied with d (or functions of d) which arise from contractions like $g^\mu_\mu = d$, $\gamma^\mu \gamma_\mu = d$ etc. For $d \to 4$ we obtain:

$$dA_0(m) = 4A_0(m) + 2m^2 \, , \quad dB_0 = 4B_0 - 2 \, . \tag{2.144}$$

The explicit evaluation of the scalar integrals (up to the scalar four–point function) is discussed in [40] (see also [41, 42]).

2.5.7 Tensor Integrals

In dimensional regularization also the calculation of tensor integrals is rather straight forward. Sign conventions are chosen in accordance with the Passarino-Veltman convention [43]. Invariant amplitudes are defined by performing covariant decompositions of the tensor integrals, which then are contracted with external vectors or with the metric tensor. A factor $i/16\pi^2$ is taken out for simplicity of notation, i.e.

$$\int_k \ldots = \frac{16\pi^2}{i} \int \frac{d^d k}{(2\pi)^d} \ldots \, . \tag{2.145}$$

1) One point integrals:
By eventually performing a shift $k \to k+p$ of the integration variable we easily find the following results:

$$\begin{aligned}
\int_k \frac{1}{(k+p)^2 - m^2} &= -A_0(m) \\
\int_k \frac{k^\mu}{(k+p)^2 - m^2} &= p^\mu A_0(m) \\
\int_k \frac{k^\mu k^\nu}{(k+p)^2 - m^2} &= -p^\mu p^\nu A_{21} + g^{\mu\nu} A_{22}
\end{aligned} \tag{2.146}$$

$$\begin{aligned}
A_{21} &= A_0(m) \\
A_{22} &= -\frac{m^2}{d} A_0(m) \stackrel{\epsilon \to 0}{\simeq} -\frac{m^2}{4} A_0(m) + \frac{m^4}{8}
\end{aligned} \tag{2.147}$$

2) Two point integrals: the defining equations here are

$$\begin{aligned}
\int_k \frac{1}{(1)(2)} &= B_0(m_1, m_2; p^2) \\
\int_k \frac{k^\mu}{(1)(2)} &= p^\mu B_1(m_1, m_2; p^2) \\
\int_k \frac{k^\mu k^\nu}{(1)(2)} &= p^\mu p^\nu B_{21} - g^{\mu\nu} B_{22} \, ,
\end{aligned} \tag{2.148}$$

where we denoted scalar propagators by $(1) \equiv k^2 - m_1^2$ and $(2) \equiv (k+p)^2 - m_2^2$. The simplest non–trivial example is B_1. Multiplying the defining equation with $2p_\mu$ we have

$$2p^2 B_1 = \int_k \frac{2pk}{k^2 - m_1^2 + i\varepsilon} \frac{1}{(p+k)^2 - m_2^2 + i\varepsilon}$$

and we may write the numerator as a difference of the two denominators plus a remainder which does not depend on the integration variable:

$$2pk = (p+k)^2 - k^2 - p^2 = [(p+k)^2 - m_2^2] - [k^2 - m_1^2] - (p^2 + m_1^2 - m_2^2)$$

After canceling the square brackets against the appropriate denominator we obtain

$$B_1(m_1, m_2; p^2) = \frac{1}{2p^2} \left\{ A_0(m_2) - A_0(m_1) - (p^2 + m_1^2 - m_2^2) B_0(m_1, m_2; p^2) \right\}$$
(2.149)

A further useful relation is

$$B_1(m, m; p^2) = -\frac{1}{2} B_0(m, m; p^2) .$$

In a similar way, by contracting the defining relation with p_ν and $g_{\mu\nu}$ we find for arbitrary dimension d

$$B_{21} = \frac{1}{(d-1)\,p^2} \left\{ (1 - d/2) A_0(m_2) - d/2 (p^2 + m_1^2 - m_2^2) B_1 - m_1^2 B_0 \right\}$$
$$B_{22} = \frac{1}{2(d-1)} \left\{ A_0(m_2) - (p^2 + m_1^2 - m_2^2) B_1 - 2 m_1^2 B_0 \right\} .$$

Expansion in $d = 4 - \epsilon, \epsilon \to 0$ yields

$$B_{21} = \frac{-1}{3p^2} \left\{ A_0(m_2) + 2(p^2 + m_1^2 - m_2^2) B_1 + m_1^2 B_0 + 1/2 (m_1^2 + m_2^2 - p^2/3) \right\}$$
$$B_{22} = \frac{1}{6} \left\{ A_0(m_2) - (p^2 + m_1^2 - m_2^2) B_1 - 2 m_1^2 B_0 - (m_1^2 + m_2^2 - p^2/3) \right\}$$

where the arguments of the B–functions are obvious.

3) Three point integrals: for the simplest cases we define the following invariant amplitudes

$$\int_k \frac{1}{(1)(2)(3)} = -C_0(m_1, m_2, m_3; p_1^2, p_2^2, p_3^2)$$
$$\int_k \frac{k^\mu}{(1)(2)(3)} = -p_1^\mu C_{11} - p_2^\mu C_{12}$$
(2.150)
$$\int_k \frac{k^\mu k^\nu}{(1)(2)(3)} = -p_1^\mu p_1^\nu C_{21} - p_2^\mu p_2^\nu C_{22} - (p_1^\mu p_2^\nu + p_2^\mu p_1^\nu) C_{23} + g^{\mu\nu} C_{24}$$

where $p_3 = -(p_1 + p_2)$, $(1) \equiv k^2 - m_1^2$, $(2) \equiv (k + p_1)^2 - m_2^2$ and $(3) \equiv (k + p_1 + p_2)^2 - m_3^2$.

The C_{1i}'s can be found using all possible independent contractions with $p_{1\mu,\nu}$, $p_{2\mu,\nu}$ and $g_{\mu\nu}$. This leads to the equations

$$\underbrace{\begin{pmatrix} p_1^2 & p_1 p_2 \\ p_1 p_2 & p_2^2 \end{pmatrix}}_{X} \begin{pmatrix} C_{11} \\ C_{21} \end{pmatrix} = \begin{pmatrix} R_1 \\ R_2 \end{pmatrix}$$

with

$$R_1 = \tfrac{1}{2}(B_0(m_2, m_3; p_2^2) - B_0(m_1, m_3; p_3^2)$$
$$- (p_1^2 + m_1^2 - m_2^2)C_0)$$
$$R_2 = \tfrac{1}{2}\left(B_0(m_1, m_3; p_3^2) - B_0(m_1, m_2; p_1^2)\right.$$
$$\left. + (p_1^2 - p_3^2 - m_2^2 + m_3^2)C_0\right) .$$

The inverse of the kinematic matrix of the equation to be solved is

$$X^{-1} = \frac{1}{DetX}\begin{pmatrix} p_2^2 & -p_1p_2 \\ -p_1p_2 & p_1^2 \end{pmatrix}, \quad DetX \doteq p_1^2 p_2^2 - (p_1 p_2)^2$$

and the solution reads

$$C_{11} = \frac{1}{DetX}\left\{p_2^2 R_1 - (p_1 p_2) R_2\right\}$$
$$C_{12} = \frac{1}{DetX}\left\{-(p_1 p_2) R_1 + p_1^2 R_2\right\} . \tag{2.151}$$

The same procedure applies to the more elaborate case of the C_{2i}'s where the solution may be written in the form

$$C_{24} = -\frac{m_1^2}{2}C_0 + \frac{1}{4}B_0(2,3) - \frac{1}{4}(f_1 C_{11} + f_2 C_{12}) + \frac{1}{4} \tag{2.152}$$

$$\begin{pmatrix} C_{21} \\ C_{23} \end{pmatrix} = X^{-1}\begin{pmatrix} R_3 \\ R_5 \end{pmatrix}; \begin{pmatrix} C_{23} \\ C_{22} \end{pmatrix} = X^{-1}\begin{pmatrix} R_4 \\ R_6 \end{pmatrix} \tag{2.153}$$

with

$$R_3 = C_{24} - \tfrac{1}{2}\left(f_1 C_{11} + B_1(1,3) + B_0(2,3)\right)$$
$$R_5 = -\tfrac{1}{2}\left(f_2 C_{11} + B_1(1,2) - B_1(1,3)\right)$$
$$R_4 = -\tfrac{1}{2}\left(f_1 C_{12} + B_1(1,3) - B_1(2,3)\right)$$
$$R_6 = C_{24} - \tfrac{1}{2}\left(f_2 C_{12} - B_1(1,3)\right)$$

and

$$f_1 = p_1^2 + m_1^2 - m_2^2 \ ; \ f_2 = p_3^2 - p_1^2 + m_2^2 - m_3^2 \ .$$

The notation used for the B–functions is as follows: $B_0(1,2)$ denotes the two point function obtained by dropping propagator $\frac{1}{(3)}$ from the form factor i.e. $\int_k \frac{1}{(1)(2)}$ and correspondingly for the other cases.

In the following sections we present an introduction to the calculation of the perturbative higher order corrections, also called *radiative corrections*, for the simplest QED processes.

2.6 One–Loop Renormalization

2.6.1 The Photon Propagator and the Photon Self–Energy

We first consider the full photon propagator

$$\mathrm{i} D_\gamma^{\mu\nu\,\prime}(x-y) = \langle 0|T\left\{A^\mu(x)A^\nu(y)\right\}|0\rangle ,$$

which includes all electromagnetic interactions, in momentum space. It is given by repeated insertion of the *one-particle irreducible* (1PI) self-energy function

$$-i\Pi_\gamma^{\mu\nu}(q) \equiv \text{〜〜⊗〜〜} = \text{〜〜◯〜〜} + \cdots$$

also called the vacuum polarization tensor. Since the external photon couples to the electromagnetic current via the vertex $ie j_{\text{em}}^\mu(x) A_\mu(x)$, the latter may also be represented as a correlator of two electromagnetic currents (2.92):

$$-i\Pi_\gamma^{\mu\nu}(q) = (ie)^2 \int d^4x \, e^{iqx} \langle 0| T\{j_{\text{em}}^\mu(x) \, j_{\text{em}}^\nu(y)\} |0\rangle \,. \tag{2.154}$$

Because the electromagnetic current is conserved[25] $\partial_\mu j_{\text{em}}^\mu = 0$ the non-trivial part of the self-energy function is transversal

$$\Pi^{\mu\nu} = -\left(q^\mu q^\nu - q^2 g^{\mu\nu}\right) \Pi'(q^2) \tag{2.156}$$

which implies $q_\nu \Pi^{\mu\nu} = 0$ automatically. Note however, that the free propagator, because of the required gauge fixing does not satisfy the transversality condition. The left over terms are gauge fixing artifacts and will drop out from physical matrix elements. An external real photon, for example, is represented by a polarization vector $\varepsilon^\mu(q, \lambda)$ which satisfy $q_\mu \varepsilon^\mu(q, \lambda) = 0$ and thus nullifies all terms proportional to q^μ.

In any case, we will need to consider the transverse part only in the following. In order to see how the splitting into transverse and longitudinal parts works, we introduce the projection tensors

$$T^{\mu\nu} = g^{\mu\nu} - \frac{q^\mu q^\nu}{q^2} \text{ (transverse projector)}, \quad L^{\mu\nu} = \frac{q^\mu q^\nu}{q^2} \text{ (longitudinal projector)}$$

which satisfy

$$T_\nu^\mu + L_\nu^\mu = \delta_\nu^\mu \,, \quad T_\rho^\mu T_\nu^\rho = T_\nu^\mu \,, \quad L_\rho^\mu L_\nu^\rho = L_\nu^\mu \,, \quad T_\rho^\mu L_\nu^\rho = L_\rho^\mu T_\nu^\rho = 0 \,.$$

Then writing

$$\Pi^{\mu\nu}(q) = \left(T_{\mu\nu} \, \Pi(q^2) + L_{\mu\nu} \, L(q^2)\right) = \left(g_{\mu\nu} \, \Pi_1(q^2) + q_\mu q_\nu \, \Pi_2(q^2)\right) \tag{2.157}$$

we have $L = q^2 \Pi_2 + \Pi_1$ and $\Pi \equiv \Pi_1$. Thus the transverse amplitude Π is uniquely given by the $g_{\mu\nu}$-term in the propagator and the longitudinal amplitude L does not mix with the transverse part.

---

[25] By Noether's theorem current conservation derives from a global symmetry, which in our case is global gauge symmetry (i.e. transformations (2.88) with gauge function $\alpha = $ constant) of the Lagrangian. Current conservation implies the existence of a conserved charge

$$Q = \int d^3x \, j^0(t, \boldsymbol{x}) \,; \quad \frac{dQ}{dt} = 0 \,, \tag{2.155}$$

the total charge of the system.

This allows us to calculate the full or dressed photon propagator by simply considering it in the *Feynman gauge* $\xi = 1$, for which the free propagator takes the simple form $iD_\gamma^{\mu\nu} = -ig^{\mu\nu}/(q^2 + i\varepsilon)$. The so called *Dyson series* of self–energy insertions then takes the form (we omit the metric tensor $g^{\mu\nu}$ which acts as a unit matrix)

$$i D'_\gamma(q^2) \equiv \frac{-i}{q^2} + \frac{-i}{q^2}(-i\Pi_\gamma)\frac{-i}{q^2} + \frac{-i}{q^2}(-i\Pi_\gamma)\frac{-i}{q^2}(-i\Pi_\gamma)\frac{-i}{q^2} + \cdots$$

$$= \frac{-i}{q^2}\left\{1 + \left(\frac{-\Pi_\gamma}{q^2}\right) + \left(\frac{-\Pi_\gamma}{q^2}\right)^2 + \cdots\right\}$$

$$= \frac{-i}{q^2}\left\{\frac{1}{1+\frac{\Pi_\gamma}{q^2}}\right\} = \frac{-i}{q^2 + \Pi_\gamma(q^2)} \,. \qquad (2.158)$$

The fact that the series of self–energy insertions represents a geometrical progression allows for a closed resummation and is called a Dyson summation. The result is very important. It shows that the full propagator indeed has a simple pole in q^2 only, as the free propagator, and no multi–poles as it might look like before the resummation has been performed.

In a more general form the dressed propagator, including an auxiliary photon mass term for a moment, reads

$$iD'^{\mu\nu}_\gamma(q) = \frac{-i}{q^2 - m_{0\gamma}^2 + \Pi_\gamma(q^2)}\left(g^{\mu\nu} - \frac{q^\mu q^\nu}{q^2}\right) + \frac{q^\mu q^\nu}{q^2}\cdots \qquad (2.159)$$

and we observe that in general the position of the pole of the propagator, at the tree level given by the mass of the particle, gets modified or renormalized by higher order corrections encoded in the self–energy function Π. The condition for the position $q^2 = s_P$ of the pole is

$$s_P - m_{0\gamma}^2 + \Pi_\gamma(s_P) = 0 \,. \qquad (2.160)$$

By $U(1)_{em}$ gauge invariance the photon necessarily is massless and must remain massless after including radiative corrections. Besides $m_{0\gamma} = 0$ this requires $\Pi_\gamma(q^2) = \Pi_\gamma(0) + q^2\, \Pi'_\gamma(q^2)$ with $\Pi_\gamma(0) \equiv 0$, in agreement with the transversality condition (2.156). As a result we obtain

$$i D'^{\mu\nu}_\gamma(q) = -ig^{\mu\nu}\, D'_\gamma(q^2) + \text{gauge terms} = \frac{-ig^{\mu\nu}}{q^2\,(1 + \Pi'_\gamma(q^2))} + \text{gauge terms} \,. \qquad (2.161)$$

The inverse full bare photon propagator is of the form

$$-iD_\gamma^{\mu\nu\,'\,-1} = \cdots$$

$$= i\left\{g^{\mu\nu}(q^2 - m_{0\gamma}^2) - \left(1 - \frac{1}{\xi}\right)q^\mu q^\nu\right\} - i\Pi_\gamma^{\mu\nu} \,. \qquad (2.162)$$

2.6 One–Loop Renormalization

After these structural considerations about the photon propagator we are ready to calculate the one–loop self–energy and to discuss the renormalization of the photon propagator. We have to calculate[26]

$$-i\Pi^{\mu\nu}(q) = \underset{q\quad\quad k+q}{\overset{k}{\mu \sim\!\!\!\bigcirc\!\!\!\sim \nu}}$$

$$= (-1)^F \, i^4 e^2 \int \frac{d^d k}{(2\pi)^d} \operatorname{Tr} \left\{ \gamma^\mu \frac{\slashed{k}+m}{k^2-m^2+i\varepsilon} \gamma^\nu \frac{\slashed{p}+\slashed{k}+m}{(q+k)^2-m^2+i\varepsilon} \right\}$$

$$= -e^2 \int_k \frac{\operatorname{Tr}\{\gamma^\mu \slashed{k} \gamma^\nu (\slashed{q}+\slashed{k})\}}{(1)(2)} - e^2 m^2 \int_k \frac{\operatorname{Tr}\{\gamma^\mu \gamma^\nu\}}{(1)(2)} \, .$$

We have used already the property that the trace of an odd number of γ–matrices is zero. F is the number of closed fermion loops, $F = 1$ in our case. As a convention the string of γ–matrices is read against the direction of the arrows. We again use the short notation

$$(1) = k^2 - m^2 + i\varepsilon \, , \quad (2) = (q+k)^2 - m^2 + i\varepsilon$$

and

$$\int_k \cdots = \int \frac{d^d k}{(2\pi)^d} \cdots \, .$$

Gauge invariance or transversality of the photon field requires

$$q_\mu \Pi^{\mu\nu} = 0$$

where $\Pi^{\mu\nu}$ is the symmetric vacuum polarization tensor. We may check transversality directly as follows

$$q_\nu \operatorname{Tr} \gamma^\mu \frac{1}{\slashed{k}-m} \gamma^\nu \frac{1}{(\slashed{q}+\slashed{k})-m} = \operatorname{Tr} \gamma^\mu \frac{1}{\slashed{k}-m} \slashed{q} \frac{1}{(\slashed{q}+\slashed{k})-m}$$

$$= \operatorname{Tr} \gamma^\mu \frac{1}{\slashed{k}-m} [(\slashed{q}+\slashed{k}-m)\} - (\slashed{k}-m)] \frac{1}{(\slashed{q}+\slashed{k})-m}$$

$$= \operatorname{Tr} \gamma^\mu \left(\frac{1}{\slashed{k}-m} - \frac{1}{(\slashed{q}+\slashed{k})-m} \right)$$

which upon integration should be zero. Indeed, in dimensional regularization, we may shift the integration variable in the second integral $q + k = k'$, and by integrating we find

---

[26] Fermion propagators are represented either as an inverse matrix $\frac{1}{\slashed{k}-m+i\varepsilon}$ or as a matrix $\frac{\slashed{k}+m-i\varepsilon}{k^2-m^2+i\varepsilon}$ with a scalar denominator. This second form is obtained from the first one by multiplying numerator and denominator from the left or from the right with $\slashed{k}+m-i\varepsilon$. In the denominator we then have $(\slashed{k}+m-i\varepsilon)(\slashed{k}-m+i\varepsilon) = \slashed{k}\slashed{k} - (m-i\varepsilon)^2 = k^2 - m^2 + i\varepsilon + O(\varepsilon^2)$ where the $O(\varepsilon^2)$ order term as well as the $O(\varepsilon)$ in the numerator in ε may be dropped as the limit $\varepsilon \to 0$ is always understood.

$$\int_k \operatorname{Tr} \gamma^\mu \frac{1}{\not{k} - m} - \int_k \operatorname{Tr} \gamma^\mu \frac{1}{(\not{q} + \not{k}) - m} = 0 \ .$$

It is understood that d is chosen such that the integrals converge to start with. The result is then analytically continued to arbitrary d. This then explicitly proves the transversality (2.156). We may exploit transversality and contract the vacuum polarization tensor with the metric tensor and consider the resulting scalar quantity

$$ig_{\mu\nu} \Pi^{\mu\nu} = -ig_{\mu\nu} \left(q^\mu q^\nu - q^2 g^{\mu\nu} \right) \Pi'(q^2) = iq^2 (d-1) \Pi'(q^2)$$

$$= e^2 \int_k \frac{\operatorname{Tr}\left(\gamma^\alpha \not{k} \gamma_\alpha (\not{q} + \not{k})\right)}{(1)(2)} + e^2 m^2 \int_k \frac{\operatorname{Tr}\left(\gamma^\alpha \gamma_\alpha\right)}{(1)(2)} \ .$$

Using the d–dimensional Dirac algebra relations (2.123) or, directly the trace relations (2.124), we have $\gamma^\alpha \not{k} \gamma_\alpha = (2-d) \not{k}$ and thus the trace in the first integral is $(2-d) \operatorname{Tr}(\not{k}(\not{q}+\not{k})) = (2-d)k(q+k)\operatorname{Tr} \mathbf{1}$. The scalar products $k^2 + kq$ in the numerator may be written as a difference of the two denominators (1) and (2) plus a term with does not depend on the integration variable k: $k^2 = (1) + m^2$ and $2kq = (q+k)^2 - m^2 - k^2 + m^2 - q^2 = (2) - (1) - q^2$ and hence $k^2 + qk = \frac{1}{2}[(2) + (1) - q^2 + 2m^2]$. The terms proportional to (1) and (2) each cancel against one of the denominators and give a momentum independent *tadpole* integral.

The point of these manipulations is that we got rid of the polynomial in k in the numerator and thus were able to reduce the integrals to a set of basic integrals of a scalar theory. In our example, with the definitions (2.146) and (2.148), we get

$$\int_k \frac{k^2 + qk}{(1)(2)} = \frac{i}{16\pi^2} \frac{1}{2} \left((2m^2 - q^2) B_0(m, m; q^2) - 2 A_0(m) \right) \ .$$

For the one–loop vacuum polarization as a result we then have[27]

$$q^2 \, \Pi'(q^2) = \frac{e^2}{16\pi^2} \frac{1}{(d-1)} \left\{ 4 (2-d) (m^2 - \frac{q^2}{2}) B_0(m, m; q^2) \right.$$
$$\left. - 4 (2-d) A_0(m) + 4 dm^2 \, B_0(m, m; q^2) \right\} \ .$$

Now we have to expand the result in $d = 4 - \epsilon$. At the one–loop level at most simple poles in ϵ are expected, thus a bare one–loop amplitude in the vicinity of $d = 4$ is of the form

$$A = a_{-1} \frac{1}{\epsilon} + a_0 + a_1 \epsilon + \cdots$$

---

[27] We adopt the scheme setting the trace of the unit matrix in spinor space $\operatorname{Tr} \mathbf{1} = 4$; it is of course mandatory to keep this convention consistently everywhere. While bare quantities obviously depend on this convention, on can prove that quantities finite in the limit $d \to 4$, like the renormalized ones, are unambiguous.

The expansions for the standard scalar integrals A_0 and B_0 are given in (2.140) and (2.142), respectively, and the singular terms read

$$A_0(m) = -m^2 \frac{2}{\epsilon} + O(1) \; , \quad B_0(m_1, m_2; q^2) = \frac{2}{\epsilon} + O(1)$$

which leads to (2.144). In addition, we have to expand

$$\frac{1}{d-1} = \frac{1}{3-\epsilon} = \frac{1}{3(1-\frac{\epsilon}{3})} \simeq \frac{1}{3} + \frac{\epsilon}{9} + O(\epsilon^2) \; .$$

As a result for the bare amplitude we obtain

$$q^2 \Pi'(q^2) = \frac{e^2}{16\pi^2} \frac{8}{3} \left\{ m^2 - \frac{q^2}{6} + A_0(m) + \left(m^2 + \frac{q^2}{2} \right) B_0(m, m; q^2) \right\} \tag{2.163}$$

an expression which exhibits regularized UV singularities, represented by the poles in ϵ present in A_0 and B_0.

We now have to discuss the renormalization of the photon propagator. Concerning mass renormalization, we first go back to the general form (2.157) of the vacuum polarization tensor and identify $\Pi_2 = -\Pi'$ and $\Pi_1 = -q^2 \Pi_2 = q^2 \Pi'(q^2)$ due to transversality. As we have shown earlier in this section, electromagnetic gauge invariance requires:

$$\lim_{q^2 \to 0} \Pi_1(q^2) = 0 \tag{2.164}$$

and we may check now explicitly whether the calculated amplitude satisfies this condition. For $q^2 = 0$ we have

$$B_0(m, m; 0) = -1 - \frac{A_0(m)}{m^2} = \text{Reg} - \ln m^2 \tag{2.165}$$

and hence, as it should be,

$$\lim_{q^2 \to 0} q^2 \Pi'(q^2) = \frac{e^2}{16\pi^2} \frac{8}{3} \left\{ m^2 + A_0(m) + m^2 B_0(m, m; 0) \right\} = 0 \; .$$

This proves the absence of a photon mass renormalization at this order as a consequence of $U(1)_{\text{em}}$ gauge invariance.

Next we consider the wavefunction renormalization. The renormalized photon propagator is $D'_{\text{ren}} = Z_\gamma^{-1} D'_0$, where the renormalized physical propagator is required to have residue unity of the pole at $q^2 = 0$. This infers that the interacting photon propagator in the vicinity of the pole behaves like a free photon (asymptotically free scattering state). From (2.161) we learn that the residue of the pole $q^2 = 0$ in the bare propagator is given by $1/(1 + \Pi'_\gamma(0))$ such that the wave function renormalization condition for the photon reads $Z_\gamma(1 + \Pi'_\gamma(0)) = 1$ or

$$Z_\gamma = \left[1 + \Pi'_\gamma(0)\right]^{-1} \simeq 1 - \Pi'_\gamma(0) \,. \tag{2.166}$$

We thus have to calculate

$$\lim_{q^2 \to 0} \Pi'_\gamma(q^2) = \frac{e^2}{16\pi^2} \frac{8}{3q^2} \left\{ \frac{m^2 - q^2}{6} + A_0(m) + \left(m^2 + \frac{q^2}{2}\right) B_0(m,m;q^2) \right\}\bigg|_{q^2 \to 0}$$

$$= \frac{e^2}{16\pi^2} \frac{8}{3} \left\{ -\frac{1}{6} + m^2 \dot{B}_0(m,m;0) + \frac{1}{2} B_0(m,m;0) \right\}$$

where we have used the expansion

$$B_0(m,m;q^2) = B_0(m,m;0) + q^2 \, \dot{B}_0(m,m;0) + O(q^4) \,.$$

Using the integral representation (2.142) it is easy to find

$$\dot{B}_0(m,m;0) = \frac{1}{6}\frac{1}{m^2}\,, \tag{2.167}$$

and together with (2.165) we obtain the simple result

$$Z_\gamma - 1 = \frac{e^2}{12\pi^2} B_0(m,m;0)$$

$$= \frac{\alpha}{3\pi} \left(\ln \frac{\mu^2}{m^2}\right) \,. \tag{2.168}$$

where the last expression in given in the $\overline{\mathrm{MS}}$ scheme (Reg $= \ln \mu^2$). We finally may write down the renormalized photon vacuum polarization which takes the form

$$\Pi'_{\gamma\,\mathrm{ren}}(q^2) = \Pi'_\gamma(q^2) - \Pi'_\gamma(0)$$

$$= \frac{e^2}{6\pi^2} \frac{1}{q^2} \left\{ m^2 - \frac{q^2}{6} + A_0(m) + \left(m^2 + \frac{q^2}{2}\right) B_0(m,m;q^2) - \frac{q^2}{2} B_0(m,m;0) \right\} \,.$$

Evaluating the integrals one obtains

$$B_0(m,m;q^2) = \mathrm{Reg} + 2 - \ln m^2 + 2\,(y-1)\,G(y) \tag{2.169}$$

where

$$y = \frac{4m^2}{q^2}$$

and

$$G(y) = \begin{cases} -\frac{1}{\sqrt{y-1}} \arctan \frac{1}{\sqrt{y-1}} & (y > 1) \\ \frac{1}{2\sqrt{1-y}} \ln \frac{\sqrt{1-y}+1}{\sqrt{1-y}-1} & (y < 1) \,. \end{cases} \tag{2.170}$$

For $0 < y < 1$, which means $q^2 > 4m^2$, the self–energy function is complex, given by

$$G(y) = \frac{1}{2\sqrt{1-y}} \left(\ln \frac{1+\sqrt{1-y}}{1-\sqrt{1-y}} - i\pi \right) . \qquad (2.171)$$

The imaginary part in the *time–like region* $q^2 > 0$ for $\sqrt{q^2} > 2m$ is a consequence of the fact that an electron–positron pair can be actually produced as real particles when the available energy exceeds the sum of the rest masses of the produced particles. The vacuum polarization function is thus an analytic function in the complex q^2–plane with a cut along the positive real axis starting at $q^2 = 4m^2$, which is the *threshold* for pair–creation[28].

The final result for the renormalized vacuum polarization then reads

$$\Pi'_{\gamma\,\mathrm{ren}}(q^2) = \frac{\alpha}{3\pi} \left\{ \frac{5}{3} + y - 2\left(1 + \frac{y}{2}\right)(1-y)\,G(y) \right\} \qquad (2.172)$$

which in fact is a function of q^2/m^2. This renormalized vacuum polarization function will play a crucial role in different places later. For later purposes

---

[28] As a rule, a cut diagram

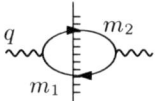

contributes to the imaginary part if the cut diagram kinematically allows physical intermediate states: $q^2 \geq (m_1 + m_2)^2$. In place of the virtual photon (a real photon requires $q^2 = 0$ and does not decay) let us consider the massive charged weak gauge boson W. The W is an unstable particle and decays predominantly as $W^- \to \ell^- \bar{\nu}_\ell$ ($\ell = e, \mu, \tau$) leptonically, and $W^- \to d\bar{u}, b\bar{c}$ hadronically. Looking at the transversal self–energy function $\Pi_W(q^2)$ of the W on the mass–shell $q^2 = M_W^2$ we have

$$\mathrm{Im}\,\Pi_W(q^2 = M_W^2) = M_W\,\Gamma_W \neq 0$$

defining the finite width Γ_W of the W–particle. Note that $W^- \to b\bar{t}$ is not allowed kinematically because the top quark t is heavier than the W ($M_W = 80.392 \pm 0.029$ GeV, $m_t = 171.4 \pm 2.1$ GeV, $m_b = 4.25 \pm 1.5$ GeV) for an on–shell W and hence does not contribute to the width.

Cutting lines means applying the substitution (see (2.139))

$$\frac{1}{p^2 - m^2 + i\varepsilon} \to -i\pi\,\delta(p^2 - m^2)$$

for the corresponding propagators. In general the imaginary part is given by cutting sets of lines of a diagram in all possible ways such that the diagram is cut into two disconnected parts. A cut contributes if the cut lines can be viewed as external lines of a real physical subprocess. Note that the imaginary part of an n–loop amplitude is given by cut diagrams exhibiting $n-1$ closed loops at most. The imaginary part therefore is less UV divergent in general. In particular, the imaginary part of a one–loop diagram is always finite.

it is useful to note that it may be written in compact form as the following integral[29]

$$\Pi'_{\gamma\,\mathrm{ren}}(q^2/m^2) = -\frac{\alpha}{\pi} \int_0^1 dz\, 2z\,(1-z)\, \ln(1 - z(1-z)q^2/m^2)$$

$$= \frac{\alpha}{\pi} \int_0^1 dt\, t^2\,(1 - t^2/3)\, \frac{1}{4m^2/q^2 - (1 - t^2)}\,. \quad (2.173)$$

The result (2.172) may be easily extended to include the other fermion contributions. In the $\overline{\mathrm{MS}}$ scheme, defined by setting $\mathrm{Reg} = \ln \mu^2$ in the bare form, we have

$$\Pi'_\gamma(q^2) = \frac{\alpha}{3\pi} \sum_f Q_f^2 N_{cf} \left[\ln \frac{\mu^2}{m_f^2} + \hat{G} \right] \quad (2.174)$$

where f labels the different fermion flavors, Q_f is the charge in units of e and N_{cf} the *color factor*, $N_{cf} = 3$ for quarks and $N_{cf} = 1$ for the leptons. We have introduced the auxiliary function

$$\hat{G} = \frac{5}{3} + y - 2\left(1 + \frac{y}{2}\right)(1 - y)\, G(y) \simeq \begin{cases} \hat{G} = 0\,, & q^2 = 0 \\ \mathrm{Re}\,\hat{G} = -\ln \frac{|q^2|}{m_f^2} + \frac{5}{3}\,, & |q^2| \gg m_f^2 \end{cases}$$

which vanishes at $q^2 = 0$. The imaginary part is given by the simple formula

$$\mathrm{Im}\,\Pi'_\gamma(q^2) = \frac{\alpha}{3} \sum_f Q_f^2 N_{cf} \left(\left(1 + \frac{y}{2}\right)\sqrt{1 - y} \right)\,. \quad (2.175)$$

Using the given low and high energy limits we get

$$\Pi'_\gamma(0) = \frac{\alpha}{3\pi} \sum_f Q_f^2 N_{cf} \ln \frac{\mu^2}{m_f^2} \quad (2.176)$$

and

$$\mathrm{Re}\,\Pi'_\gamma(q^2) = \frac{\alpha}{3\pi} \sum_f Q_f^2 N_{cf} \left(\ln \frac{\mu^2}{|q^2|} + \frac{5}{3} \right) \,;\quad |q^2| \gg m_f^2\,. \quad (2.177)$$

This concludes our derivation of the one-loop photon vacuum polarization, which will play an important role also in the calculation of the anomalous magnetic moment of the muon.

---

[29] which derives from

$$B_0(m, m; q^2) = \mathrm{Reg} - \ln m^2 - \int_0^1 dz\, \ln(1 - z(1-z)q^2/m^2)$$

(see (2.142)). The second form is obtained from the first one by a transformation of variables $z \to v = 2z - 1$, noting that $\int_0^1 dz\, \cdots = 2\int_{\frac{1}{2}}^1 dz\, \cdots$, and performing a partial integration with respect to the factor $z(1-z) = (1-v^2)/4 = \frac{d}{dv} v(1-v^2/3)/4$ in front of the logarithm.

Conformal Mapping

For numerical evaluations and for working with asymptotic expansions, it is often a big advantage to map the physical upper half $s = q^2$–plane into a bounded region as, for example, the interior of a half unit–circle as shown in Fig. 2.5. Such a conformal mapping is realized by the transformation of variables

$$s \to \xi = \frac{\sqrt{1-y} - 1}{\sqrt{1-y} + 1} \; ; \; y = \frac{4m^2}{s}$$

or

$$\frac{s}{m^2} = -\frac{(1-\xi)^2}{\xi} \; ; \; \sqrt{1-y} = \frac{1+\xi}{1-\xi} \; .$$

If we move along the real s axis from $-\infty$ to $+\infty$ we move on the half unit–circle from 0 to $+1$, then on the arc segment counter clockwise and from -1 back to 0. We distinguish the following regions:

$$\begin{array}{llll}
\text{scattering} & s < 0 & : 0 \leq \xi \leq 1 \; , & \ln \xi \\
\text{unphysical} & 0 < s < 4m^2 : & \xi = e^{i\varphi} \; , & \ln \xi = i\varphi \\
\text{production} & 4m^2 < s & : -1 \leq \xi \leq 0 \; , & \ln \xi = \ln|\xi| + i\pi
\end{array}$$

where

$$\varphi = 2 \arctan \frac{1}{\sqrt{y-1}} \; ; \; 0 \leq \varphi \leq \pi \; .$$

On the arc holds $1/y = \sin^2 \frac{\varphi}{2}$. The function $G(y)$ now has the representation

$$G(y) = \begin{cases} -\frac{1}{2}\frac{1-\xi}{1+\xi} \ln \xi \; , & 0 > s \\ -\frac{1}{2} \varphi \tan \frac{\varphi}{2} \; , & 4m^2 > s > 0 \\ -\frac{1}{2}\frac{1-\xi}{1+\xi}(\ln|\xi| + i\pi) \; , & s > 4m^2 \end{cases}$$

As an application we may write the photon vacuum polarization amplitude (2.172) in the form

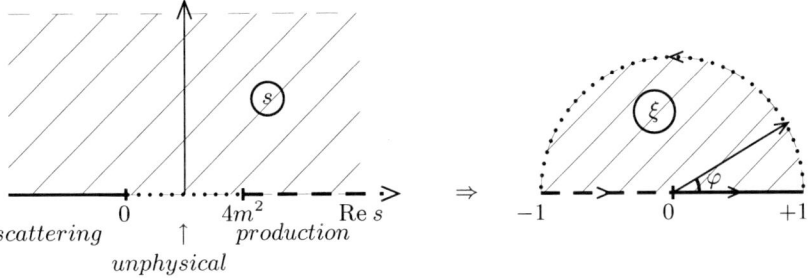

Fig. 2.5. Conformal mapping of the upper half s–plane into a half unit–circle

$$\Pi_{\gamma\,\mathrm{ren}}(q^2) = q^2 \Pi'_{\gamma\,\mathrm{ren}}(q^2)$$
$$= \frac{\alpha m^2}{3\pi} \begin{cases} -\frac{22}{3} + \frac{5}{3}\left(\xi^{-1} + \xi\right) + \left(\xi^{-1} + \xi - 4\right) \frac{1+\xi}{1-\xi} \ln\xi\,, & s < 0 \\ -\frac{5}{3}\sin^2\frac{\varphi}{2} - 4 + \left(2 + \sin^2\frac{\varphi}{2}\right) \varphi \operatorname{ctan}\frac{\varphi}{2}\,, & 0 < s < 4m^2 \end{cases}.$$

For $s > 4m^2$ the first form holds with $\ln\xi = \ln|\xi| + i\pi$. Corresponding representations are used for the vertex function as well as for the kernel function of the vacuum polarization integral contributing to $g - 2$ (see Sect. 5.2).

2.6.2 The Electron Self–Energy

Next we study the full propagator of a Dirac fermion f

$$\mathrm{i}S'_f(x - y) = \langle 0|T\left\{\psi_f(x)\bar{\psi}_f(y)\right\}|0\rangle$$

in momentum space. Again, the propagator has the structure of a repeated insertion of the 1PI self–energy $-\mathrm{i}\Sigma_f(p)$

$$\mathrm{i}\,S'_f(p) \equiv \frac{\mathrm{i}}{\not{p} - m_f} + \frac{\mathrm{i}}{\not{p} - m_f}\left(-\mathrm{i}\Sigma_f\right)\frac{\mathrm{i}}{\not{p} - m_f}$$

$$+ \frac{\mathrm{i}}{\not{p} - m_f}\left(-\mathrm{i}\Sigma_f\right)\frac{\mathrm{i}}{\not{p} - m_f}\left(-\mathrm{i}\Sigma_f\right)\frac{\mathrm{i}}{\not{p} - m_f} + \cdots$$

$$= \frac{\mathrm{i}}{\not{p} - m_f}\left\{1 + \left(\frac{\Sigma_f}{\not{p} - m_f}\right) + \left(\frac{\Sigma_f}{\not{p} - m_f}\right)^2 + \cdots\right\}$$

$$= \frac{\mathrm{i}}{\not{p} - m_f}\left\{\frac{1}{1 - \frac{\Sigma_f}{\not{p}-m_f}}\right\} = \frac{\mathrm{i}}{\not{p} - m_f - \Sigma_f}\,. \tag{2.178}$$

The Dyson series here is a geometric progression of matrix insertions which again can be summed in closed form and the inverse full fermion propagator reads

$$-\mathrm{i}S'_f{}^{-1} = \cdots = \cdots + \cdots + \cdots$$
$$= -\mathrm{i}\left\{\not{p} - m_f - \Sigma_f(p)\right\}\,. \tag{2.179}$$

The self–energy is given by an expansion in a series of 1PI diagrams

$$-\mathrm{i}\Sigma_f(p) \equiv \cdots = \cdots + \cdots\,.$$

The covariant decomposition of $\Sigma_f(p)$ for a massive fermion takes the form

$$\Sigma(p) = \not{p}\left(A(p^2, m_f, \cdots)\right) + m_f\left(B(p^2, m_f, \cdots)\right)\,, \tag{2.180}$$

where A and B are Lorentz scalar functions which depend on p^2 and on all parameters (indicated by the dots) of a given theory. In vector–like theories, like QED and QCD, no parity violating γ_5 terms are present, and the pole of the propagator, or, equivalently, the zero of the inverse propagator, is given by a multiple of the unit matrix in spinor space:

$$\not{p} = \tilde{m} \,, \quad \text{where} \quad \tilde{m}^2 = s_P \tag{2.181}$$

defines the "pole mass" of the fermion in the p^2–plane

$$\not{p} - m_f - \Sigma_f(p)|_{\not{p}=\tilde{m}} = 0 \,. \tag{2.182}$$

Among the charged leptons only the electron is stable, and hence $\tilde{m}_e = m_e$ is real and given by the physical electron mass. For the unstable fermions $s_P = \tilde{m}^2 = m^2 - im\Gamma$ is the complex *pole mass*, where the real part defines the physical mass m and the imaginary part the width Γ, which is the inverse of the life time. Looking at the full propagator

$$S'_f(p) = \frac{1}{\not{p} - m_f - \Sigma_f(p)} = \frac{\not{p}\,(1-A) + m_f\,(1+B)}{p^2\,(1-A)^2 - m_f^2\,(1+B)^2} \,. \tag{2.183}$$

the pole condition may written in a form (2.160)

$$s_P - m_0^2 - \Omega(s_P, m_0^2, \cdots) = 0 \,, \tag{2.184}$$

where

$$\Omega(p^2, m_0^2, \cdots) \equiv p^2\,(2A - A^2) + m_0^2\,(2B + B^2) \,.$$

One easily checks that the numerator matrix is non–singular at the zero of the denominator of the full Dirac propagator. Thus the solution may be obtained by iteration of (2.184) to a wanted order in perturbation theory.

Now the fermion wave function renormalization has to be considered. The renormalized propagator is obtained from the bare one by applying the appropriate wave function renormalization factor $S'_{f\mathrm{ren}} = Z_f^{-1} S'_{f0}$ (see (2.107)), where the renormalized physical propagator is required to have residue unity at the pole $\not{p} = \tilde{m}$. The interacting fermion propagator in the vicinity of the pole is supposed to behave like a free fermion (asymptotically free scattering state). In fact, this naive requirement cannot be satisfied in massless QED due to the long range nature of the electromagnetic interaction. Charged particles never become truly free isolated particles, they rather carry along a cloud of soft photons and this phenomenon is known as the infrared problem of QED. Strictly speaking the standard perturbation theory breaks down if we attempt to work with one–electron states. While the off–shell Green functions are well defined, their on–shell limit and hence the S–matrix does not exist. A way out is the so called Bloch-Nordsieck construction [56] which will be discussed below.

At intermediate stages of a calculation we may introduce an IR regulator like a tiny photon mass, which truncates the range of the electromagnetic interaction and thus allows for a perturbative treatment to start with.

In vector–like theories the fermion wave function renormalization factor $\sqrt{Z_f} = 1 + \delta Z_f$ is just a number, i.e. it is proportional to the unit matrix in spinor space[30]. Working now with a finite photon mass we may work out the on–shell wave function renormalization condition (LSZ asymptotic condition). For this purpose, we have to perform an expansion of the inverse bare propagator (2.179) about the pole $\slashed{p} = \tilde{m}$.

$$\slashed{p} - m_0 - \Sigma = \tilde{m} + (\slashed{p} - \tilde{m}) - m_0 - \tilde{m}A(\tilde{m}^2, m_0, \cdots) - m_0 B(\tilde{m}^2, m_0, \cdots)$$
$$- \tilde{m}\left(p^2 - \tilde{m}^2\right) \left.\frac{\partial A(p^2, m_0, \cdots)}{\partial p^2}\right|_{p^2=\tilde{m}^2} - m_0 \left(p^2 - \tilde{m}^2\right) \left.\frac{\partial B(p^2, m_0, \cdots)}{\partial p^2}\right|_{p^2=\tilde{m}^2}$$
$$+ \cdots,$$

where $\tilde{m}$ is the pole solution (2.182):

$$\left.\slashed{p} - m_0 - \Sigma\right|_{\slashed{p}=\tilde{m}} = \tilde{m} - m_0 - \tilde{m}A(\tilde{m}^2, m_0, \cdots) - m_0 B(\tilde{m}^2, m_0, \cdots) = 0$$

and thus using $p^2 - \tilde{m}^2 = (\slashed{p} + \tilde{m})(\slashed{p} - \tilde{m}) \simeq 2\tilde{m}(\slashed{p} - \tilde{m})$ we have

$$\slashed{p} - m_0 - \Sigma = (\slashed{p} - \tilde{m})\left(1 - \left.\frac{\partial \Sigma}{\partial \slashed{p}}\right|_{\slashed{p}=\tilde{m}}\right) + O((\slashed{p} - \tilde{m})^2)$$
$$= (\slashed{p} - \tilde{m}) Z_f^{-1} + O((\slashed{p} - \tilde{m})^2)$$

with

$$Z_f^{-1} = \left(1 - \left.\frac{\partial \Sigma}{\partial \slashed{p}}\right|_{\slashed{p}=\tilde{m}}\right)$$
$$= 1 - \left(A(\tilde{m}^2, m_0, \cdots) + 2\tilde{m}\left.\frac{\partial[\tilde{m}A(p^2, m_0, \cdots) + m_0 B(p^2, m_0, \cdots)]}{\partial p^2}\right|_{p^2=\tilde{m}^2}\right)$$
(2.185)

such that the renormalized inverse full propagator formally satisfies

$$\slashed{p} - m - \Sigma_{\mathrm{ren}} = (\slashed{p} - \tilde{m}) + O((\slashed{p} - \tilde{m})^2)$$

with residue unity of the pole.

---

[30] In the unbroken phase of the SM the left–handed and the right–handed fermion fields get renormalized independently by c–number renormalization factors $\sqrt{Z_L}$ and $\sqrt{Z_R}$, respectively. In the broken phase, a Dirac field is renormalized by $\sqrt{Z_f} = \sqrt{Z_L}\,\Pi_- + \sqrt{Z_R}\,\Pi_+$ where $\Pi_\pm = \frac{1}{2}(1 \pm \gamma_5)$ are the chiral projectors. Hence, the wave function renormalization factor, becomes a matrix $\sqrt{Z_f} = 1 + \alpha + \beta\gamma_5$ and the bare fields are related to the renormalized one's by $\psi_0(x) = \sqrt{Z_f}\psi_r(x)$, which for the adjoint field reads $\bar\psi_0(x) = \bar\psi_r(x)\gamma^0\sqrt{Z_f}\gamma^0$.

2.6 One–Loop Renormalization

We are ready now to calculate the lepton self–energy in the one–loop approximation. We have to calculate[31]

$$-i\Sigma(p) = \quad \underset{p \quad\quad k+p}{\overset{k}{\longrightarrow\!\frown\!\longrightarrow}}$$

$$= i^4 e^2 \int \frac{d^d k}{(2\pi)^d} \gamma^\rho \frac{\not{p}+\not{k}+m}{(p+k)^2 - m^2 + i\varepsilon} \gamma^\sigma D_{\rho\sigma}(k) \tag{2.187}$$

$$= -e^2 \int_k \frac{1}{k^2 - m_\gamma^2 + i\varepsilon} \frac{\gamma^\alpha(\not{p}+\not{k}+m)\gamma_\alpha}{(p+k)^2 - m^2 + i\varepsilon} + e^2(1-\xi) \int_k \text{''} \frac{1}{(k^2)^2} \text{''} \not{k} \frac{1}{\not{p}+\not{k}-m} \not{k} \,.$$

We consider the first term, applying relations (2.123) we find

$$T_1 = \int_k \frac{1}{k^2 - m_\gamma^2 + i\varepsilon} \frac{md + (2-d)(\not{p}+\not{k})}{(p+k)^2 - m^2 + i\varepsilon}$$

$$= \frac{i}{16\pi^2} \left\{ (md + (2-d)\not{p}) B_0(m_\gamma, m; p^2) + (2-d)\not{p}\, B_1(m_\gamma, m; p^2) \right\}$$

where B_1 is defined in (2.148) and may be expressed in terms of B_0 via (2.149). The limit of vanishing photon mass is regular and we may set $m_\gamma = 0$. Furthermore, expanding d about 4 using (2.144) we find

$$T_1 = \frac{i}{16\pi^2} \left\{ m(4B_0 - 2) + \not{p}\left(1 - \frac{A_0(m)}{p^2} - \frac{p^2 + m^2}{p^2} B_0 \right)\right\} \tag{2.188}$$

with

$$B_0 = B_0(0, m; p^2) = \text{Reg} + 2 - \ln m^2 + \frac{m^2 - p^2}{p^2} \ln\left(1 - \frac{p^2 + i\varepsilon}{m^2}\right).$$

We note that the first term T_1 is gauge independent. In contrast, the second term of (2.187) is gauge dependent. In the Feynman gauge $\xi = 1$ the term vanishes. In general,

$$T_2 = \int_k \frac{(1-\xi)}{(k^2 - m_\gamma^2)(k^2 - \xi m_\gamma^2)} \not{k} \frac{1}{\not{p}+\not{k}-m} \not{k}$$

where we may rewrite

---

[31] We consider the photon to have a tiny mass and thus work with a photon propagator of the form

$$D_{\rho\sigma}(k) = -\left(g_{\rho\sigma} - (1-\xi)\frac{k_\rho k_\sigma}{k^2 - \xi m_\gamma^2}\right) \frac{1}{k^2 - m_\gamma^2 + i\varepsilon}\,. \tag{2.186}$$

$$\not{k}\frac{1}{\not{p}+\not{k}-m}\not{k} = [(\not{p}+\not{k}-m)-(\not{p}-m)]\frac{1}{\not{p}+\not{k}-m}[(\not{p}+\not{k}-m)-(\not{p}-m)]$$

$$= \not{k} - (\not{p}-m) + (\not{p}-m)\frac{1}{\not{p}+\not{k}-m}(\not{p}-m) \; .$$

The first term being odd in the integration variable yields a vanishing result upon integration, while the remaining one's vanish on the mass shell $\not{p}=m$ and hence will not contribute to the mass renormalization. We obtain

$$T_2 = -(\not{p}-m)\int_k \frac{(1-\xi)}{(k^2-m_\gamma^2)(k^2-\xi m_\gamma^2)}$$
$$+(\not{p}-m)\int_k \frac{(1-\xi)}{(k^2-m_\gamma^2)(k^2-\xi m_\gamma^2)}\frac{\not{p}+\not{k}+m}{(p+k)^2-m^2+i\varepsilon}(\not{p}-m) \; ,$$

a result which affects the residue of the pole and thus contributes to the wave function renormalization. To proceed, we may use the pole decomposition

$$(1-\xi)\frac{1}{k^2-m_\gamma^2}\frac{1}{k^2-\xi m_\gamma^2} = \frac{1}{m_\gamma^2}\left(\frac{1}{k^2-m_\gamma^2}-\frac{1}{k^2-\xi m_\gamma^2}\right) \; .$$

Then all integrals are of the type we already know and the result may be worked out easily. Since these terms must cancel in physical amplitudes, we will not work them out in full detail here. Note that the second term is of order $O((\not{p}-m)^2)$ near the mass shell and hence does not contribute to the residue of the pole and hence to the wave function renormalization. The first term is very simple and given by

$$T_2 = (\not{p}-m)\left\{-(1-\xi)\frac{i}{16\pi^2}B_0(m_\gamma,\sqrt{\xi}m_\gamma;0)\right\}+O((\not{p}-m)^2) \; . \quad (2.189)$$

We now consider the mass renormalization. The latter is gauge invariant and we may start from $\Sigma = -ie^2 T_1 + ie^2 T_2$ in the Feynman gauge

$$\Sigma^{\xi=1} = -ie^2 T_1 = A(p^2)\not{p} + B(p^2) m$$
$$= \frac{e^2}{16\pi^2}\left\{\not{p}\left(1-\frac{A_0(m)}{p^2}-\frac{p^2+m^2}{p^2}B_0\right)+m\left(4B_0-2\right)\right\} \; .$$

The physical on–shell mass renormalization counter term is determined by

$$\not{p}-m_0-\Sigma|_{\not{p}=m} = \not{p}-m-\delta m-\Sigma|_{\not{p}=m}=0 \quad \text{or} \quad \delta m = -\Sigma|_{\not{p}=m}$$

and hence

$$\frac{\delta m}{m} = -\left(A(p^2)+B(p^2)\right)\big|_{p^2\to m^2}$$
$$= \frac{e^2}{16\pi^2}\left\{1+\frac{A_0(m)}{m^2}-2B_0(m_\gamma,m;m^2)\right\} = \frac{e^2}{16\pi^2}\left\{3\frac{A_0(m)}{m^2}-1\right\}$$

where we have used
$$B_0(0,m;m^2) = 1 - \frac{A_0(m)}{m^2} = \text{Reg} + 2 - \ln m^2 \;.$$

As a result the mass renormalization counter term is gauge invariant and infrared finite for $m_\gamma = 0$. The gauge dependent amplitude T_2 does not contribute. Using (2.140) we may write

$$\frac{\delta m}{m} = \frac{\alpha}{2\pi}\left\{\frac{3}{2}\ln\frac{m^2}{\mu^2} - 2\right\} \;. \tag{2.190}$$

The wave function renormalization at one–loop order is given by[32]

$$Z_f - 1 = \left(A(p^2) + 2m^2\frac{\partial(A+B)(p^2)}{\partial p^2}\right)\bigg|_{p^2 \to m^2}$$
$$= \frac{e^2}{16\pi^2}\left\{1 + \frac{A_0(m)}{m^2} + 4m^2\dot{B}_0(m_\gamma,m;m^2) + (1-\xi)\,B_0(m_\gamma,\sqrt{\xi}m_\gamma;0)\right\}\;.$$

A calculation of $\dot{B}_0$ in the limit of a small photon mass yields

$$\dot{B}_0(m_\gamma,m;m^2) \stackrel{m_\gamma \to 0}{\simeq} -\frac{1}{m^2}\left(1 + \frac{1}{2}\ln\frac{m_\gamma^2}{m^2}\right)$$

a result which exhibits an IR singularity and shows that in massless QED the residue of the pole does not exist. An asymptotically small photon mass m_γ is used as an IR regulator here. In IR regularized QED we may write the result in the form

$$Z_f - 1 = \frac{\alpha}{2\pi}\left\{\frac{1}{2}\ln\frac{m^2}{\mu^2} - 2 + 2\ln\frac{m}{m_\gamma} + \frac{1}{2}(1-\xi)\left(1 - \ln\frac{m_\gamma^2}{\mu^2}\right) + \frac{1}{2}\xi\ln\xi\right\}\;. \tag{2.192}$$

The important message here is that the residue of the pole of the bare fermion propagator is gauge dependent and infrared singular. What it means is that

---
[32] Note that with T_2 from (2.189) we have
$$\Sigma^{\xi \neq 1} = ie^2 T_2 = (\slashed{p} - m)\,A^{\xi \neq 1}$$

where
$$A^{\xi \neq 1} = (1-\xi)\,\frac{e^2}{16\pi^2}\,B_0(m_\gamma,\sqrt{\xi}m_\gamma;0)$$

and $B^{\xi \neq 1} = -A^{\xi \neq 1}$, such that $A^{\xi \neq 1} + B^{\xi \neq 1} = 0$. This leads to a contribution

$$\delta Z_f^{\xi \neq 1} = \frac{e^2}{16\pi^2}\,(1-\xi)\,B_0(m_\gamma,\sqrt{\xi}m_\gamma;0)$$
$$= \frac{e^2}{16\pi^2}\left\{(1-\xi)\left(\text{Reg} + 1 - \ln m_\gamma^2\right) + \xi\ln\xi\right\} \tag{2.191}$$

to the wave function renormalization.

the LSZ asymptotic condition for a charged particle cannot be satisfied. The cloud of soft photons accompanying any charged state would have to be included appropriately. However, usually in calculating cross sections the Bloch-Nordsieck construction is applied. This will be elaborated on below.

The renormalized fermion self–energy is given by

$$\Sigma_{f\;\mathrm{ren}} = \Sigma_f + \delta m_f - (Z_f - 1)\,(\not{p} - m_f)$$
$$= A_{\mathrm{ren}}\,(\not{p} - m_f) + C_{\mathrm{ren}}\,m_f \qquad (2.193)$$

with

$$A_{\mathrm{ren}} = A - (Z_f - 1)$$
$$C_{\mathrm{ren}} = A + B + \frac{\delta m}{m}\;.$$

In the context of $g-2$ the fermion self–energy plays a role as an insertion into higher order diagrams starting at two loops.

2.6.3 Charge Renormalization

Besides mass and wave function renormalization as a last step we have to perform a renormalization of the coupling constant, which in QED is the electric charge, or equivalently, the fine structure constant. The charge is defined via the electromagnetic vertex. The general structure of the vertex renormalization has been sketched in Sect. 2.4.1, already. Up to one–loop the diagrams to be considered are

Let us first consider the impact of *current conservation* and the resulting *Ward-Takahashi identity*. Current conservation, $\partial_\mu j^\mu_{\mathrm{em}}(x) = 0$ translates into a consideration of

$$\mathrm{i}q_\mu \Gamma^\mu = -\mathrm{i}e\not{q} - \mathrm{i}^6 e^3 \int \frac{\mathrm{d}^d k}{(2\pi^d)} D_{\rho\sigma}(k)\gamma^\rho\, S_{\mathrm{F}}(p_2 - k)\,\not{q}\, S_{\mathrm{F}}(p_1 - k)\,\gamma^\sigma + \cdots$$

with $q = p_2 - p_1$. First we note that

$$\not{q} = \not{p}_2 - \not{p}_1 = [\not{p}_2 - \not{k} - m] - [\not{p}_1 - \not{k} - m] = S_{\mathrm{F}}^{-1}(p_2 - k) - S_{\mathrm{F}}^{-1}(p_1 - k)$$

and thus

$$S_{\mathrm{F}}(p_2 - k)\not{q}S_{\mathrm{F}}(p_1 - k) = S_{\mathrm{F}}(p_2 - k)\left(S_{\mathrm{F}}^{-1}(p_2 - k) - S_{\mathrm{F}}^{-1}(p_1 - k)\right) S_{\mathrm{F}}(p_1 - k)$$
$$= S_{\mathrm{F}}(p_1 - k) - S_{\mathrm{F}}(p_2 - k)\;,$$

2.6 One–Loop Renormalization

which means that contracted with q_μ the tree–point function reduces to a difference of two two–point functions (self–energies). Therefore, for the non–trivial one–loop part, using (2.187) we obtain

$$iq_\mu \Gamma^{\mu\,(1)} = +e^3 \int_k D_{\rho\sigma}(k)\gamma^\rho\, S_F(p_1-k)\,\gamma^\sigma - e^3 \int_k D_{\rho\sigma}(k)\gamma^\rho\, S_F(p_2-k)\,\gamma^\sigma$$
$$= ie\left\{\Sigma^{(1)}(p_2) - \Sigma^{(1)}(p_1)\right\}$$

which yields the electromagnetic Ward-Takahashi (WT) identity

$$q_\mu \Gamma^\mu(p_2,p_1) = -e\left([\not{p}_2 - m - \Sigma(p_2)] - [\not{p}_1 - m - \Sigma(p_1)]\right)$$
$$= -e\left(S_F^{'-1}(p_2) - S_F^{'-1}(p_1)\right) \qquad (2.194)$$

which is the difference of the full inverse electron propagators. This relation can be shown easily to be true to all orders of perturbation theory. It has an important consequence for the renormalization of QED since it relates the vertex renormalization to the one of the charge (factor e) and the multiplicative wave function renormalization of the electron propagator. Combining the general form of the vertex renormalization (2.113) and $S'_{F0} = Z_e S'_{F\,\text{ren}}$ with the bare form of the WT identity we obtain the relationship

$$\sqrt{Z_\gamma} Z_e q_\mu \Gamma_0^\mu(p_2,p_1) = -e_0 \sqrt{Z_\gamma} Z_e \left(S_{F0}^{'-1}(p_2) - S_{F0}^{'-1}(p_1)\right)$$
$$= q_\mu \Gamma_{\text{ren}}^\mu(p_2,p_1) = -e_0 \sqrt{Z_\gamma}\left(S_{F\,\text{ren}}^{'-1}(p_2) - S_{F\,\text{ren}}^{'-1}(p_1)\right)$$
$$= -e_{\text{ren}}\left(S_{F\,\text{ren}}^{'-1}(p_2) - S_{F\,\text{ren}}^{'-1}(p_1)\right)\ .$$

We note that Z_e dropped out from the renormalized relation and we obtain the Ward-Takahashi identity

$$e_0\sqrt{Z_\gamma} = e_{\text{ren}} \quad \text{or} \quad 1 + \frac{\delta e}{e} = \frac{1}{\sqrt{1+\delta Z_\gamma}} = \sqrt{1 + \Pi'_\gamma(0)}\ . \qquad (2.195)$$

The WT identity thus has the important consequence that the charge gets renormalized only by the photon vacuum polarization! This fact will play a crucial role later, when we are going to evaluate the hadronic contributions to the effective fine structure constant.

Another important consequence of the WT identity (2.194) we obtain by taking the limit $q_\mu \to 0$:

$$\Gamma^\mu(p,p) = -e \lim_{p_2 \to p_1 = p} \frac{\left(S_F^{'-1}(p_2) - S_F^{'-1}(p_1)\right)}{(p_2-p_1)_\mu}$$
$$= -e\frac{\partial S_F^{'-1}(p)}{\partial p_\mu} = e\gamma^\mu\left(1 - \frac{\partial \Sigma}{\partial \not{p}}\right) .$$

For on–shell leptons $\not{p} = \tilde{m}$ (see (2.182)) we arrive at the electromagnetic WT identity in the form

$$\Gamma^\mu(p,p)|_{\text{on-shell}} = -e\gamma^\mu \left(1 - \left.\frac{\partial \Sigma}{\partial \not{p}}\right|_{\not{p}=\tilde{m}}\right) = -e\gamma^\mu Z_f^{-1}\;.$$

Alternatively, we may write $Z_f\, \Gamma^\mu(p,p)|_{\text{on-shell}} = -e\gamma^\mu$ or

$$-e\gamma^\mu \delta Z_f + \Gamma'^\mu(p,p)\Big|_{\text{on-shell}} = 0 \qquad (2.196)$$

where the prime denotes the non–trivial part of the vertex function. This relation tells us that some of the diagrams directly cancel. For example, we have $(V = \gamma)$

$$\text{[diagram]} + \tfrac{1}{2}\,\text{[diagram]} + \tfrac{1}{2}\,\text{[diagram]} = 0 \qquad (2.197)$$

The diagrams with the loops sitting on the external legs are contributions to the wave function renormalization and the factor $\frac{1}{2}$ has its origin in (2.108). This cancellation is the reason why the charge renormalization in QED is given by the simple relation (2.195).

We are now ready to calculate the vertex function at one–loop order. The Feynman diagram shown above translates into the Feynman integral

$$\mathrm{i}\Gamma^\mu(p_2,p_1) = -\mathrm{i}^6 e^3 \int \frac{\mathrm{d}^d k}{(2\pi^d)} D_{\rho\sigma}(k) \frac{\gamma^\rho(\not{p}_2 - \not{k} + m)\,\gamma^\mu\,(\not{p}_1 - \not{k} + m)\,\gamma^\sigma}{((p_2-k)^2 - m^2)((p_1-k)^2 - m^2)}\;. \qquad (2.198)$$

Actually, we are only interested here in the physical on–shell matrix element

$$\Gamma^\mu(p_2,p_1) \to \bar{u}(p_2,r_2)\,\Gamma^\mu(p_2,p_1)\,u(p_1,r_1)\;,$$

$p_1^2 = m^2$, $p_2^2 = m^2$, the photon being still off–shell, however. For notational simplicity we omit writing down the spinors explicitly in most cases, however, always take advantage of simplifications possible if $\Gamma^\mu(p_2,p_1)$ would be sandwiched between spinors. The first term of $D_{\rho\sigma}(k)$ (see (2.186)) produces a term proportional to

$$\gamma^\rho(\not{p}_2 - \not{k} + m)\,\gamma^\mu\,(\not{p}_1 - \not{k} + m)\,\gamma_\rho$$

and applying the Dirac algebra (2.121) and (2.123) in arbitrary dimension d together with the Dirac equation we can bring this string of γ–matrices to standard form. We anticommute $\not{p}_2$ to the left and $\not{p}_1$ to the right such that the Dirac equation $\bar{u}(p_2,r_2)\,(\not{p}_2 - m)\cdots = 0$ at the left end of the string of Dirac matrices may be used and $\cdots(\not{p}_1 - m)\,u(p_1,r_1) = 0$ at the right end. We denote $q = p_2 - p_1$ and $P = p_1 + p_2$. Furthermore we may write scalar products like $2kP = 2\,[k^2] - [(p_1-k)^2 - m^2] - [(p_2-k)^2 - m^2]$ in terms of the inverse scalar propagators which cancel against corresponding terms in the denominators. We thus obtain

$$\gamma^\mu \{(d-6)\,k^2 + 2\,([(p_1-k)^2-m^2]+[(p_2-k)^2-m^2]) + 4p_1 p_2\}$$
$$+ 4k^\alpha\,(P^\mu \gamma_\alpha - mg^\mu{}_\alpha) + 2\,(2-d)\,k^\alpha k^\mu \gamma_\alpha\;.$$

In order to stick to the definitions (2.150) we have to replace the momentum assignments as $k \to -k$, $p_1 \to p_1$ and $p_2 \to p_2 - p_1$, and we obtain

$$T_1^\mu = \frac{i}{16\pi^2}\Big\{\gamma^\mu\,\{(d-6)\,B_0(m,m,q^2)+4B_0(0,m;m^2)$$
$$+ 2\,(q^2-2m^2)\,C_0(m_\gamma,m,m)+2\,(2-d)\,C_{24}\}$$
$$+ \frac{P^\mu}{2m}\,m^2\,\{4C_{11}-2\,(2-d)\,C_{21})\}\Big\}\;.$$

An unphysical amplitude proportional to q^μ also shows up at intermediate stages of the calculation. After reduction of the tensor integrals to scalar integrals this term vanishes. On the mass shell $p_1^2 = p_2^2 = m^2$ and for $m_\gamma = 0$ the three point tensor integrals in fact are completely expressible in terms of two point functions. Evaluating the C–integrals using (2.151), (2.152) and (2.153)) we find

$$C_{11}(m_\gamma,m,m) = 2C_{12}$$
$$C_{12}(m_\gamma,m,m) = -1/(sz)\,(B_0(m,m;s)-B_0(0,m;m^2))$$
$$C_{21}(m_\gamma,m,m) = -1/(sz)\,(B_0(0,m;m^2)-B_0(m,m;s))$$
$$C_{22}(m_\gamma,m,m) = -1/(sz)[\frac{m^2}{s}\,(1+A_0(m)/m^2+B_0(m,m;s))$$
$$-\frac{1}{2}(A_0(m)/m^2+B_0(m,m;s))]$$
$$C_{23}(m_\gamma,m,m) = -1/(sz)\,\frac{1}{2}(B_0(0,m;m^2)-B_0(m,m;s))$$
$$C_{24}(m_\gamma,m,m) = \frac{1}{4}(1+B_0(m,m;s))$$

with $z = 1 - y$ where
$$y = 4m^2/q^2$$
is the kinematic variable we have encountered earlier in connection with the photon vacuum polarization.

Given the above relations we arrive at fairly simple expressions for the one–loop form factors in the Feynman gauge $\xi = 1$:

$$i\Gamma^{\mu\,\xi=1}\,(1) = -e^3\,T_1^\mu = -ie\,\Big\{\gamma^\mu A_1 + \frac{P^\mu}{2m} A_2\Big\}$$

with

$$A_1 = \frac{e^2}{16\pi^2}\left\{2\left(s-2m^2\right)C_0(m_\gamma,m,m)\right.$$
$$\left. - 3B_0(m,m;s) + 4B_0(0,m;m^2) - 2\right\}$$

$$A_2 = \frac{e^2}{16\pi^2}\left\{\frac{-y}{1-y}\left(B_0(m,m;s) - B_0(0,m;m^2)\right)\right\}. \qquad (2.199)$$

The only true vertex structure is the scalar three–point function C_0 in A_1, which may be calculated from (2.143) (see [40] Appendix E) with the result

$$C_0(m_\gamma, m, m; m^2, q^2, m^2) = -\frac{2}{q^2}\ln\frac{-q^2}{m_\gamma^2} G(y) + \frac{1}{q^2} F(y) \qquad (2.200)$$

with

$$G(y) = -\frac{1}{2\sqrt{1-y}}\ln\xi$$

$$F(y) = \frac{1}{2\sqrt{1-y}}\left\{\frac{\pi^2}{3} + 4\,\mathrm{Sp}(-\xi) + \ln^2\xi + 4\ln\xi\,\ln\frac{1+\xi}{1-\xi}\right\}.$$

The variable

$$\xi = \frac{\sqrt{1-y}-1}{\sqrt{1-y}+1}, \qquad (2.201)$$

used in this representation, was introduced in Sect. 2.6.1. The Spence function[33] or dilogarithm $\mathrm{Sp}(x)$ is defined by

$$\mathrm{Sp}(x) \equiv \mathrm{Li}_2(x) = -\int_0^1 \frac{dt}{t}\ln(1-xt). \qquad (2.203)$$

---

[33] The Spence function is an analytic function with the same cut as the logarithm. Useful relations are

$$\mathrm{Sp}(x) = -\mathrm{Sp}(1-x) + \frac{\pi^2}{6} - \ln x \ln(1-x)$$
$$\mathrm{Sp}(x) = -\mathrm{Sp}(\frac{1}{x}) - \frac{\pi^2}{6} - \frac{1}{2}\ln^2(-x)$$
$$\mathrm{Sp}(x) = -\mathrm{Sp}(-x) + \frac{1}{2}\mathrm{Sp}(x^2). \qquad (2.202)$$

For $|x| \leq 1$ it has a series expansion

$$\mathrm{Sp}(x) = \sum_{k=1}^\infty \frac{x^k}{k^2}.$$

Special values are:

$$\mathrm{Sp}(0) = 0, \quad \mathrm{Sp}(1) = \frac{\pi^2}{6}, \quad \mathrm{Sp}(-1) = -\frac{\pi^2}{12}, \quad \mathrm{Sp}(\frac{1}{2}) = \frac{\pi^2}{12} - \frac{1}{2}(\ln 2)^2.$$

2.6 One–Loop Renormalization

Looking at the standard form factor integral (2.200) for on–shell electrons, once more, we are confronted with an IR singular object. In massless QED the off–shell vertex function is regular, however, the on–shell limit does not exist. We thus again have to resort to an IR regularization by taking a small photon mass if we insist in calculating the on–shell amplitude.

Together with (2.169) the bare amplitudes may be written in a more explicit manner as in the $\overline{\text{MS}}$ scheme

$$A_1 = \frac{\alpha}{2\pi}\left\{-\frac{1}{2}\ln\frac{m^2}{\mu^2} - 2\left(1-\frac{y}{2}\right)G(y)\ln\frac{-q^2}{m_\gamma^2} + 3(1-y)G(y) + \left(1-\frac{y}{2}\right)F(y)\right\}$$

$$A_2 = \frac{\alpha}{2\pi}\left\{y\,G(y)\right\}.$$

The second term of the photon propagator in (2.198) yields a contribution

$$T_2^\mu = -(1-\xi)\int_k \frac{1}{k^2-m_\gamma^2}\frac{1}{k^2-\xi m_\gamma^2}\slashed{k}\frac{1}{\slashed{p}_2-\slashed{k}-m}\gamma^\mu\frac{1}{\slashed{p}_1-\slashed{k}-m}\slashed{k}$$

and for the on–shell vertex, applying the Dirac equation, one easily verifies that

$$\bar{u}_2\,\slashed{k}\frac{1}{\slashed{p}_2-\slashed{k}-m}\gamma^\mu\frac{1}{\slashed{p}_1-\slashed{k}-m}\slashed{k}\,u_1 = \bar{u}_2\,\gamma^\mu\,u_1$$

and hence this gauge dependent and UV divergent but q^2 independent term only contributes to the amplitude A_1 and is given by

$$i\delta\Gamma^{\mu\,\xi\neq 1\,(1)} = -e^3 T_2^\mu = -ie\gamma^\mu A_1^{\xi\neq 1} = -ie\gamma^\mu\left(-\frac{e^2}{16\pi^2}(1-\xi)\,B_0(m_\gamma,\sqrt{\xi}m_\gamma;0)\right). \tag{2.204}$$

This term exactly cancels against the gauge parameter dependent lepton part of the wave function renormalization (2.191):

$$\cdots + \cdots = -ie\gamma^\mu\delta Z_e = -ie\gamma^\mu\left(\frac{e^2}{16\pi^2}(1-\xi)\,B_0(m_\gamma,\sqrt{\xi}m_\gamma;0)\right).$$

In view of the discussion after (2.196), this cancellation is again a consequence of the WT identity. As it should be the gauge dependent term does not contribute to any physical amplitude after the appropriate wave function renormalization has been applied, i.e. the terms do not appear in the renormalized Dirac form factor A_1. The Pauli form factor in any case is not affected, it is gauge invariant and UV finite and is not subject to renormalization.

In order to discuss charge renormalization, we have to write the form factors in terms of the Dirac (electric) plus a Pauli (magnetic) term. This we may do with the help of the *Gordon identity*

$$\bar{u}(p_2)\frac{i\sigma^{\mu\nu}q_\nu}{2m}u(p_1) = \bar{u}(p_2)\left(\gamma^\mu - \frac{P^\mu}{2m}\right)u(p_1).$$

Starting from our form factor decomposition, which is more convenient from a calculational point of view, we obtain

$$i\Gamma^\mu(p_2, p_1) = -ie\left\{\gamma^\mu A_{10}(q^2) + \frac{P^\mu}{2m}A_{20}(q^2)\right\}$$
$$= -ie\left\{\gamma^\mu (A_{10} + A_{20})(q^2) - i\sigma^{\mu\alpha}\frac{q_\alpha}{2m}A_{20}(q^2)\right\}$$
$$= -ie\left\{\gamma^\mu \delta F_E(q^2) + i\sigma^{\mu\alpha}\frac{q_\alpha}{2m}F_M(q^2)\right\}.$$

Charge renormalization, according to (2.113), is fixed by the condition that $e_{\mathrm{ren}} = e$ at $q^2 = 0$ (classical charge). We therefore have to require

$$\delta F_{E\,\mathrm{ren}}(0) = A_{10}(0) + A_{20}(0) + \delta Z_e + \frac{1}{2}\delta Z_\gamma + \frac{\delta e}{e} = 0.$$

The complete Dirac form factor, including the tree level value is given by

$$F_{E\,\mathrm{ren}}(q^2) = 1 + \delta F_{E\,\mathrm{ren}}(q^2) \tag{2.205}$$

and satisfies the charge renormalization condition

$$F_{E\,\mathrm{ren}}(0) = 1. \tag{2.206}$$

However, the electromagnetic Ward-Takahashi identity (2.196) infers

$$A_{10} + A_{20} + \delta Z_e = 0$$

such that, in agreement with (2.195), the charge renormalization condition fixes the charge counter term to the wave function renormalization constant of the photon

$$\frac{\delta e}{e} = -\frac{1}{2}\delta Z_\gamma = \frac{1}{2}\Pi'_\gamma(0) = -\frac{\alpha}{2\pi}\frac{1}{3}\ln\frac{m^2}{\mu^2} \tag{2.207}$$

with the explicit result given in the $\overline{\mathrm{MS}}$ scheme $\mathrm{Reg} = \ln\mu^2$.

As a result the renormalized one–loop virtual photon contributions to the lepton electric (E) and magnetic (M) form factors read

$$\delta F_E = (A_{10} + A_{20} + \delta Z_e)$$
$$= \frac{\alpha}{2\pi}\left\{\ln\frac{m^2}{m_\gamma^2} - (2-y)\,G(y)\ln\frac{-q^2}{m_\gamma^2} - 2 + (3-2y)G(y) + \left(1 - \frac{y}{2}\right)F(y)\right\}$$
$$F_M = -A_{20} = \frac{\alpha}{2\pi}\{-y\,G(y)\}. \tag{2.208}$$

In the scattering region $q^2 < 0$ ($y < 0$) with $0 \le \xi \le 1$ the form factors are real; in the production region $q^2 > 4m^2$ ($0 < y < 1$) with $-1 \le \xi \le 0$ we have an imaginary part (using $\ln(\xi) = \ln(-\xi) + i\pi$, $\ln(-q^2/m^2 - i\varepsilon) = \ln(q^2/m^2) - i\pi$)

$$\frac{1}{\pi}\mathrm{Im}\, F_{\mathrm{E}} = \frac{\alpha}{4\pi}\frac{1}{\sqrt{1-y}}\left\{(2-y)\ln\frac{q^2-4m^2}{m_\gamma^2} - 3 + 2y\right\}$$

$$\frac{1}{\pi}\mathrm{Im}\, F_{\mathrm{M}} = \frac{\alpha}{4\pi}\frac{y}{\sqrt{1-y}} \qquad (2.209)$$

The Dirac form factor for $q^2 \neq 0$ (on–shell electron, off–shell photon) at this stage is still IR singular in the limit of vanishing photon mass and cannot be physical. Before we continue the discussion of the result we have to elaborate on the infrared problem in massless QED and the difficulties to define scattering states for charged particles.

However, the Pauli form factor, of primary interest to us turns out to be IR save. It is a perturbatively calculable quantity, which seems not to suffer from any of the usual problems of gauge dependence, UV divergences and the related renormalization scheme dependence. We thus are able to calculate the leading contribution to the anomalous magnetic moment without problems. The anomalous magnetic moment of a lepton is given by $F_{\mathrm{M}}(0)$ where $F_{\mathrm{M}}(q^2)$ is given in (2.208). We hence have to calculate $-y\, G(y)$ for $Q^2 = -q^2 > 0$ and $Q^2 \to 0$ or $y < 0$ and $|y| \to \infty$. Let $z = -y = |y|$ and z be large; the expansion yields

$$\sqrt{1-y} = \sqrt{z+1} \simeq \sqrt{z}\left(1 + \frac{1}{2z} + \cdots\right)$$

$$\ln\frac{\sqrt{1-y}-1}{\sqrt{1-y}+1} = \ln\frac{\sqrt{z+1}-1}{\sqrt{z+1}+1} \simeq -\frac{2}{\sqrt{z}} + \cdots$$

and therefore

$$-y\, G(y)\big|_{-y\to\infty} = -\frac{z}{2\sqrt{z+1}}\ln\frac{\sqrt{z+1}-1}{\sqrt{z+1}+1}\bigg|_{z\to\infty}$$

$$\simeq 1 + O\!\left(\frac{1}{\sqrt{|y|}}\right).$$

We thus arrive at

$$F_{\mathrm{M}}(0) = \frac{\alpha}{2\pi} \simeq 0.0011614\cdots \qquad (2.210)$$

which is Schwinger's classic result for the anomalous magnetic moment of the electron and which is universal for all charged leptons.

An important cross check of our calculation of F_{E} is also possible at this stage. Namely, we may check directly the WT identity (2.196), which now reads $\delta F_{\mathrm{E}}(0) = 0$. Taking the limit $q^2 \to 0$ for space–like momentum transfer $q^2 < 0$, we may use the expansion just presented for calculating $F_{\mathrm{M}}(0) = \alpha/2\pi$. For $y < 0$ and $|y| \to \infty$ we have $\xi \sim 1 - 2/\sqrt{|y|}$ and the somewhat involved expansion of $F(y)$ in (2.208) yields that $yF(y) \to 0$ in this limit. Since $-yG(y) \to 1$ we get precisely the cancellations needed to prove $\delta F_{\mathrm{E}}(q^2) \to 0$ for $q^2 \to 0$[34]. The leading term for $|q^2| \ll 4m^2$ reads

---

[34] One also may check this directly on the level of the standard scalar integrals A_0, B_0 and C_0. Denoting by $AA(m) = A_0(m)/m^2$ we have

$$\delta F_{\rm E}(q^2) = \frac{\alpha}{3\pi} \frac{q^2}{m^2} \left(\ln \frac{m}{m_\gamma} - \frac{3}{8} \right) + O(q^4/m^4)$$

and is IR singular and hence non–physical without including soft real photon emission. The leading behavior of the form factors for large $|q^2| \gg m^2$ reads

$$\delta F_{\rm E}(q^2) \sim -\frac{\alpha}{2\pi} \left(\frac{1}{2} \ln^2 \frac{|q^2|}{m^2} + 2 \ln \frac{m}{m_\gamma} \ln \frac{|q^2|}{m^2} - 2 \ln \frac{m}{m_\gamma} - \frac{3}{2} \ln \frac{|q^2|}{m^2} + 2 - \frac{\pi^2}{6} \right.$$
$$\left. - \Theta(q^2 - 4m^2) \frac{\pi^2}{2} \right) + \Theta(q^2 - 4m^2) \, i \frac{\alpha}{2} \left(\ln \frac{q^2}{m_\gamma^2} - \frac{3}{2} \right)$$

$$F_{\rm M}(q^2) \sim -\frac{\alpha}{\pi} \frac{m^2}{q^2} \ln \frac{|q^2|}{m^2} + \Theta(q^2 - 4m^2) \, i\alpha \frac{m^2}{q^2} \, .$$

As in the examples discussed so far, often we will need to know the behavior of Feynman amplitudes for large momenta or equivalently for small masses. The tools for estimating the asymptotic behavior of amplitudes are discussed next.

2.6.4 Dyson– and Weinberg–Power-Counting Theorems

Since, in momentum space, any amplitude may be obtained as a product of 1PI building blocks, the vertex functions $\Gamma(p_1, \cdots, p_n)$, it is sufficient to know the asymptotic behavior of the latter. This behavior may be obtained by considering the contributions form individual Feynman integrals $\Gamma_G(p_1, \cdots, p_n)$, the index G denoting the corresponding Feynman graph. As we know already from Sect. 2.4.2, power counting theorems play an important role for evaluating

1. the convergence of Feynman integrals (UV divergences),
2. the behavior of Feynman amplitudes for large momenta.

Weinberg's power-counting theorem is an extension of Dyson's power-counting theorem, and describes the off-shell behavior of vertex functions (amputated n–point functions with $n \geq 2$)

$$\delta F_{\rm E}(q^2) \stackrel{q^2 \to 0}{\sim} \propto \left([-4m^2 C_0 - 3B_0(m,m;0) + 4B_0(0,m;m^2) - 2]_{A_1} \right.$$
$$\left. + [B_0(m,m;0) - B_0(0,m;m^2)]_{A_2} + [1 + AA(m) + 4m^2 \dot{B}_0(m_\gamma,m;m^2)]_{\delta Z_e} \right) .$$

Using the relations

$$\begin{aligned}
C_0(m_\gamma, m, m; m^2, 0, m^2) &= \tfrac{-1}{4m^2} \left(B_0(0, m; m^2) - 1 - AA(m) + 2AA(m_\gamma) \right) \\
B_0(m, m; 0) &= -1 - AA(m) \\
B_0(0, m; m^2) &= 1 - AA(m) \\
m^2 \dot{B}_0(m_\gamma, m; m^2) &= -1 - \tfrac{1}{2} AA(m_\gamma) + \tfrac{1}{2} AA(m)
\end{aligned}$$

one easily finds that indeed $\delta F_{\rm E}(q^2) \stackrel{q^2 \to 0}{\sim} 0$. This kind of approach is usually utilized when working with computer algebra methods.

$$\Gamma(p_1,\cdots,p_n) = \sum_G \Gamma_G(p_1,\cdots,p_n)$$

for large p_i ($i = 1,\ldots,m$) in a subspace of the momenta

$$\Gamma(\lambda p_1,\cdots,\lambda p_m, p_{m+1},\cdots,p_n) \stackrel{\lambda\to\infty}{\longrightarrow} \;?$$

where $(p_1,\cdots,p_n)$ is a fixed set of momenta, $2 \leq m \leq n$ and λ a real positive stretching (dilatation) factor, which we are taking to go to infinity. The sum is over all possible Feynman graphs G which can contribute.

We first introduce some notions and notation. A set of external momenta $(p_1,\cdots,p_m)$ is called *non-exceptional* if no subsum of momenta vanishes, i.e., the set is generic. The set of external lines which carry momenta going to infinity is denoted by $\mathcal{E}_\infty$. By appropriate relabeling of the momenta we may always achieve that the first m of the momenta are the ones which go to infinity. In first place the power counting theorems hold in the Euclidean region (after Wick-rotation) or in the Minkowski region for space-like momenta, which will be sufficient for our purpose. Also for massless theories there may be additional complications [44].

Dyson's power-counting theorem [45] states that for *non-exceptional momenta* when all momenta are going to infinity a vertex function behaves as

$$\Gamma(\lambda p_1,\cdots,\lambda p_n) = \mathcal{O}(\lambda^{\alpha_\Gamma} (\ln \lambda)^{\beta_\Gamma}),$$

where $\alpha_\Gamma = \max_G d(G)$ with $d(G)$ the superficial degree of divergence of the diagram G, introduced in Sect. 2.4.2. The asymptotic coefficient β_Γ giving the leading power of the logarithm may also be characterized in terms of diagrams [47], but will not be discusse here as we will need the asymptotic behavior modulo logarithms only. For an individual 1PI diagram G the Dyson power-counting theorem says that provided all momenta go to infinity, and the set of momenta is non-exceptional the behavior is determined by the superficial degree of divergence $d(G)$ of the corresponding diagram. The crucial point is that in a renormalizable theory $d(G)$ is independent of the particular graph G and given by the dimension of the vertex function $\dim \Gamma$ which only depends on type and number of external legs as discussed before in Sect. 2.4.2. In fact, in $d = 4$ dimensions,

$$\Gamma(\lambda p_1,\cdots,\lambda p_n) = \mathcal{O}(\lambda^{4-b-\frac{3}{2}f} (\ln \lambda)^{\ell}).$$

with $b = n_B$ the number of boson lines and $f = n_F$ the number of fermion lines. ℓ is a non-negative integer depending on the order of perturbation theory. Its maximum possible value $\ell \leq L$ is given by the number L of loops.

Weinberg's power-counting theorem [46] generalizes Dyson's theorem and answers the question what happens when a subset only of all momenta is scaled to infinity. We first consider an individual Feynman integral G and 1PI subdiagrams $H \supset \mathcal{E}_\infty$ which include all lines $\mathcal{E}_\infty$ tending to infinity. A subset $H \subset G$ here is a set of lines from G (external and internal) such that at each vertex there is either no line or two or more lines[35]. Then

$$\Gamma_\Gamma(\lambda p_1, \cdots, \lambda p_m, p_{m+1}, \cdots, p_n) = \mathcal{O}(\lambda^{d(H_0)} (\ln \lambda)^{\beta(H_0)})$$

where H_0 has maximal superficial degree of divergence $d(H)$. For a characterization of the logarithmic coefficient $\beta(H)$ see [47]. The result simplifies considerably if we consider the complete vertex function. When a non-exceptional set $\mathcal{E}_\infty$ of external lines have momenta tending to infinity, then the total vertex function has as its asymptotic power a quantity $\alpha(\mathcal{E}_\infty)$

$$\Gamma(\lambda p_1, \cdots, \lambda p_m, p_{m+1}, \cdots, p_n) = \mathcal{O}(\lambda^{\alpha(\mathcal{E}_\infty)} (\ln \lambda)^\ell)$$

which depends only on the numbers and type of lines in $\mathcal{E}_\infty$, and is given by

$$\alpha(\mathcal{E}_\infty) = 4 - \frac{3}{2} f(\mathcal{E}_\infty) - b(\mathcal{E}_\infty) - \min_{\mathcal{E}'}[\frac{3}{2} f(\mathcal{E}') + b(\mathcal{E}')] \,. \qquad (2.211)$$

Here $b(\mathcal{E})$, $f(\mathcal{E})$ are the number of bosons or fermions in the set $\mathcal{E}$. The minimum in (2.211) is taken over all sets $\mathcal{E}'$ of lines such that the virtual transition $\mathcal{E}_\infty \leftrightarrow \mathcal{E}'$ is not forbidden by selection rules (charge, fermion number etc.). $\mathcal{E}'$ is the set of external lines of H which are not in $\mathcal{E}_\infty$. Again, $\ell \leq L$. For useful refinements of asymptotic expansion theorems see e.g. [48] and references therein. Another tool to study the asymptotic behavior of Green- or vertex-functions is the renormalization group which we will consider next and in particular allows to control effects due to the large UV logarithms.

2.6.5 The Running Charge and the Renormalization Group

Charge renormalization is governed by a *renormalization group* [49] (RG), which controls the response of the theory with respect to a change of the renormalization scale parameter μ in the $\overline{\text{MS}}$ scheme, like for example in the

---

[35] The following example (electrons=full lines and photons=wavy lines) may illustrate this: fat lines carry the flow of large momentum (subgraph H)

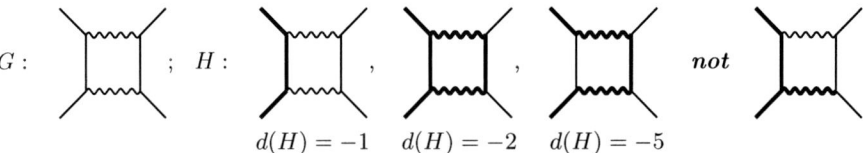

$$G: \qquad ; \quad H: \qquad , \qquad , \qquad \text{not}$$

$$d(H) = -1 \quad d(H) = -2 \quad d(H) = -5$$

The first graph in the set H determines the leading behavior $\mathcal{O}(\lambda^{-1} \ln^x \lambda)$. Note that all subgraphs H are connected and have no *dead end* lines (like the last diagram above, which is not a subgraph in the sense the term is used here). Thin lines attached to vertices of a subgraph H figure as external lines $\mathcal{E}'$, such that $\mathcal{E}_H = \mathcal{E}_\infty + \mathcal{E}'$ is the set of all external lines of H and $d(H) = 4 - \frac{3}{2} f(\mathcal{E}_H) - b(\mathcal{E}_H)$.

charge renormalization according to (2.207). It gives rise to the definition of an effective or running charge $\alpha(\mu)$ and running mass $m(\mu)$ as a function of the renormalization scale μ. However, the RG not only governs the dependence of a renormalized QFT on the renormalization scale, it yields the behavior of the theory with respect to dilatations, the simultaneous stretching of all momenta, and hence allows to discuss the asymptotic behavior for small and large momenta. The RG serves as a tool to systematically include large logarithmic radiative corrections, in fact, it permits the resummation to all orders of the perturbation expansion, of *leading logarithms* (LL), *next to leading logarithms* (NLL) etc. It thus allows to estimate leading radiative corrections of higher order without the need to actually perform elaborate calculations under the condition that *large scale changes* are involved. Besides the all orders Dyson summation of self–energy corrections and the soft photon exponentiation to be discussed in the next section, the RG is a third method which allows to predict leading higher order corrections from low order calculations. The RG generalizes the classical concept of dimensional analysis to QFT, where renormalization anomalies of the dilatation current [50] lead to a breaking of dilatation invariance by quantum effects (see Sect. 5.1.4 footnote on p. 287).

The RG may be obtained by starting from the bare vertex functions (the amputated Green functions) mentioned already briefly in Sect. 2.4.2. Note that the renormalization scale parameter μ is entering in DR by the fact that in the d–dimensional QFT the bare coupling constant $\bar{e}_0$ must have a dimension $\frac{4-d}{2}$, i.e. $\bar{e}_0 = e_0 \mu^{\varepsilon/2}$ with e_0 dimensionless (see (2.116)). This gives rise to the factors μ^{4-d} in the definitions of the standard integrals in Sect. 2.5.6 when working with the dimensionless bare coupling e_0. As a result the μ dependence formally comes in via the UV regulator term (2.141). Since μ only enters via the bare coupling $\bar{e}_0$ all bare quantities, like the vertex function Γ_0, at fixed $\bar{e}_0$ are independent of μ:

$$\mu \frac{\mathrm{d}\Gamma_0}{\mathrm{d}\mu}\bigg|_{\bar{e}_0} \equiv 0 \ . \tag{2.212}$$

The bare vertex functions in $d = 4 - \varepsilon$ dimensions

$$\Gamma_0^{(n_A, 2n_\psi)}(\{p\}; \bar{e}_0, m_0, \xi_0)_\varepsilon$$

are homogeneous under simultaneous dilatation of all momenta and all dimensionfull parameters including the scale μ. According to (2.117) we have

$$\Gamma_0^{(n_A, 2n_\psi)}\left(\{\kappa p\}; e_0 (\kappa\mu)^{\varepsilon/2}, \kappa m_0, \xi_0\right) = \kappa^{\dim\Gamma} \Gamma_0^{(n_A, 2n_\psi)}\left(\{p\}; e_0 (\mu)^{\varepsilon/2}, m_0, \xi_0\right) \tag{2.213}$$

with

$$\dim\Gamma = d - n_A \frac{d-2}{2} - 2n_\psi \frac{d-1}{2} \ .$$

The renormalized vertex functions are obtained by renormalizing parameters and fields: $A_0 = \sqrt{Z_A} A_r$, $\psi_0 = \sqrt{Z_\psi} \psi_r$, $e_0 = Z_g e_r$ and $m_0 = Z_m m_r$ and thus

$$\Gamma_0^{(n_A, 2n_\psi)}(\{p\}; \bar{e}_0, m_0, \xi_0)_\varepsilon = (Z_A)_\varepsilon^{-\frac{n_A}{2}} (Z_\psi)_\varepsilon^{-n_\psi} \Gamma_{\text{ren}}^{(n_A, 2n_\psi)}(\{p\}; e_r, m_r, \xi_r, \mu)_\varepsilon$$

where the wave function renormalization factors have the property to make the limit $\lim_{\varepsilon \to 0} \Gamma_{\text{ren}}(\{p\}; e_r, m_r, \xi_r, \mu)_\varepsilon$ exist. The trivially looking bare RG (2.212) becomes highly non–trivial if rewritten as an equation for Γ_{ren} as a function of the renormalized parameters. By applying the chain rule of differentiation we find the RG equation

$$\left\{ \mu \frac{\partial}{\partial \mu} + \beta \frac{\partial}{\partial e_r} + \omega \frac{\partial}{\partial \xi_r} + \gamma_m m_r \frac{\partial}{\partial m_r} - n_A \gamma_A - 2n_\psi \gamma_\psi \right\} \Gamma_{\text{ren}} = 0 \tag{2.214}$$

where the coefficient functions are given by

$$\beta = D_{\mu, \varepsilon} e_r = e_r \left(-\frac{\varepsilon}{2} + \frac{\varepsilon}{2} e_0 \frac{\partial}{\partial e_0} \ln Z_g \right)$$

$$\gamma_m m_r = D_{\mu, \varepsilon} m_r = \frac{\varepsilon}{2} m_0 e_0 \frac{\partial}{\partial e_0} \ln Z_m$$

$$\gamma_A = D_{\mu, \varepsilon} \ln Z_A = -\frac{\varepsilon}{4} e_0 \frac{\partial}{\partial e_0} \ln Z_A$$

$$\gamma_\psi = D_{\mu, \varepsilon} \ln Z_\psi = -\frac{\varepsilon}{4} e_0 \frac{\partial}{\partial e_0} \ln Z_\psi$$

$$\omega = D_{\mu, \varepsilon} \xi_r = -\frac{\varepsilon}{2} e_0 \frac{\partial}{\partial e_0} \xi_r = -2 \xi_r \gamma_A. \tag{2.215}$$

We have used

$$\mu \frac{\partial}{\partial \mu} F(\bar{e}_0 = e_0 \mu^{\varepsilon/2}) \Big|_{\bar{e}_0} = \left(\mu \frac{\partial}{\partial \mu} - \frac{\varepsilon}{2} e_0 \frac{\partial}{\partial e_0} \right) F(e_0, \mu) \doteq D_{\mu, \varepsilon} F(e_0, \mu)$$

$$\text{and} \quad F^{-1} D_{\mu, \varepsilon} F(e_0, \mu) = D_{\mu, \varepsilon} \ln F(e_0, \mu)$$

and the relation $\xi_0 = Z_A \xi_r$, i.e. $Z_\xi = Z_A$, which is a consequence of a WT identity, and implies $\omega = -2\xi_r \gamma_A$. Note that $\beta = \beta(e_r)$ and $\gamma_m = \gamma_m(e_r)$ are gauge invariant. In the *Landau gauge* $\xi_r = 0$ the coefficient function $\omega \equiv 0$ and $\gamma_i = \gamma_i(e_r)$ $(i = A, \psi)$. The right hand sides of (2.215) have to be rewritten in terms of the renormalized parameters by inversion of the formal power series. The renormalization factors Z_i are of the form

$$Z_i = 1 + \sum_{n=1}^{\infty} \frac{Z_{i,n}(e_r, \xi_r)}{\varepsilon^n} \tag{2.216}$$

and applying the chain rule, we observe that the coefficient functions are uniquely determined by $Z_{i,1}(e_r, \xi_r)$ alone:

$$\beta(e) = \frac{e}{2} e \frac{\partial}{\partial e} Z_{g,1}(e) = \frac{\alpha}{\pi} \frac{e}{3} + \cdots$$

$$\gamma_m(e) = \frac{1}{2} e \frac{\partial}{\partial e} Z_{m,1}(e) = \frac{\alpha}{\pi} \frac{3}{2} + \cdots$$

$$\gamma_A(e,\xi) = \frac{1}{4} e \frac{\partial}{\partial e} Z_{A,1}(e,\xi) = \frac{\alpha}{\pi} \frac{2}{3} + \cdots$$

$$\gamma_\psi(e,\xi) = \frac{1}{4} e \frac{\partial}{\partial e} Z_{\psi,1}(e,\xi) = \frac{\alpha}{\pi} \frac{\xi}{2} + \cdots \quad (2.217)$$

These are the residues of the simple ε–poles of the renormalization counter terms. The one–loop contributions we calculated above: $Z_A = Z_\gamma$ (2.168), $Z_\psi = Z_f$ (2.192), $Z_g = 1 + \frac{\delta e}{e}$ (2.207) and $Z_m = 1 + \frac{\delta m}{m}$ (2.190) with $\mathrm{Reg} = \ln \mu^2 \to \frac{2}{\varepsilon}$ (see (2.141)). Note that in QED the WT identity (2.195) implies $Z_g = 1/\sqrt{Z_\gamma}$, which is very important because it says that charge renormalization is governed by photon vacuum polarization effects. The latter will play a crucial role in calculations of $g-2$. The UV singular parts of the counter terms read

$$Z_e = 1 + \frac{e^2}{4\pi^2} \frac{1}{3} \frac{1}{\varepsilon}, \quad Z_m = 1 - \frac{e^2}{4\pi^2} \frac{3}{2} \frac{1}{\varepsilon},$$
$$Z_A = 1 + \frac{e^2}{4\pi^2} \frac{2}{3} \frac{1}{\varepsilon}, \quad Z_\psi = 1 + \frac{e^2}{4\pi^2} \frac{\xi}{2} \frac{1}{\varepsilon},$$

from which the leading terms of the RG coefficient functions given in (2.217) may be easily read off. The RG equation is a partial differential equation which is homogeneous and therefore can be solved easily along so called characteristic curves. Let s parametrize such a curve, such that als quantities become functions of a the single parameter s: $e = e(s)$, $m = m(s)$, $\mu = \mu(s)$ and

$$\frac{\mathrm{d}\Gamma}{\mathrm{d}s} (\{p\}; e(s), m(s), \mu(s)) = n\gamma \, \Gamma$$

with

$$\frac{\mathrm{d}\mu}{\mathrm{d}s} = \mu, \quad \frac{\mathrm{d}e}{\mathrm{d}s} = \beta(e), \quad \frac{\mathrm{d}m}{\mathrm{d}s} = m\gamma_m(e),$$

which is a set of ordinary differential equations the solution of which is solving the RG equation (2.215). For simplicity of notation and interpretation we have assumed the Landau gauge $\xi = 0$ and we abbreviated $n_A \gamma_A + n_\psi \gamma_\psi = n\gamma$. The successive integration then yields

1)
$$\frac{\mathrm{d}\mu}{\mathrm{d}s} = \mu \,\triangleright\, \ln \mu = s + \text{constant} \,\triangleright\, \mu = \mu_0 \, \mathrm{e}^s = \mu_0 \, \kappa$$

where $\kappa = \mathrm{e}^s$ is a scale dilatation parameter

2)
$$\frac{\mathrm{d}e}{\mathrm{d}s} = \beta(e) \,\triangleright\, \frac{\mathrm{d}e}{\beta(e)} = \mathrm{d}s = \frac{\mathrm{d}\mu}{\mu} \,\triangleright$$

$$\ln(\mu/\mu_0) = \ln \kappa = \int_e^{e(\kappa)} \frac{de'}{\beta(e')} \qquad (2.218)$$

which is the implicit definition of the running coupling $e(\kappa)$ with $e = e(1)$ the coupling at reference scale μ_0 and $e(\kappa) = e(\mu/\mu_0)$ the coupling at scale μ.

3)
$$\frac{dm}{ds} = m\gamma_m \triangleright \quad \frac{dm}{m} = \gamma_m(e)\,ds = \gamma_m(e)\frac{de}{\beta(e)} \triangleright$$

$$m(\kappa) = m \, \exp \int_e^{e(\kappa)} \frac{\gamma(e')\,de'}{\beta(e')} \qquad (2.219)$$

4)
$$\frac{d\Gamma}{ds} = n\gamma(e)\,ds = n\gamma(e)\frac{d\mu}{\mu} = n\gamma(e)\frac{de}{\beta(e)} \triangleright$$

$$\Gamma(\kappa) = \Gamma \, \exp\left\{ n \int_e^{e(\kappa)} \frac{\gamma(e')\,de'}{\beta(e')} \right\} = \Gamma \, z_A(e,\kappa)^{n_A} \, z_\psi(e,\kappa)^{2n_\psi} \qquad (2.220)$$

with $\Gamma = \Gamma(1)$, and

$$z_A(e,\kappa) = \exp \int_e^{e(\kappa)} \frac{\gamma_A(e')\,de'}{\beta(e')} \;,\quad z_\psi(e,\kappa) = \exp \int_e^{e(\kappa)} \frac{\gamma_\psi(e')\,de'}{\beta(e')}\;.$$

Altogether, we may write this as an equation which describes the response of the theory with respect to a change of the scale parameter μ:

$$\Gamma(\{p\}; e, m, \mu/\kappa) = z_A(e,\kappa)^{-n_A} \, z_\psi(e,\kappa)^{-2n_\psi} \, \Gamma(\{p\}; e(\kappa), m(\kappa), \mu) \qquad (2.221)$$

Thus **a change of the scale parameter μ is equivalent to a finite renormalization of the parameters and fields** and together with the homogeneity relation we have for the vertex functions with scaled momenta

$$\Gamma(\{\kappa p\}; e, m, \mu) = \kappa^{\dim \Gamma} \Gamma(\{p\}; e(\kappa), m(\kappa)/\kappa, \mu/\kappa)$$
$$= \kappa^{\dim \Gamma} \, z_A(e,\kappa)^{-n_A} \, z_\psi(e,\kappa)^{-2n_\psi} \, \Gamma(\{p\}; e(\kappa), m(\kappa)/\kappa, \mu) \qquad (2.222)$$

which is the basic relation for a discussion of the asymptotic behavior.

Asymptotic Behavior

Two regimes are of interest, the high energy (ultraviolet) behavior and the low energy (infrared) behavior. For the general discussion we consider a generic gauge coupling g (in place of e in QED).

1) UV behavior

The ultraviolet behavior, which determines the short distance properties, is obtained by choosing $\kappa|p| \gg m, \mu$ thus

$$\ln \kappa = \int_g^{g(\kappa)} \frac{dg'}{\beta(g')} \to +\infty \; ; \quad \kappa \to \infty .$$

However, the integral can only become divergent for finite $g(\kappa)$ if $\beta(g)$ has a zero at $\lim_{\kappa \to \infty} g(\kappa) = g^*$: more precisely, in the limit $\kappa \to \infty$ the effective coupling has to move to a fixed point $g(\kappa) \to g_-^*$ if finite, and the fixed point coupling is characterized by $\beta(g_-^*) = 0$, $\beta'(g_-^*) < 0$. Thus g_-^* is an ultraviolet fixed point coupling. Note that by dilatation of the momenta at fixed m and μ, the effective coupling is automatically driven into a fixed point, a zero of the β-function with negative slope, if it exists. If $g_-^* = 0$ we have asymptotic freedom. This is how QCD behaves, which has a β-function

$$\beta_{\text{QCD}}(g_s) = -g_s \left(\beta_0 \left(\frac{g^2}{16\pi^2} \right) + \beta_1 \left(\frac{g^2}{16\pi^2} \right)^2 + \cdots \right) \quad (2.223)$$

with $\beta_0 > 0$ (see Fig. 2.6a). QCD will be considered in more detail later on.

A possible fixed point is accessible in perturbation theory provided g^* is sufficiently small, such that perturbation theory is sufficiency "convergent" as an asymptotic series. One may then expand about g^*:

$$\beta(g) = (g - g_-^*) \, \beta'(g_-^*) + \cdots$$
$$\gamma(g) = \gamma^* + (g - g_-^*) \, \gamma'(g_-^*) + \cdots$$

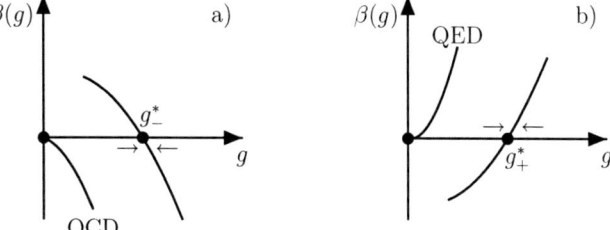

Fig. 2.6. RG fixed points are zeros of the β-function: a) UV fixed points, b) IR fixed points

and provided $\beta'(g_-^*) \neq 0$ we have

$$a(g,\kappa) = \exp \int_g^{g(\kappa)} \frac{\gamma(g')}{\beta(g')} \, \mathrm{d}g' = \exp \int_g^{g(\kappa)} \frac{\gamma(g_-^*)}{\beta(g')} \, \mathrm{d}g' \cdot r(g,\kappa)$$

$$= \kappa^{\gamma^*} \, r(g,\kappa)$$

where

$$r(g,\kappa) = \exp \int_g^{g(\kappa)} \frac{(\gamma(g') - \gamma^*)}{\beta(g')} \, \mathrm{d}g'$$

in the limit of large κ yields a finite scale independent wave function renormalization

$$\lim_{\kappa \to \infty} r(g,\kappa) = r(g,\infty) \, .$$

We thus find the asymptotic from

$$\Gamma(\{\kappa p\}; g, m, \mu) \overset{\to}{\sim} \kappa^d \left(\kappa^{d_A} \, r_A(g,\infty)\right)^{-n_A} \left(\kappa^{d_\psi} \, r_\psi(g,\infty)\right)^{-2n_\psi} \Gamma(\{p\}; g_-^*, 0, \mu) \tag{2.224}$$

which exhibits asymptotic *scaling*. In the first place it is given by the vertex functions of a *massless theory*. As expected, at high energies masses may be neglected, however on the expense that another mass scale remains in the game, the scale parameter μ. The first factor κ^d is trivial and is due to the d–momentum conservation which was factored out. Then each field exhibits a homogeneous (power–like) behavior in the dilatation factor κ, the exponent of which exhibits an *anomalous dimension* as a consequence of the dynamics of the theory:

$$d_A = \frac{d-2}{2} + \gamma_A^* \, , \quad d_\psi = \frac{d-1}{2} + \gamma_\psi^* \, . \tag{2.225}$$

The first term is the naive or engineers dimension the second part is the anomalous part which is a quantum effect, a relict of the breaking of scale invariance, when $g \neq g^*$. While naively we would expect that in $d = 4$ dimensions the massless theory has scaling: for example a scalar two–point function, the only dimensionfull physical quantity being the momentum, one would expect $G(p;g) \sim 1/p^2$ as G has dimension 2. However, if there would be a non–trivial UV fixed point one would have $G(p,g,\mu) \sim (\mu^2)^{\gamma^*}/(p^2)^{1+\gamma^*}$ ($\gamma^* > 0$) which shows the role and unavoidability of the scale parameter μ, which has to eat up the extra dimension γ^* induced by the dynamics of the theory. Otherwise only truly free theories could have scaling, called *canonical scaling* in this case. The discovery of asymptotic freedom of QCD [30] is the prime example of a dynamical theory, notabene of the theory of strong interactions, exhibiting asymptotic canonical scaling (Bjorken scaling) of liberated quarks (quark parton model) [51]. The latter was discovered before in the pioneering investigations concerning *Deep Inelastic Scattering* (DIS) [52] of electrons on protons and bound neutrons by Friedman, Kendall and Taylor (Nobel prize 1990). These experiments have been of essential importance for the development of the quark model and to the discovery of QCD as the theory of the strong interactions.

2) IR behavior

The infrared behavior corresponds to the long distance properties of a system. Here the regime of interest is $\kappa |p| \ll m, \mu$ and the discussion proceeds essentially as before: now as $\kappa \to 0$ the effective $g(\kappa) \to g_+^*$ where g_+^* is a zero of the β-function with positive slope, see Fig. 2.6b, $\beta(g_+^*) = 0$ and $\beta'(g_+^*) > 0$. This is the typical situation in the construction of low energy effective theories, particularly in the discussion of critical phenomena of statistical systems (keywords: critical behavior, critical exponents, scaling laws, universality). If $g_+^* = 0$ the effective theory is infrared free (the opposite of *asymptotic freedom*), also called Gaussian (Gaussian fixed point). Here the well known examples are QED

$$\beta_{\text{QED}}(e) = \frac{e^3}{12\pi^2} \sum_f N_{cf} Q_f^2 + \cdots \tag{2.226}$$

or the self–interacting scalar field ϕ^4–theory

$$\beta(\lambda) = -\varepsilon\lambda + \frac{3\lambda^2}{16\pi^2} + \cdots$$

in $d = 4$ dimensions. For QED the running coupling to leading order thus follows from

$$\ln \kappa = \int_e^{e(\kappa)} \frac{1}{\beta(e')} \, de' = \frac{12\pi^2}{\sum_f N_{cf} Q_f^2} \int_e^{e(\kappa)} \frac{1}{(e')^3} \, de' = \frac{24\pi^2}{\sum_f N_{cf} Q_f^2} \left(\frac{1}{e^2} - \frac{1}{e(\kappa)^2} \right)$$

where the sum extends over all light flavors $f : m_f < \mu$[36]. The running fine structure constant thus at leading order is given by

$$\alpha(\mu) = \frac{\alpha}{1 - \frac{2\alpha}{3\pi} \sum_f N_{cf} Q_f^2 \ln \mu/\mu_0} \tag{2.227}$$

---

[36]This latter restriction takes into account the decoupling of heavy flavors, valid in QED and QCD. Since in the $\overline{\text{MS}}$ scheme, i.e. renormalization by the substitution Reg $\to \ln \mu^2$, which we are considering here, decoupling is not automatic, one has to impose it by hand. At a given scale one is thus considering an effective theory, which includes only those particles with masses below the scale μ.

where μ_0 is the scale where the lightest particle starts to contribute, which is the electron $\mu_0 = m_e$. We then may identify $\alpha(\mu_0) = \alpha$ the classical low energy value of the fine structure constant, with the proviso that only logarithmic accuracy is taken into account (see below). The running α is equivalent to the Dyson summation of the transversal part of the photon self–energy to the extent that only the logs are kept. The RG running takes into account the leading radiative corrections in case the logs are dominating over constant terms, i.e. provided large scale changes are involved.

In the calculation of the contributions from electron loops in photon propagators to the muon anomaly a_μ, such large scale changes from m_e to m_μ are involved and indeed one may calculate such two–loop contributions starting from the lowest order result

$$a_\mu^{(2)} = \frac{\alpha}{2\pi} \quad \text{via the substitution} \quad \alpha \to \alpha(m_\mu) \tag{2.228}$$

where

$$\alpha(m_\mu) = \frac{\alpha}{1 - \frac{2}{3}\frac{\alpha}{\pi} \ln \frac{m_\mu}{m_e}} = \alpha \left(1 + \frac{2}{3}\frac{\alpha}{\pi} \ln \frac{m_\mu}{m_e} + \cdots \right) \tag{2.229}$$

such that we find

$$a_\mu^{(4)\,\mathrm{LL}}(\mathrm{vap}, e) = \frac{1}{3} \ln \frac{m_\mu}{m_e} \left(\frac{\alpha}{\pi}\right)^2$$

which indeed agrees with the leading log result obtained in [53] long time ago by a direct calculation. The method has been further developed and refined by Lautrup and de Rafael [54]. In the calculation of a_μ only the electron VP insertions are governed by the RG and the corresponding one–flavor QED β–function has been calculated to three loops

$$\beta(\alpha) = \frac{2}{3}\left(\frac{\alpha}{\pi}\right) + \frac{1}{2}\left(\frac{\alpha}{\pi}\right)^2 - \frac{121}{144}\left(\frac{\alpha}{\pi}\right)^3 + \cdots \tag{2.230}$$

by [55], which thus allows to calculate leading $\alpha^n (\ln m_\mu/m_e)^n$, next-to-leading $\alpha^n (\ln m_\mu/m_e)^{n-1}$ and next-to-next-to-leading $\alpha^n (\ln m_\mu/m_e)^{n-2}$ log corrections.

As $\alpha(\mu)$ is increasing with μ in the resummed perturbation theory (2.227) exhibits a pole, the so called *Landau pole* at which the coupling becomes infinite: $\lim_{\mu \lesssim \mu_L} \alpha(\mu) = \infty$ The "fixed point" very likely is an artefact of perturbation theory, which of course cease to be valid when the one–loop correction approaches 1. What this tells us is that we actually do not know what the high energy asymptotic behavior of QED is.

α in the on–shell versus α in the $\overline{\mathrm{MS}}$ scheme

In our discussion of renormalizing QED we were considering originally the *on–shell renormalization scheme*, while the RG provides α in the $\overline{\mathrm{MS}}$

2.6 One–Loop Renormalization

scheme. Here we briefly discuss the relationship between the OS and the $\overline{\text{MS}}$ fine structure constants $\alpha_{\text{OS}} = \alpha$ and $\alpha_{\overline{\text{MS}}}$, respectively. Since the bare fine structure constant

$$\alpha_0 = \alpha_{\overline{\text{MS}}} \left(1 + \left.\frac{\delta\alpha}{\alpha}\right|_{\overline{\text{MS}}}\right) = \alpha_{\text{OS}} \left(1 + \left.\frac{\delta\alpha}{\alpha}\right|_{\text{OS}}\right) \tag{2.231}$$

is independent of the renormalization scheme. The one–loop calculation in the SM yields (including the charged W contribution for completeness)

$$\left.\frac{\delta\alpha}{\alpha}\right|_{\overline{\text{MS}}} = \frac{\alpha}{3\pi} \sum Q_f^2 \ln \frac{\mu^2}{m_f^2} - \frac{\alpha}{3\pi}\frac{21}{4} \ln \frac{\mu^2}{M_W^2}$$

$$\left.\frac{\delta\alpha}{\alpha}\right|_{\text{OS}} = \Pi'_\gamma(0) + \frac{\alpha}{\pi} \ln \frac{M_W^2}{\mu^2}$$

$$= \left.\frac{\delta\alpha}{\alpha}\right|_{\overline{\text{MS}}} - \frac{\alpha}{6\pi}$$

and thus

$$\alpha_{\overline{\text{MS}}}^{-1}(0) = \alpha^{-1} + \frac{1}{6\pi} \tag{2.232}$$

as a low energy matching condition. The α–shift in the $\overline{\text{MS}}$ scheme is very simple, just the UV logs,

$$\Delta\alpha_{\overline{\text{MS}}}(\mu) = \frac{\alpha}{3\pi} \sum Q_f^2 N_{cf} \ln \frac{\mu^2}{m_f^2} + \frac{\alpha}{3\pi}\frac{21}{4} \ln \frac{\mu^2}{M_W^2} \tag{2.233}$$

such that

$$\Delta\alpha_{\overline{\text{MS}}}(\mu) = \Delta\alpha_{\text{OS}}(\mu) + \frac{\alpha}{\pi}\frac{5}{3} \sum Q_f^2 N_{cf} \tag{2.234}$$

where the sum goes over all fermions f with $N_{cf} = 1$ for leptons and $N_{cf} = 3$ for quarks.

In perturbation theory, the leading light fermion ($m_f \ll M_W, \sqrt{s}$) contribution in the OS scheme is given by

$$\Delta\alpha(s) = \frac{\alpha}{3\pi} \sum_f Q_f^2 N_{cf} (\ln \frac{s}{m_f^2} - \frac{5}{3}) \ . \tag{2.235}$$

We distinguish the contributions from the leptons, for which the perturbative expression is appropriate, the five light quarks (u, d, s, c, b) and the top

$$\Delta\alpha = \Delta\alpha_{\text{lep}} + \Delta\alpha_{\text{had}} + \Delta\alpha_{\text{top}} \ . \tag{2.236}$$

Since the top quark is heavy we cannot use the light fermion approximation for it. A very heavy top in fact decouples like

$$\Delta\alpha_{top} \simeq -\frac{\alpha}{3\pi}\frac{4}{15}\frac{s}{m_t^2} \to 0$$

when $m_t \gg s$. Since pQCD does not apply at low energies, $\Delta\alpha_{\text{had}}$ has to be evaluated via dispersion relations from e^+e^-–annihilation data.

Note that in $d = 4$ dimensions both for QCD and QED very likely there is no RG fixed point at finite value of g except $g = 0$, which always is a fixed point, either an UV one (QCD) or and IR one (QED). In QCD this could mean that $\alpha_s(\mu) \to \infty$ for $\mu \to 0$ (infrared slavery, confinement). In perturbation theory a Landau pole shows up at finite scale Λ_{QCD} when coming from higher energy scales, where $\alpha_s \to \infty$ for $\mu \overset{>}{\to} \Lambda_{\text{QCD}}$. In QED likely $\alpha(\mu) \to \infty$ for $\mu \to \infty$.

It is important to emphasize that the RG only accounts for the UV logarithms, which in DR are related to the UV poles in $d = 4-\varepsilon$ dimensions. Large logs may also be due to IR singular behavior, like the terms proportional to $\ln m_\gamma$ which we have regulated with an infinitesimally small photon mass in the on–shell lepton wave function renormalization factor $Z_\psi = Z_f$ (2.192). In spite of the fact that this term appears in the UV renormalization counter term, it has nothing to do with an UV singularity and does not contribute in the RG coefficients. In DR also IR singularities may be regularized by analytic continuation in d, however, by dimensional continuation to $d = 4 + \varepsilon_{\text{IR}}$, and corresponding IR poles at negative ε_{UV}. Also the terms proportional to $\ln \frac{-q^2}{m_\gamma^2}$ showing up in the electric form factor (2.208) is not covered by the RG analysis. As will be explained in the next section, the IR singularities have their origin in the attempt to define free charged particle states as simple isolated poles in the spectrum (by trying to impose an on–shell condition). In reality, the Coulomb potential mediated by the massless photon has infinite range and the charged states feel the interaction whatever the spatial separation in corresponding scattering states is.

2.6.6 Bremsstrahlung and the Bloch-Nordsieck Prescription

As we have seen the on–shell form factor A_1 is IR singular in the limit of physical zero mass photons at the one–loop level and beyond. As already mentioned, the problem is that we try to work with scattering states with a fixed number of free particles, while in QED due to the masslessness of the photon and the related infinite interaction range of the electromagnetic forces soft photons are emitted and eventually reabsorbed at any distance from the "interaction region", i.e. the latter extends to ∞. The basic problem in this case is the proper definition of a charged particle state as obviously the order by order treatment of a given scattering amplitude breaks down. Fortunately, as Bloch and Nordsieck [56] have observed, a simple prescription bring us back to a quasi perturbative treatment. The basic observation was that virtual and soft real photons are not distinguishable beyond the resolution of the measuring apparatus. Thus besides the virtual photons we have to include the soft real photons of energies below the resolution threshold. For a given tree level process, the Bloch-Nordsieck prescription requires to include photonic corrections at a given order $O(e^n)$ irrespective of whether the photons are virtual or real (soft). We thus are led back to a perturbative order by order scheme, on the expense that all, at the given order, possible final states which only differ by (soft) photons have to be summed over.

Thus in order to obtain a physics–wise meaningful observable quantity, in case of the electromagnetic form factor

$$e^-(p_1) + \gamma(q) \to e'^-(p_2) ,$$

at one–loop order $O(e^2)$, we have to include the corresponding process

$$e^-(p_1) + \gamma(q) \to e'^-(p_2) + \gamma'(k) ,$$

with one additional real (soft) photon attached in all possible ways to the tree diagram as shown in Fig. 2.7. The second photon is assumed to be soft, i.e. having energy $E_\gamma = |\mathbf{k}| < \omega$, where ω is the threshold of detectability of the real photon. Since the photon cannot be seen, the event looks like an "elastic" event, i.e. like one of the same final state as the tree level process. The soft photons thus factorize into the Born term of the original process times a soft photon correction, with the soft photons integrated out up to energy ω. The correction given by the bremsstrahlung cross–section is proportional to the square $|T_{\text{bre}}|^2$ of the sum of the matrix elements of the two diagrams which reads

$$T_{\text{bre}} = \mathrm{i}^3 e^2 \bar{u}(p_2) \left\{ \gamma^\rho \frac{\slashed{p}_2 + \slashed{k} + m}{(p_2+k)^2 - m^2} \gamma^\mu + \gamma^\mu \frac{\slashed{p}_1 - \slashed{k} + m}{(p_1-k)^2 - m^2} \gamma^\rho \right\} u(p_1)\, \varepsilon_\rho^*(k,\lambda) . \tag{2.237}$$

In the soft photon approximation $k \sim 0$ and hence $p_1 + q = p_2 + k \simeq p_2$ we may neglect the $\slashed{k}$ terms in the numerator. Using the Dirac–algebra and the Dirac equation we may write, in the first term, $\bar{u}(p_2)\slashed{\varepsilon}^*(\slashed{p}_2 + m) = \bar{u}(p_2)\,[2\varepsilon^* p_2 + (-\slashed{p}_2 + m)\,\slashed{\varepsilon}^*] = \bar{u}(p_2) 2\varepsilon^* p_2$, in the second term, $(\slashed{p}_1 + m)\slashed{\varepsilon}^* u(p_1) = [2\varepsilon^* p_1 + \slashed{\varepsilon}^*(-\slashed{p}_1 + m)]u(p_1) = 2\varepsilon^* p_1 u(p_1)$. Furthermore, in the bremsstrahlung integral the scalar propagators take a very special form, which comes about due to the on–shellness of the electrons and of the bremsstrahlung photon: $(p_2+k)^2 - m^2 = p_2^2 + 2(kp_2) + k^2 - m^2 = 2(kp_2)$ and $(p_1 - k)^2 - m^2 = p_1^2 - 2(kp_1) + k^2 - m^2 = -2(kp_1)$ as $p_1^2 = p_2^2 = m^2$ and $k^2 = 0$. Therefore, the soft bremsstrahlung matrix element factorizes into the Born term times a radiation factor

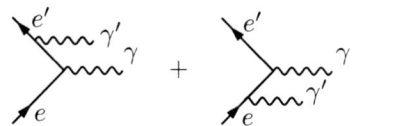

Fig. 2.7. Bremsstrahlung in $e(p_1) + \gamma(q) \to e'(p_2)$

$$T_{\text{bre}}^{\text{soft}} \simeq -ie\bar{u}(p_2)\,\gamma^\mu\,u(p_1) \left\{ -2e \left(\frac{\varepsilon^* p_1}{kp_1} - \frac{\varepsilon^* p_2}{kp_2} \right) \right\}$$

and one obtains

$$d\sigma = d\sigma_0 \frac{4e^2}{(2\pi)^3} \left| \frac{\varepsilon p_1}{kp_1} - \frac{\varepsilon p_2}{kp_2} \right|^2 \frac{d^3k}{2\omega_k}$$

where $d\sigma_0$ denotes the lowest order "cross–section" for the absorption of a virtual photon by an electron. If we sum over the two photon polarizations λ indexing the polarization vector and use the completeness relation (2.25) we find

$$d\sigma = -d\sigma_0 \frac{4e^2}{(2\pi)^3} \left(\frac{p_1}{kp_1} - \frac{p_2}{kp_2} \right)^2 \frac{d^3k}{2\omega_k}\,. \tag{2.238}$$

Actually, the integral for massless photons does not exist as it is logarithmically IR singular

$$\int_{|\mathbf{k}|<\omega} \frac{d^3k}{|\mathbf{k}|^3} \cdots = \infty\,.$$

Again an IR regularization is required and we introduce a tiny photon mass such that $\omega_k = \sqrt{k^2 + m_\gamma^2}$. As a correction to the cross–section, we may write the *inclusive cross section* for

$$e^-(p_1) + \gamma(q) \to e'^-(p_2)\,,\ \ e'^-(p_2) + \gamma'(k,\,\text{soft})$$

as

$$d\sigma_{\text{inc}} = d\sigma_0\,(1 + C_{\text{bre}})$$

which, for the vertex on the amplitude level reads

$$i\Gamma^\mu_{\text{inc}} = -ie\gamma^\mu \left(1 + \frac{1}{2} C_{\text{bre}} + \cdots \right) \simeq -ie\gamma^\mu + i\delta\Gamma^\mu_{\text{bre}}$$

where

$$i\delta\Gamma^\mu_{\text{bre}} = -ie\gamma^\mu \frac{1}{2} C_{\text{bre}} \tag{2.239}$$

with

$$C_{\text{bre}} = \frac{e^2}{2\pi^3} \int_{|\mathbf{k}|<\omega} \frac{d^3k}{2\omega_k} \left\{ \frac{2(p_1 p_2)}{(kp_1)(kp_2)} - \frac{m^2}{(kp_1)^2} - \frac{m^2}{(kp_2)^2} \right\} \tag{2.240}$$

is the $O(\alpha)$ contribution to the Dirac form factor due to bremsstrahlung. The first term is the interference from the two diagrams, the second and third correspond to the squares of the first and the second diagram, respectively. For a finite photon mass the integral is finite and may be worked out (see e.g. [40] Sect. 7.). The result may be written in the form

$$C_{\text{bre}} = \frac{\alpha}{\pi}\left\{\left(1-\frac{y}{2}\right)\left(4G'(y)\ln\frac{2\omega}{m_\gamma} - F'(y)\right) - 2\ln\frac{2\omega}{m_\gamma} + 2G'(y)\right\}$$

with

$$G'(y) = -\frac{1}{4\sqrt{1-y}}\ln(\xi^2)$$

$$F'(y) = \frac{1}{2\sqrt{1-y}}\left\{\frac{2\pi^2}{3} - 4\text{Sp}(-\xi) + \ln^2(-\xi) - 4\ln(-\xi)\ln(1+\xi)\right\}$$

where, for simplicity, F' is given for the production channel

$$\gamma(q) \to e^-(-p_1) + e'^-(p_2)\,,\ \ e^-(-p_1) + e'^-(p_2) + \gamma'(k, \text{soft})$$

where $0 < y < 1$ ($-1 < \xi < 0$). In spite of the fact that the soft bremsstrahlung factor (2.240) looks universal, the result of the evaluation of the integrals is *process dependent*: apart from the universal terms, which in particular include the IR singular ones, the function $F'(y)$ depends on the channel considered. Note that, in contrast to the form factors, like $A_{E\,\text{ren}}$, which are analytic in q^2, C_{bre} is not analytic in the same variable, because it is the integral over the absolute square $|T|^2$ of a transition matrix element. It must be real and positive. Above, we have chosen to present $F'(y)$ for the production channel as it allows us to discuss the main points of the Bloch-Nordsieck prescription, keeping the notation substantially simpler[37]. The leading behavior in this case

---

[37] In the scattering region the result is more complicated, because, there is one more kinematic variable, the scattering angle Θ, or equivalently, the electron velocity β_e. Considering, elastic scattering $|\mathbf{p}_1| = |\mathbf{p}_2|$, $E_1 = E_2$ the finite function $F'(y)$, now for $y < 0$ ($0 < \xi < 1$), reads

$$F'(y) = \frac{1}{\sqrt{1-y}}\left\{-\text{Sp}(1+\frac{2}{1+\xi}\frac{1}{1-\beta_e}) - \text{Sp}(1+\frac{2}{1+\xi}\frac{1}{1+\beta_e})\right.$$
$$\left. + \text{Sp}(1+\frac{2\xi}{1+\xi}\frac{1}{1-\beta_e}) + \text{Sp}(1+\frac{2\xi}{1+\xi}\frac{1}{1+\beta_e})\right\}$$

where $\beta_e = \sqrt{1-4m^2/s}$ is the velocity of the electron. s and $Q^2 = -q^2 > 0$ are related by $Q^2 = s\frac{1-\cos\Theta}{2}$. The asymptotic behavior $Q^2 \gg m^2$ at fixed angle requires $s \gg m^2$ with $r \equiv Q^2/s = (1-\cos\Theta)/2$ fixed. The arguments of the Spence functions behave like $1 + \frac{2}{1+\xi}\frac{1}{1-\beta_e} \simeq \frac{s}{m^2} - r^{-1} + \cdots$, $1 + \frac{2}{1+\xi}\frac{1}{1+\beta_e} \simeq 2 - \frac{m^2}{Q^2} + \frac{m^2}{s} + \cdots$, $1 + \frac{2\xi}{1+\xi}\frac{1}{1-\beta_e} \simeq 1 + r^{-1} - (1+3r^{-1})\frac{m^2}{Q^2} + \cdots$, and $1 + \frac{2\xi}{1+\xi}\frac{1}{1+\beta_e} \simeq 1 + \frac{m^2}{Q^2} + \cdots$. Utilizing the relations (2.202), one may work out the leading behavior

$$C_{\text{bre}}^{\text{scattering}} = \frac{\alpha}{\pi}\left\{2\ln\frac{Q^2}{m^2}\ln\frac{2\omega}{m_\gamma} - \frac{1}{2}\ln^2\frac{s}{m^2} - 2\ln\frac{2\omega}{m_\gamma} + \ln\frac{Q^2}{m^2} + \cdots\right\}$$

which, with $\ln^2 s/m^2 = -\ln^2 Q^2/m^2 + 2\ln Q^2/m^2 \ln s/m^2 + \ln^2 s/Q^2$ and after neglecting the last (sub leading) term, is in agreement with [2]. In the production channel with $q^2 = -Q^2 > 0$, in the center of mass frame of the produced lepton pair, the leptons are back-to-back and hence $\Theta = \pi$, or $\cos\Theta = -1$, such that s may be identified as $s = q^2$.

reads
$$C_{\text{bre}} = \frac{\alpha}{\pi}\left\{2\ln\frac{q^2}{m^2}\ln\frac{2\omega}{m_\gamma} - \frac{1}{2}\ln^2\frac{q^2}{m^2} - 2\ln\frac{2\omega}{m_\gamma} + \ln\frac{q^2}{m^2} + \cdots\right\}.$$

Now, we are able to calculate the form factor for soft photon dressed electrons. The real part of the Dirac form factor gets modified to

$$\operatorname{Re} A_{E\,\text{ren}} + \frac{1}{2}C_{\text{bre}} = \frac{\alpha}{2\pi}\left\{-2\ln\frac{2\omega}{m} + 4\left(1 - \frac{y}{2}\right)G'(y)\ln\frac{2\omega}{\sqrt{q^2}}\right.$$
$$\left. + 2\left(1 - \frac{y}{2}\right)\frac{\pi^2}{2\sqrt{1-y}} - 2 + (5 - 2y)G'(y) + \left(1 - \frac{y}{2}\right)(\operatorname{Re} F - F')(y)\right\}$$

(2.241)

where

$$(\operatorname{Re} F - F')(y) = \frac{1}{2\sqrt{1-y}}\left\{-\frac{4\pi^2}{3} + 8\operatorname{Sp}(-\xi)\right.$$
$$\left. + 4\ln(-\xi)\,(2\ln(1+\xi) - \ln(1-\xi))\right\}.$$

This is the result for the time–like region (production or annihilation) where $-1 \leq \xi \leq 0$. Here the photon mass has dropped out and we have an IR finite result, at the expense that the form factor is dependent on the experimental resolution ω, the threshold detection energy for soft photons. This is the Bloch-Nordsieck [56] solution of the IR problem. The Pauli form factor is not affected by real photon radiation. In general, as a rule, soft and collinear real photon radiation is always integral part of the radiative corrections.

When combining virtual and soft photon effects one typically observes the cancellations of large or potentially large radiative correction and the range of validity of the perturbative results must be addressed. To be more specific, the calculation has revealed terms of different type and size: typically IR sensitive soft photons logarithms of the type $\ln(m/2\omega)$, or collinear logarithms $\ln(q^2/m^2)$ show up. The latter come from photons traveling in the direction of a lepton, which again cannot be resolved in an experiment with arbitrary precision. This is the reason why the limit $m \to 0$, in which photon and lepton would travel in the same direction at the same speed (the speed of light) is singular. These logarithms can be very large (high resolution, high energy) and if the corrections $\frac{\alpha}{\pi}\ln(q^2/m^2)$ tend to be of $O(1)$ one cannot trust the perturbative expansion any longer. Even more dangerous are the double logarithmic corrections like the so called *Sudakov logarithms* $\frac{\alpha}{\pi}\ln^2(q^2/m^2)$ or the mixed IR sensitive times collinear terms $\frac{\alpha}{\pi}\ln(m/2\omega)\ln(q^2/m^2)$. There are several possibilities to deal with the large logs:

a) the leading large terms are known also in higher orders and may thus be resummed. The resummation leads to more reliable results. A typical example here is the soft photon exponentiation according to Yennie-Frautschi-Suura [57].

b) UV sensitive large logs may by resummed by the renormalization group, as discussed above.

c) Some observable quantities may have much better convergence properties in a perturbative approach than others. A typical example is the attempt of an *exclusive* measurement of a lepton, which because of the soft photon problematic per se is not a good object to look for. In fact, increasing the exclusivity by choosing the IR cut–off ω smaller and smaller, the correction becomes arbitrary large and the perturbative result becomes meaningless. Somehow the experimental question in such a situation is not well posed. In contrast, by choosing ω larger the correction gets smaller. The possibility to increase ω in the formula given above is kinematically constraint by the requirement of soft radiation *factorization*. Of course photons may be included beyond that approximation. Indeed, there is a famous theorem, the *Kinoshita-Lee-Nauenberg* (KLN) *theorem* [58] which infers the cancellations of mass singularities and infrared divergences for observables which are defined to include summation over all degenerate or quasi degenerate states. It is important that a summation over degenerate states is performed for the initial (i) and the final (f) states. Then

$$\sum_{i,f} |T_{fi}|^2 \qquad (2.242)$$

and the corresponding cross–section is free of all infrared singularities in the limit of all masses vanishing. Such observables typically are "all inclusive" cross-sections averaged over the initial spin.

In our example, the inclusive cross section is obtained by adding the hard photons of energy $E_\gamma > \omega$ up to the kinematic limit $E_{\gamma\,\mathrm{max}} = \sqrt{q^2 - 4m^2}/2$. To illustrate the point, let us consider the lepton pair creation channel $\gamma^*(q) \to \ell^-(p_-) + \ell^+(p_+) + \gamma(k)$, where the $*$ denote that the corresponding state is virtual, i.e. off–shell, with an additional real bremsstrahlung photon $\gamma(k)$ emitted from one of the final state leptons. We thus include the so called *final state radiation* (FSR). The "heavy" virtual photon γ^* of momentum $q = p_- + p_+ + k$, we may think to have been created previously in e^+e^-–annihilation, for example[38]. The center of mass energy is $E_{\mathrm{cm}} = E_- + E_+ + E_\gamma = \sqrt{q^2}$. Let $\lambda = 2\omega/E_{\mathrm{cm}}$ and $1-\lambda \gg y$ such that we may work in the approximation up to terms of order $O(\alpha\,\frac{m^2}{q^2})$, i.e. neglecting power corrections in m^2/q^2. Relaxing from the soft photon approximation which defined C_{bre} in (2.240), the hard

---

[38]The factorization into $e^+e^- \to \gamma^*$ production and subsequent decay $\gamma^* \to \ell^+\ell^-$ only makes sense at relatively low q^2, when the one–photon exchange approximation can be used. In the SM the γ^* may also be a "heavy light" particle Z of mass about $M_Z \simeq 91\,\mathrm{GeV}$ which is unstable and thus is described well by a Breit-Wigner resonance. Near the resonance energy again factorization is an excellent approximation and the following discussion applies. In e^+e^-–annihilation, the radiation of additional photons from the initial state electron or positron (Fig. 2.7 with e' an incoming e^+) is called *initial state radiation* (ISR). In the soft approximation (2.238) still holds. For details see (5.8) in Sect. 5.1.2.

bremsstrahlung integral of interest is

$$\int_\omega^{E_{cm}/2} dE_\gamma \cdots$$

with the spectral density (integrand)

$$\frac{1}{\Gamma_0(\gamma^*\to\ell\ell)} \frac{d^2\Gamma(\gamma^*\to\ell\ell\gamma)}{dudv} = P(u,v) = \frac{\alpha}{2\pi}\left\{\left(2\frac{u}{1-u}+1-u\right)\left(\frac{1}{v}+\frac{1}{1-u-v}\right)\right. \quad (2.243)$$
$$\left. -\frac{a}{2}\left(\frac{1}{v^2}+\frac{1}{(1-u-v)^2}\right)-2\right\}.$$

where $a = 4m^2/q^2$, $u = (p_-+p_+)^2/q^2$ and $v = (q-p_-)^2/q^2$. In the rest frame of the heavy photon we have $u = 1 - 2E_\gamma/M_\gamma$, $v = 1 - 2E_-/M_\gamma$ and $1-u-v = 1 - 2E_+/M_\gamma$. In the center of mass frame of the lepton pair

$$v = \frac{1}{2}(1-u)(1-\sqrt{1-y}\cos\Theta_+)$$
$$1-u-v = \frac{1}{2}(1-u)(1-\sqrt{1-y}\cos\Theta_-)$$

with $y = a/u$ and $\Theta_\pm$ the angle between the final state photon and the lepton with momentum $p_\pm$ ($\Theta_- = \pi - \Theta_+$). We have to integrate the distribution over the angles $0 \leq \Theta_\pm \leq \pi/2$ and over the hard photon $E_\gamma \geq \omega = \lambda(M_\gamma/2)$ with $1-a > \lambda > 0$ yields [59] up to $O(\alpha y)$ precision

$$\Delta C_{>\omega} = \frac{\alpha}{2\pi}\left\{\left(4\ln\frac{1}{\lambda}-(1-\lambda)(3-\lambda)\right)\ln\frac{q^2}{m^2}-4\ln\frac{1}{\lambda}\right.$$
$$\left.+4\mathrm{Sp}(\lambda)-\frac{2}{3}\pi^2-(1-\lambda)(3-\lambda)\ln(1-\lambda)+\frac{1}{2}(1-\lambda)(11-3\lambda)\right\}$$

or for $\omega \ll E_{cm}/2$

$$\Delta C_{>\omega} = \frac{\alpha}{2\pi}\left\{\left(4\ln\frac{\sqrt{q^2}}{2\omega}-3\right)\ln\frac{q^2}{m^2}-4\ln\frac{\sqrt{q^2}}{2\omega}-\frac{2}{3}\pi^2+\frac{11}{2}\right\}. \quad (2.244)$$

In this approximation the complementary soft plus virtual part (see (2.241))

$$C_{<\omega} = C_{QED}^{virtual} + C_\omega^{soft}$$
$$= \frac{\alpha}{2\pi}\left\{-\left(4\ln\frac{\sqrt{q^2}}{2\omega}-3\right)\ln\frac{q^2}{m^2}+4\ln\frac{\sqrt{q^2}}{2\omega}+\frac{2}{3}\pi^2-4\right\} \quad (2.245)$$

The total inclusive sum is

$$C^{total} = C_{<\omega} + \Delta C_{>\omega} = \frac{\alpha}{2\pi}\frac{3}{2} \simeq 1.74\times 10^{-3} \quad (2.246)$$

a truly small perturbative correction. No scale and no log involved, just a pure number. This is the KLN theorem at work. It will play a crucial role later on in this book.

The two separate contributions become large when the cut energy ω is chosen very small and in fact we get a negative cross–section, which physics wise makes no sense. The reason is that the correction gets large and one has to include other relevant higher order terms. Fortunately, the multi soft γ emission can be calculated to all orders. One can prove that the IR sensitive soft photon exponentiates: Thus,

$$1 + C_{\text{IR}} + \frac{1}{2!} C_{\text{IR}}^2 + \cdots = e^{C_{\text{IR}}}$$

$$= \exp \frac{\alpha}{2\pi} \left\{ -4 \ln \frac{\sqrt{q^2}}{2\omega} \ln \frac{q^2}{m^2} + 4 \ln \frac{\sqrt{q^2}}{2\omega} + \cdots \right\} = \left(\frac{2\omega}{\sqrt{q^2}} \right)^{\frac{2\alpha}{\pi} \left(\ln \frac{q^2}{m^2} - 1 \right)}$$

and the result is

$$1 + C_{<\omega} + \cdots = e^{C_{\text{IR}}} + \Delta C^{v+s} + \cdots \tag{2.247}$$

with

$$\Delta C^{v+s} = C_{<\omega} - C_{\text{IR}} = \frac{\alpha}{2\pi} \left\{ 3 \ln \frac{q^2}{m^2} + \frac{2}{3}\pi^2 - 4 \right\}$$

a correction which is small if q^2/m^2 is not too large. Otherwise higher order collinear logs have to be considered as well. They do not simply exponentiate. By the resummation of the leading IR sensitive terms we have obtained a result which is valid much beyond the order by order perturbative result. Even the limit $\omega \to 0$ may be taken now, with the correct result that the probability of finding a naked lepton of mass m tends to zero. In contrast $1 + C_{<\omega} \to -\infty$ as $\omega \to 0$, a nonsensical result.

For our consideration of soft photon dressed states the inspection of the complementary hard photon part is important as far as the expression (2.244) tells us which are the logs which have to be canceled for getting the log free inclusive result. Namely, the IR sensitive log terms appear with the center of mass energy scale $\sqrt{q^2}$ not with the lepton mass m. This observation allows us to write the virtual plus soft result in a slightly different form than just adding up the results.

Another consideration may be instructive about the collinear mass singularities (terms $\propto \ln(q^2/m^2)$), which are a result of integrating the propagators $2|\boldsymbol{k}|(E_i - |\boldsymbol{p}_i| \cos \Theta_i))^{-1}$ in the distribution (2.238) or (2.243). If we integrate the angular distribution over a cone $\Theta_1, \Theta_2 \leq \delta$ only, instead of over the full angular range and add up the contributions

$$C_{<\omega,\,<\delta} = C_{\text{QED}}^{\text{virtual}} + C_\omega^{\text{soft}} + \Delta C_{>\omega,\,<\delta}^{\text{hard,collinear}} \tag{2.248}$$

the collinear singularities exactly cancel in the limit $m \to 0$, provided $\delta > 0$. The result reads

$$C^{m=0}_{<\omega,<\delta} = \frac{\alpha}{2\pi}\left\{\left(4\ln\frac{1}{\lambda} - (1-\lambda)(3-\lambda)\right)\ln\frac{1-\rho}{1+\rho}\frac{3}{2} + \rho(1-\lambda^2)\right\}$$

with $\rho = \cos\delta$, $\lambda = \frac{2\omega}{M_\gamma}$ and we have assumed $\frac{1-\rho}{2} \gg \frac{m^2}{M_\gamma^2}$. Thus, in addition to the virtual plus soft photons we have included now the hard collinear photons traveling with the leptons within a cone of opening angel δ. Here the collinear cone has been defined in the c.m. frame of the lepton pair, where the two cones are directed back to back and non overlapping for arbitrary cuts $\delta \leq \pi/2$. In an experiment one would rather define the collinear cones in the c.m. frame of the incoming virtual photon. In this case a slightly more complicated formula (14) of [59] is valid, which simplifies for small angles δ_0 and $\lambda = 2\epsilon = 2\omega/M_\gamma \ll 1$ to

$$C^{m=0}_{<\omega,<\delta_0} = -\frac{\alpha}{\pi}\left\{(4\ln 2\epsilon + 3)\ln\frac{\delta_0}{2} + \frac{\pi^2}{3} - \frac{5}{2}\right\} \tag{2.249}$$

which is the QED analog of the famous Sterman-Weinberg (SW) formula [60]

$$C_{\text{SW}} = -\frac{4}{3}\frac{\alpha_s}{\pi}\left\{(4\ln 2\epsilon + 3)\ln\frac{\delta_0}{2} + \frac{\pi^2}{3} - \frac{5}{2}\right\} \tag{2.250}$$

for the two–jet event rate in QCD. The extra factor $\frac{4}{3}$ is an $SU(3)$ Casimir coefficient and α_s is the $SU(3)$ strong interaction coupling constant. The physical interpretation of this formula will be considered in Sect. 5.1.3.

Some final remarks are in order here: the IR problem of QED is a nice example of how the "theory reacts" if one is not asking the right physical questions. The degeneracy in the energy spectrum which manifests itself in particular kinematic regions (soft and/or collinear photons), at first leads to ill–defined results in a naive (assuming forces to be of finite range) scattering picture approach. At the end one learns that in QED the S–matrix as defined by the Gell-Mann Low formula does not exist, because the physical state spectrum is modified by the dynamics and is not the one suggested by the free part of the Lagrangian. Fortunately, a perturbative calculation of cross–sections is still possible, by modifications of the naive approach by accounting appropriately for the possible degeneracy of states.

As we have observed in the above discussion, the radiatively induced Pauli form factor is not affected by the IR problem. The Pauli form factor is an example of a so called *infrared save* quantity, which does not suffer from IR singularities in the naive scattering picture approach. As the anomalous magnetic moment is measured with extremely high accuracy, it nevertheless looks pretty much like a miracle how it is possible to calculate the anomalous magnetic moment in the naive approach to high orders (four loops at the moment) and confront it with an experimental result which is also measured assuming such a picture to be valid. But the states with which one formally

2.7 Pions in Scalar QED and Vacuum Polarization by Vector Mesons

The strong interaction effects in $(g-2)$ are dominated by the lightest hadrons, the isospin $SU(2)$ triplet (π^+, π^0, π^-) of pions, pseudoscalar spin 0 mesons of masses: $m_{\pi^\pm} = 139.75018(35)$ MeV, $m_{\pi^0} = 134.9766(6)$ MeV. Pions are quark–antiquark color singlet bound states $(u\bar{d}, \frac{1}{\sqrt{2}}[u\bar{u}-d\bar{d}], d\bar{u})$ and their electromagnetic interaction proceeds via the charged quarks. This is particularly pronounced in the case of the neutral π^0 which decays electromagnetically via $\pi^0 \to \gamma\gamma$ and has a much shorter life time $\tau_{\pi^0} = 8.4(6) \times 10^{-17}$ sec than the charged partners which can decay by weak interaction only according to $\pi^+ \to \mu^+ \nu_\mu$ and hence live longer by almost 10 orders of magnitude $\tau_{\pi^\pm} = 2.6033(5) \times 10^{-8}$ sec. However, at low energies, in many respects the pions behave like point particles especially what concerns soft photon emission and the Bloch-Nordsieck prescription. The effective Lagrangian for the electromagnetic interaction of a charged point–like pion described by a complex scalar field φ follows from the free Lagrangian

$$\mathcal{L}_\pi^{(0)} = (\partial_\mu \varphi)(\partial^\mu \varphi)^* - m_\pi^2 \varphi \varphi^*.$$

via *minimal substitution* $\partial_\mu \varphi \to D_\mu \varphi = (\partial_\mu + ieA_\mu(x))\varphi$ (also called *covariant derivative*), which implies the scalar QED (sQED) Lagrangian

$$\mathcal{L}_\pi^{\mathrm{sQED}} = \mathcal{L}_\pi^{(0)} - ie(\varphi^* \partial_\mu \varphi - \varphi \partial_\mu \varphi^*)A^\mu + e^2 g_{\mu\nu} \varphi \varphi^* A^\mu A^\nu. \quad (2.251)$$

Thus gauge invariance implies that the pions must couple via two different vertices to the electromagnetic field, and the corresponding Feynman rules are given in Fig. 2.8.

The bound state nature of the charged pion is taken care off by introducing a pion form factor $e \to e F_\pi(q^2)$, $e^2 \to e^2 |F_\pi(q^2)|^2$.

In sQED the contribution of a pion loop to the photon VP is given by

$$-i \Pi_\gamma^{\mu\nu\,(\pi)}(q) = \quad \text{(diagram)} \quad + \quad \text{(diagram)}.$$

The bare result for the transversal part defined by (2.156) reads

$$\Pi_\gamma^{(\pi)}(q^2) = \frac{e^2}{48\pi^2} \left\{ B_0(m,m;q^2)\,(q^2 - 4m^2) - 4A_0(m) - 4m^2 + \frac{2}{3}q^2 \right\}$$

with $\Pi_\gamma(0) = 0$. We again calculate the renormalized transversal self–energy $\Pi'_\gamma(q^2) = \Pi_\gamma(q^2)/q^2$ which is given by $\Pi'_{\gamma\mathrm{ren}}(q^2) = \Pi'_\gamma(q^2) - \Pi'_\gamma(0)$. The subtraction term

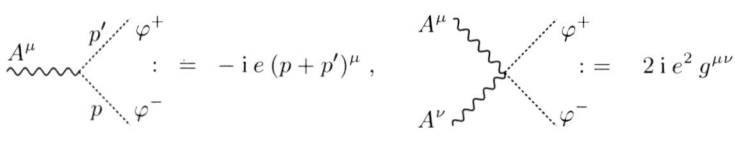

Fig. 2.8. Feynman rules for sQED. p is incoming, p' outgoing

$$\Pi_\gamma^{'(\pi)}(0) = \frac{-e^2}{48\pi^2}\left\{\frac{A_0(m)}{m^2}+1\right\}$$

is the $\pi^\pm$ contribution to the photon wavefunction renormalization and the renormalized transversal photon self–energy reads

$$\Pi_{\gamma\,\text{ren}}^{'(\pi)}(q^2) = \frac{\alpha}{6\pi}\left\{\frac{1}{3}+(1-y)-(1-y)^2\,G(y)\right\} \quad (2.252)$$

where $y = 4m^2/q^2$ and $G(y)$ given by (2.170). For $q^2 > 4m^2$ there is an imaginary or absorptive part given by substituting

$$G(y) \to \text{Im}\,G(y) = -\frac{\pi}{2\sqrt{1-y}}$$

according to (2.171)

$$\text{Im}\,\Pi_\gamma^{'(\pi)}(q^2) = \frac{\alpha}{12}(1-y)^{3/2} \quad (2.253)$$

and for large q^2 is 1/4 of the corresponding value for a lepton (2.175). According to the *optical theorem* the absorptive part may be written in terms of the $e^+e^- \to \gamma^* \to \pi^+\pi^-$ production cross–section $\sigma_{\pi^+\pi^-}(s)$ as

$$\text{Im}\,\Pi_\gamma^{'\,\text{had}}(s) = \frac{s}{4\pi\alpha}\sigma_{\text{had}}(s) \quad (2.254)$$

which hence we can read off to be

$$\sigma_{\pi^+\pi^-}(s) = \frac{\pi\alpha^2}{3s}\beta_\pi^3 \quad (2.255)$$

with $\beta_\pi = \sqrt{(1-4m_\pi^2/s)}$ the pion velocity in the CM frame. Often, one writes hadronic cross–sections as a ratio

2.7 Pions in Scalar QED and Vacuum Polarization by Vector Mesons

$$R(s) \doteq \sigma_{\text{had}}(s) / \frac{4\pi\alpha^2}{3s} \qquad (2.256)$$

in units of the high energy asymptotic form of the cross–section $\sigma(e^+e^- \to \gamma^* \to \mu^+\mu^-)$ for muon pair production in e^+e^-–annihilation. Given the cross–section or imaginary part, conversely, the real part of the renormalized vacuum polarization function may be obtained by integrating the appropriate dispersion relation (see Sect. 3.7), which reads

$$\operatorname{Re} \Pi'^{\text{had}}_{\gamma\text{ren}}(s) = \frac{s}{4\pi^2\alpha} \fint_{s_1}^{s_2} ds' \frac{\sigma_{had}(s')}{s'-s} = \frac{\alpha}{3\pi} \fint_{s_1}^{s_2} ds' \left\{ \frac{1}{s'-s} - \frac{1}{s'} \right\} R(s') . \qquad (2.257)$$

This is another way, the *dispersive approach*, to get the result (2.252) via the easier to calculate imaginary part, which here is just given by the tree level cross–section for $\gamma^* \to \pi^+\pi^-$.

As already mentioned, sometimes one has to resort to sQED in particular in connection with the soft photon radiation problem of charged particles, where sQED provides a good description of the problem. However, the photon vacuum polarization due to an elementary charged spin 0 pion, we just have been calculating, includes hard photons in the region of interest above the $\pi^+\pi^-$ production threshold to about 1 GeV, say. As we will see sQED in this case gives a rather bad approximation. In reality $e^+e^- \to \gamma^* \to \pi^+\pi^-$ is non–perturbative and exhibits a pronounced resonance, the neutral spin 1 meson ρ^0, and the hadron production cross–section is much better parametrized by a Breit-Wigner (BW) resonance shape. The relevant parameters are M_R the mass, Γ the width and $\Gamma_{e^+e^-}/\Gamma$ the branching fraction for $\rho \to e^+e^-$. We briefly present the different possible parametrizations and how a BW resonance contributes to the renormalized photon vacuum polarization when integrated over a range (s_1, s_2) with $4m_\pi^2 \leq s_1 < s_2 \leq \infty$ [62]:

- **Narrow width resonance**

The contribution from a zero width resonance

$$\sigma_{\text{NW}}(s) = \frac{12\pi^2}{M_R} \Gamma_{e^+e^-} \delta(s - M_R^2) \qquad (2.258)$$

is given by

$$\Pi'^{\text{NW}}_{\gamma\text{ren}}(s) = \frac{-3\Gamma_{e^+e^-}}{\alpha M_R} \frac{s}{s - M_R^2} \qquad (2.259)$$

which in the limit $|s| \gg M_R^2$ becomes

$$\Pi'^{\text{NW}}_{\gamma\text{ren}}(s) \simeq \frac{-3\Gamma_{e^+e^-}}{\alpha M_R} . \qquad (2.260)$$

- **Breit-Wigner resonance**

The contribution from a classical Breit-Wigner resonance

$$\sigma_{BW}(s) = \frac{3\pi}{s} \frac{\Gamma \Gamma_{e^+e^-}}{(\sqrt{s} - M_R)^2 + \frac{\Gamma^2}{4}} \tag{2.261}$$

is given by

$$\Pi'^{BW}_{\gamma \text{ren}}(s) = \frac{-3\Gamma_{e^+e^-}}{4\pi\alpha} \{I(0) - I(W)\} \tag{2.262}$$

where

$$I(W) = \frac{1}{2ic} \left\{ \frac{1}{W - M_R - ic} \left(\ln \frac{W_2 - W}{W_1 - W} - \ln \frac{W_2 - M_R - ic}{W_1 - M_R - ic} \right) \right.$$
$$\left. - \frac{1}{W + M_R + ic} \left(\ln \frac{W_2 + W}{W_1 + W} - \ln \frac{W_2 - M_R - ic}{W_1 - M_R - ic} \right) - \text{h.c.} \right\}$$

with $c = \Gamma/2$. For $W_1 \ll M_R \ll W_2$ and $\Gamma \ll M_R$ this may be approximated by

$$\Pi'^{BW}_{\gamma \text{ren}}(s) \simeq \frac{-3\Gamma_{e^+e^-}}{\alpha M_R} \frac{s(s - M_R^2 + 3c^2)}{(s - M_R^2 + c^2)^2 + M_R^2 \Gamma^2} \tag{2.263}$$

which agrees with (2.259) and (2.260) in the limits $\Gamma^2 \ll |s - M_R^2|, M_R^2$ and $|s| \gg M_R^2$, respectively.

– **Breit-Wigner resonance: field theory version**

Finally, we consider a field theoretic form of a Breit-Wigner resonance obtained by the Dyson summation of a massive spin 1 transversal part of the propagator in the approximation that the imaginary part of the self–energy yields the width by $\text{Im}\Pi_V(M_V^2) = M_V \Gamma_V$ near resonance.

$$\sigma_{BW}(s) = \frac{12\pi}{M_R^2} \frac{\Gamma_{e^+e^-}}{\Gamma} \frac{s\Gamma^2}{(s - M_R^2)^2 + M_R^2 \Gamma^2} \tag{2.264}$$

which yields

$$\Pi'^{BW}_{\gamma \text{ren}}(s) = \frac{-3\Gamma_{e^+e^-}}{\pi\alpha M_R} \frac{s(s - M_R^2 - \Gamma^2)}{(s - M_R^2)^2 + M_R^2 \Gamma^2} \left\{ \left(\pi - \arctan \frac{\Gamma M_R}{s_2 - M_R^2} \right. \right.$$
$$\left. \left. - \arctan \frac{\Gamma M_R}{M_R^2 - s_1} \right) - \frac{\Gamma}{M_R} \frac{s}{(s - M_R^2 - \Gamma^2)} \left(\ln \left| \frac{s_2 - s}{s_1 - s} \right| - \ln \left| \frac{s_2 - M_R^2 - iM_R\Gamma}{s_1 - M_R^2 - iM_R\Gamma} \right| \right) \right\} \tag{2.265}$$

and reduces to

$$\Pi'^{BW}_{\gamma \text{ren}}(s) \simeq \frac{-3\Gamma_{e^+e^-}}{\alpha M_R} \frac{s(s - M_R^2 - \Gamma^2)}{(s - M_R^2)^2 + M_R^2 \Gamma^2} \tag{2.266}$$

for $s_1 \ll M_R^2 \ll s_2$ and $\Gamma \ll M_R$. Again we have the known limits for small Γ and for large $|s|$.

For broad resonances the different parametrizations of the resonance in general yield very different results. Therefore, it is important to know how a resonance was parametrized to get the resonance parameters like M_R and Γ. For narrow resonances, which we will have to deal with later, results are not affected in a relevant way by using different parametrizations. Note that for the broad non–relativistic ρ meson only the classical BW parametrization works. In fact, due to isospin breaking of the strong interactions ($m_d - m_u$ mass difference as well as electromagnetic effects $Q_u = 2/3 \neq Q_d = -1/3$) the ρ and ω mix and more sophisticated parametrizations must be applied, like the Gounaris-Sakurai parametrization [63] based on the vector meson dominance (VMD) model (see Sect. 5.4). More appropriate is a parametrization which relies on first principle concepts only, the description by unitarity, analyticity and constrained by chiral perturbation theory (χPT), which is the low energy effective form of QCD (see [64] and references therein).

We will use the results presented here later for the evaluation of the contributions to $g-2$ from hadron–resonances. In e^+e^-–annihilation a large number of resonances, like ρ, ω, ϕ, J/ψ series and the Υ series, show up and will have to be taken into account.

2.8 Note on QCD: The Feynman Rules and the Renormalization Group

Quantum chromodynamics, the modern theory of the strong interactions, is a non–Abelian gauge theory with gauge group $SU(3)_c$ consisting of unitary 3×3 matrices of determinant unity. The corresponding internal degrees of freedom are called *color*. The generators are given by the basis of hermitian traceless 3×3 matrices T_i, $i = 1, \cdots 8$. Quarks transform under the fundamental 3–dimensional representation 3 (quark triplets) antiquarks under the complex conjugate 3* (antiquark anti–triplets). The requirement of local gauge invariance with respect to $SU(3)_c$ transformations implies that quark fields $\psi_i(x)$ must couple to an octet of gauge fields, the gluon fields $G_{\mu j}$, $j = 1, \cdots, 8$, and together with the requirement of renormalizability this fixes the form of the interactions of the quarks completely: in the free quark Dirac–Lagrangian we have to replace the derivative by the *covariant derivative*

$$\partial_\mu \psi(x) \to D_\mu \psi(x) \,, \quad (D_\mu)_{ik} = \partial_\mu \delta_{ik} - \mathrm{i}\, g_s \sum_j (T_j)_{ik} G_{\mu j}(x) \qquad (2.267)$$

where g_s is the $SU(3)_c$ gauge coupling constant. The dynamics of the gluon fields is controlled by the non–Abelian *field strength tensor*

$$G_{\mu\nu i} = \partial_\mu G_{\nu i} - \partial_\nu G_{\mu i} + g_s c_{ijk} G_{\mu j} G_{\nu k} \qquad (2.268)$$

where c_{ijk} are the $SU(3)$ structure constants obtained from the commutator of the generators $[T_i, T_j] = \mathrm{i}\, c_{ijk}\, T_k$. The locally gauge invariant Lagrangian density is then given by

$$\mathcal{L}_{\mathrm{inv}} = -\frac{1}{4}\sum_i G_{i\mu\nu}\, G_i^{\mu\nu} + \bar{\psi}\left(i\gamma^\mu D_\mu - m\right)\psi \ . \tag{2.269}$$

We split $\mathcal{L}_{\mathrm{inv}}$ into a free part $\mathcal{L}_0$ and an interaction part $\mathcal{L}_{\mathrm{int}}$ which is taken into account as a formal power series expansion in the gauge coupling g_s. The perturbation expansion is an expansion in terms of the free fields described by $\mathcal{L}_0$. The basic problem of quantizing massless spin 1 fields is familiar from QED. Since $\mathcal{L}_{\mathrm{YM}}$ is gauge invariant, the gauge potentials $G_{i\mu}$ cannot be uniquely determined from the gauge invariant field equations. Again one has to break the gauge invariance, now, for a $SU(n)$ gauge group, by a sum of $r = n^2 - 1$ gauge fixing conditions

$$C_i(G) = 0\ , \quad i = 1, \cdots, r \ .$$

It is known from QED that the only relativistically invariant condition linear in the gauge potential which we can write is the Lorentz condition. Correspondingly we require

$$C_i(G) = -\partial_\mu G_i^\mu(x) = 0\ , \quad i = 1, \cdots, r \ . \tag{2.270}$$

It should be stressed that a covariant formulation is mandatory for calculations beyond the tree level. We are thus lead to break the gauge invariance of the Lagrangian by adding the **gauge fixing** term

$$\mathcal{L}_{\mathrm{GF}} = -\frac{1}{2\xi}\sum_i \left(\partial_\mu G_i^\mu(x)\right)^2 \tag{2.271}$$

with ξ a free gauge parameter. Together with the term $\mathcal{L}_0^G$ from $\mathcal{L}_{\mathrm{inv}}$ we obtain for the bilinear gauge field part

$$\mathcal{L}_{0,i}^{G,\xi} = -\frac{1}{4}\left(\partial_\mu G_{i\,\nu} - \partial_\nu G_{i\,\mu}\right)^2 - \frac{1}{2\xi}\left(\partial_\mu G_i^\mu(x)\right)^2 \tag{2.272}$$

which now uniquely determines a free gauge field propagator. Unlike in QED, however, $\mathcal{L}_{\mathrm{GF}}$ breaks local gauge invariance explicitly and one has to restore gauge invariance by a compensating **Faddeev-Popov term** (Faddeev and Popov 1967). The Faddeev-Popov trick consists in adding further charged ghost fields $\bar{\eta}_i(x)$ and $\eta_i(x)$, the so called **Faddeev-Popov ghosts**, which conspire with the other ghosts in such a way that physical matrix elements remain gauge invariant. Unitarity and renormalizability are then restored. The FP–ghosts must be **massless spin 0 fermions**. For the unphysical ghosts this wrong spin–statistics assignment is no obstacle. The Faddeev-Popov term must be of the form

$$\mathcal{L}_{\mathrm{FP}} = \bar{\eta}_i(x) M_{ik} \eta_k(x)$$

where

2.8 Note on QCD: The Feynman Rules and the Renormalization Group

$$M_{ik} = \frac{\partial C_i(G)}{\partial G_{j\mu}(x)} (D_\mu)_{jk} = -\partial^\mu \left(\partial_\mu \delta_{ik} - g c_{ikj} G_{j\mu}(x)\right)$$
$$= -\Box \delta_{ik} + g c_{ikj} G_{j\mu}(x) \partial^\mu + g c_{ikj} \left(\partial^\mu G_{j\mu}(x)\right) \;.$$

By partial integration of $S_{FP} = \int d^4x \, \mathcal{L}_{FP}(x)$ we may write

$$\mathcal{L}_{\text{FP}} = \partial_\mu \bar{\eta}_i \partial^\mu \eta_i - g c_{ikj} \left(\partial^\mu \bar{\eta}_i\right) C_{j\mu} \eta_k \tag{2.273}$$

which describes massless scalar fermions in interaction with the gauge fields. The complete Lagrangian for a quantized Yang-Mills theory is

$$\mathcal{L}_{\text{eff}} = \mathcal{L}_{\text{inv}} + \mathcal{L}_{\text{GF}} + \mathcal{L}_{\text{FP}} \;. \tag{2.274}$$

The free (bilinear) part

$$\mathcal{L}_0 = \mathcal{L}_0(G) + \mathcal{L}_0(\psi) + \mathcal{L}_0(\eta)$$

with

$$\mathcal{L}_0(G) = \frac{1}{2} G_{i\mu} \left[\left(\Box g^{\mu\nu} - \left(1 - \frac{1}{\xi}\right) \partial^\mu \partial^\nu\right) \delta_{ik}\right] G_{k\nu}$$
$$\mathcal{L}_0(\psi) = \bar{\psi}_{\alpha a} \left[\left((i\gamma^\mu)_{\alpha\beta} \partial_\mu - m\delta_{\alpha\beta}\right) \delta_{ab}\right] \psi_{\beta b}$$
$$\mathcal{L}_0(\eta) = \bar{\eta}_i \left[(-\Box) \delta_{ik}\right] \eta_k$$

determines the free propagators, the differential operators in the square brackets being the inverses of the propagators. By Fourier transformation the free propagators are obtained in algebraic form (i.e. the differential operators are represented by c–numbers) in momentum space. Inverting these c–number matrices we obtain the results depicted in Fig. 2.9.

The interaction part of the Lagrangian is given by

$$\mathcal{L}_{\text{int}} = g_s \bar{\psi} \gamma^\mu T_i \psi G_{i\mu} - \frac{1}{2} g_s c_{ikl} \left(\partial^\mu G_i^\nu - \partial^\nu G_i^\mu\right) G_{k\mu} G_{l\nu}$$
$$- \frac{1}{4} g_s^2 c_{ikl} c_{ik'l'} G_k^\mu G_l^\nu G_{k'\mu} G_{l'\nu} - g_s c_{ikj} \left(\partial^\mu \bar{\eta}_i\right) G_{j\mu} \eta_k \tag{2.275}$$

with a single coupling constant g_s for the four different types of vertices.

While the formal argumentation which leads the the construction of local gauge theories looks not too different for Abelian and non–Abelian gauge groups, the physical consequences are very different and could not be more dramatic: in contrast to Abelian theories where the gauge field is neutral and exhibits no self–interaction, non–Abelian gauge fields necessarily carry non–Abelian charge and must be self–interacting. These Yang-Mills self–interactions are responsible for the *anti–screening* of the non–Abelian charge, known as *asymptotic freedom* (AF) (see end of section). It implies that the strong interaction force gets weaker the higher the energy, or equivalently,

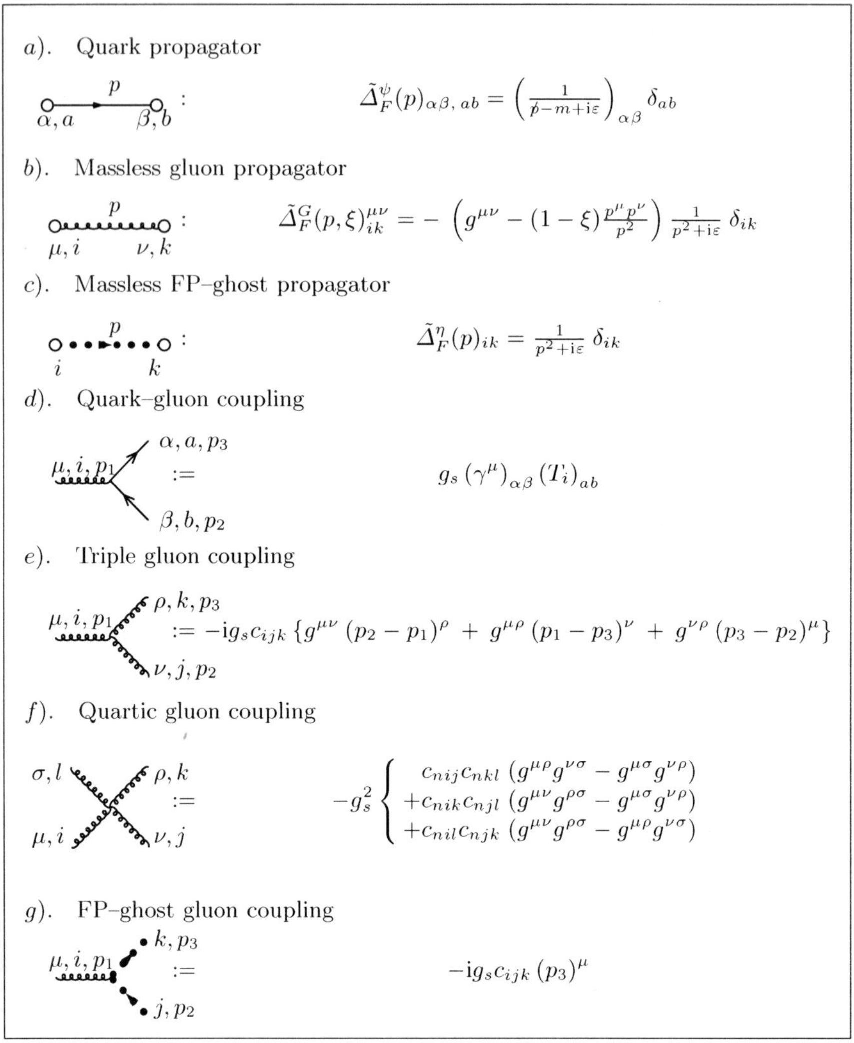

$a)$. Quark propagator
p α, a β, b : $\tilde{\Delta}^{\psi}_{F}(p)_{\alpha\beta,\,ab} = \left(\frac{1}{\not p - m + \mathrm{i}\varepsilon}\right)_{\alpha\beta} \delta_{ab}$
$b)$. Massless gluon propagator
p μ, i ν, k : $\tilde{\Delta}^{G}_{F}(p,\xi)^{\mu\nu}_{ik} = -\left(g^{\mu\nu} - (1-\xi)\frac{p^{\mu}p^{\nu}}{p^2}\right)\frac{1}{p^2+\mathrm{i}\varepsilon}\,\delta_{ik}$
$c)$. Massless FP–ghost propagator
p i k : $\tilde{\Delta}^{\eta}_{F}(p)_{ik} = \frac{1}{p^2+\mathrm{i}\varepsilon}\,\delta_{ik}$
$d)$. Quark–gluon coupling
μ, i, p_1 α, a, p_3 β, b, p_2 := $g_s\,(\gamma^{\mu})_{\alpha\beta}\,(T_i)_{ab}$
$e)$. Triple gluon coupling
μ, i, p_1 ρ, k, p_3 ν, j, p_2 := $-\mathrm{i}g_s c_{ijk}\,\{g^{\mu\nu}\,(p_2-p_1)^{\rho} + g^{\mu\rho}\,(p_1-p_3)^{\nu} + g^{\nu\rho}\,(p_3-p_2)^{\mu}\}$
$f)$. Quartic gluon coupling
σ, l ρ, k μ, i ν, j := $-g_s^2\left\{\begin{array}{l} c_{nij}c_{nkl}\,(g^{\mu\rho}g^{\nu\sigma} - g^{\mu\sigma}g^{\nu\rho}) \\ +c_{nik}c_{njl}\,(g^{\mu\nu}g^{\rho\sigma} - g^{\mu\sigma}g^{\nu\rho}) \\ +c_{nil}c_{njk}\,(g^{\mu\nu}g^{\rho\sigma} - g^{\mu\rho}g^{\nu\sigma}) \end{array}\right.$
$g)$. FP–ghost gluon coupling
μ, i, p_1 k, p_3 j, p_2 := $-\mathrm{i}g_s c_{ijk}\,(p_3)^{\mu}$

Fig. 2.9. Feynman rules for QCD. Momenta at vertices are chosen ingoing

the shorter the distance. While it appears most natural to us that particles interact the less the farther apart they are, non–Abelian forces share the opposite property, the forces get the stronger the farer away we try to separate the quarks. In QCD this leads to the *confinement* of the constituents within hadrons. The latter being quark bound states which can never be broken up into free constituents. This makes QCD an intrinsically non–perturbative theory, the fields in the Lagrangian, quarks and gluons, never appear in scattering states, which define the physical state space and the S–matrix. QED is very different, it has an perturbative S-matrix, its proper definition being complicated by the existence of the long range Coulomb forces (see Sect. 2.6.6 above). Nevertheless, the fields in the QED Lagrangian as interpolating fields are closely related to the physical states, the photons and leptons. This extends to the electroweak SM, where the weak non–Abelian gauge bosons, the $W^\pm$ and the Z particles, become massive as a consequence of the breakdown of the $SU(2)_L$ gauge symmetry by the Higgs mechanism. Also the weak gauge bosons cannot be seen as scattering states in a detector, but this time because of their very short lifetime. Due to its non–perturbative nature, precise predictions in strong interaction physics are often difficult, if not impossible. Fortunately, besides perturbative QCD which applies to hard subprocesses, non–perturbative methods have been developed to a high level of sophistication, like *chiral perturbation theory* (CHPT) [68] and QCD on a Euclidean space–time lattice (*lattice QCD*) [69]. Chiral perturbation theory is based on the low energy structure of QCD: in the limit of vanishing quark masses QCD has a global $SU(N_f)_V \otimes SU(N_f)_A \otimes U(1)_V$ symmetry (*chiral symmetry*). Thereby the $SU(N_f)_A$ subgroup turns out broken spontaneously, which, in the isospin limit $N_f = 2$, $m_u = m_d = 0$, implies the existence of a triplet of massless pions (Goldstone bosons). $U(1)_V$ is responsible for *baryon number conservation*, whereas in contrast $U(1)_A$ is broken by the *Adler-Bell-Jackiw anomaly* (see p. 233 below).

The RG of QCD in Short

The renormalization group, introduced in Sect. 2.6.5, for QCD plays a particularly important role for a quantitative understanding of AF as well as a tool for improving the convergence of the perturbative expansion [30, 65]. For QCD the RG is given by

$$\mu \frac{\mathrm{d}}{\mathrm{d}\mu} g_s(\mu) = \beta\left(g_s(\mu)\right)$$
$$\mu \frac{\mathrm{d}}{\mathrm{d}\mu} m_i(\mu) = -\gamma\left(g_s(\mu)\right) m_i(\mu) \qquad (2.276)$$

with

$$\beta(g) = -\beta_0 \frac{g^3}{16\pi^2} - \beta_1 \frac{g^5}{(16\pi^2)^2} + O(g^7)$$
$$\gamma(g) = \gamma_0 \frac{g^2}{4\pi^2} + \gamma_1 \frac{g^4}{(4\pi^2)^2} + O(g^6) \qquad (2.277)$$

where, in the $\overline{\mathrm{MS}}$ scheme (Sect. 2.5.6),

$$\begin{aligned}\beta_0 &= 11 - \tfrac{2}{3} N_f \; ; \; \gamma_0 = \quad 2 \\ \beta_1 &= 102 - \tfrac{38}{3} N_f \; ; \; \gamma_1 = \tfrac{101}{12} - \tfrac{5}{18} N_f \end{aligned} \qquad (2.278)$$

and N_f is the number of quark flavors. The RG for QCD is known to 4 loops [66, 67]. It allows to define effective parameters in QCD, which incorporate the summation of leading logarithmic (1–loop), next–to–leading logarithmic (2–loop), $\cdots$ corrections (RG improved perturbation theory). The solution of (2.276) for the running coupling constant $\alpha_s(\mu) = g_s^2(\mu)/(4\pi)$ yields (see (2.218))

$$\frac{4\pi}{\beta_0 \alpha_s(\mu)} - \frac{\beta_1}{\beta_0^2} \ln\left(\frac{4\pi}{\beta_0 \alpha_s(\mu)} + \frac{\beta_1}{\beta_0^2}\right) =$$
$$\ln \mu^2/\mu_0^2 + \frac{4\pi}{\beta_0 \alpha_s(\mu_0)} - \frac{\beta_1}{\beta_0^2} \ln\left(\frac{4\pi}{\beta_0 \alpha_s(\mu_0)} + \frac{\beta_1}{\beta_0^2}\right) \equiv \ln \mu^2/\Lambda^2 \quad (2.279)$$

with reference scale (integration constant)

$$\Lambda_{\text{QCD}} = \Lambda_{\overline{\text{MS}}}^{(N_f)} = \mu \exp\left\{-\frac{4\pi}{2\beta_0 \alpha_s(\mu)}\left(1 + \frac{\alpha_s(\mu)}{4\pi}\frac{\beta_1}{\beta_0}\ln\frac{\beta_0 \alpha_s(\mu)}{4\pi + \frac{\beta_1}{\beta_0}\alpha_s(\mu)}\right)\right\}$$
$$(2.280)$$

which can be shown easily to be independent of the reference scale μ. It is RG invariant

$$\mu \frac{\mathrm{d}}{\mathrm{d}\mu} \Lambda_{\text{QCD}} = 0,$$

and thus QCD has its own intrinsic scale Λ_{QCD} which is related directly to the coupling strength (dimensional transmutation). This is most obvious at the one–loop level where we have the simple relation

$$\alpha_s(\mu) = \frac{1}{\frac{\beta_0}{4\pi} \ln \frac{\mu^2}{\Lambda^2}}. \quad (2.281)$$

Thus Λ_{QCD} incorporates the reference coupling $\alpha_s(\mu_0)$ measured at scale μ_0 in a scale invariant manner, i.e. each experiment measures the same Λ_{QCD} irrespective of the reference energy μ_0 at which the measurement of $\alpha_s(\mu_0)$ is performed.

The solution of (2.276) for the effective masses $m_i(\mu)$ reads (see (2.219))

$$m_i(\mu) = m_i(\mu_0) \frac{r(\mu)}{r(\mu_0)} \equiv \bar{m}_i r(\mu) \quad (2.282)$$

with

$$r(\mu) = \exp -2\left\{\frac{\gamma_0}{\beta_0} \ln \frac{4\pi}{\beta_0 \alpha_s(\mu)} + \left(\frac{\gamma_0}{\beta_0} - \frac{4\gamma_1}{\beta_1}\right) \ln(1 + \frac{\beta_1}{\beta_0}\frac{\alpha_s(\mu)}{4\pi})\right\}. \quad (2.283)$$

Note that also the $\bar{m}_i$ are RG invariant masses (integration constants) and for the masses play a role similar to Λ_{QCD} for the coupling. The solution of the RG equation may be expanded in the large log $L \equiv \ln \frac{\mu^2}{\Lambda^2}$, which of course only makes sense if L is large ($\mu \gg \Lambda$),

$$\alpha_s(\mu) = \frac{4\pi}{\beta_0 L}\left(1 - \frac{\beta_1}{\beta_0^2}\frac{\ln(L + \frac{\beta_1}{\beta_0^2})}{L} + \cdots\right)$$

$$m_i(\mu) = \bar{m}_i\left(\frac{L}{2}\right)^{-\frac{\gamma_0}{\beta_0}}\left(1 - \frac{2\beta_1\gamma_0}{\beta_0^3}\frac{\ln L + 1}{L} + \frac{8\gamma_1}{\beta_0^2 L} + \cdots\right). \quad (2.284)$$

If L is not large one should solve (2.279) or its higher order version numerically by iteration for $\alpha_s(\mu)$. For the experimental prove of the running of the strong coupling constant [70] see Fig. 3.3 in Sect. 3.2.1. The non-perturbative calculations in lattice QCD are able to demonstrate a surprisingly good agreement with perturbative results (see [71] and references therein).

References

1. J. D. Bjorken, S. D. Drell, *Relativistic Quantum Mechanics*, 1st edn (McGraw-Hill, New York 1964) 300 p; *Relativistic Quantum Fields*, 1st edn (McGraw-Hill, New York 1965) 396 p
2. V. B. Berestetskii, E. M. Lifshitz, L. P. Pitaevskii, Quantum Electrodynamics, *Landau and Lifshitz Course of Theoretical Physics Vol. 4* 2nd edn (Pergamon, London 1982) 652 p
3. P. A. M. Dirac, Proc. Roy. Soc. A **114** (1927) 243; Proc. Roy. Soc. A **126** (1930) 360; Proc. Roy. Soc. A **136** (1932) 453
4. E. P. Wigner, Ann. Math. **40** (1939) 149
5. P. Jordan, E. P. Wigner, Zeits. Phys. **47** (1928) 631
6. W. Heisenberg, W. Pauli, Zeits. Phys. **56** (1929) 1; Zeits. Phys. **59** (1930) 168; P. A. M. Dirac, V. A. Fock, B. Podolsky, Phys. Zeits. Sowjetunion **3** (1932) 64
7. H. Joos, Fortsch. Phys. **10** (1962) 65; S. Weinberg, Phys. Rev. **133** (1964) B1318; Phys. Rev. **134** (1964) B882
8. P. A. M. Dirac, Proc. Roy. Soc. A **114** (1927) 243; A **117** (1928) 610
9. W. Pauli, Zeits. Physik **31** (1925) 765
10. W. Pauli, Phys. Rev. **58** (1940) 716
11. G. Lüders, K. Danske Vidensk. Selsk. Mat.-Fys. Medd. **28** (1954) No. 5; W. Pauli, Exclusion principle, Lorentz group and reflection of space-time and charge. In: *Niels Bohr and the Development of Physics*, ed by W. Pauli (Pergamon Press London 1955, reissued 1962) pp. 30–51; W. Pauli, Il Nuovo Cim. **6** (1957) 204; G. Lüders, Ann. Phys. N. Y. **2** (1957) 1; G. Lüders, B. Zumino, Phys. Rev. **106** (1957) 345; R. Jost, Helv. Phys. Acta **30** (1957) 409
12. S. Gasiorowicz, *Elementary particle physics*, (John Wiley and Sons, Inc., New York, 1966), p. 513
13. S. Eidelman et al. [Particle Data Group], Phys. Lett. B **592** (2004) 1
14. B. C. Regan, E. D. Commins, C. J. Schmidt, D. DeMille, Phys. Rev. Lett. **88** (2002) 071805
15. W. Gerlach, O. Stern, Zeits. Physik **8** (1924) 110
16. G. E. Uhlenbeck, S. Goudsmit, Naturwissenschaften **13** (1925) 953; Nature **117** (1926) 264
17. W. Pauli, Zeits. Phys. **43** (1927) 601
18. S. Weinberg, Phys. Rev. **134** (1964) B882

19. R. Frisch, O. Stern, Zeits. Physik **85** (1933) 4; I. Estermann, O. Stern, ibid **85** (1933) 17
20. G. Charpak, F. J. M. Farley, R. L. Garwin, T. Muller, J. C. Sens, A. Zichichi, Phys. Lett. **1B** (1962) 16
21. H. Weyl, Zeits. Phys. **56** (1929) 330
22. C. N. Yang, R. L. Mills, Phys. Rev. **96** (1954) 191
23. M. Gell-Mann, F. Low, Phys. Rev. **95** (1954) 1300
24. G. C. Wick, Phys. Rev. **80** (1950) 268
25. W. Pauli, F. Villars, Rev. Mod. Phys. **21** (1949) 434
26. S. Tomonaga, Riken Iho, Progr. Theor. Phys. **1** (1946) 27; J. Schwinger, Phys. Rev. **74** (1948) 1439; R. P. Feynman, Phys. Rev. **76** (1949) 749; F. Dyson, Phys. Rev. **75** (1949) 486, ibid. 1736
27. N. N. Bogoliubov, D. V. Shirkov, *Introduction to the Theory of Quantized Fields*, 1st & 2nd edn (John Wiley & Sons, Inc., New York 1957, 1980) 720 p
28. G. 't Hooft, Nucl. Phys. B **33** (1971) 173; **35** (1971) 167; G. 't Hooft, M. Veltman, Nucl. Phys. B **50** (1972) 318
29. H. Fritzsch, M. Gell-Mann, H. Leutwyler, Phys. Lett. **47**B (1973) 365
30. H. D. Politzer, Phys. Rev. Lett. **30** (1973) 1346; D. Gross, F. Wilczek, Phys. Rev. Lett. **30** (1973) 1343
31. S. L. Glashow, Nucl. Phys. B **22** (1961) 579; S. Weinberg, Phys. Rev. Lett. **19** (1967) 1264; A. Salam, Weak and electromagnetic interactions. In: *Elementary Particle Theory*, ed by N. Svartholm, (Amquist and Wiksells, Stockholm 1969) pp. 367–377
32. G. t' Hooft, M. Veltman, Nucl. Phys. B **44** (1972) 189
33. W. H. Furry, Phys. Rev. **51** (1937) 125
34. H. Lehmann, K. Symanzik, W. Zimmermann, Nuovo Cimento **1** (1955) 205; Nuovo Cimento **6** (1957) 319
35. F. Jegerlehner, Eur. Phys. J. C **18** (2001) 673
36. D. Akyeampong, R. Delbourgo, Nuovo Cim. **17A** (1973) 47; W. A. Bardeen, R. Gastmans, B. Lautrup, Nucl. Phys. B **46** (1972) 319; M. Chanowitz, M. Furman, I. Hinchliffe, Nucl. Phys. B **159** (1979) 225
37. R. F. Streater, A. S. Wightman, *CPT, spin, statistics and all that* (Benjamin, New York 1964)
38. K. Osterwalder, R. Schrader, Commun. Math. Phys. **31** (1973) 83; ibid. **42** (1975) 281
39. K. G. Wilson, Phys. Rev. D **10** (1974) 2445
40. G. t'Hooft, M. Veltman, Nucl. Phys. B **153** (1979) 365
41. A. I. Davydychev, M. Y. Kalmykov, Nucl. Phys. B **605** (2001) 266
42. J. Fleischer, F. Jegerlehner, O. V. Tarasov, Nucl. Phys. B **672** (2003) 303
43. G. Passarino, M. Veltman, Nucl. Phys. B **160** (1979) 151
44. Y. Hahn, W. Zimmermann, Commun. Math. Phys. **10** (1968) 330; W. Zimmermann, Commun. Math. Phys. **11** (1968) 1; J. H. Lowenstein, W. Zimmermann, Commun. Math. Phys. **44** (1975) 73
45. F. J. Dyson, Phys. Rev. **75** (1949) 1736
46. S. Weinberg, Phys. Rev. **118** (1960) 838
47. J. P. Fink, J. Math. Phys. **9** (1968) 1389; E. B. Manoukian, J. Math. Phys. **19** (1978) 917
48. V. A. Smirnov, Mod. Phys. Lett. A **10** (1995) 1485 and references therein

49. E. C. G. Stueckelberg, A. Petermann, Helv. Phys. Acta **26** (1953) 499; M. Gell-Mann, F. E. Low, Phys. Rev. **95** (1954) 1300
50. G. Mack, Nucl. Phys. B **5** (1968) 499
51. J. D. Bjorken, Phys. Rev. **179** (1969) 1547; J. D. Bjorken, E. A. Paschos, Phys. Rev. **185** (1969) 1975
52. D. H. Coward et al., Phys. Rev. Lett. **20** (1968) 292; E. D. Bloom et al., Phys. Rev. Lett. **23** (1969) 930; G. Miller et al., Phys. Rev. D **5** (1972) 528
53. H. Suura, E. Wichmann, Phys. Rev. **105** (1957) 1930; A. Petermann, Phys. Rev. **105** (1957) 1931
54. B. E. Lautrup, E. de Rafael, Nucl. Phys. B **70** (1974) 317
55. E. de Rafael, J. L. Rosner, Ann. Phys. (N. Y.) **82** (1974) 369
56. F. Bloch, A. Nordsieck, Phys. Rev. D **52** (1937) 54
57. D. R. Yennie, S. C. Frautschi, H. Suura, Ann. Phys. **13** (1961) 379; D. R. Yennie, Phys. Lett. **34**B (1975) 239; J. D. Jackson, D. L. Scharre, Nucl. Instr. **128** (1975) 13; M. Greco, G. Pancheri, Y. Srivastava, Nucl. Phys. B **101** (1975) 234
58. T. Kinoshita, J. Math. Phys. **3** (1962) 650; T. D. Lee, M. Nauenberg, Phys. Rev. D **133** (1964) B1549
59. J. Fleischer, F. Jegerlehner, Z. Phys. C **26** (1985) 629
60. G. Sterman, S. Weinberg, Phys. Rev. Lett. **39** (1977) 1436
61. O. Steinmann, Commun. Math. Phys. **237** (2003) 181
62. F. Jegerlehner, Nucl. Phys. B (Proc. Suppl.) **51C** (1996) 131
63. G. J. Gounaris, J. J. Sakurai, Phys. Rev. Lett. **21** (1968) 244; A. Quenzer et al, Phys. Lett. B **76** (1978) 512
64. H. Leutwyler, *Electromagnetic form factor of the pion*, hep-ph/0212324
65. W. E. Caswell, Phys. Rev. Lett. **33** (1974) 244; D. R. T. Jones, Nucl. Phys. B **75** (1974) 531; E. Egorian, O. V. Tarasov, Theor. Math. Phys. **41** (1979) 863 [Teor. Mat. Fiz. **41** (1979) 26]
66. O. V. Tarasov, A. A. Vladimirov, A. Y. Zharkov, Phys. Lett. B **93** (1980) 429; S. A. Larin, J. A. M. Vermaseren, Phys. Lett. B **303** (1993) 334
67. T. van Ritbergen, J. A. M. Vermaseren, S. A. Larin, Phys. Lett. B **400** (1997) 379; M. Czakon, Nucl. Phys. B **710** (2005) 485; K. G. Chetyrkin, Nucl. Phys. B **710** (2005) 499
68. J. Gasser, H. Leutwyler, Annals Phys. **158** (1984) 142; Nucl. Phys. B **250** (1985) 465
69. K. G. Wilson, Phys. Rev. D **10** (1974) 2445; M. Creutz, Phys. Rev. D **21** (1980) 2308
70. S. Bethke, Phys. Rept. **403–404** (2004) 203
71. M. Della Morte, R. Frezzotti, J. Heitger, J. Rolf, R. Sommer, U. Wolff, Nucl. Phys. B **713** (2005) 378

3

Lepton Magnetic Moments: Basics

3.1 Equation of Motion for a Lepton in an External Field

For the measurement of the anomalous magnetic moment of a lepton we have to investigate the motion of a relativistic point–particle of charge $Q_\ell e$ (e the positron charge) and mass m_ℓ in an external electromagnetic field $A^{\text{ext}}_\mu(x)$. The equation of motion of a charged Dirac particle in an external field is given by

$$\begin{aligned}\left(\mathrm{i}\hbar\gamma^\mu\partial_\mu + Q_\ell \tfrac{e}{c}\gamma^\mu(A_\mu + A^{\text{ext}}_\mu(x)) - m_\ell c\right)\psi_\ell(x) &= 0 \\ \left(\Box g^{\mu\nu} - (1-\xi^{-1})\partial^\mu\partial^\nu\right) A_\nu(x) &= -Q_\ell e \bar\psi_\ell(x)\gamma^\mu\psi_\ell(x)\ .\end{aligned} \quad (3.1)$$

What we are looking for is the solution of the Dirac equation with an external field as a relativistic one–particle problem, neglecting the radiation field in a first step. We thus are interested in a solution of the first of the above equations, which we may write as

$$\mathrm{i}\hbar\frac{\partial\psi_\ell}{\partial t} = \left(-c\,\boldsymbol{\alpha}\left(\mathrm{i}\hbar\boldsymbol{\nabla} - Q_\ell\frac{e}{c}\boldsymbol{A}\right) - Q_\ell\,e\,\Phi + \beta\,m_\ell c^2\right)\psi_\ell\ , \quad (3.2)$$

with $\beta = \gamma^0$, $\boldsymbol{\alpha} = \gamma^0\boldsymbol{\gamma}$ and $A^{\mu\,\text{ext}} = (\Phi, \boldsymbol{A})$. For the interpretation of the solution the non–relativistic limit plays an important role, because many relativistic problems in QED may be most easily understood in terms of the non–relativistic problem as a starting point, which usually is easier to solve. We will consider a lepton e^-, μ^- or τ^- with $Q_\ell = -1$ in the following and drop the index ℓ.

1. Non–relativistic limit

For studying the non–relativistic limit of the motion of a Dirac particle in an external field it is helpful and more transparent to work in natural units[1].

---

[1] The general rules of translation read: $p^\mu \to p^\mu$, $\mathrm{d}\mu(p) \to \hbar^{-3}\mathrm{d}\mu(p)$, $m \to mc$, $e \to e/(\hbar c)$, $e^{\mathrm{i}px} \to e^{\mathrm{i}\frac{px}{\hbar}}$, spinors: $u, v \to u/\sqrt{c}, v/\sqrt{c}$.

3 Lepton Magnetic Moments: Basics

In order to get from the Dirac spinor ψ the two component Pauli spinors in the non–relativistic limit, one has to perform an appropriate unitary transformation, called Foldy-Wouthuysen transformation. Looking at the Dirac equation (3.2)

$$i\hbar \frac{\partial \psi}{\partial t} = H\psi, \quad H = c\,\boldsymbol{\alpha}\left(\boldsymbol{p} - \frac{e}{c}\boldsymbol{A}\right) + \beta\,mc^2 + e\,\Phi$$

with

$$\beta = \gamma^0 = \begin{pmatrix} 1 & 0 \\ 0 & -1 \end{pmatrix}, \quad \boldsymbol{\alpha} = \gamma^0 \boldsymbol{\gamma} = \begin{pmatrix} 0 & \boldsymbol{\sigma} \\ \boldsymbol{\sigma} & 0 \end{pmatrix},$$

we note that H has the form

$$H = \beta\,mc^2 + c\,\mathcal{O} + e\,\Phi$$

where $[\beta, \Phi] = 0$ is commuting and $\{\beta, \mathcal{O}\} = 0$ anti–commuting. In the absence of an external field spin is a conserved quantity in the rest frame, i.e. the Dirac equation must be equivalent to the Pauli equation. This fixes the unitary transformation to be performed in the case $A_\mu^{\text{ext}} = 0$:

$$\psi' = U\psi, \quad H' = U\left(H - i\hbar\frac{\partial}{\partial t}\right)U^{-1} = UHU^{-1} \tag{3.3}$$

where the time–independence of U has been used, and we obtain

$$i\hbar\frac{\partial \psi'}{\partial t} = H'\psi'; \quad \psi' = \begin{pmatrix} \varphi' \\ 0 \end{pmatrix}, \tag{3.4}$$

where φ' is the Pauli spinor. In fact U is a Lorentz boost matrix

$$U = \mathbf{1}\cosh\theta + \boldsymbol{n}\,\boldsymbol{\gamma}\sinh\theta = e^{\theta \boldsymbol{n}\boldsymbol{\gamma}} \tag{3.5}$$

with

$$\boldsymbol{n} = \frac{\boldsymbol{p}}{|\boldsymbol{p}|}, \quad \theta = \frac{1}{2}\mathrm{arccosh}\frac{p^0}{mc} = \mathrm{arcsinh}\frac{|\boldsymbol{p}|}{mc}$$

and we obtain, with $p^0 = \sqrt{\boldsymbol{p}^2 + m^2c^2}$,

$$H' = cp^0\beta; \quad [H', \boldsymbol{\Sigma}] = 0, \quad \boldsymbol{\Sigma} = \boldsymbol{\alpha}\,\gamma_5 = \begin{pmatrix} \boldsymbol{\sigma} & 0 \\ 0 & \boldsymbol{\sigma} \end{pmatrix} \tag{3.6}$$

where $\boldsymbol{\Sigma}$ is the spin operator. Actually, there exist two projection operators U one to the upper and one to the lower components:

$$U_+\psi = \begin{pmatrix} \varphi' \\ 0 \end{pmatrix}, \quad U_-\psi = \begin{pmatrix} 0 \\ \chi \end{pmatrix},$$

given by
$$U_+ = \frac{(p^0 + mc)\mathbf{1} + \boldsymbol{p}\boldsymbol{\gamma}}{\sqrt{2p^0}\sqrt{p^0 + mc}}, \quad U_- = \frac{(p^0 + mc)\mathbf{1} - \boldsymbol{p}\boldsymbol{\gamma}}{\sqrt{2p^0}\sqrt{p^0 + mc}}.$$

For the spinors we have
$$U_+ u(p, r) = \sqrt{\frac{2p^0}{c}} \begin{pmatrix} U(r) \\ 0 \end{pmatrix}, \quad U_- v(p, r) = \sqrt{\frac{2p^0}{c}} \begin{pmatrix} 0 \\ V(r) \end{pmatrix}$$

with $U(r)$ and $V(r) = i\sigma_2 U(r)$ the two component spinors in the rest system. We now look at the lepton propagator. The Feynman propagator reads
$$iS_{F\alpha\beta}(x - y) \equiv \langle 0|T\{\psi_\alpha(x)\bar{\psi}_\beta(y)\}|0\rangle$$
$$= c\sum_r \int d\mu(p)\, u_\alpha(p, r)\, \bar{u}_\beta(p, r)\, e^{-ip(x-y)} = \int d\mu(p)\, (\slashed{p} + mc)\, e^{-ip(x-y)}$$

where[2]
$$S_{F\alpha\beta}(z; m^2) = (i\hbar\gamma^\mu \partial_\mu + mc)\, \Delta_F(z; m^2) = \Theta(z^0)\, S^+(z) + \Theta(-z^0)\, S^-(z)$$

where the retarded positive frequency part is represented by
$$\Theta(z^0)\, S^+(z) = \int \frac{d^4p}{(2\pi)^4} \frac{c}{2\omega_p} \frac{\sum_r u_\alpha(p, r)\, \bar{u}_\beta(p, r)}{p^0 - \omega_p + i0}\, e^{-ipz}$$

and the advanced negative frequency part by
$$\Theta(-z^0)\, S^-(z) = -\int \frac{d^4p}{(2\pi)^4} \frac{c}{2\omega_p} \frac{\sum_r v_\alpha(p, r)\, \bar{v}_\beta(p, r)}{p^0 - \omega_p + i0}\, e^{ipz}.$$

Using
$$\sum_r u_\alpha(p, r)\, \bar{u}_\beta(p, r) = \frac{2\omega_p}{c} U(\boldsymbol{p})\, \gamma_+\, U(\boldsymbol{p})$$
$$\sum_r v_\alpha(p, r)\, \bar{v}_\beta(p, r) = \frac{2\omega_p}{c} U(\boldsymbol{p})\, \gamma_-\, U(\boldsymbol{p})$$

---
[2] The positive frequency part is given by
$$iS^+_{\alpha\beta}(x - y) \equiv \langle 0|\psi_\alpha(x)\bar{\psi}_\beta(y)|0\rangle$$
$$= c\sum_r \int d\mu(p)\, u_\alpha(p, r)\, \bar{u}_\beta(p, r)\, e^{-ip(x-y)} = \int d\mu(p)\, (\slashed{p} + mc)\, e^{-ip(x-y)}$$

and the negative frequency part by
$$-iS^-_{\alpha\beta}(x - y) \equiv \langle 0|\bar{\psi}_\beta(y)\psi_\alpha(x)|0\rangle$$
$$= c\sum_r \int d\mu(p)\, v_\alpha(p, r)\, \bar{v}_\beta(p, r)\, e^{ip(x-y)} = \int d\mu(p)\, (\slashed{p} - mc)\, e^{ip(x-y)}.$$

with
$$\gamma_\pm = \frac{1}{2}\left(1\pm\gamma^0\right)\ ;\ \gamma^0\gamma_\pm = \pm\gamma_\pm\ ,\ \gamma_+\gamma_- = \gamma_-\gamma_+ = 0$$
the projection matrices for the upper and lower components, respectively. We thus arrive at our final representation which allows a systematic expansion in $1/c$:
$$S_F(x-y) = \int \frac{d^4p}{(2\pi)^4}\, e^{-ip(x-y)}\, \boldsymbol{U}(\boldsymbol{p})\left(\frac{\gamma_+}{p^0-\omega_p+i0} - \frac{\gamma_-}{p^0+\omega_p-i0}\right)\boldsymbol{U}(\boldsymbol{p}). \tag{3.7}$$

The $1/c$-expansion simply follows by expanding the matrix $\boldsymbol{U}$:
$$\boldsymbol{U}(\boldsymbol{p}) = \exp\theta\frac{\boldsymbol{p}}{|\boldsymbol{p}|}\gamma = \exp\theta\frac{\boldsymbol{p}\gamma}{2mc}\ ;\ \theta = \sum_{n=0}^{\infty}\frac{(-1)^n}{2n+1}\left(\frac{\boldsymbol{p}^2}{m^2c^2}\right)^n.$$

The non–relativistic limit thus reads:
$$S_F(x-y)_{\text{NR}} = \int \frac{d^4p}{(2\pi)^4}\, e^{-ip(x-y)}\left(\frac{\gamma_+}{p^0-(mc^2+\frac{p^2}{2m})+i0} - \frac{\gamma_-}{p^0+(mc^2+\frac{p^2}{2m})-i0}\right)$$

i.e.
$$S_F(x-y) = S_F(x-y)_{\text{NR}} + O(1/c).$$

2. Non–relativistic lepton with $A_\mu^{\text{ext}} \neq 0$

Again we start from the Dirac equation (3.2). In order to get the non–relativistic representation for small velocities we have to split off the phase of the Dirac field, which is due to the rest energy of the lepton:
$$\psi = \hat{\psi}\, e^{-i\frac{mc^2}{\hbar}t}\quad \text{with}\quad \hat{\psi} = \begin{pmatrix}\hat{\varphi}\\\hat{\chi}\end{pmatrix}.$$

Consequently, the Dirac equation takes the form
$$i\hbar\frac{\partial\hat{\psi}}{\partial t} = \left(\boldsymbol{H} - mc^2\right)\hat{\psi}$$
and describes the coupled system of equations
$$\left(i\hbar\frac{\partial}{\partial t} - e\Phi\right)\hat{\varphi} = c\,\boldsymbol{\sigma}\left(\boldsymbol{p} - \frac{e}{c}\boldsymbol{A}\right)\hat{\chi}$$
$$\left(i\hbar\frac{\partial}{\partial t} - e\Phi + 2mc^2\right)\hat{\chi} = c\,\boldsymbol{\sigma}\left(\boldsymbol{p} - \frac{e}{c}\boldsymbol{A}\right)\hat{\varphi}.$$

For $c \to \infty$ we obtain

3.1 Equation of Motion for a Lepton in an External Field

$$\hat{\chi} \simeq \frac{1}{2mc}\boldsymbol{\sigma}\left(\boldsymbol{p} - \frac{e}{c}\boldsymbol{A}\right)\hat{\varphi} + O(1/c^2)$$

and hence

$$\left(i\hbar\frac{\partial}{\partial t} - e\,\Phi\right)\hat{\varphi} \simeq \frac{1}{2m}\left(\boldsymbol{\sigma}\left(\boldsymbol{p} - \frac{e}{c}\boldsymbol{A}\right)\right)^2 \hat{\varphi}\;.$$

As $\boldsymbol{p}$ does not commute with $\boldsymbol{A}$, we may use the relation

$$(\boldsymbol{\sigma}\boldsymbol{a})(\boldsymbol{\sigma}\boldsymbol{b}) = \boldsymbol{ab} + i\boldsymbol{\sigma}\,(\boldsymbol{a}\times\boldsymbol{b})$$

to obtain

$$\left(\boldsymbol{\sigma}\left(\boldsymbol{p} - \frac{e}{c}\boldsymbol{A}\right)\right)^2 = \left(\boldsymbol{p} - \frac{e}{c}\boldsymbol{A}\right)^2 - \frac{e\hbar}{c}\boldsymbol{\sigma}\cdot\boldsymbol{B}\;;\quad \boldsymbol{B} = \mathrm{rot}\,\boldsymbol{A}\;.$$

This leads us to the *Pauli equation* (W. Pauli 1927)

$$i\hbar\frac{\partial\hat{\varphi}}{\partial t} = \hat{H}\,\hat{\varphi} = \left(\frac{1}{2m}(\boldsymbol{p} - \frac{e}{c}\boldsymbol{A})^2 + e\,\Phi - \frac{e\hbar}{2mc}\boldsymbol{\sigma}\cdot\boldsymbol{B}\right)\hat{\varphi} \qquad (3.8)$$

which up to the spin term is nothing but the non–relativistic Schrödinger equation. The last term is the one this book is about: it has the form of a potential energy of a magnetic dipole in an external field. In leading order in $1/c$ the lepton behaves as a particle which has besides a charge also a magnetic moment

$$\boldsymbol{\mu} = \frac{e\hbar}{2mc}\boldsymbol{\sigma} = \frac{e}{mc}\boldsymbol{S}\;;\quad \boldsymbol{S} = \hbar\,\boldsymbol{s} = \hbar\frac{\boldsymbol{\sigma}}{2} \qquad (3.9)$$

with $\boldsymbol{S}$ the angular momentum. For comparison: the orbital angular momentum reads

$$\boldsymbol{\mu}_{\mathrm{orbital}} = \frac{Q}{2M}\boldsymbol{L} = g_l\frac{Q}{2M}\boldsymbol{L}\;;\quad \boldsymbol{L} = \boldsymbol{r}\times\boldsymbol{p} = -i\hbar\,\boldsymbol{r}\times\boldsymbol{\nabla} = \hbar\boldsymbol{l}$$

and thus the total magnetic moment is

$$\boldsymbol{\mu}_{\mathrm{total}} = \frac{Q}{2M}(g_l\,\boldsymbol{L} + g_s\,\boldsymbol{S}) = \frac{m_e}{M}\mu_{\mathrm{B}}\,(g_l\,\boldsymbol{l} + g_s\,\boldsymbol{s}) \qquad (3.10)$$

where

$$\mu_{\mathrm{B}} = \frac{e\hbar}{2m_e c} \qquad (3.11)$$

is Bohr's magneton. As a result for the electron: $Q = -e$, $M = m_e$, $g_l = -1$ and $g_s = -2$. The last remarkable result is due to Dirac (1928) and tells us that the gyromagnetic ratio ($\frac{e}{mc}$) is twice as large as the one from the orbital motion.

The Foldy-Wouthuysen transformation for arbitrary A_μ cannot be performed in closed analytic form. However, the expansion in $1/c$ can be done in a systematic way (see e.g. [1]) and yields the effective Hamiltonian

$$H' = \beta \left(mc^2 + \frac{(\boldsymbol{p} - \frac{e}{c}\boldsymbol{A})^2}{2m} - \frac{\boldsymbol{p}^4}{8m^3c^2} \right) + e\Phi - \beta \frac{e\hbar}{2mc} \boldsymbol{\sigma} \cdot \boldsymbol{B}$$

$$- \frac{e\hbar^2}{8m^2c^2} \operatorname{div} \boldsymbol{E} - \frac{e\hbar}{4m^2c^2} \boldsymbol{\sigma} \cdot \left[(\boldsymbol{E} \times \boldsymbol{p} + \frac{\mathrm{i}}{2} \operatorname{rot} \boldsymbol{E}) \right]$$

$$+ O(1/c^3). \tag{3.12}$$

The additional terms are $\frac{\boldsymbol{p}^4}{8m^3c^2}$ originating from the relativistic kinematics, $\frac{e\hbar^2}{8m^2c^2} \operatorname{div} \boldsymbol{E}$ is the Darwin term as a result of the fluctuations of the electrons position and $\frac{e\hbar}{4m^2c^2} \boldsymbol{\sigma} \cdot \left[(\boldsymbol{E} \times \boldsymbol{p} + \frac{1}{2} \operatorname{rot} \boldsymbol{E}) \right]$ is the spin–orbit interaction energy. The latter plays an important role in setting up a muon storage ring in the $g - 2$ experiment (magic energy tuning). As we will see, however, in such an experiment the muons are required to be highly relativistic such that relativistic kinematics is required. The appropriate modifications will be discussed in Chap. 6.

3.2 Magnetic Moments and Electromagnetic Form Factors

3.2.1 Main Features: An Overview

Our particular interest is the motion of a lepton in an external field under consideration of the full relativistic quantum behavior. It is controlled by the QED equations of motion (3.1) with an external field added (3.2), specifically a constant magnetic field. For slowly varying fields the motion is essentially determined by the generalized Pauli equation (3.12), which also serves as a basis for understanding the role of the magnetic moment of a lepton on the classical level. As we will see, in the absence of electrical fields $\boldsymbol{E}$ the quantum correction miraculously may be subsumed in a single number the anomalous magnetic moment, which is the result of relativistic quantum fluctuations, usually simply called *radiative corrections* (RC).

To study radiative corrections we have to extend the discussion of the preceding section and consider the full QED interaction Lagrangian

$$\mathcal{L}_{\text{int}}^{\text{QED}} = -e\bar{\psi}\gamma^\mu\psi\, A_\mu \tag{3.13}$$

in case the photon field is part of the dynamics but has an external classical component A_μ^{ext}

$$A_\mu \to A_\mu + A_\mu^{\text{ext}}. \tag{3.14}$$

We are thus dealing with QED exhibiting an additional external field insertion "vertex":

$$\otimes = -\mathrm{i}e\,\gamma^\mu\,\tilde{A}_\mu^{\text{ext}}.$$

Gauge invariance (2.88) requires that a gauge transformation of the external field

3.2 Magnetic Moments and Electromagnetic Form Factors

$$A_\mu^{\text{ext}}(x) \to A_\mu^{\text{ext}}(x) - \partial_\mu \alpha(x) , \qquad (3.15)$$

for an arbitrary scalar classical field $\alpha(x)$, leaves physics invariant.

The motion of the lepton in the external field is described by a simultaneous expansion in the fine structure constant $\alpha = \frac{e^2}{4\pi}$ and in the external field A_μ^{ext} assuming the latter to be weak

In the following we will use the more customary graphic representation

of the external vertex, just as an amputated photon line at zero momentum.

The gyromagnetic ratio of the muon is defined by the ratio of the magnetic moment which couples to the magnetic field in the Hamiltonian and the spin operator in units of $\mu_0 = e\hbar/2m_\mu c$

$$\boldsymbol{\mu} = g_\mu \frac{e\hbar}{2m_\mu c} \boldsymbol{s} \; ; \quad g_\mu = 2(1 + a_\mu) \qquad (3.16)$$

and as indicated has a tree level part, the Dirac moment $g_\mu^{(0)} = 2$ [2], and a higher order part the muon anomaly or anomalous magnetic moment

$$a_\mu = \frac{1}{2}(g_\mu - 2) . \qquad (3.17)$$

In general, the anomalous magnetic moment of a lepton is related to the gyromagnetic ratio by

$$a_\ell = \mu_\ell/\mu_B - 1 = \frac{1}{2}(g_\ell - 2) \qquad (3.18)$$

where the precise value of the Bohr magneton is given by

$$\mu_B = \frac{e\hbar}{2m_e c} = 5.788381804(39) \times 10^{-11} \text{ MeVT}^{-1} . \qquad (3.19)$$

Here T as a unit stands for 1 Tesla = 10^4 Gauss. It is the unit in which the magnetic field B usually is given. In QED a_μ may be calculated in perturbation theory by considering the matrix element

$$\mathcal{M}(x;p) = \langle \mu^-(p_2, r_2) | j_{\text{em}}^\mu(x) | \mu^-(p_1, r_1) \rangle$$

of the electromagnetic current for the scattering of an incoming muon $\mu^-(p_1, r_1)$ of momentum p_1 and 3rd component of spin r_1 to a muon $\mu^-(p_2, r_2)$ of momentum p_2 and 3rd component of spin r_2, in the classical limit of zero momentum transfer $q^2 = (p_2 - p_1)^2 \to 0$. In momentum space, by virtue of space–time translational invariance $j^\mu_{\text{em}}(x) = \mathrm{e}^{\mathrm{i}Px} j^\mu_{\text{em}}(0)\mathrm{e}^{-\mathrm{i}Px}$ and the fact that the lepton states are eigenstates of four–momentum $\mathrm{e}^{-\mathrm{i}Px}|\mu^-(p_i, r_i)\rangle = \mathrm{e}^{-\mathrm{i}p_i x}|\mu^-(p_i, r_i)\rangle$ $(i = 1, 2)$, we find

$$\tilde{\mathcal{M}}(q;p) = \int \mathrm{d}^4 x\, \mathrm{e}^{-\mathrm{i}qx} \langle \mu^-(p_2, r_2)|j^\mu_{\text{em}}(x)|\mu^-(p_1, r_1)\rangle$$

$$= \int \mathrm{d}^4 x\, \mathrm{e}^{\mathrm{i}(p_2-p_1-q)x} \langle \mu^-(p_2, r_2)|j^\mu_{\text{em}}(0)|\mu^-(p_1, r_1)\rangle$$

$$= (2\pi)^4\, \delta^{(4)}(q - p_2 + p_1) \langle \mu^-(p_2, r_2)|j^\mu_{\text{em}}(0)|\mu^-(p_1, r_1)\rangle\,,$$

proportional to the δ–function of four–momentum conservation. The T–matrix element is then given by

$$\langle \mu^-(p_2)|j^\mu_{\text{em}}(0)|\mu^-(p_1)\rangle\,.$$

In QED it has a relativistically covariant decomposition of the form

$$\gamma(q)\quad \mu(p_2)\quad \mu(p_1) \quad = (-\mathrm{i}e)\, \bar{u}(p_2) \left[\gamma^\mu F_{\mathrm{E}}(q^2) + \mathrm{i}\frac{\sigma^{\mu\nu} q_\nu}{2m_\mu} F_{\mathrm{M}}(q^2)\right] u(p_1)\,, \quad (3.20)$$

where $q = p_2 - p_1$ and $u(p)$ denote the Dirac spinors. $F_{\mathrm{E}}(q^2)$ is the electric charge or Dirac form factor and $F_{\mathrm{M}}(q^2)$ is the magnetic or Pauli form factor. Note that the matrix $\sigma^{\mu\nu} = \frac{\mathrm{i}}{2}[\gamma^\mu, \gamma^\nu]$ represents the spin 1/2 angular momentum tensor. In the static (classical) limit we have (see (2.205))

$$F_{\mathrm{E}}(0) = 1\,, \quad F_{\mathrm{M}}(0) = a_\mu\,, \quad (3.21)$$

where the first relation is the *charge renormalization condition* (in units of the physical positron charge e, which by definition is taken out as a factor in (3.20)), while the second relation is the finite prediction for a_μ, in terms of the form factor F_{M} the calculation of which will be described below. The leading contribution (2.210) we have been calculating already in Sect. 2.6.3.

Note that in higher orders the form factors in general acquire an imaginary part. One may write therefore an effective dipole moment Lagrangian with complex "coupling"

$$\mathcal{L}^{\mathrm{DM}}_{\mathrm{eff}} = -\frac{1}{2}\left\{\bar\psi \sigma^{\mu\nu}\left[D_\mu \frac{1+\gamma_5}{2} + D^*_\mu \frac{1-\gamma_5}{2}\right]\psi\right\} F_{\mu\nu} \quad (3.22)$$

with ψ the muon field and

$$\mathrm{Re}\, D_\mu = a_\mu \frac{e}{2m_\mu}\,, \quad \mathrm{Im}\, D_\mu = d_\mu = \frac{\eta_\mu}{2}\frac{e}{2m_\mu}\,, \quad (3.23)$$

(see (3.78) and (3.79) below). Thus the imaginary part of $F_{\mathrm{M}}(0)$ corresponds to an electric dipole moment. The latter is non–vanishing only if we have T violation. For some more details we refer to Sect. 3.3.

As illustrated in Fig. 3.1, when polarized muons travel on a circular orbit in a constant magnetic field, then a_μ is responsible for the *Larmor precession* of the direction of the spin of the muon, characterized by the angular frequency ω_a. At the magic energy of about ~ 3.1 GeV, the latter is directly proportional to a_μ:

$$\omega_a = \frac{e}{m}\left[a_\mu \boldsymbol{B} - \left(a_\mu - \frac{1}{\gamma^2-1}\right)\boldsymbol{\beta}\times\boldsymbol{E}\right]_{\text{at "magic }\gamma\text{"}}^{E\sim 3.1\text{GeV}} \simeq \frac{e}{m}\left[a_\mu \boldsymbol{B}\right]\,. \quad (3.24)$$

Electric quadrupole fields $\boldsymbol{E}$ are needed for focusing the beam and they affect the precession frequency in general. $\gamma = E/m_\mu = 1/\sqrt{1-\beta^2}$ is the relativistic Lorentz factor with $\beta = v/c$ the velocity of the muon in units of the speed of light c. The magic energy $E_{\text{mag}} = \gamma_{\text{mag}} m_\mu$ is the energy E for which $\frac{1}{\gamma_{\text{mag}}^2-1} = a_\mu$. The existence of a solution is due to the fact that a_μ is a positive constant in competition with an energy dependent factor of opposite sign (as $\gamma \geq 1$). The second miracle, which is crucial for the feasibility of the experiment, is the fact that $\gamma_{\text{mag}} = \sqrt{(1+a_\mu)/a_\mu} \simeq 29.378$ is large enough to provide the time dilatation factor for the unstable muon boosting the life time $\tau_\mu \simeq 2.197 \times 10^{-6}$ sec to $\tau_{\text{in flight}} = \gamma \tau_\mu \simeq 6.454 \times 10^{-5}$ sec, which allows the muons, traveling at $v/c = 0.99942\cdots$, to be stored in a ring of reasonable size (diameter ~ 14 m).

This provided the basic setup for the $g-2$ experiments at the *Muon Storage Rings* at CERN and at BNL. The oscillation frequency ω_a can be measured very precisely. Also the precise tuning to the magic energy is not the major problem. The most serious challenge is to manufacture a precisely known con-

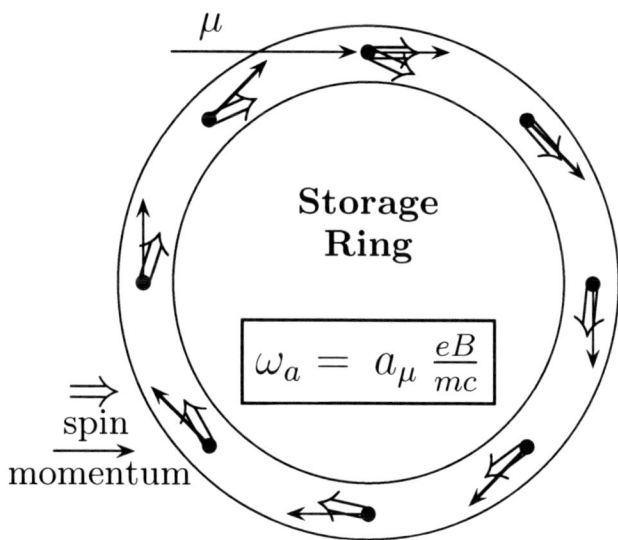

Fig. 3.1. Spin precession in the $g-2$ ring ($\sim 12°$/circle)

stant magnetic field B (magnetic flux density), as the latter directly enters the experimental extraction of a_μ (3.24). Of course one also needs high enough statistics to get sharp values for the oscillation frequency. The basic principle of the measurement of a_μ is a measurement of the "anomalous" frequency difference $\omega_a = |\boldsymbol{\omega}_a| = \omega_s - \omega_c$, where $\omega_s = g_\mu \left(e\hbar/2m_\mu\right) B/\hbar = g_\mu/2 \; e/m_\mu B$ is the muon spin–flip *precession frequency* in the applied magnetic field and $\omega_c = e/m_\mu B$ is the muon *cyclotron frequency*. Instead of eliminating the magnetic field by measuring ω_c, B is determined from proton *Nuclear Magnetic Resonance* (NMR) measurements. This procedure requires the value of μ_μ/μ_p to extract a_μ from the data. Fortunately, a high precision value for this ratio is available from the measurement of the hyperfine structure in muonium. One obtains

$$a_\mu = \frac{\bar{R}}{|\mu_\mu/\mu_p| - \bar{R}}, \qquad (3.25)$$

where $\bar{R} = \omega_a/\bar{\omega}_p$ and $\bar{\omega}_p = (e/m_\mu c)\langle B \rangle$ is the free–proton NMR frequency corresponding to the average magnetic field seen by the muons in their orbits in the storage ring. We mention that for the electron a *Penning trap* is employed to measure a_e rather than a storage ring. The B field in this case can be eliminated via a measurement of the cyclotron frequency.

On the theory side, the crucial point is that a_ℓ is dimensionless, just a number, and must vanish at tree level in any renormalizable theory. As an effective interaction it would look like

$$\delta \mathcal{L}_{\text{eff}}^{\text{AMM}} = -\frac{\delta g}{2}\frac{e}{4m}\left\{\bar{\psi}_L(x)\,\sigma^{\mu\nu}\,\psi_R(x) + \bar{\psi}_R(x)\,\sigma^{\mu\nu}\,\psi_L(x)\right\}F_{\mu\nu}(x) \qquad (3.26)$$

where ψ_L and ψ_R are Dirac fields of negative (left–handed L) and positive (right–handed R) chirality and $F_{\mu\nu} = \partial_\mu A_\nu - \partial_\nu A_\mu$ is the electromagnetic field strength tensor. This Pauli term has dimension 5 (=2×3/2 for the two Dirac fields plus 1 for the photon plus 1 for the derivative included in F) and thus would spoil renormalizability. In a renormalizable theory, however, a_μ is a finite unambiguous prediction of that theory. It is testing the rate of helicity flip transition and is one of the most precisely measured electroweak observables. Of course the theoretical prediction only may agree with the experimental result to the extend that we know the complete theory of nature, within the experimental accuracy.

Before we start discussing the theoretical prediction for the magnetic moment anomaly, we will specify the parameters which we will use for the numerical evaluations below.

Since the lowest order result for a_ℓ is proportional to α, obviously, the most important basic parameter for calculating a_μ is the fine structure constant α. It is best determined now from the very recent extraordinary precise measurement of the electron anomalous magnetic moment [3]

$$a_e^{\text{exp}} = 0.001\,159\,652\,180\,85(076) \qquad (3.27)$$

3.2 Magnetic Moments and Electromagnetic Form Factors

which, confronted with its theoretical prediction as a series in α (see Sect. 3.2.2 below) determines [4, 5]

$$\alpha^{-1}(a_e) = 137.035999069(96)[0.70\,\text{ppb}]\,. \tag{3.28}$$

This is now the by far most precise determination of α and we will use it throughout in the calculation of a_μ.

All QED contributions associated with diagrams with lepton–loops, where the "internal" lepton has mass different from the mass of the external one, depend on the corresponding mass ratio. These mass–dependent contributions differ for a_e, a_μ and a_τ, such that lepton universality is broken: $a_e \neq a_\mu \neq a_\tau$. Lepton universality is broken in any case by the difference in the masses and whatever depends on them. Such mass–ratio dependent contributions start at two loops. For the evaluation of these contributions precise values for the lepton masses are needed. We will use the following values for the muon–electron mass ratio, the muon and the tau mass [6, 7]

$$\begin{aligned} m_\mu/m_e &= 206.768\,2838\,(54)\,, \quad m_\mu/m_\tau = 0.059\,4592\,(97) \\ m_e &= 0.510\,9989\,918(44)\,\text{MeV}\,, \quad m_\mu = 105.658\,3692\,(94)\,\text{MeV} \\ m_\tau &= 1776.99\,(29)\,\text{MeV}\,. \end{aligned} \tag{3.29}$$

Note that the primary determination of the electron and muon masses come from measuring the ratio with respect to the mass of a nucleus and the masses are obtained in atomic mass units (amu). The conversion factor to MeV is more uncertain than the mass of the electron and muon in amu. The ratio of course does not suffer from the uncertainty of the conversion factor.

Other physical constants which we will need later for evaluating the weak contributions are the Fermi constant

$$G_\mu = 1.16637(1) \times 10^{-5}\ \text{GeV}^{-2}\,, \tag{3.30}$$

the weak mixing parameter (here defined by $\sin^2\Theta_W = 1 - M_W^2/M_Z^2$)

$$\sin^2\Theta_W = 0.22276(56) \tag{3.31}$$

and the masses of the intermediate gauge bosons Z and W

$$M_Z = 91.1876 \pm 0.0021\ \text{GeV}\,,\quad M_W = 80.392 \pm 0.029\ \text{GeV}\,. \tag{3.32}$$

For the not yet discovered SM Higgs particle the mass is constrained by LEP data to the range

$$114\ \text{GeV} < m_H < 200\ \text{GeV}\,(\text{at } 95\%\ \text{CL})\,. \tag{3.33}$$

We also mention here that virtual pion–pair production is an important contribution to the photon vacuum polarization and actually yields the leading hadronic contribution to the anomalous magnetic moment. For the dominating $\pi^+\pi^-$ channel, the threshold is at $2m_\pi$ with the pion mass given by

$$m_{\pi^\pm} = 139.570\,18\,(35)\,\text{MeV}\,. \tag{3.34}$$

There is also a small contribution from $\pi^0\gamma$ with threshold at m_{π}^0 which has the value

$$m_{\pi^0} = 134.976\,6\,(6)\,\text{MeV}\,. \tag{3.35}$$

For the quark masses needed in some cases we use running current quark masses in the $\overline{\text{MS}}$ scheme [6] with renormalization scale parameter μ. For the light quarks $q = u, d, s$ we give $m_q = \bar{m}_q(\mu = 2 \text{ GeV})$, for the heavier $q = c, b$ the values at the mass as a scale $m_q = \bar{m}_q(\mu = \bar{m}_q)$ and for q = t the pole mass:

$$\begin{aligned} m_u &= 3 \pm 1 \text{ MeV} & m_d &= 6 \pm 2 \text{ MeV} & m_s &= 105 \pm 25 \text{ MeV} \\ m_c &= 1.25 \pm 0.10 \text{ GeV} & m_b &= 4.25 \pm 0.15 \text{ GeV} & m_t &= 171.4 \pm 2.1 \text{ GeV}\,. \end{aligned} \tag{3.36}$$

This completes the list of the most relevant parameters and we may discuss the various contributions in turn now.

The profile of the most important contributions may be outlined as follows:

1) QED universal part:

The by far largest QED/SM contribution comes from the one–loop QED diagram [8]

$$: \quad a_e^{(2)} = a_\mu^{(2)} = a_\tau^{(2)} = \frac{\alpha}{2\pi} \qquad \text{(Schwinger 1948)}$$

which we have calculated in Sect. 2.6.3, and which is universal for all charged leptons. As it is customary we indicate the perturbative order in powers of e, i.e. $a^{(n)}$ denotes an $O(e^n)$ term, in spite of the fact that the perturbation expansion is usually represented as an expansion in $\alpha = e^2/4\pi$. Typically, analytic results for higher order terms may be expressed in terms of the Riemann zeta function

$$\zeta(n) = \sum_{k=1}^{\infty} \frac{1}{k^n} \tag{3.37}$$

and of the polylogarithmic integrals[3]

$$\text{Li}_n(x) = \frac{(-1)^{n-1}}{(n-2)!} \int_0^1 \frac{\ln^{n-2}(t)\ln(1-tx)}{t} dt = \sum_{k=1}^{\infty} \frac{x^k}{k^n}\,, \tag{3.38}$$

---

[3] The appearance of transcendental numbers like $\zeta(n)$ and higher order polylogarithms $\text{Li}_n(x)$ or so called harmonic sums is directly connected to the number of loops of a Feynman diagram. Typically, 2–loop results exhibit $\zeta(3)$ 3–loop ones $\zeta(5)$ etc of increasing transcendentality [9].

3.2 Magnetic Moments and Electromagnetic Form Factors

where $\text{Li}_2(x)$ is often referred to as the Spence function $\text{Sp}(x)$ (see (2.203) in Sect. 2.6.3 and [10] and references therein). Special $\zeta(n)$ values we will need are

$$\zeta(2) = \frac{\pi^2}{6}, \ \zeta(3) = 1.202\ 056\ 903\cdots, \ \zeta(4) = \frac{\pi^4}{90}, \ \zeta(5) = 1.036\ 927\ 755\cdots. \tag{3.39}$$

Also the constants

$$\text{Li}_n(1) = \zeta(n), \ \text{Li}_n(-1) = -[1 - 2^{1-n}]\zeta(n)$$

$$a_4 \equiv \text{Li}_4(\frac{1}{2}) = \sum_{n=1}^{\infty} 1/(2^n n^4) = 0.517\ 479\ 061\ 674\cdots, \tag{3.40}$$

related to polylogarithms, will be needed later for the evaluation of analytical results. Since a_μ is a number all QED contributions calculated in "one flavor QED", with just one species of lepton, which exhibits *one* physical mass scale only, equal to the mass of the external lepton, are universal. The following universal contributions (one flavor QED) are known:

- 2–loop diagrams [7 diagrams] with one type of fermion lines yield

$$a_\ell^{(4)} = \left[\frac{197}{144} + \frac{\pi^2}{12} - \frac{\pi^2}{2}\ln 2 + \frac{3}{4}\zeta(3)\right] \left(\frac{\alpha}{\pi}\right)^2. \tag{3.41}$$

The first calculation performed by Karplus and Kroll (1950) [11] later was recalculated and corrected by Peterman (1957) [12] and, independently, by Sommerfield (1957) [13]. An instructive compact calculation based on the dispersion theoretic approach is due to Terentev (1962) [14].

- 3–loop diagrams [72 diagrams] with common fermion lines

$$a_\ell^{(6)} = \left[\frac{28259}{5184} + \frac{17101}{810}\pi^2 - \frac{298}{9}\pi^2 \ln 2 + \frac{139}{18}\zeta(3) \right.$$

$$+ \frac{100}{3}\left\{\text{Li}_4(\frac{1}{2}) + \frac{1}{24}\ln^4 2 - \frac{1}{24}\pi^2 \ln^2 2\right\}$$

$$\left. - \frac{239}{2160}\pi^4 + \frac{83}{72}\pi^2\zeta(3) - \frac{215}{24}\zeta(5)\right] \left(\frac{\alpha}{\pi}\right)^3 \tag{3.42}$$

This is the famous analytical result of Laporta and Remiddi (1996) [15], which largely confirmed an earlier numerical result of Kinoshita [16]. For the evaluation of (3.42) one needs the constants given in (3.39) and (3.40) before.

- 4–loop diagrams [891 diagrams] with common fermion lines so far have been calculated by numerical methods mainly by Kinoshita and collaborators. The status had been summarized by Kinoshita and Marciano (1990) [17] some time ago. Since then, the result has been further improved by Kinoshita and his collaborators (2002/2005/2007) [18, 19, 5]. They find

$$- 1.9144(35) \left(\frac{\alpha}{\pi}\right)^4 ,$$

which is correcting an older result shifting the coefficient of the $\left(\frac{\alpha}{\pi}\right)^4$ term by $-$ 0.19. The missing higher order terms contribute to the theoretical uncertainty. Adopting the estimation [20], that the size of the coefficient of the missing 5–loop term should not exceed 3.8, we will account for this by a term

$$0.0(3.8) \left(\frac{\alpha}{\pi}\right)^5 ,$$

in our further discussion.

Collecting the universal terms we have

$$\begin{aligned} a_\ell^{\text{univ}} &= 0.5 \left(\frac{\alpha}{\pi}\right) - 0.328\,478\,965\,579\,193\,78\ldots \left(\frac{\alpha}{\pi}\right)^2 \\ &\quad + 1.181\,241\,456\,587\ldots \left(\frac{\alpha}{\pi}\right)^3 - 1.9144(35) \left(\frac{\alpha}{\pi}\right)^4 + 0.0(3.8) \left(\frac{\alpha}{\pi}\right)^5 \\ &= 0.001\,159\,652\,176\,42(81)(10)(26)[86] \cdots \end{aligned} \qquad (3.43)$$

for the one–flavor QED contribution. The three errors are from the error of α given in (3.28), from the numerical uncertainty of the α^4 coefficient and for the missing higher order terms, respectively.

It is interesting to note that the first term $a_\ell^{(2)} \simeq 0.00116141 \cdots$ contributes the first three significant digits. Thus the anomalous magnetic moment of a lepton is an effect of about 0.12%, $g_\ell/2 \simeq 1.00116 \cdots$, but in spite of the fact that it is so small we know a_e and a_μ more precisely than most other precision observables.

2) QED mass dependent part:

Since fermions, as demanded by the SM[4], only interact via photons or other spin one gauge bosons, mass dependent corrections only may show up at the two–loop level via photon *vacuum polarization* effects. There are two different regimes for the mass dependent effects [21, 22]:

- LIGHT internal masses give rise to potentially large logarithms of mass ratios which get singular in the limit $m_{\text{light}} \to 0$

$$a_\mu^{(4)}(\text{vap}, e) = \left[\frac{1}{3} \ln \frac{m_\mu}{m_e} - \frac{25}{36} + O\left(\frac{m_e}{m_\mu}\right)\right] \left(\frac{\alpha}{\pi}\right)^2 .$$

---

[4]Interactions are known to derive from a local gauge symmetry principle, which implies the structure of gauge couplings, which must be of vector (V) or axial–vector (A) type.

Here we have a typical result for a light field which produces a large logarithm $\ln \frac{m_\mu}{m_e} \simeq 5.3$, such that the first term ~ 2.095 is large relative to a typical constant second term -0.6944. Here[5] the exact two–loop result is

$$a_\mu^{(4)}(\text{vap}, e) \simeq 1.094\,258\,3111(84) \left(\frac{\alpha}{\pi}\right)^2 = 5.90406007(5) \times 10^{-6} \ .$$

The error is due to the uncertainty in the mass ratio (m_e/m_μ).

The kind of leading short distance log contribution just discussed, which is related to the UV behavior[6], in fact may be obtained from a renormalization group type argument. In Sect. 2.6.5 (2.228) we have shown that if we replace in the one–loop result $\alpha \to \alpha(m_\mu)$ we obtain

$$a_\mu = \frac{1}{2}\frac{\alpha}{\pi}\left(1 + \frac{2}{3}\frac{\alpha}{\pi}\ln\frac{m_\mu}{m_e}\right) \ , \tag{3.44}$$

which reproduces precisely the leading term of the two–loop result. RG type arguments, based on the related Callan-Symanzik (CS) equation approach, were further developed and refined in [23]. The CS equation is a differential equation which quantifies the response of a quantity to a change of a physical mass like m_e relative to the renormalization scale which is m_μ if we consider a_μ. For the leading m_e–dependence of a_μ, neglecting all terms which behave like powers of m_e/m_μ for $m_e \to 0$ at fixed m_μ, the CS equation takes the simple homogeneous form

$$\left(m_e\frac{\partial}{\partial m_e} + \beta(\alpha)\,\alpha\frac{\partial}{\partial \alpha}\right) a_\mu^{(\infty)}(\frac{m_\mu}{m_e}, \alpha) = 0 \ , \tag{3.45}$$

where $a_\mu^{(\infty)}$ denotes the contribution to a_μ from powers of logarithms $\ln\frac{m_\mu}{m_e}$ and constant terms and $\beta(\alpha)$ is the QED β–function. The latter governs the charge screening of the electromagnetic charge, which will be discussed below. The charge is running according to

$$\alpha(\mu) = \frac{\alpha}{1 - \frac{2}{3}\frac{\alpha}{\pi}\ln\frac{\mu}{m_e}} \tag{3.46}$$

which in linear approximation yields (3.44).

We continue with the consideration of the other contributions. For comparison we also give the result for the

- EQUAL internal masses case which yields a pure number and has been included in the $a_\ell^{(4)}$ universal part (3.41) already:

---

[5]The leading terms shown yield $5.84199477 \times 10^{-6}$.

[6]The muon mass m_μ here serves as an UV cut–off, the electron mass as an IR cut–off, and the relevant integral reads

$$\int_{m_e}^{m_\mu} \frac{\mathrm{d}E}{E} = \ln\frac{m_\mu}{m_e} \ .$$

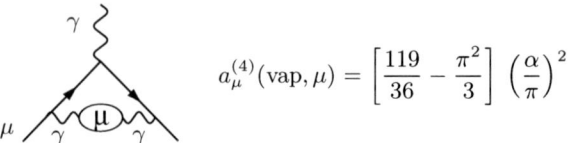

$$a_\mu^{(4)}(\text{vap}, \mu) = \left[\frac{119}{36} - \frac{\pi^2}{3}\right] \left(\frac{\alpha}{\pi}\right)^2.$$

This no scale result shows another typical aspect of perturbative answers. There is a rational term of size 3.3055... and a transcendental π^2 term of very similar size 3.2899... but of opposite sign which yields as a sum a result which is only 0.5% of the individual terms:

$$a_\mu^{(4)}(\text{vap}, \mu) \simeq 0.015\,687\,4219 \left(\frac{\alpha}{\pi}\right)^2 = 8.464\,13320 \times 10^{-8}. \quad (3.47)$$

- HEAVY internal masses decouple[7] in the limit $m_{\text{heavy}} \to \infty$ and thus only yield small power correction

$$a_\mu^{(4)}(\text{vap}, \tau) = \left[\frac{1}{45}\left(\frac{m_\mu}{m_\tau}\right)^2 + O\left(\frac{m_\mu^4}{m_\tau^4}\ln\frac{m_\tau}{m_\mu}\right)\right]\left(\frac{\alpha}{\pi}\right)^2.$$

Note that "heavy physics" contributions, from mass scales $M \gg m_\mu$, typically are proportional to m_μ^2/M^2. This means that besides the order in α there is an extra suppression factor, e.g. $O(\alpha^2) \to Q(\alpha^2 \frac{m_\mu^2}{M^2})$ in our case. To unveil new heavy states thus requires a corresponding high precision in theory and experiment. For the τ the contribution is relatively tiny[8]

$$a_\mu^{(4)}(\text{vap}, \tau) \simeq 0.000\,078\,064(25) \left(\frac{\alpha}{\pi}\right)^2 = 4.2120(13) \times 10^{-10},$$

with error from the mass ratio (m_μ/m_τ). However, at the level of accuracy reached by the Brookhaven experiment (63×10^{-11}), the contribution is non-negligible.

At the next higher order, in $a^{(6)}$ up to two internal closed fermion loops show up. The photon vacuum polarization (VP) insertions into photon lines again yield mass dependent effects if one or two of the μ loops of the universal contributions are replaced by an electron or a τ. These contributions will be discussed in more detail in Chap. 4. Here we just give the numerical results for the coefficients of $\left(\frac{\alpha}{\pi}\right)^3$ [24, 25, 26]:

---

[7]The decoupling–theorem infers that in theories like QED or QCD, where couplings and masses are independent parameters of the Lagrangian, a heavy particle of mass M decouples from physics at lower scales E_0 as E_0/M for $M \to \infty$.

[8]The leading approximation shown in the formula yields 4.2387×10^{-10}.

3.2 Magnetic Moments and Electromagnetic Form Factors

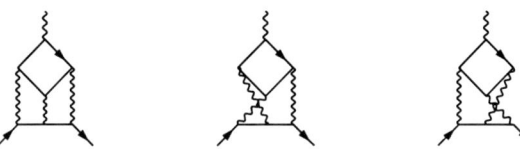

$$A_\mu^{(6)}(\text{vap},e) = 1.920\,455\,130(33)$$
$$A_\mu^{(6)}(\text{vap},\tau) = -0.001\,782\,33(48)$$
$$A_\mu^{(6)}(\text{vap},e,\tau) = 0.000\,527\,66(17).$$

Besides these photon self–energy corrections, a new kind of contributions are the so called *light–by–light scattering* (LbL) insertions: closed fermion loops with four photons attached. Light–by–light scattering $\gamma\gamma \to \gamma\gamma$ is a fermion–loop induced process between real on–shell photons. There are 6 diagrams which follow from the first one below, by permutation of the photon vertices on the external muon line:

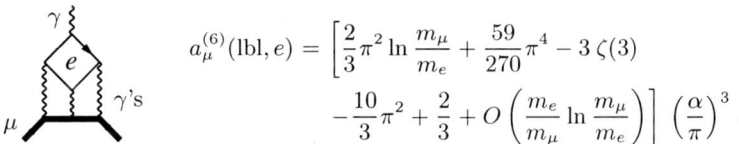

plus the one's obtained by reversing the direction of the fermion loop. Remember that closed fermion loops with three photons vanish by Furry's theorem. Again, besides the equal mass case $m_{\text{loop}} = m_\mu$ there are two different regimes [27, 28]:

- LIGHT internal masses also in this case give rise to potentially large logarithms of mass ratios which get singular in the limit $m_{\text{light}} \to 0$

$$a_\mu^{(6)}(\text{lbl},e) = \left[\frac{2}{3}\pi^2 \ln \frac{m_\mu}{m_e} + \frac{59}{270}\pi^4 - 3\,\zeta(3) \right.$$
$$\left. - \frac{10}{3}\pi^2 + \frac{2}{3} + O\left(\frac{m_e}{m_\mu} \ln \frac{m_\mu}{m_e}\right)\right]\left(\frac{\alpha}{\pi}\right)^3.$$

This again is a light loop which yields an unexpectedly large contribution

$$a_\mu^{(6)}(\text{lbl},e) \simeq 20.947\,924\,89(16)\left(\frac{\alpha}{\pi}\right)^3 = 2.625\,351\,02(2) \times 10^{-7},$$

with error from the (m_e/m_μ) mass ratio. Historically, it was calculated first numerically by Aldins et al. [29], after a 1.7σ discrepancy with the CERN measurement [30] in 1968 showed up[9].

For comparison we also present the

- EQUAL internal masses case which yields a pure number which is included in the $a_\ell^{(6)}$ universal part (3.42) already:

where a_4 is the constant defined in (3.40). The single scale QED contribution is much smaller

---

[9] The result of [29] was $2.30 \pm 0.14 \times 10^{-7}$ pretty close to the "exact" answer above. The occurrence of such large terms of course has a physical interpretation [31].

$$a_\mu^{(6)}(\text{lbl},\mu) = \left[\frac{5}{6}\,\zeta(5) - \frac{5}{18}\,\pi^2\,\zeta(3) - \frac{41}{540}\pi^4 - \frac{2}{3}\pi^2\ln^2 2 \right.$$
$$\left. + \frac{2}{3}\ln^4 2 + 16 a_4 - \frac{4}{3}\,\zeta(3) - 24\pi^2\ln 2 + \frac{931}{54}\pi^2 + \frac{5}{9}\right]\left(\frac{\alpha}{\pi}\right)^3 ,$$

$$a_\mu^{(6)}(\text{lbl},\mu) \simeq 0.371005293 \left(\frac{\alpha}{\pi}\right)^3 = 4.64971652 \times 10^{-9} \qquad (3.48)$$

but is still a substantial contributions at the required level of accuracy.
- HEAVY internal masses again decouple in the limit $m_{\text{heavy}} \to \infty$ and thus only yield small power correction

$$a_\mu^{(6)}(\text{lbl},\tau) = \left[\left[\frac{3}{2}\,\zeta(3) - \frac{19}{16}\right]\left(\frac{m_\mu}{m_\tau}\right)^2 \right.$$
$$\left. + O\left(\frac{m_\mu^4}{m_\tau^4}\ln^2\frac{m_\tau}{m_\mu}\right)\right]\left(\frac{\alpha}{\pi}\right)^3 .$$

As expected this heavy contribution is power suppressed yielding

$$a_\mu^{(6)}(\text{lbl},\tau) \simeq 0.00214283(69) \left(\frac{\alpha}{\pi}\right)^3 = 2.68556(86) \times 10^{-11} ,$$

and therefore would play a significant role at a next level of precision experiments only. Again the error is from the (m_μ/m_τ) mass ratio.

We mention that except for the mixed term $A_\mu^{(6)}(\text{vap},e,\tau)$, which has been worked out as a series expansion in the mass ratios [25, 26], all contributions are known analytically in exact form [24, 27][10] up to 3 loops. At 4 loops only a few terms are known analytically [33]. Again the relevant 4-loop contributions have been evaluated by numerical integration methods by Kinoshita and Nio [18]. The 5-loop term has been estimated to be $A_2^{(10)}(m_\mu/m_e) = 663(20)$ in [34, 35, 36].

Combining the universal and the mass dependent terms discussed so far wa arrive at the following QED result for a_μ

Firstly, the large logs $\ln(m_\mu/m_e)$ are due to a logarithmic UV divergence in the limit $m_\mu \to \infty$, i.e. m_μ serves as an UV cut-off, in conjunction with an IR singularity in the limit $m_e \to 0$, i.e. m_e serves as an IR cut-off: $\int_{m_e}^{m_\mu} \frac{dE}{E} = \ln\frac{m_\mu}{m_e}$. The integral is large because of the large range $[m_e, m_\mu]$ and an integrand with the property that it is contributing equally at all scales. Secondly, and this is the new point here, there is an unusual $\pi^2 \sim 10$ factor in the coefficient of the large log. It becomes possible in conjunction with the LbL scattering subdiagram where the electron is moving in the field of an almost static non-relativistic muon. A non-relativistic spin-flip interaction (required to contribute to a_μ) gets dressed by Coulomb interactions between muon and electron, which produces the large π^2 factor.

[10] Explicitly, the papers only present expansions in the mass ratios; some result have been extended in [28] and cross checked against the full analytic result in [32].

3.2 Magnetic Moments and Electromagnetic Form Factors

$$a_\mu^{\mathrm{QED}} = \frac{\alpha}{2\pi} + 0.765\,857\,410(26) \left(\frac{\alpha}{\pi}\right)^2$$
$$+ 24.050\,509\,65(46) \left(\frac{\alpha}{\pi}\right)^3 + 130.8105(85) \left(\frac{\alpha}{\pi}\right)^4 + 663(20) \left(\frac{\alpha}{\pi}\right)^5. \quad (3.49)$$

Growing coefficients in the α/π expansion reflect the presence of large $\ln \frac{m_\mu}{m_e} \simeq 5.3$ terms coming from electron loops. In spite of the strongly growing expansion coefficients the convergence of the perturbation series is excellent

# n of loops	$C_i\,[(\alpha/\pi)^n]$	$a_\mu^{\mathrm{QED}} \times 10^{11}$
1	+0.5	116140973.30 (0.08)
2	+0.765 857 410(26)	413217.62 (0.01)
3	+24.050 509 65(46)	30141.90 (0.00)
4	+130.8105(85)	380.81 (0.03)
5	+663.0(20.0)	4.48 (0.14)
tot		116584718.11 (0.16)

because α/π is a truly small expansion parameter.

Now we have to address the question what happens beyond QED. What is measured in an experiment includes effects from the real world and we have to include the contributions from all known particles and interactions such that from a possible deviation between theory and experiment we may get a hint of the yet unknown physics.

Going from QED of leptons to the SM the most important step is to include the hadronic effects mediated by the quarks, which in the SM sit in families together with the leptons and neutrinos. The latter being electrically neutral do not play any role, in contrast the charged quarks. The *strong interaction effects* are showing up in particular through the hadronic structure of the photon via vacuum polarization starting at $O(\alpha^2)$ or light–by–light scattering starting at $O(\alpha^3)$.

3) Hadronic VP effects:

Formally, these are the contributions obtained by replacing lepton–loops by quark–loops (see Fig. 3.2), however, the quarks are strongly interacting via gluons as described by the $SU(3)_{\mathrm{color}}$ gauge theory QCD [37] (see Sect. 2.8). While electromagnetic and weak interactions are weak in the sense that they allow us to perform perturbation expansions in the coupling constants, strong interactions are weak only at high energies as inferred by the property of asymptotic freedom (anti–screening)[11]. At energies above about 2 GeV

---

[11] Asymptotic freedom, discovered in 1973 by Politzer, Gross and Wilczek [38] (Nobel Prize 2004), is one of the key properties of QCD and explains why at high enough energies one observes quasi–free quarks. Thus, while quarks remain imprisoned inside color neutral hadrons (quark confinement), at high enough energies (so called hard subprocesses) the *quark parton model* (QPM) of free quarks may be a reasonable approximation, which may be systematically improved by including the perturbative corrections.

154   3 Lepton Magnetic Moments: Basics

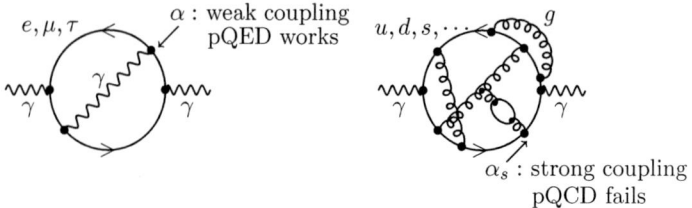

Fig. 3.2. The hadronic analog of the lepton loops

perturbative QCD (pQCD) may be applied as well. In the regime of interest to us here, however, perturbative QCD fails. The strength of the strong coupling "constant" increases dramatically as we approach lower energies. This is firmly illustrated by Fig. 3.3, which shows a compilation of measured strong coupling constants as a function of energy in comparison to perturbative QCD. The latter seems to describes very well the running of α_s down to 2 GeV.

Fortunately the leading hadronic effects are vacuum polarization type corrections, which can be safely evaluated by exploiting causality (analyticity) and unitarity (optical theorem) together with experimental low energy data. The imaginary part of the photon self–energy function $\Pi'_\gamma(s)$ (see Sect. 2.6.1) is determined via the optical theorem by the total cross–section of hadron production in electron–positron annihilation:

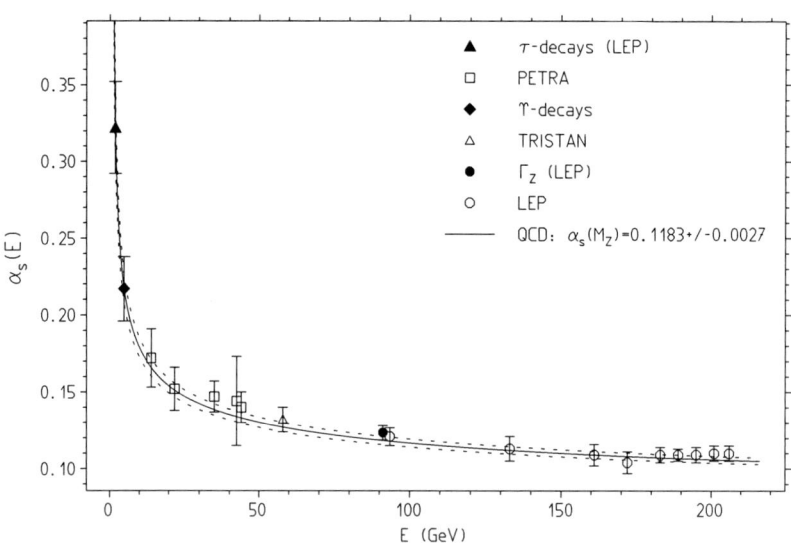

Fig. 3.3. A compilation of α_s measurements from [39]. The lowest point shown is at the τ lepton mass $M_\tau = 1.78$ GeV where $\alpha_s(M_\tau) = 0.322 \pm 0.030$

3.2 Magnetic Moments and Electromagnetic Form Factors

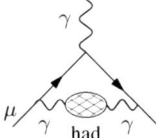

Fig. 3.4. The leading order (LO) hadronic vacuum polarization diagram

$$\sigma(s)_{e^+e^- \to \gamma^* \to \text{hadrons}} = \frac{4\pi^2 \alpha}{s} \frac{1}{\pi} \operatorname{Im} \Pi_\gamma^{\prime\text{had}}(s) \,. \tag{3.50}$$

The leading hadronic contribution is represented by the diagram Fig. 3.4, which has a representation as a dispersion integral

$$a_\mu = \frac{\alpha}{\pi} \int_0^\infty \frac{ds}{s} \frac{1}{\pi} \operatorname{Im} \Pi_\gamma^{\prime\text{had}}(s) K(s) \,, \quad K(s) \equiv \int_0^1 dx \frac{x^2(1-x)}{x^2 + \frac{s}{m_\mu^2}(1-x)} \,. \tag{3.51}$$

As a result the leading non–perturbative hadronic contributions a_μ^{had} can be obtained in terms of $R_\gamma(s) \equiv \sigma^{(0)}(e^+e^- \to \gamma^* \to \text{hadrons})/\frac{4\pi\alpha^2}{3s}$ data via the dispersion integral:

$$a_\mu^{\text{had}} = \left(\frac{\alpha m_\mu}{3\pi}\right)^2 \left(\int_{m_{\pi^0}^2}^{E_{\text{cut}}^2} ds \, \frac{R_\gamma^{\text{data}}(s) \hat{K}(s)}{s^2} + \int_{E_{\text{cut}}^2}^\infty ds \, \frac{R_\gamma^{\text{pQCD}}(s) \hat{K}(s)}{s^2} \right), \tag{3.52}$$

where the rescaled kernel function $\hat{K}(s) = 3s/m_\mu^2 \, K(s)$ is a smooth bounded function, increasing from 0.63... at $s = 4m_\pi^2$ to 1 as $s \to \infty$. The $1/s^2$ enhancement at low energy implies that the $\rho \to \pi^+\pi^-$ resonance is dominating the dispersion integral ($\sim 75\,\%$). Data can be used up to energies where $\gamma - Z$ mixing comes into play at about $E_{\text{cut}} = 40$ GeV. However, by the virtue of asymptotic freedom, perturbative Quantum Chromodynamics (see p. 129) (pQCD) becomes the more reliable the higher the energy and, in fact, it may be used safely in regions away from the flavor thresholds, where resonances show up: ρ, ω, ϕ, the J/ψ series and the Υ series. We thus use perturbative QCD [40, 41] from 5.2 to 9.6 GeV and for the high energy tail above 13 GeV, as recommended in [40, 41, 42].

Hadronic cross section measurements $e^+e^- \to$ hadrons at electron–positron storage rings started in the early 1960's and continued up to date. Since our analysis [43] in 1995 data from MD1 [44], BES-II [45] and from CMD-2 [46] have lead to a substantial reduction in the hadronic uncertainties on a_μ^{had}. More recently, KLOE [47], SND [48] and CMD-2 [49] published new measurements in the region below 1.4 GeV. My up–to–date evaluation of the leading order hadronic VP yields [50]

Table 3.1. Contributions and errors from different energy ranges

Energy range	$a_\mu^{\text{had}}[\%](\text{error}) \times 10^{10}$	rel. err.	abs. err.
ρ, ω ($E < 2M_K$)	538.33 [77.8](3.65)	0.7 %	42.0 %
$2M_K < E < 2$ GeV	102.31 [14.8](4.07)	4.0 %	52.1 %
2 GeV $< E < M_{J/\psi}$	22.13 [3.2](1.23)	5.6 %	4.8 %
$M_{J/\psi} < E < M_\Upsilon$	26.40 [3.8](0.59)	2.2 %	1.1 %
$M_\Upsilon < E < E_{\text{cut}}$	1.40 [0.2](0.09)	6.2 %	0.0 %
$E_{\text{cut}} < E$ pQCD	1.53 [0.2](0.00)	0.1 %	0.0 %
$E < E_{\text{cut}}$ data	690.57 [99.8](5.64)	0.8 %	100.0 %
total	692.10 [100.0](5.64)	0.8 %	100.0 %

$$a_\mu^{\text{had}(1)} = (692.1 \pm 5.6) \times 10^{-10} \, . \tag{3.53}$$

Table 3.1 gives more details about the origin of contributions and errors from different regions. Some other recent evaluations are collected in Table 3.2. Differences in errors come about mainly by utilizing more "theory–driven" concepts : use of selected data sets only, extended use of perturbative QCD in place of data [assuming local duality], sum rule methods, low energy effective methods [59]. Only the last three (**) results include the most recent data from SND, CMD-2, and BaBar[12].

In principle, the $I = 1$ iso–vector part of $e^+e^- \to$ hadrons can be obtained in an alternative way by using the precise vector spectral functions from hadronic τ–decays $\tau \to \nu_\tau +$ hadrons which are related by an isospin

Table 3.2. Some recent evaluations of $a_\mu^{\text{had}(1)}$

$a_\mu^{\text{had}(1)} \times 10^{10}$	data	Ref.		Reference
696.3[7.2]	e^+e^-	[51]		Davier et al. (03) (e^+e^-)
711.0[5.8]	$e^+e^- + \tau$	[51]		Davier et al. (03) (e^+e^-, τ)
694.8[8.6]	e^+e^-	[52]		Ghozzi, Jegerlehner (03) (e^+e^-)
684.6[6.4]	e^+e^- TH	[53]		Narison (03) (e^+e^-)
699.6[8.9]	e^+e^-	[54]		Ezhela et al. (03) (e^+e^-)
692.4[6.4]	e^+e^-	[55]		Hagiwara et al. (03) (e^+e^-)(incl)
693.5[5.9]	e^+e^-	[56]		Troconiz, Yndurain (04) (e^+e^-)
701.8[5.8]	$e^+e^- + \tau$	[56]		Troconiz, Yndurain (04) (e^+e^-, τ)
690.9[4.4]	e^+e^- **	[57]		Davier et al. (06) (e^+e^-)
689.4[4.6]	e^+e^- **	[58]		Hagiwara et al. (06) (e^+e^-)(incl)
692.1[5.6]	e^+e^- **	[50]		Jegerlehner (06) (e^+e^-)

[12]The analysis [58] does not include exclusive data in a range from 1.43 to 2 GeV; therefore also the new BaBar data are not included in that range. In [51, 57] pQCD is used in the extended ranges 1.8–3.7 GeV and above 5.0 GeV and in [57] KLOE data are not included. See also the comments to Fig. 7.1.

rotation [60]. After isospin violating corrections, due to photon radiation and the mass splitting $m_d - m_u \neq 0$, have been applied, there remains an unexpectedly large discrepancy between the e^+e^-- and the τ-based determinations of a_μ [51], as may be seen in Table 3.2. Possible explanations are so far unaccounted isospin breaking [52] or experimental problems with the data. Since the e^+e^--data are more directly related to what is required in the dispersion integral, one usually advocates to use the e^+e^- data only.

At order $O(\alpha^3)$ diagrams of the type shown in Fig. 3.5 have to be calculated, where the first diagram stands for a class of higher order hadronic contributions obtained if one replaces in any of the first 6 two–loop diagrams of Fig. 4.2, below, one internal photon line by a dressed one. The relevant kernels for the corresponding dispersion integrals have been calculated analytically in [61] and appropriate series expansions were given in [62] (for earlier estimates see [63, 64]). Based on my recent compilation of the e^+e^- data [50] I obtain

$$a_\mu^{(6)}(\text{vap}, \text{had}) = -100.3(2.2) \times 10^{-11},$$

in accord with previous evaluations [55, 60, 62, 64]. The errors include statistical and systematic errors added in quadrature.

4) Hadronic LbL effects:

A much more problematic set of hadronic corrections are those related to hadronic light–by–light scattering, which sets in only at order $O(\alpha^3)$, fortunately. However, we already know from the leptonic counterpart that this contribution could be dramatically enhanced. It was estimated for the first time in [63]. Even for real–photon light–by–light scattering, perturbation theory is far from being able to describe reality, as the reader may convince himself by a glance at Fig. 3.6, showing sharp spikes of π^0, η and η' production, while pQCD predicts a smooth continuum[13]. As a contribution to the anomalous magnetic moment three of the four photons are virtual and to be integrated over all four–momentum space, such that a direct experimental input for the non–perturbative dressed four–photon cor-

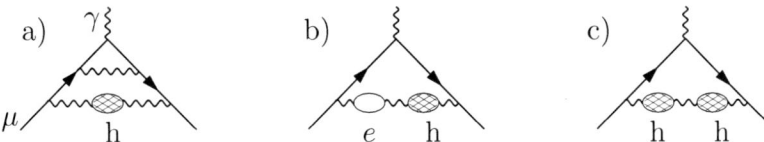

Fig. 3.5. Higher order (HO) vacuum polarization contributions

---

[13] The pion which gives the by far largest contribution is a quasi Goldstone boson. In the chiral limit of vanishing light quark masses $m_u = m_d = m_s = 0$ pions and kaons are true Goldstone bosons which exist due to the spontaneous breakdown of the chiral $U(N_f)_V \otimes U_A(N_f)$ ($N_f = 3$) symmetry, which is a non–perturbative phenomenon, absent in pQCD.

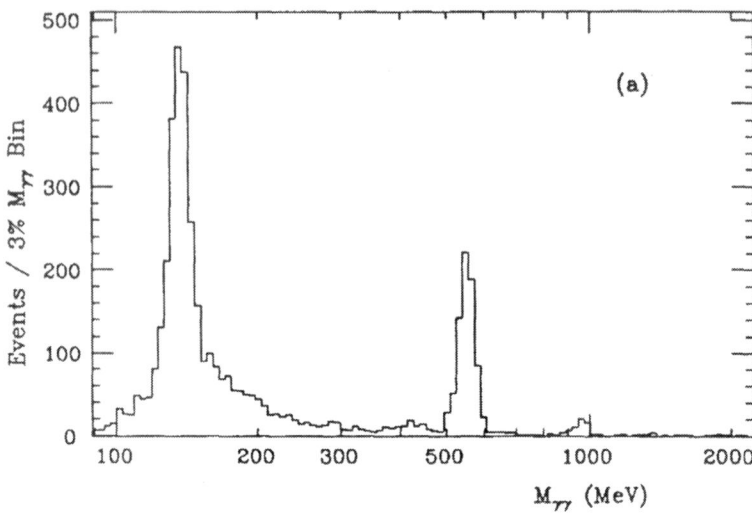

Fig. 3.6. The spectrum of invariant $\gamma\gamma$ masses obtained with the Crystal Ball detector [65]. The three rather pronounced spikes seen are the $\gamma\gamma \to$ pseudoscalar (PS) $\to \gamma\gamma$ excitations: PS $= \pi^0, \eta, \eta'$

relator is not available. In this case one has to resort to the low energy effective descriptions of QCD like *chiral perturbation theory* (CHPT) extended to include vector–mesons. This Resonance Lagrangian Approach (RLA) is realizing vector–meson dominance model (VMD) ideas in accord with the low energy structure of QCD [66]. Other effective theories are the extended Nambu-Jona-Lasinio (ENJL) model [67] (see also [70]) or the very similar hidden local symmetry (HLS) model [71, 72]; approaches more or less accepted as a framework for the evaluation of the hadronic LbL effects. The amazing fact is that the interactions involved in the hadronic LbL scattering process are the parity conserving QED and QCD interactions while the process is dominated by the parity odd pseudoscalar meson–exchanges. This means that the effective $\pi^0\gamma\gamma$ interaction vertex exhibits the parity violating γ_5 coupling, which of course in $\gamma\gamma \to \pi^0 \to \gamma\gamma$ must appear twice (an even number of times). The process indeed is associated with the parity odd Wess-Zumino-Witten (WZW) effective interaction term

$$\mathcal{L}^{(4)} = -\frac{\alpha N_c}{12\pi F_0}\varepsilon_{\mu\nu\rho\sigma}F^{\mu\nu}A^\rho\partial^\sigma\pi^0 + \cdots \qquad (3.54)$$

which reproduces the Adler-Bell-Jackiw anomaly and which plays a key role in estimating the leading hadronic LbL contribution. F_0 denotes the pion decay constant F_π in the chiral limit of massless light quarks ($F_\pi \simeq 92.4$ MeV). The constant WZW form factor yields a divergent result, applying a cut–off Λ one obtains the leading term

$$a_\mu^{(6)}(\text{lbl},\pi^0) = \left[\frac{N_c^2}{48\pi^2}\frac{m_\mu^2}{F_\pi^2}\ln^2\frac{\Lambda}{m} + \cdots\right]\left(\frac{\alpha}{\pi}\right)^3,$$

with a universal coefficient $\mathcal{C} = N_c^2 m_\mu^2/(48\pi^2 F_\pi^2)$ [73]; in the VMD dressed cases M_V represents the cut–off $\Lambda \to M_V$ if $M_V \to \infty$[14].

Based on refined effective field theory (EFT) models, two major efforts in evaluating the full a_μ^{LbL} contribution were made by Hayakawa, Kinoshita and Sanda (HKS 1995) [71], Bijnens, Pallante and Prades (BPP 1995) [67] and Hayakawa and Kinoshita (HK 1998) [72] (see also Kinoshita, Nizic and Okamoto (KNO 1985) [64]). Although the details of the calculations are quite different, which results in a different splitting of various contributions, the results are in good agreement and essentially given by the π^0-pole contribution, which was taken with the wrong sign, however. In order to eliminate the cut–off dependence in separating L.D. and S.D. physics, more recently it became favorable to use quark–hadron duality, as it holds in the large N_c limit of QCD [68, 69], for modeling of the hadronic amplitudes [70]. The infinite series of narrow vector states known to show up in the large N_c limit is then approximated by a suitable lowest meson dominance (LMD+V) ansatz [74], assumed to be saturated by known low lying physical states of appropriate quantum numbers. This approach was adopted in a reanalysis by Knecht and Nyffeler (KN 2001) [73, 75, 76, 77] in 2001, in which they discovered a sign mistake in the dominant π^0, η, η' exchange contribution, which changed the central value by $+167 \times 10^{-11}$, a 2.8σ shift, and which reduces a larger discrepancy between theory and experiment. More recently Melnikov and Vainshtein (MV 2004) [78] found additional problems in previous calculations, this time in the short distance constraints (QCD/OPE) used in matching the high energy behavior of the effective models used for the π^0, η, η' exchange contribution. We will elaborate on this in much more detail in Sect. 5.4 below.

We advocate to use consistently dressed form factors as inferred from the resonance Lagrangian approach. However, other effects which were first considered in [78] must be taken into account: **i)** the constraint on the twist four $(1/q^4)$-term in the OPE requires $h_2 = -10$ GeV^2 in the Knecht-Nyffeler form factor [75]: $\delta a_\mu \simeq +5 \pm 0$ relative to $h_2 = 0$, **ii)** the contributions from the f_1 and f_1' isoscalar axial-vector mesons: $\delta a_\mu \simeq +10 \pm 4$ (using dressed photons), **iii)** for the remaining effects, scalars (f_0) + dressed $\pi^\pm, K^\pm$ loops + dressed quark loops: $\delta a_\mu \simeq -5 \pm 13$. Note that this last group of terms have been evaluated in [67, 71] only. The splitting into the different terms is model dependent and only the sum should be considered; the results read -5 ± 13

---

[14] Since the leading term is divergent and requires UV subtraction, we expect this term to drop from the physical result, unless a physical cut–off tames the integral, like the physical ρ in effective theories which implement the VMD mechanism.

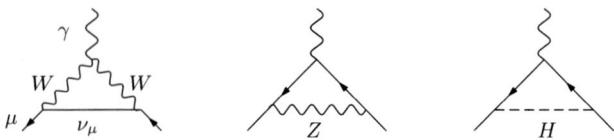

Fig. 3.7. The leading weak contributions to a_ℓ; diagrams in the physical unitary gauge

(BPP) and 5.2 ± 13.7 and (HKS) hence contribution the remains unclear[15]. As an estimate based on [71, 67, 75, 79, 78] we adopt π^0, η, η' [88 ± 12] + axial–vector [10 ± 4] + scalar [-7 ± 2] + π, K loops [-19 ± 13] + quark loops [21 ± 3] which yields

$$a_\mu^{(6)}(\text{lbl, had}) = (93 \pm 34) \times 10^{-11} \ .$$

5) Weak interaction corrections:
The last set of corrections are due to the weak interaction as described by the electroweak SM. The weak corrections are those mediated by the weak currents which couple to the heavy spin 1 gauge bosons, the charged $W^\pm$ or the neutral "heavy light" particle Z or by exchange of a Higgs particle H (see Fig. 3.7; masses are given in (3.32, 3.33)). What is most interesting is the occurrence of the first diagram of Fig. 3.7, which exhibits a non–Abelian triple gauge vertex and the corresponding contribution provides a test of the Yang-Mills structure involved. It is of course not surprising that the photon couples to the charged W boson the way it is dictated by electromagnetic gauge invariance. In spite of the fact that the contribution is of leading one-loop order, it is vastly suppressed by the fact that the corrections are mediated by the exchange of very heavy states which makes them suppressed by $O(2\frac{\alpha}{\pi}\frac{m_\mu^2}{M^2}) \sim 5 \times 10^{-9}$ for M of about 100 GeV. The gauge boson contributions up to negligible terms of order $O(\frac{m_\mu^2}{M_{W,Z}^2})$ are given by (the Higgs contribution is negligible) [80]

$$a_\mu^{(2)\,\text{EW}} = [5 + (-1 + 4\sin^2\Theta_W)^2]\,\frac{\sqrt{2}G_\mu m_\mu^2}{48\pi^2} \simeq (194.82 \pm 0.02) \times 10^{-11}$$

(3.55)

The error comes from the uncertainty in $\sin^2\Theta_W$ [see (3.31)].

Electroweak two–loop calculations started 1992 with Kukhto et al. [81], who observed potentially large terms proportional to $\sim G_F m_\mu^2 \frac{\alpha}{\pi} \ln \frac{M_Z}{m_\mu}$ enhanced by a large logarithm. The most important diagrams are triangle fermion–loops:
where T_{3f} is the 3rd component of the weak isospin, Q_f the charge and N_{cf} the color factor, 1 for leptons, 3 for quarks. The mass $m_{f'}$ is m_μ if $m_f < m_\mu$

---

[15] We adopt the result of [67] as the sign must be negative in any case (see [76])

$$a_\mu^{(4)\,\text{EW}}([f]) \simeq \frac{\sqrt{2}G_\mu m_\mu^2}{16\pi^2}\frac{\alpha}{\pi}\,2T_{3f}N_{cf}Q_f^2\left[3\ln\frac{M_Z^2}{m_{f'}^2}+C_f\right]$$

and m_f if $m_f > m_\mu$, and $C_e = 5/2$, $C_\mu = 11/6 - 8/9\,\pi^2$, $C_\tau = -6$ [81]. Note that triangle fermion–loops cannot contribute in QED due to Furry's theorem. However, the weak interactions are parity violating and if one of the three vector vertices $V^\mu = \gamma^\mu$ is replaced by an axial vertex $A^\mu = \gamma^\mu\gamma_5$ one gets a non–vanishing contribution. This is what happens if we replace one of the photons by a "heavy light" particle Z. However, these diagrams are responsible for the Adler-Bell-Jackiw anomaly [82] which is leading to a violation of axial current conservation and would spoil renormalizability. The anomalous terms must cancel and in the SM this happens by lepton quark duality: leptons and quarks have to live in families and for each family $\sum_f N_{cf}Q_f^2 T_{3f} = 0$, which is the anomaly cancellation condition in the $SU(3)_c \otimes SU(2)_L \otimes U(1)_Y$ gauge theory. This is again one of the amazing facts, that at the present level of precision one starts to be sensitive to the anomaly cancellation mechanism. This anomaly cancellation leads to substantial cancellations between the individual fermion contributions. The original results therefore get rectified by taking into account the family structure of SM fermions [83, 84, 85].

For more sophisticated analyses we refer to [83, 84, 86] which was corrected and refined in [85, 87]. Including subleading effects yields -5.0×10^{-11} for the first two families. The 3rd family of fermions including the heavy top quark can be treated in perturbation theory and was worked out to be -8.2×10^{-11} in [88]. Subleading fermion loops contribute -5.3×10^{-11}. There are many more diagrams contributing, in particular the calculation of the bosonic contributions (1678 diagrams) is a formidable task and has been performed 1996 by Czarnecki, Krause and Marciano as an expansion in $(m_\mu/M_V)^2$ and $(M_V/m_H)^2$ [89]. Later complete calculations, valid also for lighter Higgs masses, were performed [90, 91], which confirmed the previous result -22.3×10^{-11}. The 2–loop result reads[16]

$$a_\mu^{(4)\,\text{EW}} = -41(3) \times 10^{-11}\,.$$

The complete weak contribution may be summarized by [87]

$$a_\mu^{\text{EW}} = \frac{\sqrt{2}G_\mu\,m_\mu^2}{16\pi^2}\left\{\frac{5}{3} + \frac{1}{3}\left(1 - 4\sin^2\Theta_W\right)^2 - \frac{\alpha}{\pi}[155.5(4)(2)]\right\}$$

---

[16]The authors of [81] reported

$$a_\mu^{(4)\,\text{EW}} = -42 \times 10^{-11}$$

for what they thought was the leading correction, which is very close to the complete weak two–loop corrections, however, this coincidence looks to be a mere accident. Nevertheless, the sign and the order of magnitude turned out to be correct.

$$= (154 \pm 1[\text{had}] \pm 2[m_H, m_t, 3-\text{loop}]) \times 10^{-11} \quad (3.56)$$

with errors from triangle quark–loops and from variation of the Higgs mass in the range $m_H = 150^{+100}_{-40}$ GeV. The 3–loop effect has been estimated to be negligible [85, 87].

This closes our overview of the various contributions to the anomalous magnetic moment of the muon. More details about the higher order QED corrections as well as the weak and strong interaction corrections will be discussed in detail in the next Chap. 4. First we give a brief account of the status of the theory in comparison to the experiments. We will consider the electron and the muon in turn.

3.2.2 The Anomalous Magnetic Moment of the Electron

The electron magnetic moment anomaly likely is the experimentally most precisely known quantity. For almost 20 years the value was based on the extraordinary precise measurements of electron and positron anomalous magnetic moments

$$\begin{aligned} a_{e^-}^{\text{exp}} &= 0.001\,159\,652\,188\,4(43), \\ a_{e^+}^{\text{exp}} &= 0.001\,159\,652\,187\,9(43), \end{aligned} \quad (3.57)$$

by Van Dyck et al. (1987) [92]. The experiment used the ion trap technique, which has made it possible to study a single electron with extreme precision[17]. The result impressively confirms the conservation of CPT: $a_{e^+} = a_{e^-}$. Being a basic prediction of any QFT, CPT symmetry will be assumed to hold in the following. This allows us to average the electron and positron values with the result [7]

$$a_e = \mu_e/\mu_B - 1 = (g_e - 2)/2 = 1.159\,652\,1883(42) \times 10^{-3}\,. \quad (3.58)$$

The relative standard uncertainty is 3.62 ppb. Since recently, a new substantially improved result for a_e is available. It was obtained by Gabrielse et al. [3] in an experiment at Harvard University using a one–electron quantum cyclotron. The new result is

$$a_e = 1.159\,652\,180\,85(76)[.66\text{ppb}] \times 10^{-3}\,, \quad (3.59)$$

with an accuracy nearly 6 times better than (3.58) and shifting up a_e by 1.7 standard deviations.

---

[17] The ion trap technique was introduced and developed by Paul and Dehmelt, whom was awarded the Nobel Prize in 1989. The ion traps utilize electrical quadrupole fields obtained with hyperboloid shaped electrodes. The Paul trap works with dynamical trapping using r.f. voltage, the Penning trap used by Dehmelt works with d.c. voltage and a magnetic field in z-direction.

The measurements of a_e not only played a key role in the history of precision tests of QED in particular, and of QFT concepts in general, today we may use the anomalous magnetic moment of the electron to get the most precise indirect measurement of the fine structure constant α. This possibility of course hangs on our ability to pin down the theoretical prediction with very high accuracy. Indeed a_e is much saver to predict reliably than a_μ. The reason is that non–perturbative hadronic effects as well as the sensitivity to unknown physics beyond the SM are suppressed by the large factor $m_\mu^2/m_e^2 \simeq 42\,753$ in comparison to a_μ. This suppression has to be put into perspective with the 829 times higher precision with which we know a_e. We thus can say that effectively a_e is a factor 52 less sensitive to model dependent physics than a_μ.

The reason why it is so interesting to have such a precise measurement of a_e of course is that it can be calculated with comparable accuracy in theory. The prediction is given by a perturbation expansion of the form

$$a_e = \sum_{n=1}^{N} A^{(2n)} (\alpha/\pi)^n \;, \tag{3.60}$$

with terms up to five loops, $N = 5$, under consideration. The experimental precision of a_e requires the knowledge of the coefficients with accuracies $\delta A^{(4)} \sim 1 \times 10^{-7}$, $\delta A^{(6)} \sim 6 \times 10^{-5}$, $\delta A^{(8)} \sim 2 \times 10^{-2}$ and $\delta A^{(10)} \sim 10$. For what concerns the universal terms one may conclude by inspecting the convergence of (3.43) that one would expect the completely unknown coefficient $A^{(10)}$ to be $O(1)$ and hence negligible at present accuracy. In reality it is one of the main uncertainties, which is already accounted for in (3.43). Concerning the mass–dependent contributions, the situation for the electron is quite different from the muon. Since the electron is the lightest of the leptons a potentially large "light internal loop" contribution is absent. For a_e the muon is a heavy particle $m_\mu \gg m_e$ and its contribution is of the type "heavy internal loops" which is suppressed by an extra power of m_e^2/m_μ^2. In fact the μ–loops tend to decouple and therefore only yield small terms. We may evaluate them by just replacing $m_\mu \to m_e$ and $m_\tau \to m_\mu$ in the formula for the τ–loop contributions to a_μ. Corrections due to internal μ–loops are suppressed as $O(2\alpha/\pi \, m_e^2/m_\mu^2) \simeq 1.1 \times 10^{-7}$ relative to the leading term and the τ–loops practically play no role at all. The fact that muons and tau leptons tend to decouple is also crucial for the unknown five–loop contribution, since we can expect that corresponding contributions can be safely neglected.

The results we obtain[18] may be written in the form

---

[18]The order α^3 terms are given by two parts which cancel partly

$$A_2^{(6)}(m_e/m_\mu) = -7.373\,941\,64(29) \times 10^{-6}$$
$$= -2.17684015(11) \times 10^{-5}\big|_{\mu-\mathrm{vap}} + 1.439445989(77) \times 10^{-5}\big|_{\mu-\mathrm{lbl}}$$
$$A_2^{(6)}(m_e/m_\tau) = -6.5819(19) \times 10^{-8}$$

$$a_e^{\text{QED}} = a_e^{\text{univ}} + a_e(\mu) + a_e(\tau) + a_e(\mu, \tau) \tag{3.61}$$

with universal term given by (3.43) and

$$a_e(\mu) = 5.197\,386\,70(27) \times 10^{-7} \left(\frac{\alpha}{\pi}\right)^2 - 7.373\,941\,64(29) \times 10^{-6} \left(\frac{\alpha}{\pi}\right)^3$$

$$a_e(\tau) = 1.83763(60) \times 10^{-9} \left(\frac{\alpha}{\pi}\right)^2 - 6.5819(19) \times 10^{-8} \left(\frac{\alpha}{\pi}\right)^3$$

$$a_e(\mu, \tau) = 0.190945(62) \times 10^{-12} \left(\frac{\alpha}{\pi}\right)^3 .$$

Altogether the perturbative expansion for the QED prediction of a_e is given by

$$\begin{aligned}
a_e^{\text{QED}} &= \frac{\alpha}{2\pi} - 0.328\,478\,444\,002\,90(60) \left(\frac{\alpha}{\pi}\right)^2 \\
&+ 1.181\,234\,016\,828(19) \left(\frac{\alpha}{\pi}\right)^3 \\
&- 1.9144(35) \left(\frac{\alpha}{\pi}\right)^4 \\
&+ 0.0(3.8) \left(\frac{\alpha}{\pi}\right)^5 .
\end{aligned} \tag{3.62}$$

The largest uncertainty comes from 518 diagrams without fermion loops contributing to the universal term $A_1^{(8)}$ (3.43). As mentioned before, completely unknown is the universal five–loop term $A_1^{(10)}$, which is leading for a_e and is included by the uncertainty estimate (3.43).

What still is missing are the hadronic and weak contributions, which both are suppressed by the $(m_e/m_\mu)^2$ factor relative to a_μ. For a_e they are small: $a_e^{\text{had}} = 1.67(3) \times 10^{-12}$ and $a_e^{\text{weak}} = 0.036 \times 10^{-12}$, respectively[19]. The hadronic

$$= -1.16723(36) \times 10^{-7}\big|_{\tau-\text{vap}} + 0.50905(17) \times 10^{-7}\big|_{\tau-\text{lbl}} .$$

The errors are due to the errors in the mass ratios. They are completely negligible in comparison to the other errors.

[19] The precise procedure of evaluating the hadronic contributions will be discussed extensively in Chap. 5 for the muon, for which the effects are much more sizable. For the electron, as expected, one finds a small contribution. We use an updated value $a_e^{\text{had(l.o.)}} = 1.8588(156) \times 10^{-12}$ [50] for the leading hadronic vacuum polarization contribution $a_e^{\text{had(h.o.)}} = -0.223(2) \times 10^{-12}$ [50] for the higher order VP contributions and a value rescaled by $(m_e/m_\mu)^2$ of the light–by–light scattering contribution $a_\mu^{\text{had(lbl)}} = 100(39) \times 10^{-11}$ [79] yields $a_e^{\text{had(lbl)}} = 2.34(92) \times 10^{-14}$. The more reliable direct evaluation of this contribution for a_e in the approach by Knecht and Nyffeler [79] yields a slightly larger result: [93] $a_e^{\text{lbl};\pi^0} \sim 2.5(1.2) \times 10^{-14}$ for the π^0 exchange diagram. For a_μ the corresponding result is $a_\mu^{\text{lbl};\pi^0} \sim 5.8(1.0) \times 10^{-10}$ and for the sum of all pseudoscalars $a_\mu^{\text{lbl};\pi^0,\eta,\eta'} \sim 8.3(1.2) \times 10^{-10}$. Assuming that the contributions scale in the same proportion we find $a_e^{\text{lbl};\pi^0,\eta,\eta'} \sim 3.6(1.7) \times 10^{-14}$ as an estimate we will use.

contribution now just starts to be significant, however, unlike in a_μ^{had} for the muon, a_e^{had} is known with sufficient accuracy and is not the limiting factor here. The theory error is dominated by the missing 5–loop QED term. As a result a_e essentially only depends on perturbative QED, while hadronic, weak and new physics (NP) contributions are suppressed by $(m_e/M)^2$, where M is a weak, hadronic or new physics scale. As a consequence a_e at this level of accuracy is theoretically well under control (almost a pure QED object) and therefore is an excellent observable for extracting α_{QED} based on the SM prediction

$$a_e^{\text{SM}} = a_e^{\text{QED}}[\text{Eq. (3.62)}] + 1.706(30) \times 10^{-12} \text{ (hadronic \& weak)}. \quad (3.63)$$

We now compare this result with the very recent extraordinary precise measurement of the electron anomalous magnetic moment [3]

$$a_e^{\text{exp}} = 0.001\,159\,652\,180\,85(76) \quad (3.64)$$

which yields

$$\alpha^{-1}(a_e) = 137.035999069(90)(12)(30)(3),$$

which is the value (3.28) [4, 5] given earlier. The first error is the experimental one of a_e^{exp}, the second and third are the numerical uncertainties of the α^4 and α^5 terms, respectively. The last one is the hadronic uncertainty, which is completely negligible. Note that the largest theoretical uncertainty comes from the almost completely missing information concerning the 5–loop contribution. This is now the by far most precise determination of α and we will use it throughout in the calculation of a_μ, below.

A different strategy is to use a_e for a precision test of QED. For a theoretical prediction of a_e we then need the best determinations of α which do not depend on a_e. These are [94, 95][20]

$$\alpha^{-1}(\text{Cs}) = 137.03600000(110)[8.0\,\text{ppb}], \quad (3.65)$$
$$\alpha^{-1}(\text{Rb}) = 137.03599878(091)[6.7\,\text{ppb}]. \quad (3.66)$$

In terms of $\alpha(\text{Cs})$ one predicts $a_e = 0.00115965217298(930)$ which agrees well with the experiment value $a_e^{\text{exp}} - a_e^{\text{the}} = 7.87(9.33) \times 10^{-12}$; similarly, using $\alpha(\text{Rb})$ the prediction is $a_e = 0.00115965218279(769)$, again in good agreement with experiment $a_e^{\text{exp}} - a_e^{\text{the}} = -1.94(7.73) \times 10^{-12}$. Errors are completely dominated by the uncertainties in α. The following Table 3.3 collects the typical contributions to a_e evaluated in terms of (3.65, 3.66).

---

[20]The results rely upon a number of other experimental quantities. One is the measured Rydberg constant [96], others are the Cesium (Cs) and Rubidium (Rb) masses in atomic mass units (amu) [97] and the electron mass in amu [98, 99]. The $\hbar/M_{\text{Cs}}$ needed comes from an optical measurement of the Cs D1 line [94, 101] and the preliminary recoil shift in an atom interferometer [100], while $\hbar/M_{\text{Rb}}$ comes from a measurement of an atom recoil of a Rb atom in an optical lattice [94].

Table 3.3. Contributions to $a_e(h/M)$ in units 10^{-6}

contribution	$\alpha(h/M_{\text{Cs}})$	$\alpha(h/M_{\text{Rb}})$
universal	1159.652 16856(929)(15)(26)	1159.652 17067(769)(15)(26)
μ–loops	0.000 00271 (0)	0.000 00271 (0)
τ–loops	0.000 00001 (0)	0.000 00001 (0)
hadronic	0.000 00167 (3)	0.000 00167 (3)
weak	0.000 000036 (0)	0.000 000036 (0)
theory	1159.652 17298(930)	1159.652 18279(769)
experiment	1159.652 180 85 (76)	1159.652 180 85 (76)

An improvement of α by a factor 10 would allow a much more stringent test of QED, and therefore is urgently needed [4]. The sensitivity of future QED tests may be illustrated as follows: if one assumes that $\left|\Delta a_e^{\text{New Physics}}\right| \simeq m_e^2/\Lambda^2$ where Λ approximates the scale of "New Physics", then the agreement between $\alpha^{-1}(a_e)$ and $\alpha^{-1}(\text{Rb06})$ currently probes $\Lambda \lesssim O(250 \text{ GeV})$. To access the much more interesting $\Lambda \sim O(1 \text{ TeV})$ region would also require a reliable estimate of the first significant digit of the 5–loop QED contribution, and an improved calculation of the 4–loop QED contribution to a_e^{SM}.

3.2.3 The Anomalous Magnetic Moment of the Muon

The muon magnetic moment anomaly is defined by

$$a_\mu = \frac{1}{2}(g_\mu - 2) = \frac{\mu_\mu}{e\hbar/2m_\mu} - 1 , \quad (3.67)$$

where $g_\mu = 2\mu_\mu/(e\hbar/2m_\mu)$ is the g–factor and μ_μ the magnetic moment of the muon. The different higher order QED contributions are collected in Table 3.4. We thus arrive at a QED prediction of a_μ given by

$$a_\mu^{\text{QED}} = 116\,584\,718.113(.082)(.014)(.025)(.137)[.162] \times 10^{-11} \quad (3.68)$$

where the first error is the uncertainty of α in (3.28), the second one combines in quadrature the uncertainties due to the errors in the mass ratios (3.29), the

Table 3.4. QED contributions to a_μ in units 10^{-6}

term	universal	e–loops	τ–loops	$e \times \tau$–loops
$a^{(4)}$	$-1.772\,305\,06$ (0)	$5.904\,060\,07$ (5)	$0.000\,421\,20$ (13)	–
$a^{(6)}$	$0.014\,804\,20$ (0)	$0.286\,603\,69$ (0)	$0.000\,004\,52$ (1)	$0.000\,006\,61$ (0)
$a^{(8)}$	$-0.000\,055\,73$ (10)	$0.003\,862\,56$ (21)	$0.000\,000\,15$ (9)	$0.000\,001\,09$ (0)
$a^{(10)}$	$0.000\,000\,00$ (26)	$0.000\,044\,83$ (135)	?	?

3.2 Magnetic Moments and Electromagnetic Form Factors

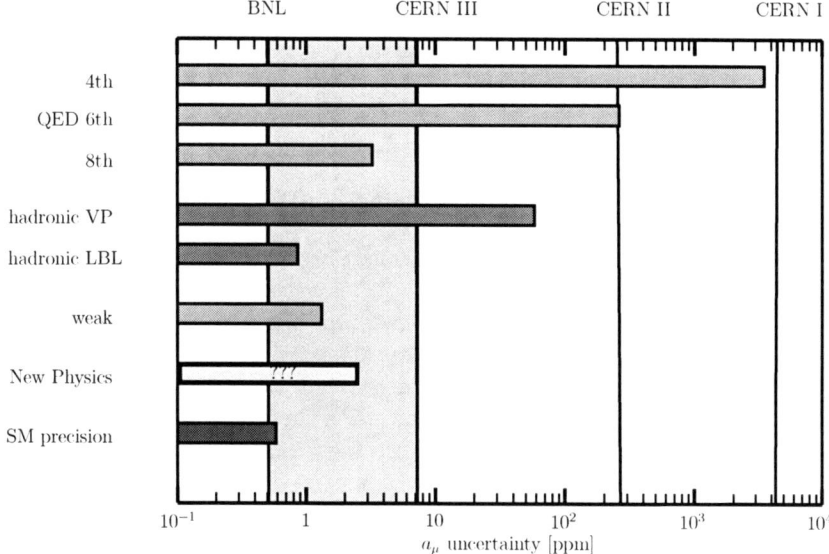

Fig. 3.8. Sensitivity of $g-2$ experiments to various contributions. The increase in precision with the BNL $g-2$ experiment is shown as a gray vertical band. New Physics is illustrated by the deviation $(a_\mu^{\text{exp}} - a_\mu^{\text{the}})/a_\mu^{\text{exp}}$

third is due to the numerical uncertainty and the last stands for the missing $O(\alpha^5)$ terms. With the new value of $\alpha[a_e]$ the combined error is dominated by limited knowledge of the 5–loop term.

The following Table 3.5 collects the typical contributions to a_μ evaluated in terms of α (3.28) determined via a_e.

Table 3.5. The various types of contributions to a_μ in units 10^{-6}, ordered according to their size (L.O. lowest order, H.O. higher order, LbL. light–by–light)

L.O. universal	1161.409 73	(0)
e-loops	6.194 57	(0)
H.O. universal	$-1.757\,55$	(0)
L.O. hadronic	0.069 21	(56)
L.O. weak	0.001 95	(0)
H.O. hadronic	$-0.001\,00$	(2)
LbL. hadronic	0.000 93	(34)
τ-loops	0.000 43	(0)
H.O. weak	$-0.000\,41$	(2)
$e+\tau$-loops	0.000 01	(0)
theory	1165.917 86	(66)
experiment	1165.920 80	(63)

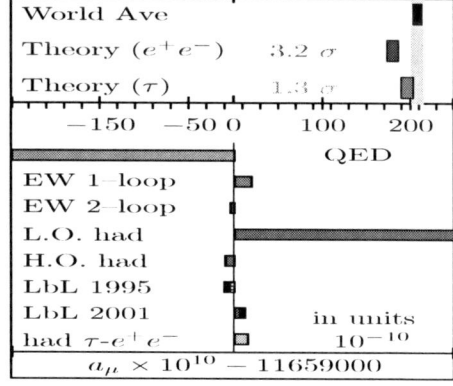

The world average experimental muon magnetic anomaly, dominated by the very precise BNL result, now is [102]

$$a_\mu^{\text{exp}} = 1.16592080(63) \times 10^{-3} \tag{3.69}$$

(relative uncertainty 5.4×10^{-7}), which confronts the SM prediction

$$a_\mu^{\text{the}} = 1.16591786(66) \times 10^{-3} \ . \tag{3.70}$$

Fig. 3.8 illustrates the sensitivity to various contributions and how it developed in history. The high sensitivity of a_μ to physics from not too high scales M above m_μ, which is scaling like $(m_\mu/M)^2$, and the more than one order of magnitude improvement of the experimental accuracy has brought many SM effects into the focus of the interest. Not only are we testing now the 4–loop QED contribution, higher order hadronic VP effects, the infamous hadronic LbL contribution and the weak loops, we are reaching or limiting possible New Physics at a level of sensitivity which caused and still causes a lot of excitement. "New Physics" is displayed in the figure as the ppm deviation of

$$a_\mu^{\text{exp}} - a_\mu^{\text{the}} = (294 \pm 89) \times 10^{-11} \ , \tag{3.71}$$

which is 3.3 σ. We note that the theory error is somewhat larger than the experimental one. It is fully dominated by the uncertainty of the hadronic low energy cross–section data, which determine the hadronic vacuum polarization and, partially, from the uncertainty of the hadronic light–by–light scattering contribution.

As we notice, the enhanced sensitivity to "heavy" physics is somehow good news and bad news at the same time: the sensitivity to "New Physics" we are always hunting for at the end is enhanced due to

$$a_\ell^{\text{NP}} \sim \left(\frac{m_\ell}{M_{\text{NP}}}\right)^2$$

by the mentioned mass ratio square, but at the same time also scale dependent SM effects are dramatically enhanced, and the hadronic ones are not easy to estimate with the desired precision.

The perspectives for future developments will be discussed at the end of Chap. 7.

After this summary of the current status of a_μ and a_e, we will now go on and present basic techniques and tools used in calculating the various effects, before we are going to present a more detailed account of the individual contributions in the next chapter.

3.3 Structure of the Electromagnetic Vertex in the SM

Here we want to discuss the lepton moments beyond QED in the more general context of the SM, in which parity P as well as CP are broken by the

3.3 Structure of the Electromagnetic Vertex in the SM

weak interactions. We again start from the relevant matrix element of the electromagnetic current between lepton states

$$i\Gamma^\mu_{\gamma\ell\ell}(p_1, p_2; r_1, r_2) = \langle \ell^-(p_2, r_2)|j^\mu_{\text{em}}(0)|\ell^-(p_1, r_1)\rangle = i\bar{u}(p_2, r_2)\Pi^\mu_{\gamma\ell\ell}u(p_1, r_1) \tag{3.72}$$

and look for the additional form factors showing up if P and C are violated. Again $q = p_2 - p_1$ is the momentum transfer. $u(p_1, r_1)$ is the Dirac spinor, the wave function of the incoming lepton, with momentum p_1 and 3rd component of spin $r_1 (=\pm\frac{1}{2})$, and $\bar{u} = u^+\gamma^0$ is the adjoint spinor representing the wave function of the outgoing lepton. $\Pi^\mu_{\gamma\ell\ell}$ is a hermitian 4×4 matrix in spinor space and a Lorentz four–vector.

Besides the Dirac matrix γ^μ we have two further independent four–vectors, the momenta p_1 and p_2 or linear combinations of them. It is convenient to choose the orthogonal vectors $P = p_1 + p_2$ and $q = p_2 - p_1$ (with $Pq = 0$). The general covariant decomposition for on shell leptons in the SM then may be written in the form

$$\Pi^\mu_{\gamma\ell\ell} = \gamma^\mu A_1 + \frac{P^\mu}{2m} A_2 + \frac{q^\mu}{2m} A_3 + \gamma^\mu \gamma_5 A_4 + \frac{q^\mu}{2m}\gamma_5 A_5 + i\frac{P^\mu}{2m}\gamma_5 A_6 \tag{3.73}$$

where the scalar amplitudes $A_i(p_1, p_2)$ are functions of the scalar products p_1^2, p_2^2 and $p_1 p_2$. Since the lepton is on the mass shell $p_1^2 = p_2^2 = m^2$ and using $q^2 = 2m^2 - 2p_1 p_2$, the dimensionless amplitudes depend on the single kinematic variable q^2 and on all the parameters of the theory: the fine structure constant $\alpha = e^2/4\pi$ and all physical particle masses. We will simply write $A_i = A_i(q^2)$ in the following.

When writing (3.73) we already have made use of the Gordon identities

$$\begin{array}{ll} i\sigma^{\mu\nu} q_\nu = -P^\mu + 2m\gamma^\mu, & i\sigma^{\mu\nu} P_\nu = -q^\mu, \\ i\sigma^{\mu\nu} q_\nu \gamma_5 = -P^\mu \gamma_5, & i\sigma^{\mu\nu} P_\nu \gamma_5 = -q^\mu \gamma_5 + 2m\gamma^\mu \gamma_5, \end{array} \tag{3.74}$$

which hold if sandwiched between the spinors like $\bar{u}(p_2)\cdots u(p_1)$. In QED due to parity conservation the terms proportional to γ_5 are absent.

The electromagnetic current still is conserved:

$$\partial_\mu j^\mu_{\text{em}} = 0. \tag{3.75}$$

On a formal level, this may be considered as a trivial consequence of the inhomogeneous Maxwell equation (see [103] for a manifestly gauge invariant formulation in the SM)

$$\partial_\mu F^{\mu\nu} = -e\, j^\nu_{\text{em}} \quad \text{with} \quad F_{\mu\nu} = \partial_\mu A_\nu - \partial_\nu A_\mu$$

since $\partial_\nu\partial_\mu F^{\mu\nu} = -e\,\partial_\nu j^\nu_{\text{em}} \equiv 0$ as $\partial_\nu\partial_\mu$ is symmetric in $\mu \leftrightarrow \nu$ while $F^{\mu\nu}$ is antisymmetric. As a consequence we must have $q_\mu \bar{u}_2 \Pi^\mu_{\gamma\ell\ell} u_1 = 0$. By the Dirac equations $\slashed{p}_i u_i = m u_i$ ($i = 1, 2$) we have $\bar{u}_2 \slashed{q} u_1 = 0$, while $\bar{u}_2 \slashed{q} \gamma_5 u_1 = -2m \bar{u}_2 \gamma_5 u_1$, furthermore, $qP = 0$ while keeping $q^2 \neq 0$ at first. Hence current conservation requires $A_3 = 0$ and $A_5 = -4m^2/q^2 A_4$ such that we remain with four physical form factors[21]

$$\bar{u}_2 \Pi^\mu_{\gamma\ell\ell} u_1 = \bar{u}_2 \left(\gamma^\mu A_1 + \frac{P^\mu}{2m} A_2 + \left(\gamma^\mu - \frac{2mq^\mu}{q^2} \right) \gamma_5 A_4 + i\frac{P^\mu}{2m} \gamma_5 A_6 \right) u_1 \, .$$

This shows that the two amplitudes A_3 and A_6 are redundant for physics, however, they show up in actual calculations at intermediate steps and/or for contributions from individual Feynman diagrams. By virtue of the Gordon decomposition

$$\bar{u}(p_2) \frac{P^\mu}{2m} u(p_1) \equiv \bar{u}(p_2) \left(\gamma^\mu - i\frac{\sigma^{\mu\nu} q_\nu}{2m} \right) u(p_1)$$

we finally obtain for the form factor

$$\Pi^\mu_{\gamma\ell\ell} = \gamma^\mu F_\mathrm{E}(q^2) + \left(\gamma^\mu - \frac{2mq^\mu}{q^2} \right) \gamma_5 F_\mathrm{A} + i\sigma^{\mu\nu} \frac{q_\nu}{2m} F_\mathrm{M}(q^2) + \sigma^{\mu\nu} \frac{q_\nu}{2m} \gamma_5 F_\mathrm{D}(q^2) \quad (3.76)$$

With $F_\mathrm{E} = A_1 + A_2$ the electric charge form factor, normalized by charge renormalization to $F_\mathrm{E}(0) = 1$, $F_\mathrm{A} = A_4$ the *anapole moment* [104, 105, 106, 107, 108] which is P violating and vanishing at $q^2 = 0$: $F_\mathrm{A}(0) = 0$. The magnetic form factor is $F_\mathrm{M} = -A_2$ which yields the anomalous magnetic moment as $a_\ell = F_\mathrm{M}(0)$. The last term with $F_\mathrm{D} = A_6$ represents the CP violating *electric dipole moment* (EDM)

$$d_\ell = -\frac{F_\mathrm{D}(0)}{2m} \, . \quad (3.77)$$

Note that (3.76) is the most general Lorentz covariant answer, which takes into account current conservation (3.75) and the on–shell conditions for the leptons (Dirac equation for the spinors).

In the SM at the tree level $F_\mathrm{E}(q^2) = 1$, while $F_\mathrm{i}(q^2) = 0$ for ($i = $ M, A, D).

The anomalous magnetic moment a_ℓ is a dimensionless quantity, just a number, and corresponds to an effective interaction term

$$\delta \mathcal{L}^\mathrm{AMM}_\mathrm{eff} = -\frac{e_\ell a_\ell}{4 m_\ell} \bar{\psi}(x) \, \sigma^{\mu\nu} \, \psi(x) \, F_{\mu\nu}(x) \, , \quad (3.78)$$

with classical low energy limit

$$-\delta \mathcal{L}^\mathrm{AMM}_\mathrm{eff} \Rightarrow \mathcal{H}_m \simeq \frac{e_\ell a_\ell}{2 m_\ell} \, \sigma B \, ,$$

---

[21] In the SM the proper definition of the form factors is highly non–trivial. The conventional definition of the photon field has to be replaced by one which satisfies Maxwell's equations to all orders. This has been investigated extensively in [103]. Since we are interested only in the form factors in the classical limit here, we need not go further into this discussion.

written as a Hamiltonian in 2–spinor space à la Pauli. Note that a term (3.78), if present in the fundamental Lagrangian, would spoil renormalizability of the theory and contribute to $F_i(q^2)$ ($i = $ M, D) at the tree level. In addition it is not $SU(2)_L$ gauge invariant, because gauge invariance only allows minimal couplings via the covariant derivative: vector and/or axial–vector terms. The emergence of an anomalous magnetic moment term in the SM is a consequence of the symmetry breaking by the Higgs mechanism[22], which provides the mass to the physical particles and allows for helicity flip processes like the anomalous magnetic moment transitions. In any renormalizable theory the anomalous magnetic moment term must vanish at tree level, which also means that there is no free parameter associated with it. It is thus a finite prediction of the theory to all orders of the perturbation expansion.

The EDM term only can be non–vanishing if both parity P and time–reversal T are violated [109, 110]. It corresponds to an effective interaction

$$\delta \mathcal{L}_{\text{eff}}^{\text{EDM}} = -\frac{d_\ell}{2} \, \bar{\psi}(x) \, \mathrm{i} \, \sigma^{\mu\nu} \gamma_5 \, \psi(x) \, F_{\mu\nu}(x) \;, \qquad (3.79)$$

which in the non–relativistic limit becomes

$$-\delta \mathcal{L}_{\text{eff}}^{\text{EDM}} \Rightarrow \mathcal{H}_e \simeq -d_\ell \, \boldsymbol{\sigma} \boldsymbol{E} \;, \qquad (3.80)$$

again written as a Hamiltonian in 2–spinor space. Again a term (3.79) is non–renormalizable and it is not $SU(2)_L$ gauge invariant and thus can be there only because the symmetry is broken in the Higgs phase. In the framework of a QFT where CPT is conserved T violation directly corresponds to CP violation, which is small (0.3%) in the light particle sector and can come in at second order at best [111][23]. This is the reason why the EDM is so much smaller than its magnetic counter part. The experimental limit for the

---

[22] Often the jargon *spontaneously broken gauge symmetry* (or the like) is used for the Higgs mechanism. The formal similarity to true *spontaneous symmetry breaking*, like in the Goldstone model, which requires the existence of physical zero mass Goldstone bosons, only shows up on an unphysical state space which is including the Higgs ghosts (would be Goldstone bosons). In fact it is the discrete Z_2 symmetry $H \leftrightarrow -H$ of the physical Higgs field (in the unitary gauge) which is spontaneously broken. This also explains the absence of physical Goldstone bosons.

[23] CP-violation in the SM arises from the complex phase δ in the CKM matrix, which enters the interactions of the quarks with the $W^\pm$ gauge bosons. The magnitude in the 3 family SM is given by the Jarlskog invariant [112]

$$J = \cos\theta_1 \cos\theta_2 \cos\theta_3 \sin^2\theta_1 \sin\theta_2 \sin\theta_3 \sin\delta = (2.88 \pm 0.33) \times 10^{-5} \qquad (3.81)$$

where the θ_i are the 3 mixing angles and δ is the phase in the CKM matrix. Note that J is very small. In addition, only diagrams with at least one quark–loop with at least four CC vertices can give a contribution. This requires 3–loop diagrams exhibiting four virtual W–boson lines inside. Such contributions are highly suppressed. Expected CP violation in the neutrino mixing matrix are expected to yield even much smaller effects.

electron is $|d_e| < 1.6 \times 10^{-27} e \cdot \text{cm}$ at 90% C.L. [113]. The direct test for the muon gave $d_\mu = 3.7 \pm 3.4 \times 10^{-19} e \cdot \text{cm}$ at 90% C.L. [114]. New much more precise experiments for d_μ are under discussion [115]. Theory expects $d_e^{\text{the}} \sim 10^{-28} e \cdot \text{cm}$ [111], 10 times smaller than the present limit. For a theoretical review I refer to [116]. If we assume that $\eta_\mu \sim (m_\mu/m_e)^2 \eta_e$ (see (1.5)), i.e. η_ℓ scales like heavy particle (X) effects in $\delta a_\ell(X) \propto (m_\ell/M_X)^2$, as they do in many new physics scenarios, we expect that $d_\mu \sim (m_\mu/m_e) d_e$, and thus $d_\mu \sim 3.2 \times 10^{-25} e \cdot \text{cm}$. This is too small to affect the extraction of a_μ, for example, as we will see.

3.4 Dipole Moments in the Non–Relativistic Limit

Here we are interested in the non–relativistic limits of the effective dipole moment interaction terms (3.78)

$$\delta \mathcal{L}_{\text{eff}}^{\text{AMM}} = -\frac{e_\ell a_\ell}{4 m_\ell} \bar\psi(x) \, \sigma^{\mu\nu} \, \psi(x) \, F_{\mu\nu}(x) \,,$$

and (3.79)

$$\delta \mathcal{L}_{\text{eff}}^{\text{EDM}} = -\frac{d_\ell}{2} \bar\psi(x) \, i \sigma^{\mu\nu} \gamma_5 \, \psi(x) \, F_{\mu\nu}(x) \,,$$

when the electron is moving in a classical external field described by $F_{\mu\nu}^{\text{ext}}$. The relevant expansion may be easily worked out as follows: since the antisymmetric electromagnetic field strength tensor $F_{\mu\nu}$ exhibits the magnetic field in the spatial components F_{ik}: $B^l = \frac{1}{2}\epsilon^{ikl} F_{ik}$ and the electric field in the mixed time–space part: $E^i = F_{0i}$, we have to work out $\sigma^{\mu\nu}$ for the corresponding components:

$$\sigma^{ik} = \frac{i}{2} \left(\gamma^i \gamma^k - \gamma^k \gamma^i\right)$$
$$= \frac{i}{2} \left(\begin{pmatrix} 0 & \sigma^i \\ -\sigma^i & 0 \end{pmatrix} \begin{pmatrix} 0 & \sigma^k \\ -\sigma^k & 0 \end{pmatrix} - \begin{pmatrix} 0 & \sigma^k \\ -\sigma^k & 0 \end{pmatrix} \begin{pmatrix} 0 & \sigma^i \\ -\sigma^i & 0 \end{pmatrix} \right)$$
$$= -\frac{i}{2} \begin{pmatrix} [\sigma^i, \sigma^k] & 0 \\ 0 & [\sigma^i, \sigma^k] \end{pmatrix} = \epsilon^{ikl} \begin{pmatrix} \sigma^l & 0 \\ 0 & \sigma^l \end{pmatrix}$$

$$\sigma^{0i} \gamma_5 = \frac{i}{2} \left(\gamma^0 \gamma^i - \gamma^i \gamma^0\right) \gamma_5$$
$$= \frac{i}{2} \left(\begin{pmatrix} 1 & 0 \\ 0 & -1 \end{pmatrix} \begin{pmatrix} 0 & \sigma^i \\ -\sigma^i & 0 \end{pmatrix} - \begin{pmatrix} 0 & \sigma^i \\ -\sigma^i & 0 \end{pmatrix} \begin{pmatrix} 1 & 0 \\ 0 & -1 \end{pmatrix} \right) \gamma_5$$
$$= i \begin{pmatrix} 0 & \sigma^i \\ \sigma^i & 0 \end{pmatrix} \begin{pmatrix} 0 & 1 \\ 1 & 0 \end{pmatrix} = i \begin{pmatrix} \sigma^i & 0 \\ 0 & \sigma^i \end{pmatrix}$$

Note that the γ_5 here is crucial to make the matrix block diagonal, because, only block diagonal terms contribute to the leading order in the non–relativistic expansion, as we will see now.

In the rest frame of the electron the spinors have the form
$$u(p,r) = \frac{1}{\sqrt{p^0 + m}} (\not{p} + m)\, \tilde{u}(0,r) \simeq \tilde{u}(0,r)$$
with
$$\tilde{u}(0,r) = \begin{pmatrix} U(r) \\ 0 \end{pmatrix}, \quad U(\tfrac{1}{2}) = \begin{pmatrix} 1 \\ 0 \end{pmatrix}, \quad U(-\tfrac{1}{2}) = \begin{pmatrix} 0 \\ 1 \end{pmatrix}.$$

We first work out the magnetic dipole term
$$\begin{aligned}
\bar{u}_2 \sigma^{\mu\nu} u_1 F_{\mu\nu} &\simeq (U^T(r_2), 0)\, \sigma^{\mu\nu} \begin{pmatrix} U(r_1) \\ 0 \end{pmatrix} F_{\mu\nu} \\
&= (U^T(r_2), 0)\, \sigma^{ik} \begin{pmatrix} U(r_1) \\ 0 \end{pmatrix} F_{ik} \\
&= \epsilon^{ikl} (U^T(r_2), 0) \begin{pmatrix} \sigma^l & 0 \\ 0 & \sigma^l \end{pmatrix} \begin{pmatrix} U(r_1) \\ 0 \end{pmatrix} F_{ik} \\
&= 2 U^T(r_2)\, \boldsymbol{\sigma}\, U(r_1)\, \boldsymbol{B} = 2 (\boldsymbol{\sigma})_{r_2, r_1}\, \boldsymbol{B}\,.
\end{aligned}$$

The other non–diagonal terms do not contribute in this static limit. Similarly, for the electric dipole term
$$\begin{aligned}
\bar{u}_2 \sigma^{\mu\nu} \gamma_5 u_1 F_{\mu\nu} &\simeq (U^T(r_2), 0)\, \sigma^{\mu\nu} \gamma_5 \begin{pmatrix} U(r_1) \\ 0 \end{pmatrix} F_{\mu\nu} \\
&= 2\, (U^T(r_2), 0)\, \sigma^{0i} \gamma_5 \begin{pmatrix} U(r_1) \\ 0 \end{pmatrix} F_{0i} \\
&= 2\mathrm{i} (U^T(r_2), 0) \begin{pmatrix} \sigma^i & 0 \\ 0 & \sigma^i \end{pmatrix} \begin{pmatrix} U(r_1) \\ 0 \end{pmatrix} F_{0i} \\
&= 2\mathrm{i} U^T(r_2)\, \boldsymbol{\sigma}\, U(r_1)\, \boldsymbol{E} = 2\mathrm{i} (\boldsymbol{\sigma})_{r_2, r_1}\, \boldsymbol{E}\,.
\end{aligned}$$

In the low energy expansion matrix elements of the form $\bar{v}_2 \Gamma_i u_1$ or $\bar{u}_2 \Gamma_i v_1$ pick out off–diagonal 2×2 sub–matrices mediating electron–positron creation or annihilation processes, which have thresholds $\sqrt{s} \geq 2m$ and thus are genuinely relativistic effects. The leading terms are the known classical low energy effective terms
$$-\delta \mathcal{L}_{\mathrm{eff}}^{\mathrm{AMM}} \Rightarrow \mathcal{H}_m \simeq \frac{e_\ell a_\ell}{2 m_\ell}\, \boldsymbol{\sigma} \boldsymbol{B}\,,$$
and
$$-\delta \mathcal{L}_{\mathrm{eff}}^{\mathrm{EDM}} \Rightarrow \mathcal{H}_e \simeq -d_\ell\, \boldsymbol{\sigma} \boldsymbol{E}\,,$$
written as 2×2 matrix Hamiltonian, as given before.

3.5 Projection Technique

Especially the calculations of the anomalous magnetic moment in higher orders require most efficient techniques to perform such calculations. As we have seen in Chap. 2 the straight forward calculation of the electromagnetic form factors turns out to be quite non–trivial at the one–loop level already. In particular the occurrence of higher order tensor integrals (up to second rank) makes such calculations rather tedious. Here we outline a projection operator technique which appears to be a much more clever set up for such calculations. Calculations turn out to simplify considerably as we will see.

The tensor integrals showing up in the direct evaluation of the Feynman integrals may be handled in a different way, which allows us to deal directly with the individual amplitudes appearing in the covariant decomposition (3.73). With the matrix element of the form (3.72) we may construct projection operators $\mathcal{P}_{\mu i}$ such that the amplitudes A_i are given by the trace

$$A_i = \mathrm{Tr}\left\{\mathcal{P}_{\mu i}\Pi^\mu_{\gamma \ell \ell}\right\}. \tag{3.82}$$

Since we assume parity P and CP symmetry here (QED) and we have to form a scalar amplitude, a projection operator has to be of a form like (3.73) but with different coefficients which have to be chosen such that the individual amplitudes are obtained. An additional point we have to take into account is the following: since we are working on the physical mass shell (off–shell there would be many more amplitudes), we have to enforce that contributions to $\Pi^\mu_{\gamma \ell \ell}$ of the form $\delta \Pi^\mu_{\gamma \ell \ell} = \cdots (\slashed{p}_1 - m) + (\slashed{p}_2 - m) \cdots$ give vanishing contribution as $\bar{u}_2 \delta \Pi^\mu_{\gamma \ell \ell} u_1 = 0$. This is enforced by applying the projection matrices $\slashed{p}_1 + m$ from the right and $\slashed{p}_2 + m$ from the left, respectively, such that the general form of the projector of interest reads

$$\mathcal{P} = (\slashed{p}_1 + m)\left(\gamma^\mu c_1 + \frac{P^\mu}{2m}c_2 + \frac{q^\mu}{2m}c_3 + \gamma^\mu \gamma_5 c_4 + \frac{q^\mu}{2m}\gamma_5 c_5 - \mathrm{i}\frac{P^\mu}{2m}\gamma_5 c_6\right)(\slashed{p}_2 + m). \tag{3.83}$$

It indeed yields

$$\mathrm{Tr}\left\{\mathcal{P}_\mu \delta \Pi^\mu_{\gamma \ell \ell}\right\} = 0$$

for arbitrary values of the constants c_i, because $(\slashed{p}_2 + m)(\slashed{p}_2 - m) = p_2^2 - m^2 = 0$ if we set $p_2^2 = m^2$ and making use of the cyclicity of the trace, similarly, $(\slashed{p}_1 - m)(\slashed{p}_1 + m) = p_1^2 - m^2 = 0$ if we set $p_1^2 = m^2$. In order to find the appropriate sets of constants which allow us to project to the individual amplitudes we compute $\mathrm{Tr}\,\mathcal{P}_\mu \Pi^\mu_{\gamma \ell \ell}$ and obtain

$$\mathrm{Tr}\left\{\mathcal{P}_\mu \Pi^\mu_{\gamma \ell \ell}\right\} = \sum_{i=1}^{6} g_i A_i \tag{3.84}$$

$\sum_{i=1}^{6} g_i A_i = A_1 \left[c_1(2ds - 4s + 8m^2) + c_2(-2s + 8m^2)\right]$
$\quad + A_2 \left[c_1(-2s + 8m^2) + c_2(-4s + 1/2s^2 m^{-2} + 8m^2)\right]$
$\quad + A_3 \left[c_3(2s - 1/2s^2 m^{-2})\right]$
$\quad + A_4 \left[c_4(2ds - 8dm^2 - 4s + 8m^2) + c_5(2s)\right]$
$\quad + A_5 \left[c_4(-2s) + c_5(1/2s^2 m^{-2})\right]$
$\quad + A_6 \left[c_6(2s - 1/2s^2 m^{-2})\right]$

where $s = q^2$. We observe, firstly, that each of the amplitudes A_3 and A_6 does not mix with any other amplitude and hence may be projected out in a trivial way setting $c_3 = 1$ or $c_6 = 1$, respectively, with all others zero in (3.83). Secondly, the parity violating amplitudes A_i $i = 4, 5, 6$ do not interfere of course with the parity conserving ones A_i $i = 1, 2, 3$ which are the only ones present in QED. To disentangle A_1 and A_2 we have to choose c_1/c_2 such that the coefficient of A_2 or the one of A_1 vanish, and correspondingly for A_4 and A_5. The coefficient of the projected amplitude A_i has to be normalized to unity, such that the requested projector yields (3.82).

Thus, $\mathcal{P}_i$ is obtained by choosing c_j such that $g_i = 1$ and $g_j = 0$ for all $j \neq i$. The following table lists the non–zero coefficients required for the corresponding projector:

$$\mathcal{P}_1 : c_1 = c_2 \frac{s-4m^2}{4m^2} \qquad c_2 = \frac{1}{(d-2)f_1(d)} \frac{2m^2}{s(s-4m^2)}$$

$$\mathcal{P}_2 : c_2 = c_1 \frac{(d-2)s+4m^2}{s-4m^2} \qquad c_1 = \frac{1}{(d-2)f_1(d)} \frac{2m^2}{s(s-4m^2)}$$

$$\mathcal{P}_3 : \qquad c_3 = \frac{1}{f_1(d)} \frac{2m^2}{s(s-4m^2)}$$

$$\mathcal{P}_4 : c_4 = c_5 \frac{s}{4m^2} \qquad c_5 = \frac{1}{(d-2)f_1(d)} \frac{2m^2}{s(s-4m^2)}$$

$$\mathcal{P}_5 : c_5 = -c_4 \frac{(d-2)(s-4m^2)-4m^2}{s} \qquad c_4 = \frac{1}{(d-2)f_1(d)} \frac{2m^2}{s(s-4m^2)}$$

$$\mathcal{P}_6 : \qquad c_6 = -i\frac{1}{f_1(d)} \frac{2m^2}{s(s-4m^2)}$$

with $f_1(d)$ we denote $f(d)/f(d=4)$ where $f(d) \doteq \text{Tr}\,1 = 2^{(d/2)}$ ($\lim_{d\to 4} f(d) = 4$). As discussed in Sect. 2.4.2 p. 62 physics is not affected by the way $f(d=4) = 4$ is extrapolated to $d \neq 4$, provided one sticks to a given convention, like setting $f(d) = 4$ for arbitrary d which means we may take $f_1(d) = 1$ everywhere as a convention. For the amplitudes we are interested in the following we have

$$\mathcal{P}_1^\mu = \frac{1}{2f_1(d)(d-2)}(\not{p}_1 + m)\left(\gamma^\mu + \frac{4m^2}{s(s-4m^2)}\frac{P^\mu}{2m}\right)(\not{p}_2 + m),$$

$$\mathcal{P}_2^\mu = \frac{2m^2/s}{f_1(d)(d-2)(s-4m^2)}(\not{p}_1 + m)\left(\gamma^\mu + \frac{(d-2)s+4m^2}{(s-4m^2)}\frac{P^\mu}{2m}\right)(\not{p}_2 + m),$$

$$\mathcal{P}_3^\mu = \frac{1}{f_1(d)}\frac{2m^2/s}{(s-4m^2)}(\not{p}_1 + m)\left(\frac{q^\mu}{2m}\right)(\not{p}_2 + m). \qquad (3.85)$$

All projectors are of the form

$$\mathcal{P}_i^\mu = (\not{p}_1 + m)\Lambda_i^\mu(p_2, p_1)(\not{p}_2 + m), \qquad (3.86)$$

for example, in the projector for A_2 taking $f_1(d) = 1$ we have

$$\Lambda_2^\mu(p_2, p_1) = \frac{2m^2}{(d-2)s(s-4m^2)}\left(\gamma^\mu + \frac{(d-2)s+4m^2}{(s-4m^2)}\frac{P^\mu}{2m}\right). \qquad (3.87)$$

This projector we will need later for calculating higher order contributions to the anomalous magnetic moment in an efficient manner.

The amplitudes A_i at one–loop are now given by the integrals

$$A_i = e^2 \int \frac{d^d k}{(2\pi)^d} \frac{f_i(k)}{((p_2 - k)^2 - m^2)((p_1 - k)^2 - m^2)(k^2)} \qquad (3.88)$$

with

$$f_1(k) = (4m^2 - 2s) - 4kP + (d-4)k^2 + \frac{2(kq)^2}{s} - \frac{2(kP)^2}{(s - 4m^2)}$$

$$f_2(k) = -\frac{8m^2}{s - 4m^2}\left(kP + k^2 + (d-1)\frac{(kP)^2}{(s-4m^2)} - \frac{(kq)^2}{s}\right)$$

$$f_3(k) = \frac{8m^2}{s} kq \left(1 - (d-2)\frac{kP}{(s-4m^2)}\right). \qquad (3.89)$$

Again we use the relations $2kP = 2[k^2] - [(p_1-k)^2 - m^2] - [(p_2-k)^2 - m^2]$ and $2kq = [(p_1-k)^2 - m^2] - [(p_2-k)^2 - m^2]$ when it is possible to cancel against the scalar propagators $\frac{1}{(1)}$, $\frac{1}{(2)}$ and $\frac{1}{(3)}$ where $(1) \doteq (p_1 - k)^2 - m^2$, $(2) \doteq (p_2 - k)^2 - m^2$, $(3) \doteq k^2$:

$$f_1(k) = (4m^2 - 2s) + (d-8)(3) + 2(1) + 2(2)$$
$$+ \frac{(kq)}{s}[(1) - (2)] - \frac{(kP)}{(s-4m^2)}[2(3) - (1) - (2)]$$

$$f_2(k) = -\frac{4m^2}{s - 4m^2}\bigg(4(3) - (1) - (2)$$
$$+ (d-1)\frac{(kP)}{(s-4m^2)}[2(3) - (1) - (2)] - \frac{(kq)}{s}[(1) - (2)]\bigg)$$

$$f_3(k) = \frac{4m^2}{s}[(1) - (2)]\left(1 - (d-2)\frac{kP}{(s-4m^2)}\right). \qquad (3.90)$$

We observe that besides the first term in f_1 which yields a true vertex correction (three point function) all other terms have at least one scalar propagator (1), (2) or (3) in the numerator which cancels against one of the denominators and hence only yields a much simpler two point function. In particular f_i $i = 2, 3$ are completely given by two point functions and the remaining k dependence in the numerator is at most linear (first rank tensor) and only in combination of two point functions. This is a dramatic simplification in comparison to the most frequently applied direct method presented before. With $\int_k \frac{1}{(1)(2)(3)} = -C_0$, $\int_k \frac{1}{(1)(2)} = B_0(m, m; s)$, $\int_k \frac{1}{(1)(3)} = \int_k \frac{1}{(2)(3)} = B_0(0, m; m^2)$, $\int_k \frac{k^\mu}{(1)(3)} = p_1^\mu \frac{A_0(m)}{2m^2}$, $\int_k \frac{k^\mu}{(2)(3)} = p_2^\mu \frac{A_0(m)}{2m^2}$, $\int_k \frac{k^\mu}{(1)(2)} = 0$, $\int_k \frac{1}{(1)} = \int_k \frac{1}{(3)} = -A_0(m)$ and $\int_k \frac{1}{(3)} = 0$ we easily find

$$A_1 = \frac{e^2}{16\pi^2}\left\{(2s - 4m^2)\,C_0(m_\gamma, m, m)\right.$$
$$\left. - 3\,B_0(m, m; s) + 4\,B_0(0, m; m^2) - 2\right\}$$
$$A_2 = \frac{e^2}{16\pi^2}\left\{\frac{-4m^2}{s - 4m^2}\left(B_0(m, m; s) - B_0(0, m; m^2)\right)\right\}$$
$$A_3 = 0 \tag{3.91}$$

in agreement with (2.199).

For our main goal of calculating the muon anomaly $a_\mu = F_M(0) = -A_2(0)$ we may work out the classical limit $s = q^2 \to 0$

$$a_\mu = \lim_{q^2 \to 0} \text{Tr}\left\{(\not{p}_1 + m)\,\Lambda_2^\mu(p_2, p_1)\,(\not{p}_2 + m)\,\Pi_\mu(P, q)\right\} \tag{3.92}$$

explicitly. Because of the singular factor $1/q^2$ in front of the projector Λ_2 (3.87) we are required to expand the amplitude $\Pi^\mu(P, q)$ to first order for small q,

$$\Pi_\mu(P, q) \simeq \Pi_\mu(P, 0) + q^\nu \frac{\partial}{\partial q^\nu}\,\Pi_\mu(P, q)|_{q=0} \equiv V_\mu(p) + q^\nu\,T_{\nu\mu}(p)\,, \tag{3.93}$$

where for $q = 0$ we have $p = P/2 = p_1$. Other factors of q come from expanding the other factors in the trace by setting $p_2 = (P+q)/2$ and $p_1 = (P-q)/2$ and performing an expansion in $q = p_2 - p_1$ for fixed $P = p_2 + p_1$. We note that due to the on–shell condition $p_2^2 = p_1^2 = m^2$ we have $Pq = 2pq + q^2 = 0$. The only relevant q^μ dependence left are the terms linear and quadratic in q, proportional to q^μ and $q^\mu q^\nu$. Since the trace under consideration projects to a scalar, we may average the residual q dependence over all spatial directions without changing the result. Since P and q are two independent and orthogonal vectors, averaging is relative to the direction of P. For the linear term we have

$$\overline{q^\mu} \equiv \int \frac{d\Omega(P, q)}{4\pi}\,q^\mu = 0 \tag{3.94}$$

because the integrand is an odd function, while

$$\overline{q^\mu q^\nu} \equiv \int \frac{d\Omega(P, q)}{4\pi}\,q^\mu q^\nu = \alpha g^{\mu\nu} + \beta\,\frac{P^\mu P^\nu}{P^2}$$

must be a second rank tensor in P. Since $Pq = 0$, the contraction with P_μ is vanishing, which requires

$$\beta = -\alpha\,.$$

The other possible contraction with $g_{\mu\nu}$ yields q^2:

178   3 Lepton Magnetic Moments: Basics

$$\int \frac{\mathrm{d}\Omega(P,q)}{4\pi} q^2 = q^2 \int \frac{\mathrm{d}\Omega(P,q)}{4\pi} = q^2 = \alpha\, d + \beta = (d-1)\,\alpha$$

and hence

$$\alpha = \frac{q^2}{(d-1)}$$

or

$$\overline{q^\mu q^\nu} = \frac{q^2}{(d-1)} \left(g^{\mu\nu} - \frac{P^\mu P^\nu}{P^2} \right). \tag{3.95}$$

Using these averages we may work out the limit which yields

$$a_\mu = \frac{1}{8(d-2)(d-1)\,m} \operatorname{Tr}\left\{ (\not{p}+m)\,[\gamma^\mu,\gamma^\nu]\,(\not{p}+m)\,T_{\nu\mu}(p) \right\} \tag{3.96}$$

$$+ \frac{1}{4(d-1)\,m^2} \operatorname{Tr}\left\{ \left[m^2\,\gamma^\mu - (d-1)\,m\,p^\mu - d\,\not{p}\,p^\mu \right] V_\mu(p) \right\}\Big|_{p^2=m^2}$$

as a master formula for the calculation of a_μ [61]. The form of the first term is obtained upon anti–symmetrization in the indices $[\mu\nu]$. The amplitudes $V_\mu(p)$ and $T_{\nu\mu}(p)$ depend on one on–shell momentum p, only, and thus the problem reduces to the calculation of on–shell self–energy type diagrams shown in Fig. 3.9.

In $T_{\nu\mu}$ the extra vertex is generated by taking the derivative of the internal muon propagators

$$\frac{\partial}{\partial q^\nu}\,\frac{\mathrm{i}}{\not{p}-\not{k}\mp\not{q}/2-m}\bigg|_{q=0} = \mp\frac{1}{2}\,\frac{\mathrm{i}}{\not{p}-\not{k}-m}\,(-\mathrm{i}\gamma_\nu)\,\frac{\mathrm{i}}{\not{p}-\not{k}-m}.$$

Usually, writing the fermion propagators in terms of scalar propagators

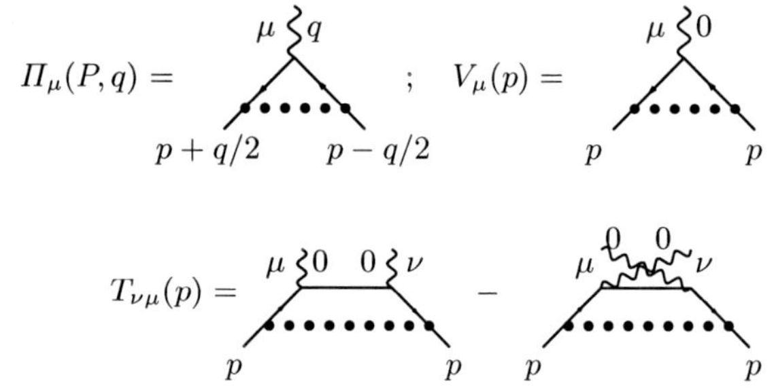

$$\Pi_\mu(P,q) = \begin{array}{c}\mu\,\zeta\,q\\ \triangle\\ p+q/2 \quad p-q/2\end{array} \quad ; \quad V_\mu(p) = \begin{array}{c}\mu\,\zeta\,0\\ \triangle\\ p \quad p\end{array}$$

$$T_{\nu\mu}(p) = \begin{array}{c}\mu\,\zeta\,0 \quad 0\,\zeta\,\nu\\ \triangle\\ p \quad p\end{array} - \begin{array}{c}\mu\,\zeta\,0 \quad 0\,\zeta\,\nu\\ \triangle\\ p \quad p\end{array}$$

Fig. 3.9. To calculate a_μ one only needs the on–shell vertex $V_\mu(p) = \Pi_\mu(P,q)|_{q=0}$ and its $\mu \leftrightarrow \nu$ anti–symmetrized derivative $T_{\nu\mu} = \frac{\partial}{\partial q^\nu} \Pi_\mu(P,q)|_{q=0}$ at zero momentum transfer. Illustrated here for the lowest order diagram; the dotted line may be a photon or a heavy "photon" as needed in the dispersive approach to be discussed below

$$\frac{i}{\not{p}_i - \not{k} - m} = \frac{i(\not{p}_i - \not{k} + m)}{(p_i - k)^2 - m^2}$$

as done in (2.198), only the expansion of the numerators contributes to $T_{\nu\mu}$, while expanding the product of the two scalar propagators

$$\frac{1}{(p_2 - k)^2 - m^2} \frac{1}{(p_1 - k)^2 - m^2} = \frac{1}{((p - k)^2 - m^2)^2} + Q(q^2)$$

gives no contribution linear in q, as the linear terms coming from the individual propagators cancel in the product. Looking at (2.198), for the lowest order contribution we thus have to calculate the trace (3.97) with

$$V_\mu \to v_\mu = \gamma^\rho (\not{p} - \not{k} + m) \gamma_\mu (\not{p} - \not{k} + m) \gamma_\rho$$
$$T_{\nu\mu} \to t_{\nu\mu} = \frac{1}{2} \gamma^\rho \left(\gamma_\nu \gamma_\mu (\not{p} - \not{k} + m) - (\not{p} - \not{k} + m) \gamma_\mu \gamma_\nu \right) \gamma_\rho \, .$$

The trace yields

$$2k^2 \left(\frac{1}{d-1} - 1 \right) - 4kp + \frac{(2kp)^2}{2m^2} \left(d - 1 - \frac{1}{d-1} \right)$$

which is to be integrated as in (2.198). The result is (see Sect. 2.6.3 p.99)

$$a_\mu = \frac{e^2}{16\pi^2} \frac{2}{3} \{ B_0(0, m; m^2) - B_0(m, m; 0) + 1 \} = \frac{\alpha}{\pi} \frac{1}{2}$$

as it should be. Note that the result differs in structure from (3.91) because integration and taking the limit is interchanged. Since we are working throughout with dimensional regularization, it is crucial to take the dimension d generic until after integration. In particular setting $d = 4$ in the master formula (3.97) would lead to a wrong constant term in the above calculation. In fact, the constant term would just be absent.

The projection technique just outlined provides an efficient tool for calculating individual on-shell amplitudes directly. One question may be addressed here, however. The muon is an unstable particle and mass and width are defined via the resonance pole in the complex p^2–plane. In this case the projection technique as presented above has its limitation. However, the muon width is so many orders of magnitude smaller than the muon mass, that at the level of accuracy which is of any practical interest, this is not a matter of worry, i.e. the muon as a quasi stable particle may be safely approximated to be stable in calculations of a_μ.

3.6 Properties of the Form Factors

We again consider the interaction of a lepton in an external field: the relevant T–matrix element is

$$T_{fi} = e J_{fi}^\mu \tilde{A}_\mu^{\text{ext}}(q) \tag{3.97}$$

with

$$J_{fi}^{\mu} = \bar{u}_2 \Gamma^{\mu} u_1 = \langle f | j_{em}^{\mu}(0) | i \rangle = \langle \ell^-(p_2) | j_{em}^{\mu}(0) | \ell^-(p_1) \rangle \,. \tag{3.98}$$

By the crossing property we have the following channels:

- Elastic ℓ^- scattering: $s = q^2 = (p_2 - p_1)^2 \leq 0$
- Elastic ℓ^+ scattering: $s = q^2 = (p_1 - p_2)^2 \leq 0$
- Annihilation (or pair–creation) channel: $s = q^2 = (p_1 + p_2)^2 \geq 4m_\ell^2$

The domain $0 < s < 4m_\ell^2$ is unphysical. A look at the unitarity condition

$$i\{T_{if}^* - T_{fi}\} = \sum_n (2\pi)^4 \, \delta^{(4)}(P_n - P_i) \, T_{nf}^* T_{ni}, \tag{3.99}$$

which derives from (2.94), (2.101), taking $\langle f | S^+ S | i \rangle$ and using (3.120) below, tells us that for $s < 4m_\ell^2$ there is no physical state $|n\rangle$ allowed by energy and momentum conservation and thus

$$T_{fi} = T_{if}^* \quad \text{for} \quad s < 4m_\ell^2, \tag{3.100}$$

which means that the current matrix element is hermitian. As the electromagnetic potential $A_\mu^{\text{ext}}(x)$ is real its Fourier transform satisfies

$$\tilde{A}_\mu^{\text{ext}}(-q) = \tilde{A}_\mu^{*\,\text{ext}}(q) \tag{3.101}$$

and hence

$$J_{fi}^{\mu} = J_{if}^{\mu *} \quad \text{for} \quad s < 4m_\ell^2. \tag{3.102}$$

If we interchange initial and final state, the four–vectors p_1 and p_2 are interchanged such that q changes sign: $q \to -q$. The unitarity relation for the form factor decomposition of $\bar{u}_2 \, \Pi_{\gamma\ell\ell}^{\mu} \, u_1$ (3.76) thus reads ($u_i = u(p_i, r_i)$)

$$\bar{u}_2 \left(\gamma^\mu F_E(q^2) + [\gamma^\mu - \frac{2mq^\mu}{q^2}] \gamma_5 F_A + i\sigma^{\mu\nu} \frac{q_\nu}{2m} F_M(q^2) + \sigma^{\mu\nu} \frac{q_\nu}{2m} \gamma_5 F_D \right) u_1$$

$$= \left\{ \bar{u}_1 \left(\gamma^\mu F_E(q^2) + [\gamma^\mu + \frac{2mq^\mu}{q^2}] \gamma_5 F_A - i\sigma^{\mu\nu} \frac{q_\nu}{2m} F_M(q^2) - \sigma^{\mu\nu} \frac{q_\nu}{2m} \gamma_5 F_D \right) u_2 \right\}^*$$

$$= u_2^+ \left(\gamma^{\mu+} F_E^*(q^2) + \gamma_5^+ [\gamma^{\mu+} + \frac{2mq^\mu}{q^2}] F_A^* + i\sigma^{\mu\nu+} \frac{q_\nu}{2m} F_M^*(q^2) - \gamma_5^+ \sigma^{\mu\nu+} \frac{q_\nu}{2m} F_D^* \right) \bar{u}_1^+$$

$$= \bar{u}_2 \left(\gamma^\mu F_E^*(q^2) + [\gamma^\mu - \frac{2mq^\mu}{q^2}] \gamma_5 F_A^* + i\sigma^{\mu\nu} \frac{q_\nu}{2m} F_M^*(q^2) + \sigma^{\mu\nu} \frac{q_\nu}{2m} \gamma_5 F_D^* \right) u_1 \,.$$

The last equality follows using $u_2^+ = \bar{u}_2 \gamma^0$, $\bar{u}_1^+ = \gamma^0 u_1$, $\gamma_5^+ = \gamma_5$, $\gamma^0 \gamma_5 \gamma^0 = -\gamma_5$, $\gamma^0 \gamma^{\mu+} \gamma^0 = \gamma^\mu$ and $\gamma^0 \sigma^{\mu\nu+} \gamma^0 = \sigma^{\mu\nu}$. Unitarity thus implies that the form factors are real

$$\text{Im}\, F(s)_i = 0 \quad \text{for} \quad s < 4m_e^2 \tag{3.103}$$

below the threshold of pair production $s = 4m_e^2$. For $s \geq 4m_e^2$ the form factors are complex; they are analytic in the complex s–plane with a cut along the positive axis starting at $s = 4m_e^2$ (see Fig. 3.10). In the annihilation channel $(p_- = p_2, p_+ = -p_1)$

$$\langle 0|j_{em}^\mu(0)|p_-, p_+\rangle = \sum_n \langle 0|j_{em}^\mu(0)|n\rangle\langle n|p_-, p_+\rangle , \qquad (3.104)$$

where the lowest state $|n\rangle$ contributing to the sum is an e^+e^- state at threshold : $E_+ = E_- = m_e$ and $\boldsymbol{p}_+ = \boldsymbol{p}_- = 0$ such that $s = 4m_e^2$. Because of the causal $i\varepsilon$–prescription in the time–ordered Green functions the amplitudes change sign when $s \to s^*$ and hence

$$F_i(s^*) = F_i^*(s) , \qquad (3.105)$$

which is the Schwarz reflection principle.

3.7 Dispersion Relations

Causality together with unitarity imply analyticity of the form factors in the complex s–plane except for the cut along the positive real axis starting at $s \geq 4m_i^2$. Cauchy's integral theorem tells us that the contour integral, for the contour $\mathcal{C}$ shown in Fig. 3.10, satisfies

$$F_i(s) = \frac{1}{2\pi i} \oint_\mathcal{C} \frac{ds' F(s')}{s' - s} . \qquad (3.106)$$

Since $F^*(s) = F(s^*)$ the contribution along the cut may be written as

$$\lim_{\varepsilon \to 0} (F(s + i\varepsilon) - F(s - i\varepsilon)) = 2\,i\,\mathrm{Im}\,F(s) ; \quad s \text{ real}, s > 0$$

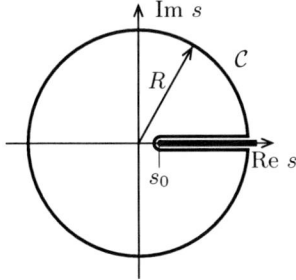

Fig. 3.10. Analyticity domain and Cauchy contour $\mathcal{C}$ for the lepton form factor (vacuum polarization). $\mathcal{C}$ is a circle of radius R with a cut along the positive real axis for $s > s_0 = 4m^2$ where m is the mass of the lightest particles which can be pair–produced

and hence for $R \to \infty$

$$F(s) = \lim_{\varepsilon \to 0} F(s+i\varepsilon) = \frac{1}{\pi} \lim_{\varepsilon \to 0} \int_{4m^2}^{\infty} ds' \frac{\mathrm{Im}\, F(s')}{s'-s-i\varepsilon} + \mathcal{C}_{\infty}\,.$$

In all cases where $F(s)$ falls off sufficiently rapidly as $|s| \to \infty$ the boundary term $\mathcal{C}_{\infty}$ vanishes and the integral converges. This may be checked order by order in perturbation theory. In this case the "un–subtracted" dispersion relation (DR)

$$F(s) = \frac{1}{\pi} \lim_{\varepsilon \to 0} \int_{4m^2}^{\infty} ds' \frac{\mathrm{Im}\, F(s')}{s'-s-i\varepsilon} \tag{3.107}$$

uniquely determines the function by its imaginary part. A technique based on DRs is frequently used for the calculation of Feynman integrals, because the calculation of the imaginary part is simpler in general. The real part which actually is the object to be calculated is given by the principal value ($\mathcal{P}$) integral

$$\mathrm{Re}\, F(s) = \frac{1}{\pi} \fint_{4m^2}^{\infty} ds' \frac{\mathrm{Im}\, F(s')}{s'-s}\,, \tag{3.108}$$

which is also known under the name Hilbert transform.

For our form factors the fall off condition is satisfied for the Pauli form factor F_{M} but not for the Dirac form factor F_{E}. In the latter case the fall off condition is not satisfied because $F_{\mathrm{E}}(0) = 1$ (charge renormalization condition = subtraction condition). However, performing a subtraction of $F_{\mathrm{E}}(0)$ in (3.107), one finds that $(F_{\mathrm{E}}(s) - F_{\mathrm{E}}(0))/s$ satisfies the "subtracted" dispersion relations

$$\frac{F(s) - F(0)}{s} = \frac{1}{\pi} \lim_{\varepsilon \to 0} \int_{4m^2}^{\infty} ds' \frac{\mathrm{Im}\, F(s')}{s'(s'-s-i\varepsilon)}\,, \tag{3.109}$$

which exhibits one additional power of s' in the denominator and hence improves the damping of the integrand at large s' by one additional power. Order by order in perturbation theory the integral (3.109) is convergent for the Dirac form factor. A very similar relation is satisfied by the vacuum polarization amplitude which we will discuss in the following section.

3.7.1 Dispersion Relations and the Vacuum Polarization

Dispersion relations play an important role for taking into account the photon propagator contributions. The related photon self–energy, obtained from the photon propagator by the amputation of the external photon lines, is given by the correlator of two electromagnetic currents and may be interpreted as vacuum polarization for the following reason: as we have seen in Sect. 2.6.3

charge renormalization in QED, according to (2.207), is caused solely by the photon self–energy correction; the fundamental electromagnetic fine structure constant α in fact is a function of the energy scale $\alpha \to \alpha(E)$ of a process due to charge screening. The latter is a result of the fact that a naked charge is surrounded by a cloud of virtual particle–antiparticle pairs (dipoles mostly) which line up in the field of the central charge and such lead to a vacuum polarization which screens the central charge. This is illustrated in Fig. 3.11. From long distances (classical charge) one thus sees less charge than if one comes closer, such that one sees an increasing charge with increasing energy. Figure 3.12 shows the usual diagrammatic representation of a vacuum polarization effect.

As discussed in Sect. 2.6.1 the vacuum polarization affects the photon propagator in that the full or dressed propagator is given by the geometrical progression of self–energy insertions $-i\Pi_\gamma(q^2)$. The corresponding Dyson summation implies that the free propagator is replaced by the dressed one

$$\mathrm{i}D^{\mu\nu}_\gamma(q) = \frac{-\mathrm{i}g^{\mu\nu}}{q^2 + \mathrm{i}\varepsilon} \to \mathrm{i}D^{'\mu\nu}_\gamma(q) = \frac{-\mathrm{i}g^{\mu\nu}}{q^2 + \Pi_\gamma(q^2) + \mathrm{i}\varepsilon} \qquad (3.110)$$

modulo unphysical gauge dependent terms. By $U(1)_{\mathrm{em}}$ gauge invariance the photon remains massless and hence we have $\Pi_\gamma(q^2) = \Pi_\gamma(0) + q^2\,\Pi'_\gamma(q^2)$ with $\Pi_\gamma(0) \equiv 0$. As a result we obtain

$$\mathrm{i}D^{'\mu\nu}_\gamma(q) = \frac{-\mathrm{i}g^{\mu\nu}}{q^2\,(1 + \Pi'_\gamma(q^2))} + \text{gauge terms} \qquad (3.111)$$

where the "gauge terms" will not contribute to gauge invariant physical quantities, and need not be considered further.

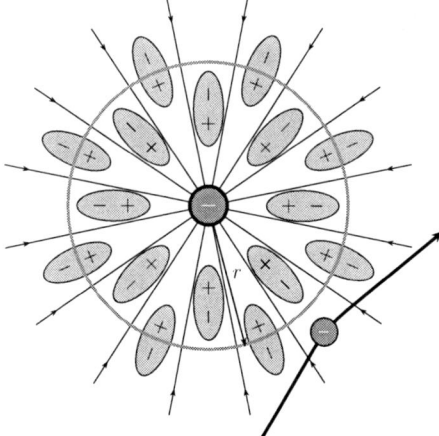

Fig. 3.11. Vacuum polarization causing charge screening by virtual pair creation and re–annihilation

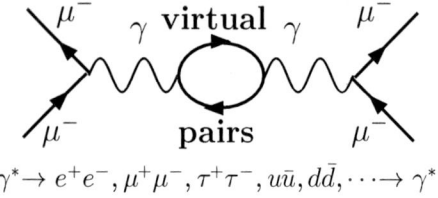

$$\gamma^* \to e^+e^-, \mu^+\mu^-, \tau^+\tau^-, u\bar{u}, d\bar{d}, \cdots \to \gamma^*$$

Fig. 3.12. Feynman diagram describing the vacuum polarization in muon scattering

Including a factor e^2 and considering the renormalized propagator (wave function renormalization factor Z_γ) we have

$$i e^2 D'^{\mu\nu}_\gamma(q) = \frac{-ig^{\mu\nu} e^2 Z_\gamma}{q^2 \left(1 + \Pi'_\gamma(q^2)\right)} + \text{gauge terms} \qquad (3.112)$$

which in effect means that the charge has to be replaced by a **running charge**

$$e^2 \to e^2(q^2) = \frac{e^2 Z_\gamma}{1 + \Pi'_\gamma(q^2)} \;. \qquad (3.113)$$

The wave function renormalization factor Z_γ is fixed by the condition that at $q^2 \to 0$ one obtains the classical charge (charge renormalization in the *Thomson limit*; see also (2.166)). Thus the renormalized charge is

$$e^2 \to e^2(q^2) = \frac{e^2}{1 + (\Pi'_\gamma(q^2) - \Pi'_\gamma(0))} \qquad (3.114)$$

where the lowest order diagram in perturbation theory which contributes to $\Pi'_\gamma(q^2)$ is

and describes the virtual creation and re–absorption of fermion pairs $\gamma^* \to e^+e^-, \mu^+\mu^-, \tau^+\tau^-, u\bar{u}, d\bar{d}, \cdots \to \gamma^*$.

In terms of the fine structure constant $\alpha = \frac{e^2}{4\pi}$ (3.114) reads

$$\alpha(q^2) = \frac{\alpha}{1 - \Delta\alpha} \;\;;\;\; \Delta\alpha = -\text{Re}\left(\Pi'_\gamma(q^2) - \Pi'_\gamma(0)\right) \;. \qquad (3.115)$$

The various contributions to the shift in the fine structure constant come from the leptons (lep = e, μ and τ) the 5 light quarks (u, d, s, c, and b and the corresponding hadrons = had) and from the top quark:

$$\Delta\alpha = \Delta\alpha_{\text{lep}} + \Delta^{(5)}\alpha_{\text{had}} + \Delta\alpha_{\text{top}} + \cdots \quad (3.116)$$

Also W–pairs contribute at $q^2 > M_W^2$. While the other contributions can be calculated order by order in perturbation theory the hadronic contribution $\Delta^{(5)}\alpha_{\text{had}}$ exhibits low energy strong interaction effects and hence cannot be calculated by perturbative means. Here the dispersion relations play a key role. This will be discussed in detail in Sect. 5.2 below.

The leptonic contributions are calculable in perturbation theory. Using our result (2.172) for the renormalized photon self–energy, at leading order the free lepton loops yield

$$\begin{aligned} \Delta\alpha_{\text{lep}}(q^2) &= \\ &= \sum_{\ell=e,\mu,\tau} \frac{\alpha}{3\pi} \left[-\frac{5}{3} - y_\ell + \left(1 + \frac{y_\ell}{2}\right) \sqrt{1 - y_\ell} \ln\left(\left|\frac{\sqrt{1-y_\ell}+1}{\sqrt{1-y_\ell}-1}\right|\right) \right] \\ &= \sum_{\ell=e,\mu,\tau} \frac{\alpha}{3\pi} \left[-\frac{8}{3} + \beta_\ell^2 + \frac{1}{2}\beta_\ell(3 - \beta_\ell^2) \ln\left(\left|\frac{1+\beta_\ell}{1-\beta_\ell}\right|\right) \right] \quad (3.117) \\ &= \sum_{\ell=e,\mu,\tau} \frac{\alpha}{3\pi} \left[\ln\left(|q^2|/m_\ell^2\right) - \frac{5}{3} + O\left(m_\ell^2/q^2\right) \right] \text{ for } |q^2| \gg m_\ell^2 \\ &\simeq 0.03142 \text{ for } q^2 = M_Z^2 \end{aligned}$$

where $y_\ell = 4m_\ell^2/q^2$ and $\beta_\ell = \sqrt{1-y_\ell}$ are the lepton velocities. This leading contribution is affected by small electromagnetic corrections only in the next to leading order. The leptonic contribution is actually known to three loops [117, 118] at which it takes the value

$$\Delta\alpha_{\text{leptons}}(M_Z^2) \simeq 314.98 \times 10^{-4}. \quad (3.118)$$

As already mentioned, in contrast, the corresponding free quark loop contribution gets substantially modified by low energy strong interaction effects, which cannot be calculated reliably by perturbative QCD. The evaluation of the hadronic contribution will be discussed later.

Vacuum polarization effects are large when large scale changes are involved (large logarithms) and because of the large number of light fermionic degrees of freedom (see (2.177)) as we infer from the asymptotic form in perturbation theory

$$\Delta\alpha^{\text{pert}}(q^2) \simeq \frac{\alpha}{3\pi} \sum_f Q_f^2 N_{cf} \left(\ln \frac{|q^2|}{m_f^2} - \frac{5}{3} \right) \; ; \; |q^2| \gg m_f^2 . \quad (3.119)$$

Fig. 3.13 illustrates the running of the effective charges at lower energies in the space–like region[24]. Typical values are $\Delta\alpha(5\text{GeV}) \sim 3\%$ and $\Delta\alpha(M_Z) \sim 6\%$,

---

[24] A direct measurement is difficult because of the normalizing process involved in any measurement which itself depends on the effective charge. Measurements of the evolution of the electromagnetic coupling are possible in any case with an offset energy scale and results have been presented in [119] (see also [120]).

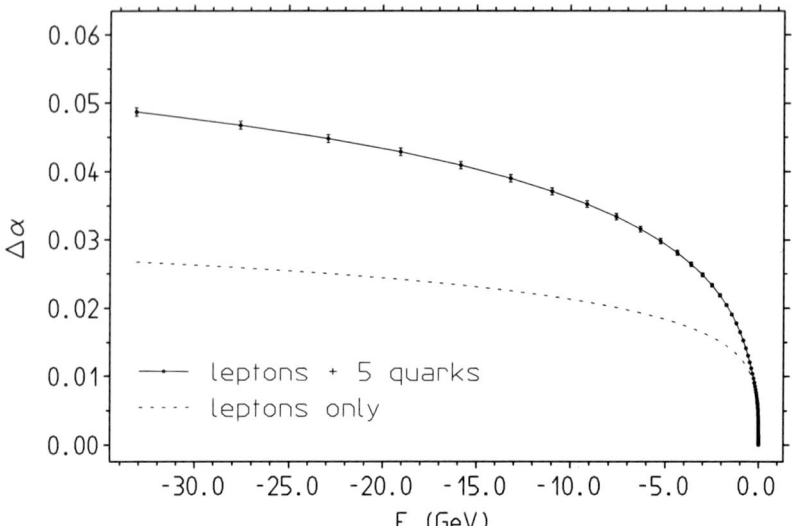

Fig. 3.13. Shift of the effective fine structure constant $\Delta\alpha$ as a function of the energy scale in the space–like region $q^2 < 0$ ($E = -\sqrt{-q^2}$). The vertical bars at selected points indicate the uncertainty

where about $\sim 50\%$ of the contribution comes from leptons and about $\sim 50\%$ from hadrons. Note the sharp increase of the screening correction at relatively low energies.

The vacuum polarization may be described alternatively as the vacuum expectation value of the time ordered product of two electromagnetic currents, which follows by amputation of the external photon lines of the photon propagator: at one loop order

One may represent the current correlator as a Källen-Lehmann representation [121] in terms of spectral densities. To this end, let us consider first the Fourier transform of the vacuum expectation value of the product of two currents. Using translation invariance and inserting a complete set of states n of momentum p_n[25], satisfying the completeness relation

$$\int \frac{d^4 p_n}{(2\pi)^3} \sum_n |n\rangle\langle n| = 1 \qquad (3.120)$$

---

[25] Note that the intermediate states are multi–particle states, in general, and the completeness integral includes an integration over p_n^0, since p_n is not on the mass shell $p_n^0 \neq \sqrt{m_n^2 + \boldsymbol{p}_n^2}$. In general, in addition to a possible discrete part of the spectrum we are dealing with a continuum of states.

where $\sumint_n$ includes, for fixed total momentum p_n, integration over the phase space available to particles of all possible intermediate physical states $|n\rangle$, we have

$$i \int d^4x \, e^{iqx} \langle 0|j^\mu(x) \, j^\nu(0)|0\rangle$$

$$= i \int \frac{d^4 p_n}{(2\pi)^3} \int d^4x \, e^{i(q-p_n)x} \sumint_n \langle 0|j^\mu(0)|n\rangle\langle n|j^\nu(0)|0\rangle$$

$$= i \int \frac{d^4 p_n}{(2\pi)^3} \sumint_n (2\pi)^4 \, \delta^{(4)}(q-p_n) \langle 0|j^\mu(0)|n\rangle\langle n|j^\nu(0)|0\rangle$$

$$= i \, 2\pi \sumint_n \langle 0|j^\mu(0)|n\rangle\langle n|j^\nu(0)|0\rangle|_{p_n=q} \, .$$

Key ingredient of the representation we are looking for is the **spectral function** tensor $\rho^{\mu\nu}(q)$ defined by

$$\rho^{\mu\nu}(q) \doteq \sumint_n \langle 0|j^\mu(0)|n\rangle\langle n|j^\nu(0)|0\rangle|_{p_n=q} \, . \tag{3.121}$$

Taking into account that q is the momentum of a physical state (spectral condition $q^2 \geq 0$, $q^0 \geq 0$), the relativistic covariant decomposition may be written as

$$\rho^{\mu\nu}(q) = \Theta(q^0)\Theta(q^2) \left\{ \left[q^\mu q^\nu - q^2 g^{\mu\nu}\right] \rho_1(q^2) + q^\mu q^\nu \rho_0(q^2) \right\} \tag{3.122}$$

and current conservation $\partial_\mu j^\mu = 0 \Leftrightarrow q_\mu \rho^{\mu\nu} = 0$ implies $\rho_0 \equiv 0$, which is the transversality condition. For non–conserved currents, like the one's of the weak interactions, a longitudinal component ρ_0 exists in addition to the transversal one ρ_1. Note that $\Theta(p^2)$ may be represented as

$$\Theta(q^2) = \int_0^\infty dm^2 \delta(q^2 - m^2)$$

and therefore we may write

$$i \int d^4x \, e^{iqx} \langle 0|j^\mu(x) \, j^\nu(0)|0\rangle \tag{3.123}$$

$$= \int_0^\infty dm^2 \left\{ \left[m^2 g^{\mu\nu} - q^\mu q^\nu\right] \rho_1(m^2) - q^\mu q^\nu \rho_0(m^2) \right\} \left(-2\pi i \Theta(q^0) \delta(q^2 - m^2)\right),$$

which is the Källen-Lehmann representation for the positive frequency part of the current correlator. The latter, according to (2.139), is twice the imaginary part of the corresponding time–ordered current correlation function

$$\mathrm{i}\int \mathrm{d}^4x\, \mathrm{e}^{\mathrm{i}qx} \langle 0|Tj^\mu(x)\, j^\nu(0)|0\rangle \qquad (3.124)$$

$$= \int_0^\infty \mathrm{d}m^2 \left\{ [m^2 g^{\mu\nu} - q^\mu q^\nu]\, \rho_1(m^2) - q^\mu q^\nu \rho_0(m^2) \right\} \left(\frac{1}{q^2 - m^2 + \mathrm{i}\varepsilon} \right)$$

constrained to positive q^0.

In our case, for the conserved electromagnetic current, only the transversal amplitude is present: thus $\rho_0 \equiv 0$ and we denote ρ_1 by ρ, simply[26]. Thus, formally, in Fourier space we have

$$\mathrm{i}\int \mathrm{d}^4x\, \mathrm{e}^{\mathrm{i}qx} \langle 0|Tj^\mu_{\mathrm{em}}(x)\, j^\nu_{\mathrm{em}}(0)|0\rangle$$

$$= \int_0^\infty \mathrm{d}m^2\, \rho(m^2)\, (m^2 g^{\mu\nu} - q^\mu q^\nu)\, \frac{1}{q^2 - m^2 + \mathrm{i}\varepsilon}$$

$$= -\left(q^2 g^{\mu\nu} - q^\mu q^\nu\right) \hat{\Pi}'_\gamma(q^2) \qquad (3.125)$$

where $\hat{\Pi}'_\gamma(q^2)$ up to a factor e^2 is the **photon vacuum polarization function** introduced before (see (2.154) and (2.156)):

$$\Pi'_\gamma(q^2) = e^2 \hat{\Pi}'_\gamma(q^2)\,. \qquad (3.126)$$

With this bridge to the photon self–energy function Π'_γ we can get its imaginary part by substituting

$$\frac{1}{q^2 - m^2 + \mathrm{i}\varepsilon} \to -\pi\,\mathrm{i}\,\delta(q^2 - m^2)$$

in (3.125), which if constrained to positive q^0 yields back half of (3.123) with $\rho_0 = 0$. Thus contracting (3.123) with $2\Theta(q^0) g_{\mu\nu}$ and dividing by $g_{\mu\nu}(q^2 g^{\mu\nu} - q^\mu q^\nu) = 3q^2$ we obtain

$$2\Theta(q^0)\, \mathrm{Im}\, \hat{\Pi}'_\gamma(q^2) = \Theta(q^0)\, 2\pi\, \rho(q^2) \qquad (3.127)$$

$$= -\frac{1}{3q^2}\, 2\pi \sum_n\!\!\!\!\!\!\!\int \langle 0|j^\mu_{\mathrm{em}}(0)|n\rangle\langle n|j_{\mu\,\mathrm{em}}(0)|0\rangle\big|_{p_n=q}\,.$$

---

[26] In case of a conserved current, where $\rho_0 \equiv 0$, we may formally derive that $\rho_1(s)$ is real and positive $\rho_1(s) \geq 0$. To this end we consider the element ρ^{00}

$$\rho^{00}(q) = \sum_n\!\!\!\!\!\!\!\int \langle 0|j^0(0)|n\rangle\langle n|j^0(0)|0\rangle\big|_{q=p_n}$$

$$= \sum_n\!\!\!\!\!\!\!\int |\langle 0|j^0(0)|n\rangle|^2_{q=p_n} \geq 0$$

$$= \Theta(q^0)\, \Theta(q^2)\, \boldsymbol{q}^2\, \rho_1(q^2)$$

from which the statement follows.

3.7 Dispersion Relations

Again causality implies analyticity and the validity of a dispersion relation. In fact the electromagnetic current correlator exhibits a logarithmic UV singularity and thus requires one subtraction such that from (3.125) we find

$$\Pi'_\gamma(q^2) - \Pi'_\gamma(0) = \frac{q^2}{\pi} \int_0^\infty ds \, \frac{\text{Im} \, \Pi'_\gamma(s)}{s(s - q^2 - i\varepsilon)} . \tag{3.128}$$

Unitarity (3.99) implies the *optical theorem*, which is obtained from this relation in the limit of elastic forward scattering $|f\rangle \to |i\rangle$ where

$$2\text{Im} \, T_{ii} = \sum_n (2\pi)^4 \, \delta^{(4)}(P_n - P_i) \, |T_{ni}|^2 . \tag{3.129}$$

which tells us that the imaginary part of the photon propagator is proportional to the total cross section $\sigma_{\text{tot}}(e^+e^- \to \gamma^* \to \text{anything})$ ("anything" means any possible state). The precise relationship reads (see Sect. 5.1.3 below)

$$\text{Im} \, \hat{\Pi}'_\gamma(s) = \frac{1}{12\pi} R(s) \tag{3.130}$$

$$\text{Im} \, \Pi'_\gamma(s) = e(s)^2 \, \text{Im} \, \hat{\Pi}'_\gamma(s) = \frac{s}{e(s)^2} \, \sigma_{\text{tot}}(e^+e^- \to \gamma^* \to \text{anything}) = \frac{\alpha(s)}{3} R(s)$$

where

$$R(s) = \sigma_{\text{tot}} / \frac{4\pi\alpha(s)^2}{3s} . \tag{3.131}$$

The normalization factor is the point cross section (tree level) $\sigma_{\mu\mu}(e^+e^- \to \gamma^* \to \mu^+\mu^-)$ in the limit $s \gg 4m_\mu^2$. Taking into account the mass effects the $R(s)$ which corresponds to the production of a lepton pair reads

$$R_\ell(s) = \sqrt{1 - \frac{4m_\ell^2}{s}} \left(1 + \frac{2m_\ell^2}{s} \right) , \quad (\ell = e, \mu, \tau) \tag{3.132}$$

which may be read of from the imaginary part given in (2.175). This result provides an alternative way to calculate the renormalized vacuum polarization function (2.172), namely, via the DR (3.128) which now takes the form

$$\Pi'^\ell_{\gamma \, \text{ren}}(q^2) = \frac{\alpha q^2}{3\pi} \int_{4m_\ell^2}^\infty ds \, \frac{R_\ell(s)}{s(s - q^2 - i\varepsilon)} \tag{3.133}$$

yielding the vacuum polarization due to a lepton–loop.

In contrast to the leptonic part, the hadronic contribution cannot be calculated analytically as a perturbative series, but it can be expressed in terms of the cross section of the reaction $e^+e^- \to$ hadrons, which is known from experiments. Via

$$R_{\text{had}}(s) = \sigma(e^+e^- \to \text{hadrons}) / \frac{4\pi\alpha(s)^2}{3s} . \tag{3.134}$$

we obtain the relevant hadronic vacuum polarization

$$\Pi'^{had}_{\gamma\,ren}(q^2) = \frac{\alpha q^2}{3\pi} \int_{4m_\pi^2}^{\infty} ds \, \frac{R_{had}(s)}{s(s - q^2 - i\varepsilon)} \,. \quad (3.135)$$

At low energies, where the final state necessarily consists of two pions, the cross section is given by the square of the electromagnetic form factor of the pion (undressed from VP effects),

$$R_{had}(s) = \frac{1}{4}\left(1 - \frac{4m_\pi^2}{s}\right)^{\frac{3}{2}} |F_\pi^{(0)}(s)|^2 \,, \qquad s < 9 m_\pi^2 \,, \quad (3.136)$$

which directly follows from the corresponding imaginary part (2.253) of the photon vacuum polarization. There are three differences between the pionic loop integral and those belonging to the lepton loops:

- the masses are different
- the spins are different
- the pion is composite – the Standard Model leptons are elementary

The compositeness manifests itself in the occurrence of the form factor $F_\pi(s)$, which generates an enhancement: at the ρ peak, $|F_\pi(s)|^2$ reaches values about 45, while the quark parton model would give about 7. The remaining difference in the expressions for the quantities $R_\ell(s)$ and $R_h(s)$ in (3.132) and (3.136), respectively, originates in the fact that the leptons carry spin $\frac{1}{2}$, while the spin of the pion vanishes. Near threshold, the angular momentum barrier suppresses the function $R_h(s)$ by three powers of momentum, while $R_\ell(s)$ is proportional to the first power. The suppression largely compensates the enhancement by the form factor – by far the most important property is the mass.

3.8 Dispersive Calculation of Feynman Diagrams

Dispersion relations (DR) may be used to calculate Feynman integrals in a way different from the Feynman parametric approach described in Sect. 2.5. The reason is simply because the imaginary part of an amplitude in general is much easier to calculate than the amplitude itself, which then follows from the imaginary part by a one–fold integral. The imaginary part in principle may be obtained by the unitarity relation of the form (3.99) which translate into *Cutkosky rules* [122], which may be obtained using Veltman's [123] largest time equation in coordinate space. The latter make use of the splitting of the Feynman propagator into real and imaginary part (2.139) and contributes to the imaginary part of a Feynman integral if the substitution

$$\frac{1}{p^2 - m^2 + i\varepsilon} \to -\pi\, i\, \delta(p^2 - m^2)$$

replacing a virtual particle (un–cut line) by a physical state (cut line) is made for an odd number of propagators, and provided the corresponding state is physical, i.e. is admissible by energy–momentum conservation and all other physical conservation laws (charge, lepton number etc.). With a diagram we may associate a specific physical channel by specifying which external lines are in–coming and which are out–going. For a given channel then the imaginary part of the diagram is given by cutting internal lines of the diagram between the in–coming and the out–going lines in all possible ways into two disconnected parts. A cut contributes if the cut lines can be viewed as external lines of a real physical subprocess. On the right hand side of the cut the amplitude has to be taken complex conjugate, since the out–going state produced by the cut on the left hand side becomes the in–coming state on the right hand side. Due to the many extra δ–functions (on–shell conditions) part of the integrations become phase space integrations, which in general are easier to do. As a rule, the complexity is reduced from n–loop to a $n-1$–loop problem, on the expense that the last integration, a dispersion integral, still has to be done. A very instructive non–trivial example has been presented by Terentev [14] for the complete two–loop calculation of $g-2$ in QED.

Cut diagrams in conjunction with DRs play a fundamental role also beyond being just a technical trick for calculating Feynman integrals. They not only play a key role for the evaluation of non–perturbative hadronic effect but allow to calculate numerically or sometimes analytically all kinds of VP effects in higher order diagrams as we will see. Before we discuss this in more detail, let us summarize the key ingredients of the method, which we have considered before, once more:

- *Optical theorem* implied by *unitarity:* maybe most familiar is its application to scattering processes: the imaginary part of the forward scattering amplitude of an elastic process $A + B \to A + B$ is proportional to the sum over all possible final states $A + B \to$ "anything" (see Fig. 3.14)

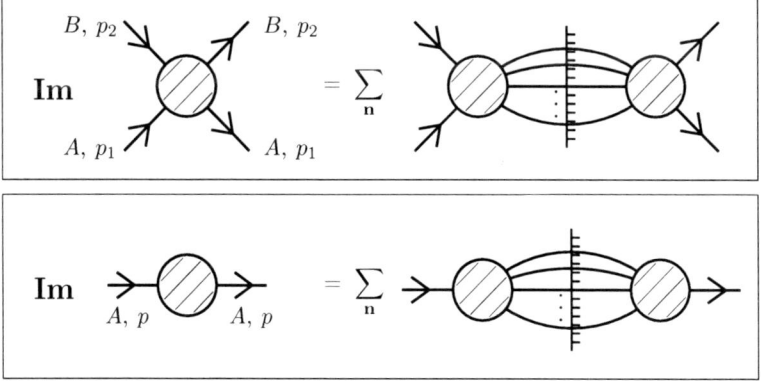

Fig. 3.14. Optical theorem for scattering and propagation

192   3 Lepton Magnetic Moments: Basics

$$\text{Im } T_{\text{forward}} (A + B \to A + B) = \sqrt{\lambda(s, m_1^2, m_2^2)}\, \sigma_{tot}(A + B \to \text{anything})$$

for the photon propagator it implies

$$\text{Im}\Pi'_\gamma(s) = \frac{s}{4\pi\alpha}\, \sigma_{tot}(e^+e^- \to \text{anything})$$

which we have been proving in the last section already.

- *Analyticity*, implied by *causality*, may be expressed in form of a so–called (subtracted) dispersion relation

$$\Pi'_\gamma(k^2) - \Pi'_\gamma(0) = \frac{k^2}{\pi} \int_0^\infty ds\, \frac{\text{Im}\Pi'_\gamma(s)}{s(s - k^2 - i\varepsilon)}\,. \qquad (3.137)$$

The latter, together with the optical theorem, directly implies the validity of (3.135). Note that its validity is based on general principles and holds beyond perturbation theory. It is the basis of all non–perturbative evaluations of hadronic vacuum polarization effects in terms of experimental data. But more than that.

Within the context of calculating $g - 2$ in the SM the maybe most important application of DRs concerns the vacuum polarization contribution related to diagrams of the type

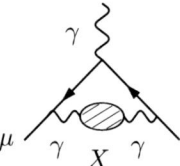

where the "blob" is the full photon propagator, including all kinds of contributions as predicted by the SM and maybe additional yet unknown contributions from physics beyond the SM. The vacuum polarization amplitude satisfies a dispersion relation (3.128) and the spectral function is given by (3.131).

The contribution to the anomalous magnetic moment from graphs of the photon vacuum polarization type shown above can be obtained in a straight forward way as follows: The physics wise relevant $g^{\mu\nu}$–term of the full photon propagator, carrying loop momentum k, reads

$$\frac{-ig^{\mu\nu}}{k^2(1 + \Pi'_\gamma(k^2))} \simeq \frac{-ig^{\mu\nu}}{k^2}\left(1 - \Pi'_\gamma(k^2) + \left(\Pi'_\gamma(k^2)\right)^2 - \cdots\right) \qquad (3.138)$$

and the renormalized photon self–energy may be written as

$$-\frac{\Pi'_{\gamma\,\text{ren}}(k^2)}{k^2} = \int_0^\infty \frac{ds}{s}\, \frac{1}{\pi}\text{Im}\,\Pi'_\gamma(s)\, \frac{1}{k^2 - s}\,. \qquad (3.139)$$

Note that the only k dependence under the convolution integral shows up in the last factor. Thus, the free photon propagator in the one–loop vertex graph discussed in Sect. 2.6.3 in the next higher order is replaced by

$$-ig_{\mu\nu}/k^2 \to -ig_{\mu\nu}/(k^2 - s)$$

which is the exchange of a photon of mass square s. The result afterward has to be convoluted with the imaginary part of the photon vacuum polarization. In a first step we have to calculate the contribution from the massive photon which may be calculated exactly as in the massless case. As discussed above $F_M(0)$ most simply may be calculated using the projection method directly at $q^2 = 0$. The result is [124, 125][27]

$$K_\mu^{(2)}(s) \equiv a_\mu^{(2)\ \text{heavy } \gamma} = \frac{\alpha}{\pi} \int_0^1 dx\, \frac{x^2(1-x)}{x^2 + (s/m_\mu^2)(1-x)}\,, \qquad (3.140)$$

which is the second order contribution to a_μ from an exchange of a photon with square mass s. Note that for $s = 0$ we get the known Schwinger result.

Utilizing this result and (3.139), the contribution from the "blob" to $g-2$ reads

$$a_\mu^{(X)} = \frac{1}{\pi} \int_0^\infty \frac{ds}{s}\, \text{Im}\, \Pi_\gamma^{'(X)}(s)\, K_\mu^{(2)}(s)\,. \qquad (3.141)$$

If we exchange integrations and evaluating the DR we arrive at

$$a_\mu^{(X)} = \frac{\alpha}{\pi} \int_0^1 dx\, (1-x) \int_0^\infty \frac{ds}{s}\, \frac{1}{\pi}\, \text{Im}\, \Pi_\gamma^{'(X)}(s)\, \frac{x^2}{x^2 + (s/m_\mu^2)(1-x)}$$

$$= \frac{\alpha}{\pi} \int_0^1 dx\, (1-x)\, \left[-\Pi_\gamma^{'(X)}(s_x)\right] \qquad (3.142)$$

where

$$s_x = -\frac{x^2}{1-x}\, m_\mu^2\,.$$

The last simple representation in terms of $\Pi_\gamma^{'(X)}(s_x)$ follows using

$$\frac{x^2}{x^2 + (s/m_\mu^2)(1-x)} = -s_x\, \frac{1}{s - s_x}\,.$$

In this context a convenient one–fold integral representation of the VP function is (2.173)

---

[27] Replacing the heavy vector exchange by a heavy scalar exchange leads to the substitution

$$x^2(1-x) \text{ (vector)} \to x^2(1-x/2) \text{ (scalar)}$$

in the numerator of (3.140).

$$\Pi'^{\ell}_{\gamma\,\text{ren}}\left(\frac{-x^2}{1-x}m_\mu^2\right) = -\frac{\alpha}{\pi}\int_0^1 dz\, 2z\,(1-z)\,\ln\left(1+\frac{x^2}{1-x}\frac{m_\mu^2}{m_\ell^2}z\,(1-z)\right),\tag{3.143}$$

which together with (3.142) leads to a two–fold integral representation of the VP contribution by lepton loops at two–loop order.

This kind of dispersion integral representation can be generalized to higher order and sequential VP insertions corresponding to the powers of $\Pi'(k^2)$ in (3.138). Denoting $\rho(s) = \text{Im}\,\Pi'_{\gamma\,\text{ren}}(s)/\pi$ we may write (3.139) in the form $-\Pi'_{\gamma\,\text{ren}}(k^2) = \int_0^\infty \frac{ds}{s}\,\rho(s)\,\frac{k^2}{k^2-s}$ such that the n–th term of the propagator expansion (3.138) is given by

$$(-\Pi'_{\gamma\,\text{ren}}(k^2))^n / k^2 = \frac{1}{k^2}\prod_{i=1}^n \int_0^\infty \frac{ds_i}{s_i}\,\rho(s_i)\,\frac{k^2}{k^2-s_i}$$

$$= \sum_{j=1}^n \int_0^\infty \frac{ds_j}{s_j}\,\rho(s_j)\,\frac{1}{k^2-s_j}\prod_{i\neq j}\int_0^\infty \frac{ds_i}{s_i}\,\rho(s_i)\,\frac{s_j}{s_j-s_i},$$

where we have been applying the partial fraction decomposition

$$\frac{1}{k^2}\prod_{i=1}^n \frac{k^2}{k^2-s_i} = \sum_{k=1}^n \frac{1}{k^2-s_j}\prod_{i\neq j}\frac{s_j}{s_j-s_i}.$$

We observe that the integration over the loop momentum k of the one–loop muon vertex proceeds exactly as before, with the photon replaced by a single heavy photon of mass s_j. Thus, the contribution to a_μ reads

$$a_\mu^{(X)} = \frac{\alpha}{\pi}\int_0^1 dx\,(1-x)\sum_{j=1}^n \int_0^\infty \frac{ds_j}{s_j}\,\rho(s_j)\,\frac{-s_x}{s_j-s_x}\prod_{i\neq j}\int_0^\infty \frac{ds_i}{s_i}\,\rho(s_i)\,\frac{s_j}{s_j-s_i}$$

$$= \frac{\alpha}{\pi}\int_0^1 dx\,(1-x)\left(\prod_{k=1}^n \int_0^\infty \frac{ds_k}{s_k}\,\rho(s_k)\right)s_x\left(\sum_{j=1}^n \frac{1}{s_x-s_j}\prod_{i\neq j}\frac{s_j}{s_j-s_i}\right).$$

Under the integral, to the last factor, we may apply the above partial fraction decomposition backward

$$\sum_{j=1}^n \frac{1}{s_x-s_j}\prod_{i\neq j}\frac{s_j}{s_j-s_i} = \frac{1}{s_x}\prod_{i=1}^n \frac{s_x}{s_x-s_i}$$

which proves that the s_i-integrals factorize and we find [126]

$$a_\mu^{(X)} = \frac{\alpha}{\pi} \int_0^1 dx \, (1-x) \left(\int_0^\infty \frac{ds}{s} \rho(s) \frac{-s_x}{s-s_x} \right)^n$$

$$= \frac{\alpha}{\pi} \int_0^1 dx \, (1-x) \left(-\Pi'_{\gamma \, \text{ren}}(s_x) \right)^n \tag{3.144}$$

We are thus able to write formally the result for the one–loop muon vertex when we replace the free photon propagator by the full transverse propagator as [127]

$$a_\mu^{(X)} = \frac{\alpha}{\pi} \int_0^1 dx \, (1-x) \left(\frac{1}{1+\Pi'_{\gamma \, \text{ren}}(s_x)} \right)$$

$$= \frac{1}{\pi} \int_0^1 dx \, (1-x) \, \alpha(s_x) \,, \tag{3.145}$$

which according to (3.114) is equivalent to the contribution of a free photon interacting with dressed charge (effective fine structure constant). However, since $\Pi'_{\gamma \, \text{ren}}(k^2)$ is negative and grows logarithmically with k^2 the full photon propagator develops a so called *Landau pole* where the effective fine structure constant becomes infinite. Thus resumming the perturbation expansion under integrals may produce a problem and one better resorts to the order by order approach, by expanding the full propagator into its geometrical progression. Nevertheless (3.145) is a very useful bookkeeping device, collecting effects from different contributions and different orders. In particular if we expand the 1PI photon self–energy into order by order contributions

$$\Pi'_{\gamma \, \text{ren}}(k^2) = \Pi'^{(2)}_{\gamma \, \text{ren}}(k^2) + \Pi'^{(4)}_{\gamma \, \text{ren}}(k^2) + \cdots$$

and also write $\rho = \rho^{(2)} + \rho^{(4)} + \cdots$ for the spectral densities.

Coming back to the single VP insertion formula (3.142) we may use (3.143) as well as the second form given in (2.173) which reads

$$\Pi'^{\ell}_{\gamma \, \text{ren}} \left(q^2/m^2 \right) = -\frac{\alpha}{\pi} \frac{q^2}{m^2} \int_0^1 dt \, \frac{\rho_2(t)}{\frac{q^2}{m^2} - 4/(1-t^2)}, \tag{3.146}$$

with[28]

---

[28] We adopt the notation of Kinoshita [126] and mention that the densities $\rho(t)$ used here are not to be confused with the $\rho(s)$ used just before, although they are corresponding to each other.

$$\rho_2(t) = \frac{t^2\,(1-t^2/3)}{1-t^2}\,, \tag{3.147}$$

and we may write

$$a_\mu^{(X)} = \left(\frac{\alpha}{\pi}\right)^2 \int_0^1 dx\,(1-x) \int_0^1 dt\,\frac{\rho_2(t)}{W_t(x)}\,, \tag{3.148}$$

where

$$1/W_t(x) = \frac{q^2}{m^2}\,\frac{1}{\frac{q^2}{m^2} - \frac{4}{1-t^2}}\bigg|_{\frac{q^2}{m^2} = -\frac{x^2}{1-x}\frac{m_\mu^2}{m^2}}$$

and hence

$$W_t(x) = 1 + \frac{4m^2}{(1-t^2)\,m_\mu^2}\,\frac{1-x}{x^2}\,. \tag{3.149}$$

If n equal loops are inserted we have

$$a_\mu^{(X)} = \frac{\alpha}{\pi} \int_0^1 dx\,(1-x)\,\left(\frac{\alpha}{\pi} \int_0^1 dt\,\frac{\rho(t)}{W_t(x)}\right)^n \tag{3.150}$$

according to the factorization theorem demonstrated before. This formula is suitable for calculating the contribution to the lepton anomalous magnetic moment once the spectral function $\rho(t)$ is known. For the one–loop 1PI self–energy we have $\rho_2(t)$ given by (3.147) and the corresponding density for the two–loop case reads [117]

$$\rho_4(t) = \frac{2t}{3\,(1-t^2)}\left\{\frac{(3-t^2)\,(1+t^2)}{2}\left(\mathrm{Li}_2(1) + \ln\frac{1+t}{2}\,\ln\frac{1+t}{1-t}\right.\right.$$
$$+ 2\left(\mathrm{Li}_2(\frac{1-t}{1+t}) + \mathrm{Li}_2(\frac{1+t}{2}) - \mathrm{Li}_2(\frac{1-t}{2})\right) - 4\,\mathrm{Li}_2(t) + \mathrm{Li}_2(t^2)\right)$$
$$+ \left(\frac{11}{16}(3-t^2)\,(1+t^2) + \frac{1}{4}t^4 - \frac{3}{2}t\,(3-t^2)\right)\ln\frac{1+t}{1-t}$$
$$\left. + t\,(3-t^2)\left(3\ln\frac{1+t}{2} - 2\ln t\right) + \frac{3}{8}t\,(5-3t^2)\right\}\,. \tag{3.151}$$

The corresponding result for the three–loop photon self–energy has been calculated in [128]. For four loops an approximate result is available [129]. Generally, the contribution to a_μ which follow from the lowest order lepton (ℓ) vertex diagram by modifying the photon propagator with l electron loops of order $2i$, m muon loops of order $2j$ and n tau loops of order $2k$ is given by

$$a_\ell = \left(\frac{\alpha}{\pi}\right)^{(1+il+jm+kn)} \int_0^1 dx\,(1-x) \left(\int_0^1 dt_1 \frac{\rho_{2i}(t_1)}{1+\frac{4}{1-t_1^2}\frac{1-x}{x^2}\left(\frac{m_e}{m_\ell}\right)^2}\right)^l$$

$$\cdot \left(\int_0^1 dt_2 \frac{\rho_{2j}(t_2)}{1+\frac{4}{1-t_2^2}\frac{1-x}{x^2}\left(\frac{m_\mu}{m_\ell}\right)^2}\right)^m \cdot \left(\int_0^1 dt_3 \frac{\rho_{2k}(t_3)}{1+\frac{4}{1-t_3^2}\frac{1-x}{x^2}\left(\frac{m_\tau}{m_\ell}\right)^2}\right)^n.$$

(3.152)

The same kind of approach works for the calculation of diagrams with VP insertions not only for the lowest order vertex. For any group of diagrams we may calculate in place of the true QED contribution the one obtained in massive QED with a photon of mass $\sqrt{s}$, and then convolute the result with the desired density of the photon VP analogous to (3.141) where (3.140) gets replaced by a different more complicated kernel function (see e.g. [61, 130] and below). It also should be noted that the representation presented here only involve integration over finite intervals ([0, 1]) and hence are particularly suited for numerical integration of higher order contributions when analytic results are not available.

The formalism developed here also is the key tool to evaluate the *hadronic contributions*, for which perturbation theory fails because of the strong interactions. In this case we represent Im $\Pi_\gamma^{'\,\mathrm{had}}(s)$ via (3.131) in terms of

$$\sigma_{\mathrm{had}}(s) = \sigma(e^+e^- \to \mathrm{hadrons})$$

where

$$\sigma_{\mathrm{had}}(s) = \frac{4\pi^2\alpha}{s}\frac{1}{\pi}\mathrm{Im}\,\Pi_\gamma^{'\,\mathrm{had}}(s) \qquad (3.153)$$

or in terms of the cross section ratio $R(s)$ defined by (3.131) where both $\sigma_{\mathrm{had}}(s)$ or equivalently $R_{\mathrm{had}}(s)$ will be taken from experiment, since they are not yet calculable reliably from first principles at present.

Starting point is the basic integral representation (from (3.141) using (3.131))

$$a_\mu^{\mathrm{had}} = \frac{\alpha}{\pi}\int_0^\infty \frac{ds}{s}\int_0^1 dx\,\frac{x^2(1-x)}{x^2+(1-x)s/m_\mu^2}\,\frac{\alpha}{3\pi}R(s)\,. \qquad (3.154)$$

If we first integrate over x we find the well known standard representation

$$a_\mu^{\mathrm{had}} = \frac{\alpha}{3\pi}\int_0^\infty \frac{ds}{s}\,K_\mu^{(2)}(s)\,R(s) \qquad (3.155)$$

as an integral along the cut of the vacuum polarization amplitude in the time-like region, while an interchange of the order of integrations yields the analog of (3.142): an integral over the hadronic shift of the fine structure constant (3.115) in the space–like domain [131]:

$$a_\mu^{\text{had}} = \frac{\alpha}{\pi} \int_0^1 dx\,(1-x)\,\Delta\alpha_{\text{had}}^{(5)}(-Q^2(x)) \tag{3.156}$$

where $Q^2(x) \equiv \frac{x^2}{1-x} m_\mu^2$ is the space–like square momentum–transfer or

$$x = \frac{Q^2}{2m_\mu^2}\left(\sqrt{1+\frac{4m_\mu^2}{Q^2}}-1\right).$$

Alternatively, by writing $(1-x) = -\frac{1}{2}\frac{d}{dx}(1-x)^2$ and performing a partial integration in (3.156) one finds

$$a_\mu^{\text{had}} = \frac{\alpha}{\pi} m_\mu^2 \int_0^1 dx\, x\,(2-x)\,\left(D(Q^2(x))/Q^2(x)\right) \tag{3.157}$$

where $D(Q^2)$ is the *Adler–function* [132] defined as a derivative of the shift of the fine structure constant

$$D(-s) = -(12\pi^2)\, s\, \frac{d\Pi_\gamma'(s)}{ds} = \frac{3\pi}{\alpha}\, s\, \frac{d}{ds}\Delta\alpha_{\text{had}}(s)\,. \tag{3.158}$$

The Adler–function is represented by

$$D(Q^2) = Q^2 \left(\int_{4m_\pi^2}^\infty \frac{R(s)}{(s+Q^2)^2}ds\right) \tag{3.159}$$

in terms of $R(s)$. i.e. in case of hadrons it can be evaluated in terms of experimental e^+e^-–data. The Adler–function is discussed in [133] and in Fig. 5.13 a comparison between theory and experiment is shown. The Adler–function is an excellent monitor for checking where pQCD works in the Euclidean region (see also [50]), and, in principle, it allows one to calculate a_μ^{had} relying more on pQCD and less on e^+e^-–data, in a well controllable manner. The advantage of this method at present is limited by the inaccuracies of the quark masses, in particular of the charm mass [134, 135].

References

1. J. D. Bjorken, S. D. Drell, *Relativistic Quantum Mechanics*, 1st edn (McGraw-Hill, New York 1964) p. 300 ; *Relativistic Quantum Fields*, 1st edn (McGraw-Hill, New York 1965) p. 396
2. P. A. M. Dirac, Proc. Roy. Soc. A **117** (1928) 610; A **118** (1928) 351

3. B. Odom, D. Hanneke, B. D'Urso, G. Gabrielse Phys. Rev. Lett. **97** (2006) 030801
4. G. Gabrielse, D. Hanneke, T. Kinoshita, M. Nio, B. Odom, Phys. Rev. Lett. **97** (2006) 030802
5. T. Aoyama, M. Hayakawa, T. Kinoshita, M. Nio, arXiv:0706.3496 [hep-ph]
6. S. Eidelman et al. [Particle Data Group], Phys. Lett. B **592** (2004) 1
7. P. J. Mohr, B. N. Taylor, Rev. Mod. Phys. **72** (2000) 351; **77** (2005) 1
8. J. S. Schwinger, Phys. Rev. **73** (1948) 416
9. D. J. Broadhurst, D. Kreimer, J. Symb. Comput. **27** (1999) 581; Int. J. Mod. Phys. C **6** (1995) 519; Phys. Lett. B **393** (1997) 403;
   D. J. Broadhurst, J. A. Gracey, D. Kreimer, Z. Phys. C **75** (1997) 559
10. A. Devoto, D. W. Duke, Riv. Nuovo Cim. **7N6** (1984) 1
11. R. Karplus, N. M. Kroll, Phys. Rep. C **77** (1950) 536
12. A. Petermann, Helv. Phys. Acta **30** (1957) 407; Nucl. Phys. **5** (1958) 677
13. C. M. Sommerfield, Phys. Rev. **107** (1957) 328; Ann. Phys. (N.Y.) **5** (1958) 26
14. M. V. Terentev, Sov. Phys. JETP **16** (1963) 444 [Zh. Eksp. Teor. Fiz. **43** (1962) 619]
15. S. Laporta, E. Remiddi, Phys. Lett. B **379** (1996) 283
16. T. Kinoshita, Phys. Rev. Lett. **75** (1995) 4728
17. T. Kinoshita, W. J. Marciano, In: *Quantum Electrodynamics*, ed. T. Kinoshita (World Scientific, Singapore 1990) pp. 419–478
18. T. Kinoshita, M. Nio, Phys. Rev. Lett. **90** (2003) 021803; Phys. Rev. D **70** (2004) 113001
19. T. Kinoshita, M. Nio, Phys. Rev. D **73** (2006) 013003
20. P. J. Mohr, B. N. Taylor, Rev. Mod. Phys. **77** (2005) 1
21. H. Suura, E. Wichmann, Phys. Rev. **105** (1957) 1930;
    A. Petermann, Phys. Rev. **105** (1957) 1931
22. H. H. Elend, Phys. Lett. **20** (1966) 682; Erratum-ibid. **21** (1966) 720
23. B. E. Lautrup, E. de Rafael, Nucl. Phys. B **70** (1974) 317
24. S. Laporta, Nuovo Cim. A **106** (1993) 675
25. A. Czarnecki, M. Skrzypek, Phys. Lett. B **449** (1999) 354
26. S. Friot, D. Greynat, E. De Rafael, Phys. Lett. B **628** (2005) 73
27. S. Laporta, E. Remiddi, Phys. Lett. B **301** (1993) 440
28. J. H. Kühn, A. I. Onishchenko, A. A. Pivovarov, O. L. Veretin, Phys. Rev. D **68** (2003) 033018
29. J. Aldins, T. Kinoshita, S. J. Brodsky, A. J. Dufner, Phys. Rev. Lett. **23** (1969) 441; Phys. Rev. D **1** (1970) 2378
30. J. Bailey et al., Phys. Lett. B **28** (1968) 287
31. A. S. Elkhovsky, Sov. J. Nucl. Phys. **49** (1989) 656 [Yad. Fiz. **49** (1989) 1059]
32. M. Passera, J. Phys. G **31** (2005) R75; Phys. Rev. D **75** (2007) 013002
33. M. Caffo, S. Turrini, E. Remiddi, Phys. Rev. D **30** (1984) 483; E. Remiddi, S. P. Sorella, Lett. Nuovo Cim. **44** (1985) 231; D. J. Broadhurst, A. L. Kataev, O. V. Tarasov, Phys. Lett. B **298** (1993) 445; S. Laporta, Phys. Lett. B **312** (1993) 495; P. A. Baikov, D. J. Broadhurst, hep-ph/9504398
34. S. G. Karshenboim, Phys. Atom. Nucl. **56** (1993) 857 [Yad. Fiz. **56N6** (1993) 252]
35. T. Kinoshita, M. Nio, Phys. Rev. D **73** (2006) 053007
36. A. L. Kataev, Nucl. Phys. Proc. Suppl. **155** (2006) 369; hep-ph/0602098; Phys. Rev. D **74** (2006) 073011

37. H. Fritzsch, M. Gell-Mann, H. Leutwyler, Phys. Lett. **47**B (1973) 365
38. H. D. Politzer, Phys. Rev. Lett. **30** (1973) 1346;
    D. Gross, F. Wilczek, Phys. Rev. Lett. **30** (1973) 1343
39. S. Bethke, Phys. Rept. **403–404** (2004) 203
40. S. G. Gorishnii, A. L. Kataev, S. A. Larin, Phys. Lett. B **259** (1991) 144; L. R. Surguladze, M. A. Samuel, Phys. Rev. Lett. **66** (1991) 560 [Erratum-ibid. **66** (1991) 2416]; K. G. Chetyrkin, Phys. Lett. B **391** (1997) 402
41. K. G. Chetyrkin, J. H. Kühn, Phys. Lett. B **342** (1995) 356; K. G. Chetyrkin, R. V. Harlander, J. H. Kühn, Nucl. Phys. B **586** (2000) 56 [Erratum-ibid. B **634** (2002) 413]
42. R. V. Harlander, M. Steinhauser, Comput. Phys. Commun. **153** (2003) 244
43. S. Eidelman, F. Jegerlehner, Z. Phys. C **67** (1995) 585
44. A. E. Blinov et al. [MD-1 Collaboration], Z. Phys. C **70** (1996) 31
45. J. Z. Bai et al. [BES Collaboration], Phys. Rev. Lett. **84** (2000) 594; Phys. Rev. Lett. **88** (2002) 101802
46. R. R. Akhmetshin et al. [CMD-2 Collaboration], Phys. Lett. B **578** (2004) 285; Phys. Lett. B **527** (2002) 161
47. A. Aloisio et al. [KLOE Collaboration], Phys. Lett. B **606** (2005) 12
48. M. N. Achasov et al. [SND Collaboration], J. Exp. Theor. Phys. **103** (2006) 380 [Zh. Eksp. Teor. Fiz. **130** (2006) 437]
49. V. M. Aulchenko et al. [CMD-2], JETP Lett. **82** (2005) 743 [Pisma Zh. Eksp. Teor. Fiz. **82** (2005) 841]; R. R. Akhmetshin et al., JETP Lett. **84** (2006) 413 [Pisma Zh. Eksp. Teor. Fiz. **84** (2006) 491]; hep-ex/0610021
50. F. Jegerlehner, Nucl. Phys. Proc. Suppl. **162** (2006) 22 [hep-ph/0608329]
51. M. Davier, S. Eidelman, A. Höcker, Z. Zhang, Eur. Phys. J. C **27** (2003) 497; ibid. **31** (2003) 503
52. S. Ghozzi, F. Jegerlehner, Phys. Lett. B **583** (2004) 222
53. S. Narison, Phys. Lett. B **568** (2003) 231
54. V. V. Ezhela, S. B. Lugovsky, O. V. Zenin, hep-ph/0312114
55. K. Hagiwara, A. D. Martin, D. Nomura, T. Teubner, Phys. Lett. B **557** (2003) 69; Phys. Rev. D **69** (2004) 093003
56. J. F. de Troconiz, F. J. Yndurain, Phys. Rev. D **71** (2005) 073008
57. S. Eidelman, Proceedings of the XXXIII International Conference on High Energy Physics, July 27 – August 2, 2006, Moscow (Russia), World Scientific, to appear; M. Davier, Nucl. Phys. Proc. Suppl. **169** (2007) 288
58. K. Hagiwara, A. D. Martin, D. Nomura, T. Teubner, Phys. Lett. B **649** (2007) 173
59. H. Leutwyler, *Electromagnetic form factor of the pion,* hep-ph/0212324; G. Colangelo, Nucl. Phys. Proc. Suppl. 131 (2004) 185; ibid. **162** (2006) 256
60. R. Alemany, M. Davier, A. Höcker, Eur. Phys. J. C **2** (1998) 123
61. R. Barbieri, E. Remiddi, Phys. Lett. B **49** (1974) 468; Nucl. Phys. B **90** (1975) 233
62. B. Krause, Phys. Lett. B **390** (1997) 392
63. J. Calmet, S. Narison, M. Perrottet, E. de Rafael, Phys. Lett. B **61** (1976) 283
64. T. Kinoshita, B. Nizic, Y. Okamoto, Phys. Rev. Lett. **52** (1984) 717; Phys. Rev. D **31** (1985) 2108
65. H. Kolanoski, P. Zerwas, Two-Photon Physics, In: *High Energy Electron-Positron Physics,* ed. A. Ali, P. Söding, (World Scientific, Singapore 1988) pp. 695–784; D. Williams et al. [Crystal Ball Collaboration], SLAC-PUB-4580, 1988, unpublished

66. G. Ecker, J. Gasser, A. Pich, E. de Rafael, Nucl. Phys. B **321** (1989) 311;
    G. Ecker, J. Gasser, H. Leutwyler, A. Pich, E. de Rafael, Phys. Lett. B **223** (1989) 425
67. J. Bijnens, E. Pallante, J. Prades, Phys. Rev. Lett. **75** (1995) 1447 [Erratum-ibid. **75** (1995) 3781]; Nucl. Phys. B **474** (1996) 379; [Erratum-ibid. **626** (2002) 410]
68. G. 't Hooft, Nucl. Phys. B **72** (1974) 461; ibid. **75** (1974) 461
69. A. V. Manohar, Hadrons in the $1/N$ Expansion, In: *At the frontier of Particle Physics*, ed M. Shifman, (World Scientific, Singapore 2001) Vol. 1, pp. 507–568
70. E. de Rafael, Phys. Lett. B **322** (1994) 239
71. M. Hayakawa, T. Kinoshita, A. I. Sanda, Phys. Rev. Lett. **75** (1995) 790; Phys. Rev. D **54** (1996) 3137
72. M. Hayakawa, T. Kinoshita, Phys. Rev. D **57** (1998) 465 [Erratum-ibid. D **66** (2002) 019902]
73. M. Knecht, A. Nyffeler, M. Perrottet, E. De Rafael, Phys. Rev. Lett. **88** (2002) 071802
74. S. Peris, M. Perrottet, E. de Rafael, JHEP **9805** (1998) 011; M. Knecht, S. Peris, M. Perrottet, E. de Rafael, Phys. Rev. Lett. **83** (1999) 5230; M. Knecht, A. Nyffeler, Eur. Phys. J. C **21** (2001) 659
75. M. Knecht, A. Nyffeler, Phys. Rev. D **65**, 073034 (2002)
76. I. Blokland, A. Czarnecki, K. Melnikov, Phys. Rev. Lett. **88** (2002) 071803
77. M. Ramsey-Musolf, M. B. Wise, Phys. Rev. Lett. **89** (2002) 041601
78. K. Melnikov, A. Vainshtein, Phys. Rev. D **70** (2004) 113006
79. A. Nyffeler, Nucl. Phys. B (Proc. Suppl.) **131** (2004) 162
80. R. Jackiw, S. Weinberg, Phys. Rev. D **5** (1972) 2396;
    I. Bars, M. Yoshimura, Phys. Rev. D **6** (1972) 374;
    G. Altarelli, N. Cabibbo, L. Maiani, Phys. Lett. B **40** (1972) 415;
    W. A. Bardeen, R. Gastmans, B. Lautrup, Nucl. Phys. B **46** (1972) 319;
    K. Fujikawa, B. W. Lee, A. I. Sanda, Phys. Rev. D **6** (1972) 2923
81. E. A. Kuraev, T. V. Kukhto and A. Schiller, Sov. J. Nucl. Phys. **51** (1990) 1031 [Yad. Fiz. **51** (1990) 1631];
    T. V. Kukhto, E. A. Kuraev, A. Schiller, Z. K. Silagadze, Nucl. Phys. B **371** (1992) 567
82. S. L. Adler, Phys. Rev. **177** (1969) 2426;
    J. S. Bell, R. Jackiw, Nuovo Cim. **60A** (1969) 47;
    W. A. Bardeen, Phys. Rev. **184** (1969) 1848;
    C. Bouchiat, J. Iliopoulos, P. Meyer, Phys. Lett. **38**B (1972) 519;
    D. Gross, R. Jackiw, Phys. Rev. D **6** (1972) 477;
    C. P. Korthals Altes, M. Perrottet, Phys. Lett. **39**B (1972) 546
83. S. Peris, M. Perrottet, E. de Rafael, Phys. Lett. B **355** (1995) 523
84. A. Czarnecki, B. Krause, W. Marciano, Phys. Rev. D **52** (1995) R2619
85. G. Degrassi, G. F. Giudice, Phys. Rev. **58D** (1998) 053007
86. M. Knecht, S. Peris, M. Perrottet, E. de Rafael, JHEP **0211** (2002) 003
87. A. Czarnecki, W. J. Marciano, A. Vainshtein, Phys. Rev. D **67** (2003) 073006
88. E. D'Hoker, Phys. Rev. Lett. **69** (1992) 1316
89. A. Czarnecki, B. Krause, W. J. Marciano, Phys. Rev. Lett. **76** (1996) 3267
90. S. Heinemeyer, D. Stöckinger, G. Weiglein, Nucl. Phys. B **699** (2004) 103
91. T. Gribouk, A. Czarnecki, Phys. Rev. D **72** (2005) 053016
92. R. S. Van Dyck, P. B. Schwinberg, H. G. Dehmelt, Phys. Rev. Lett. **59** (1987) 26

93. A. Nyffeler, private communication
94. P. Cladé et al., Phys. Rev. Lett. **96** (2006) 033001
95. V. Gerginov et al., Phys. Rev. A **73** (2006) 032504
96. C. Schwob et al., Phys. Rev. Lett. **82** (1999) 4960
97. M. P. Bradley et al., Phys. Rev. Lett. **83** (1999) 4510
98. T. Beier et al., Phys. Rev. Lett. **88** (2002) 011603
99. D. L. Farnham, R. S. Van Dyck, P. B. Schwinberg, Phys. Rev. Lett. **75** (1995) 3598
100. A. Wicht et al.: In: *Proc. of the 6th Symp. on Freq. Standards and Metrology* (World Scientific, Singapore 2002) pp. 193–212, Phys. Scr. **T102** (2002) 82
101. T. Udem et al., Phys. Rev. Lett. **82** (1999) 3568
102. G. W. Bennett et al. [Muon (g-2) Collaboration], Phys. Rev. Lett. **92** (2004) 161802
103. F. Jegerlehner, J. Fleischer, Phys. Lett. B **151** (1985) 65; Acta Phys. Polon. B **17** (1986) 709
104. Ya. B. Zeldovich, Soviet Phys. JETP **6** (1958) 1184
105. Ya. B. Zeldovich, A. M. Perelomov, Soviet Phys. JETP **12** (1961) 777
106. R. E. Marshak, Riazuddin, C. P. Ryan, *Theory of Weak Interactions in Particle Physics*, (Wiley-Interscience, New York 1969) p. 776
107. H. Czyż, K. Kołodziej, M. Zrałek, P. Khristova, Can. J. Phys. **66** (1988) 132; H. Czyż, M. Zrałek, Can. J. Phys. **66** (1988) 384
108. A. Gongora, R. G. Stuart, Z. Phys. C **55** (1992) 101
109. L. D. Landau, Nucl. Phys. **3** (1957) 127; Soviet Phys. JETP **5** (1957) 336 [Zh. Eksp. Teor. Fiz. **32** (1957) 405]
110. Ya. B. Zeldovich, Soviet Phys. JETP **12** (1960) 1030 [Zh. Eksp. Teor. Fiz. **39** (1960) 1483]
111. F. Hoogeveen, Nucl. Phys. B **341** (1990) 322
112. C. Jarlskog, Phys. Rev. Lett. **55** (1985) 1039
113. B. C. Regan, E. D. Commins, C. J. Schmidt, D. DeMille, Phys. Rev. Lett. **88** (2002) 071805
114. J. Bailey et al., Nucl. Phys. B **150** (1979) 1
115. F. J. M. Farley et al., Phys. Rev. Lett. **93** (2004) 052001; M. Aoki et al. [J-PARC Letter of Intent]: *Search for a Permanent Muon Electric Dipole Moment at the* $\times 10^{-24}$ *e· cm Level*, http://www-ps.kek.jp/jhf-np/LOIlist/pdf/L22.pdf
116. W. Bernreuther, M. Suzuki, Rev. Mod. Phys. **63** (1991) 313 [Erratum-ibid. **64** (1992) 633]
117. G. Källén, A. Sabry, K. Dan. Vidensk. Selsk. Mat.-Fys. Medd. **29** (1955) No. 17
118. M. Steinhauser, Phys. Lett. B**429** (1998) 158
119. M. Acciarri et al. [L3 Collaboration], Phys. Lett. B **476** (2000) 40; G. Abbiendi et al. [OPAL Collaboration], Eur. Phys. J. C **45** (2006) 1
120. L. Trentadue, Nucl. Phys. Proc. Suppl. **162** (2006) 73
121. G. Källén, Helv. Phys. Acta **25** (1952) 417; H. Lehmann, Nuovo Cim. **11** (1954) 342
122. L. D. Landau, Nucl. Phys. **13** (1959) 181; S. Mandelstam, Phys. Rev. **112** (1958) 1344, **115** (1959) 1741; R. E. Cutkosky, J. Math. Phys. **1** (1960) 429
123. M. J. G. Veltman, Physica **29** (1963) 186

124. V. B. Berestetskii, O. N. Krokhin, A. K. Khelbnikov, Sov. Phys. JETP **3** (1956) 761 [Zh. Eksp. Teor. Fiz. **30** (1956) 788]
125. S. J. Brodsky, E. De Rafael, Phys. Rev. **168** (1968) 1620
126. T. Kinoshita, Nuovo Cim. B **51** (1967) 140;
    T. Kinoshita, W. B. Lindquist, Phys. Rev. D **27** (1983) 867
127. B. Lautrup, Phys. Lett. B **69** (1977) 109
128. A. H. Hoang et al., Nucl. Phys. B **452** (1995) 173
129. D. J. Broadhurst, A. L. Kataev, O. V. Tarasov, Phys. Lett. B **298** (1993) 445; P. A. Baikov, D. J. Broadhurst, In: Proceedings of the 4th International Workshop on Software Engineering and Artificial Intelligence for High Energy and Nuclear Physics (AIHENP95), Pisa, Italy, 1995, p.167 [hep-ph/9504398]
130. M. Caffo, E. Remiddi, S. Turrini, Nucl. Phys. B **141** (1978) 302; J. A. Mignaco, E. Remiddi, 1969, unpublished
131. B. E. Lautrup, A. Peterman, E. de Rafael, Phys. Reports 3C (1972) 193
132. S. L. Adler, Phys. Rev. D **10** (1974) 3714;
    A. De Rujula, H. Georgi, Phys. Rev. D **13** (1976) 1296
133. S. Eidelman, F. Jegerlehner, A. L. Kataev, O. Veretin, Phys. Lett. B **454** (1999) 369
134. F. Jegerlehner, J. Phys. G **29** (2003) 101
135. F. Jegerlehner, In: *Radiative Corrections*, ed by J. Solà (World Scientific, Singapore 1999) pp. 75–89

Part II

A Detailed Account of the Theory, Outline of Concepts of the Experiment, Status and Perspectives

4

Electromagnetic and Weak Radiative Corrections

4.1 $g - 2$ in Quantum Electrodynamics

The by far largest contribution to the anomalous magnetic moment is of pure QED origin. This is of course the reason why the measurements of a_e and a_μ until not so long time ago may have been considered as precision tests of QED. This clear dominance of just one type of interaction, the interaction of the charged leptons e, μ and τ with the photon, historically, was very important for the development of QFT and QED, since it allowed to test QED as a model theory under very simple, clean and unambiguous conditions. This was very crucial in strengthening our confidence in QFT as a basic theoretical framework. We should remember that it took about 20 years from the invention of QED (Dirac 1928 [$g_e = 2$]) until the first reliable results could be established (Schwinger 1948 [$a_e^{(1)} = \alpha/2\pi$]) after a covariant formulation and renormalization was understood and settled in its main aspects. As precision of experiments improved, the QED part by itself became a big challenge for theorists, because higher order corrections are sizable, and as the order of perturbation theory increases, the complexity of the calculations grow dramatically. Thus experimental tests were able to check QED up to 7 digits in the prediction which requires to evaluate the perturbation expansion up to 5 loops (5 terms in the expansion). The anomalous magnetic moment as a dimensionless quantity exhibits contributions which are just numbers expanded in powers of α, what one would get in QED with just one species of leptons, and contributions depending on the mass ratios if different leptons come into play. Thus taking into account all three leptons we obtain functions of the ratios of the lepton masses m_e, m_μ and m_τ. Considering a_μ, we can cast it into the following form [1, 2]

$$a_\mu^{\text{QED}} = A_1 + A_2(m_\mu/m_e) + A_2(m_\mu/m_\tau) + A_3(m_\mu/m_e, m_\mu/m_\tau) \quad (4.1)$$

The term A_1 in QED is universal for all leptons. Only those internal fermion loops count here where the fermion is the muon (=external lepton). The contribution A_2 has one scale and only shows up if an additional lepton loop of a lepton different from the external one is involved. This requires at least one more loop, thus two at least: for the muon as external lepton we have two possibilities: an additional electron–loop (light–in–heavy) $A_2(m_\mu/m_e)$ or an additional τ–loop (heavy–in–light) $A_2(m_\mu/m_\tau)$ two contributions of quite different character. The first produces large logarithms $\propto \ln(m_\mu/m_e)^2$ and accordingly large effects while the second, because of the *decoupling* of heavy particles in QED like theories, produces only small effects of order $\propto (m_\mu/m_\tau)^2$. The two–scale contribution requires a light as well as a heavy extra loops and hence starts at three loop order. We will discuss the different types of contributions in the following. Each of the terms is given in renormalized perturbation theory by an appropriate expansion in α:

$$A_1 = A_1^{(2)} \left(\frac{\alpha}{\pi}\right) + A_1^{(4)} \left(\frac{\alpha}{\pi}\right)^2 + A_1^{(6)} \left(\frac{\alpha}{\pi}\right)^3 + A_1^{(8)} \left(\frac{\alpha}{\pi}\right)^4 + A_1^{(10)} \left(\frac{\alpha}{\pi}\right)^5 + \cdots$$
$$A_2 = \phantom{A_1^{(2)} \left(\frac{\alpha}{\pi}\right) +} A_2^{(4)} \left(\frac{\alpha}{\pi}\right)^2 + A_2^{(6)} \left(\frac{\alpha}{\pi}\right)^3 + A_2^{(8)} \left(\frac{\alpha}{\pi}\right)^4 + A_2^{(10)} \left(\frac{\alpha}{\pi}\right)^5 + \cdots$$
$$A_3 = \phantom{A_1^{(2)} \left(\frac{\alpha}{\pi}\right) + A_1^{(4)} \left(\frac{\alpha}{\pi}\right)^2 +} A_3^{(6)} \left(\frac{\alpha}{\pi}\right)^3 + A_3^{(8)} \left(\frac{\alpha}{\pi}\right)^4 + A_3^{(10)} \left(\frac{\alpha}{\pi}\right)^5 + \cdots$$

and later we will denote by

$$C_L = \sum_{k=1}^{3} A_k^{(2L)},$$

the total L-loop confident of the $(\alpha/\pi)^L$ term.

In collecting various contributions we should always keep in mind the precision of the present experimental result [3]

$$a_\mu^{\mathrm{exp}} = 116592080(63) \times 10^{-11}$$

and the future prospects of possible improvements [4] which could reach an ultimate precision

$$\delta a_\mu^{\mathrm{fin}} \sim 10 \times 10^{-11}. \tag{4.2}$$

For the n–loop coefficients multiplying $(\alpha/\pi)^n$ this translates into the required accuracies given in Table 4.1. To match the current accuracy one has to multiply each entry with a factor 6, which is the experimental error in units of 10^{-10}.

As we will see many contributions are enhancement by large short–distance logarithms of the type $\ln m_\mu/m_e$. These terms are controlled by the RG equation of QED or equivalently by the homogeneous Callan-Symanzik (CS) equation [5]

Table 4.1. Numerical precision of coefficients up to five loops

$\delta A^{(2)}$	$\delta A^{(4)}$	$\delta A^{(6)}$	$\delta A^{(8)}$	$\delta A^{(10)}$
4×10^{-8}	1×10^{-5}	7×10^{-3}	3	1×10^3

4.1 $g-2$ in Quantum Electrodynamics

$$\left(m_e \frac{\partial}{\partial m_e} + \beta(\alpha)\,\alpha \frac{\partial}{\partial \alpha}\right) a_\mu^{(\infty)}\left(\frac{m_\mu}{m_e},\alpha\right) = 0$$

where $\beta(\alpha)$ is the QED β–function associated with charge renormalization. $a_\mu^{(\infty)}(\frac{m_\mu}{m_e},\alpha)$ is the leading form of a_μ in the sense that it includes powers of logs of mass ratios and constant terms but powers of m_e/m_μ are dropped[1]. The solution of the CS equation amounts to replace α by the running fine structure constant $\alpha(m_\mu)$ in $a_\mu^{(\infty)}(\frac{m_\mu}{m_e},\alpha)$, which implies taking into account the leading logs of higher orders. Since β is known to three loops and also a_μ is known analytically at three loops, it is possible to obtain the important higher leading logs quite easily. The basic RG concepts have been discussed in Sect. 2.6.5.

4.1.1 One–Loop QED Contribution

For completeness we mention this contribution represented by Fig. 4.1 here once more. According to (3.145) the leading order contribution may be written in the form

$$a_\mu^{(2)\,\text{QED}} = \frac{\alpha}{\pi}\int_0^1 \mathrm{d}x\,(1-x) = \frac{\alpha}{\pi}\frac{1}{2} \tag{4.3}$$

which is trivial to evaluate.

4.1.2 Two–Loop QED Contribution

At two loops in QED the 9 diagrams shown in Fig. 4.2 are contributing to $g-2$. The (within QED) universal contribution comes from the first 6 diagrams, which besides the external muon string of lines have attached two virtual photons. They form a gauge invariant subset of diagrams and yield the result

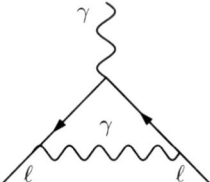

Fig. 4.1. The universal lowest order QED contribution to a_ℓ

---

[1] a_μ itself satisfies an exact CS equation which is inhomogeneous, the inhomogeneity being a mass insertion (m_e) on a_μ; this inhomogeneous part is $O(m_e/m_\mu)$ and thus drops from the CS equation for the asymptotic approximation $a_\mu^{(\infty)}$.

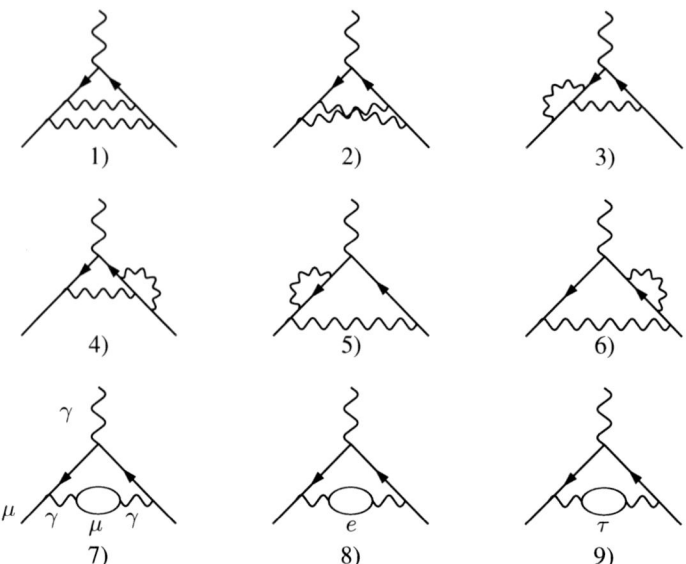

Fig. 4.2. Diagrams 1–7 represent the universal second order contribution to a_ℓ, diagram 8 yields the "light", diagram 9 the "heavy" mass dependent corrections

$$A^{(4)}_{1\,[1-6]} = -\frac{279}{144} + \frac{5\pi^2}{12} - \frac{\pi^2}{2}\ln 2 + \frac{3}{4}\zeta(3)\,.$$

The last 3 diagrams include photon vacuum polarization (vap) due to the lepton loops. The one with the muon loop is also universal in the above sense (one flavor = single scale QED) and contributes the mass independent correction

$$A^{(4)}_{1\,\mathrm{vap}}(m_\mu/m_\ell = 1) = \frac{119}{36} - \frac{\pi^2}{3}\,.$$

The complete "universal" part yields the coefficient $A^{(4)}_1$ calculated first by Petermann [6] and by Sommerfield [7] in 1957:

$$A^{(4)}_{1\,\mathrm{uni}} = \frac{197}{144} + \frac{\pi^2}{12} - \frac{\pi^2}{2}\ln 2 + \frac{3}{4}\zeta(3) = -0.328\,478\,965\,579\,193\,78... \quad (4.4)$$

where $\zeta(n)$ is the Riemann zeta function of argument n. The mass dependent non–universal contribution is due to the last two diagrams of Fig. 4.2. The coefficient now is a function of the mass m_ℓ of the lepton forming the closed loop. Using the representation (3.142) together with (2.173) we see that the coefficient of $(\alpha/\pi)^2$ may be written as double integral [8]

$$A^{(4)}_{2\,\mathrm{vap}}(1/x_\ell) = \int_0^1 du \int_0^1 dv\,\frac{u^2\,(1-u)\,v^2\,(1-v^2/3)}{u^2\,(1-v^2) + 4x_\ell^2\,(1-u)}\,, \quad (4.5)$$

where $x_\ell = m_\ell/m_\mu$ and m_ℓ is the mass of the virtual lepton in the vacuum polarization subgraph[2]. It was computed in the late 1950s [9] for $m_\ell = m_e$ and neglecting terms of $O(m_e/m_\mu)$. Its exact expression was calculated in 1966 [10]. The first integration yields logarithms, the second one double logarithms (products of logarithms) and a new type of integrals, the dilogarithms or Spence functions $\text{Li}_2(x) = -\int_0^1 dt \ln(1-xt)/t$ defined earlier in (2.203). Actually, by taking advantage of the properties of the dilogarithm (2.202), the full analytic result of [10] can be simplified to [11]

$$A_{2\,\text{vap}}^{(4)}(1/x) = -\frac{25}{36} - \frac{\ln x}{3} + x^2(4 + 3\ln x) + x^4\left[\frac{\pi^2}{3} - 2\ln x \ln\left(\frac{1}{x} - x\right) - \text{Li}_2(x^2)\right]$$
$$+ \frac{x}{2}(1 - 5x^2)\left[\frac{\pi^2}{2} - \ln x \ln\left(\frac{1-x}{1+x}\right) - \text{Li}_2(x) + \text{Li}_2(-x)\right]$$
$$= -\frac{25}{36} - \frac{\ln x}{3} + x^2(4 + 3\ln x) + x^4\left[2\ln^2(x) - 2\ln x \ln\left(x - \frac{1}{x}\right) + \text{Li}_2(1/x^2)\right]$$
$$+ \frac{x}{2}(1 - 5x^2)\left[-\ln x \ln\left(\frac{x-1}{x+1}\right) + \text{Li}_2(1/x) - \text{Li}_2(-1/x)\right] \quad (x > 1). \quad (4.7)$$

The first version of the formula is valid for arbitrary x. However, for $x > 1$ some of the logs as well as $\text{Li}_2(x)$ develop a cut and a corresponding imaginary part like the one of $\ln(1-x)$. Therefore, for the numerical evaluation in terms of a series expansion[3], it is an advantage to rewrite the $\text{Li}_2(x)$'s in terms of $\text{Li}_2(1/x)$'s, according to (2.202), which leads to the second form. For $x = 1$ (muon loop), using $\text{Li}_2(1) = \zeta(2) = \frac{\pi^2}{6}$ and $\text{Li}_2(-1) = -\frac{1}{2}\zeta(2) = -\frac{\pi^2}{12}$ the evaluation of (4.7) yields $A_{2\,\text{vap}}^{(4)}(1) = 119/36 - \pi^2/3$ the contribution already included in $A_{1\,\text{uni}}^{(4)}$ given by (4.4).

---

[2] We remind that the above integral representation is obtained by applying the method presented in Sect. 3.8. To start with, $a_\mu^{(4)}(\text{vap}, \ell)$ is given by a dispersion integral of the form (3.155) with $R(s) \to R_\ell(s)$ given by (3.132). Thus

$$a_\mu^{(4)}(\text{vap}, \ell) = \frac{\alpha}{3\pi} \int_{4m_\ell^2}^{\infty} \frac{ds}{s} K_\mu^{(2)}(s) R_\ell(s) \quad (4.6)$$

where $K_\mu^{(2)}(s)$ represents the contribution to a_μ from the one–loop diagram Fig. 4.1, where the photon has been replaced by a "heavy photon" of mass $\sqrt{s}$. The convolution with R_ℓ accomplishes the insertion of the corresponding lepton loop into the photon line of the one-loop vertex.

[3] A frequently used rapidly converging series expansion is

$$\text{Li}_2(x) = \sum_0^{\infty} B_n \frac{u^{n+1}}{(n+1)!}$$

where $u = -\ln(1-x)$ and B_n are the Bernoulli numbers.

For numerical calculations it is often convenient to work with asymptotic expansions. For a τ-loop an expansion for large arguments x gives formula (12) of [12]:

$$A_{2\,\text{vap}}^{(4)}(1/x_\tau \equiv l = \frac{m_\mu}{m_\tau}) = \frac{l^2}{45} + \frac{l^4 \ln l}{70} + \frac{9}{19600}l^4 - \frac{131}{99225}l^6 + \frac{4l^6}{315}\ln l$$
$$- \sum_{n=3}^{\infty} \frac{(8n^3 + 28n^2 - 45)l^{2n+2}}{[(n+3)(2n+3)(2n+5)]^2} + 2\ln l \sum_{n=3}^{\infty} \frac{nl^{2n+2}}{(n+3)(2n+3)(2n+5)} .$$

For the electron–loop an expansion for small x leads to formula (11) of [12]:

$$A_{2\,\text{vap}}^{(4)}(1/x_e \equiv 1/k = \frac{m_\mu}{m_e}) = -\frac{25}{36} + \frac{\pi^2}{4}k - \frac{1}{3}\ln k + (3 + 4\ln k)k^2 - \frac{5}{4}\pi^2 k^3$$
$$+ \left[\frac{\pi^2}{3} + \frac{44}{9} - \frac{14}{3}\ln k + 2\ln^2 k\right] k^4 + \frac{8}{15}k^6 \ln k - \frac{109}{225}k^6$$
$$+ \sum_{n=2}^{\infty} \left[\frac{2(n+3)}{n(2n+1)(2n+3)}\ln k - \frac{8n^3 + 44n^2 + 48n + 9}{n^2(2n+1)^2(2n+3)^2}\right] k^{2n+4} .$$

Evaluations of (4.7) or of the appropriate series expansions yields

$$A_{2\,\text{vap}}^{(4)}(m_\mu/m_e) = 1.094\,258\,3111\,(84)$$
$$A_{2\,\text{vap}}^{(4)}(m_\mu/m_\tau) = 0.000\,078\,064\,(25),$$

where the errors are solely due to the experimental uncertainties of the mass ratios.

According to Table 4.1 the τ yields a non–negligible contribution. At the two–loop level a $e - \tau$ mixed contribution is not possible, and hence $A_3^{(4)}(m_\mu/m_e, m_\mu/m_\tau) = 0$.

The complete two–loop QED contribution from the diagrams displayed in Fig. 4.2 is given by

$$C_2 = A_{1\,\text{uni}}^{(4)} + A_{2\,\text{vap}}^{(4)}(m_\mu/m_e) + A_{2\,\text{vap}}^{(4)}(m_\mu/m_\tau) = 0.765\,857\,410\,(26) .$$

and we have

$$a_\mu^{(4)\,\text{QED}} = 0.765\,857\,410\,(26)\left(\frac{\alpha}{\pi}\right)^2 \simeq 413217.621(14) \times 10^{-11} \quad (4.8)$$

for the complete 2–loop QED contribution to a_μ. The errors of $A_2^{(4)}(m_\mu/m_e)$ and $A_2^{(4)}(m_\mu/m_\tau)$ have been added in quadrature as the errors of the different measurements of the lepton masses may be treated as independent. The combined error $\delta C_2 = 2.6 \times 10^{-8}$ is negligible by the standards 1×10^{-5} of Table 4.1.

4.1.3 Three–Loop QED Contribution

At three loops in QED there are the 72 diagrams shown in Fig. 4.3 contributing to $g - 2$ of the muon. In closed fermion loops any of the SM fermions may circulate. The gauge invariant subset of 72 diagrams where all closed fermion loops are muon–loops yield the universal one–flavor QED contribution $A_{1\,\text{uni}}^{(6)}$. This set has been calculated analytically mainly by Remiddi and his collaborators [13], and Laporta and Remiddi obtained the final result in 1996 after finding a trick to calculate the non–planar "triple cross" topology diagram (diagram 25) of Fig. 4.3) [14] (see also [15]). The result, presented in (3.42) before, turned out to be surprisingly compact. All other corrections follow from Fig. 4.3 by replacing at least one muon in a loop by another lepton or quark. The such obtained mass dependent corrections are of particular interest because the light electron loops typically yield contributions, enhanced by large logarithms. Results for $A_2^{(6)}$ have been obtained in [16, 17, 18, 19, 20], for $A_3^{(6)}$ in [12, 21, 22, 23]. The leading terms of the expansion in the appropriate mass ratios have been discussed in Sect. 3.2.1 before. For the light–by–light contribution, graphs 1) to 6) of Fig. 4.3, the exact analytic result is known [19], but because of its length has not been published. The following asymptotic expansions are simple enough and match the requirement of the precision needed at the time:

$$A_{2\,\text{lbl}}^{(6)}(m_\mu/m_e) = \frac{2}{3}\pi^2 \ln\frac{m_\mu}{m_e} + \frac{59}{270}\pi^4 - 3\zeta(3) - \frac{10}{3}\pi^2 + \frac{2}{3}$$

$$+ \left(\frac{m_e}{m_\mu}\right)\left[\frac{4}{3}\pi^2\ln\frac{m_\mu}{m_e} - \frac{196}{3}\pi^2\ln 2 + \frac{424}{9}\pi^2\right]$$

$$+ \left(\frac{m_e}{m_\mu}\right)^2 \left[-\frac{2}{3}\ln^3\frac{m_\mu}{m_e} + \left(\frac{\pi^2}{9} - \frac{20}{3}\right)\ln^2\frac{m_\mu}{m_e} - \left(\frac{16}{135}\pi^4 + 4\zeta(3) - \frac{32}{9}\pi^2\right.\right.$$

$$\left.\left. + \frac{61}{3}\right)\ln\frac{m_\mu}{m_e} + \frac{4}{3}\pi^2\zeta(3) - \frac{61}{270}\pi^4 + 3\zeta(3) + \frac{25}{18}\pi^2 - \frac{283}{12}\right]$$

$$+ \left(\frac{m_e}{m_\mu}\right)^3 \left[\frac{10}{9}\pi^2\ln\frac{m_\mu}{m_e} - \frac{11}{9}\pi^2\right]$$

$$+ \left(\frac{m_e}{m_\mu}\right)^4 \left[\frac{7}{9}\ln^3\frac{m_\mu}{m_e} + \frac{41}{18}\ln^2\frac{m_\mu}{m_e} + \left(\frac{13}{9}\pi^2 + \frac{517}{108}\right)\ln\frac{m_\mu}{m_e}\right.$$

$$\left. + \frac{1}{2}\zeta(3) + \frac{191}{216}\pi^2 + \frac{13283}{2592}\right] + O\left((m_e/m_\mu)^5\right),$$

$$= 20.947\,924\,89(16) \tag{4.9}$$

214   4 Electromagnetic and Weak Radiative Corrections

Fig. 4.3. The universal third order contribution to a_μ. All fermion loops here are muon–loops. Graphs 1) to 6) are the light–by–light scattering diagrams. Graphs 7) to 22) include photon vacuum polarization insertions. All non–universal contributions follow by replacing at least one muon in a closed loop by some other fermion

where here and in the following we use m_e/m_μ as given in (3.29). The leading term in the (m_e/m_μ) expansion turns out to be surprisingly large. It has been calculated first in [24]. Prior to the exact calculation in [19] good numerical estimates 20.9471(29) [25] and 20.9469(18) [26] have been available.

$$
\begin{aligned}
A^{(6)}_{2\,\text{vap}}(m_\mu/m_e) &= \frac{2}{9}\ln^2\frac{m_\mu}{m_e} + \left(\zeta(3) - \frac{2}{3}\pi^2\ln 2 + \frac{1}{9}\pi^2 + \frac{31}{27}\right)\ln\frac{m_\mu}{m_e} \\
&+ \frac{11}{216}\pi^4 - \frac{2}{9}\pi^2\ln^2 2 - \frac{8}{3}a_4 - \frac{1}{9}\ln^4 2 - 3\zeta(3) + \frac{5}{3}\pi^2\ln 2 - \frac{25}{18}\pi^2 + \frac{1075}{216} \\
&+ \left(\frac{m_e}{m_\mu}\right)\left[-\frac{13}{18}\pi^3 - \frac{16}{9}\pi^2\ln 2 + \frac{3199}{1080}\pi^2\right] \\
&+ \left(\frac{m_e}{m_\mu}\right)^2\left[\frac{10}{3}\ln^2\frac{m_\mu}{m_e} - \frac{11}{9}\ln\frac{m_\mu}{m_e} - \frac{14}{3}\pi^2\ln 2 - 2\zeta(3) + \frac{49}{12}\pi^2 - \frac{131}{54}\right] \\
&+ \left(\frac{m_e}{m_\mu}\right)^3\left[\frac{4}{3}\pi^2\ln\frac{m_\mu}{m_e} + \frac{35}{12}\pi^3 - \frac{16}{3}\pi^2\ln 2 - \frac{5771}{1080}\pi^2\right] \\
&+ \left(\frac{m_e}{m_\mu}\right)^4\left[-\frac{25}{9}\ln^3\left(\frac{m_\mu}{m_e}\right) - \frac{1369}{180}\ln^2\left(\frac{m_\mu}{m_e}\right) + \left(-2\zeta(3) + 4\pi^2\ln 2 - \frac{269}{144}\pi^2\right.\right. \\
&\left.\left.- \frac{7496}{675}\right)\ln\frac{m_\mu}{m_e} - \frac{43}{108}\pi^4 + \frac{8}{9}\pi^2\ln^2 2 + \frac{80}{3}a_4 + \frac{10}{9}\ln^4 2\right. \\
&\left.+ \frac{411}{32}\zeta(3) + \frac{89}{48}\pi^2\ln 2 - \frac{1061}{864}\pi^2 - \frac{274511}{54000}\right] + O\left((m_e/m_\mu)^5\right), \\
&= 1.920\,455\,130(33) \quad (4.10)
\end{aligned}
$$

The leading and finite terms were first given in [27], the correct (m_e/m_μ) terms have been given in [21]. In contrast to the LbL contribution the leading logs of the VP contribution may be obtained relatively easy by renormalization group considerations using the running fine structure constant [5, 28]. In place of the known but lengthy exact result only the expansion shown was presented in [18]. Despite the existence of large leading logs the VP contribution is an order of magnitude smaller than the one from the LbL graphs.

$$
\begin{aligned}
A^{(6)}_{2\,\text{lbl}}(m_\mu/m_\tau) &= \frac{m_\mu^2}{m_\tau^2}\left[\frac{3}{2}\zeta_3 - \frac{19}{16}\right] \\
&+ \frac{m_\mu^4}{m_\tau^4}\left[\frac{13}{18}\zeta_3 - \frac{161}{1620}\zeta_2 - \frac{831931}{972000} - \frac{161}{3240}L^2 - \frac{16189}{97200}L\right] \\
&+ \frac{m_\mu^6}{m_\tau^6}\left[\frac{17}{36}\zeta_3 - \frac{13}{224}\zeta_2 - \frac{1840256147}{3556224000} - \frac{4381}{120960}L^2 - \frac{24761}{317520}L\right] \\
&+ \frac{m_\mu^8}{m_\tau^8}\left[\frac{7}{20}\zeta_3 - \frac{2047}{54000}\zeta_2 - \frac{453410778211}{1200225600000} - \frac{5207}{189000}L^2 - \frac{41940853}{952560000}L\right]
\end{aligned}
$$

216   4 Electromagnetic and Weak Radiative Corrections

$$+\frac{m_\mu^{10}}{m_\tau^{10}}\left[\frac{5}{18}\zeta_3-\frac{1187}{44550}\zeta_2-\frac{86251554753071}{287550049248000}-\frac{328337}{14968800}L^2-\frac{640572781}{23051952000}L\right]$$
$$+O\left((m_\mu/m_\tau)^{12}\right)=0.002\,142\,832(691) \tag{4.11}$$

where $L=\ln(m_\tau^2/m_\mu^2)$, $\zeta_2=\zeta(2)=\pi^2/6$ and $\zeta_3=\zeta(3)$. The expansion given in [19] in place of the exact formula has been extended in [20] with the result presented here.

$$A_{2\,\text{vap}}^{(6)}(m_\mu/m_\tau)=\left(\frac{m_\mu}{m_\tau}\right)^2\left[-\frac{23}{135}\ln\frac{m_\tau}{m_\mu}-\frac{2}{45}\pi^2+\frac{10117}{24300}\right]$$
$$+\left(\frac{m_\mu}{m_\tau}\right)^4\left[\frac{19}{2520}\ln^2\frac{m_\tau}{m_\mu}-\frac{14233}{132300}\ln\frac{m_\tau}{m_\mu}+\frac{49}{768}\zeta(3)-\frac{11}{945}\pi^2+\frac{2976691}{296352000}\right]$$
$$+\left(\frac{m_\mu}{m_\tau}\right)^6\left[\frac{47}{3150}\ln^2\frac{m_\tau}{m_\mu}-\frac{805489}{11907000}\ln\frac{m_\tau}{m_\mu}+\frac{119}{1920}\zeta(3)-\frac{128}{14175}\pi^2\right.$$
$$\left.+\frac{102108163}{30005640000}\right]+O\left((m_\mu/m_\tau)^8\right)=-0.001\,782\,326(484) \tag{4.12}$$

Again, in place of exact result obtained in [18] only the expansion shown was presented in the paper. All the expansions presented are sufficient for numerical evaluations at the present level of accuracy. This has been cross checked recently against the exact results in [11].

At three loops for the first time a contribution to $A_3(m_\mu/m_e, m_\mu/m_\tau)$, depending on two mass ratios, shows up. It is represented by diagram 22) of Fig. 4.3 with one fermion loop an electron–loop and the other a τ–loop. In view of the general discussion of VP contributions in Sect. 3.8 it is obvious to write

$$a_\mu^{(6)}(\text{vap},e,\tau)\Big|_{\text{dia 22}}=\frac{\alpha}{\pi}\int_0^1 dx(1-x)\,2\left[-\Pi'^{\,e}_{\gamma\,\text{ren}}\left(\frac{-x^2}{1-x}m_\mu^2\right)\right]$$
$$\times\left[-\Pi'^{\,\tau}_{\gamma\,\text{ren}}\left(\frac{-x^2}{1-x}m_\mu^2\right)\right], \tag{4.13}$$

which together with (3.146) or (2.173) leads to a three–fold integral representation, which we may try to integrate. Since $\Pi'^{\,\ell}_{\gamma\,\text{ren}}$ given by (2.172) is analytically known, in fact (4.13) is a one–fold integral representation. It has been calculated as an expansion in the two mass ratios in [21, 22] and was extended to $O((m_\mu^2/m_\tau^2)^5)$ recently in [23]. The result reads

$$A_{3\,\text{vap}}^{(6)}(m_\mu/m_e,m_\mu/m_\tau)=\left(\frac{m_\mu^2}{m_\tau^2}\right)\left[\frac{2}{135}\ln\frac{m_\mu^2}{m_e^2}-\frac{1}{135}\right]$$

$$+\left(\frac{m_\mu^2}{m_\tau^2}\right)^2\left[-\frac{1}{420}\ln\frac{m_\tau^2}{m_\mu^2}\ln\frac{m_\tau^2 m_\mu^2}{m_e^4}-\frac{37}{22050}\ln\frac{m_\tau^2}{m_e^2}+\frac{1}{504}\ln\frac{m_\mu^2}{m_e^2}+\frac{\pi^2}{630}\right.$$
$$\left.-\frac{229213}{12348000}\right]$$
$$+\left(\frac{m_\mu^2}{m_\tau^2}\right)^3\left[-\frac{2}{945}\ln\frac{m_\tau^2}{m_\mu^2}\ln\frac{m_\tau^2 m_\mu^2}{m_e^4}-\frac{199}{297675}\ln\frac{m_\tau^2}{m_e^2}-\frac{1}{4725}\ln\frac{m_\mu^2}{m_e^2}+\frac{4\pi^2}{2835}\right.$$
$$\left.-\frac{1102961}{75014100}\right]$$
$$+\left(\frac{m_\mu^2}{m_\tau^2}\right)^4\left[-\frac{1}{594}\ln\frac{m_\tau^2}{m_\mu^2}\ln\frac{m_\tau^2 m_\mu^2}{m_e^4}-\frac{391}{2058210}\ln\frac{m_\tau^2}{m_e^2}-\frac{19}{31185}\ln\frac{m_\mu^2}{m_e^2}+\frac{\pi^2}{891}\right.$$
$$\left.-\frac{161030983}{14263395300}\right]$$
$$+\frac{2}{15}\frac{m_e^2}{m_\tau^2}-\frac{4\pi^2}{45}\frac{m_e^3}{m_\tau^2 m_\mu}+\mathcal{O}\left[\left(\frac{m_\mu^2}{m_\tau^2}\right)^5\ln\frac{m_\tau^2}{m_\mu^2}\ln\frac{m_\tau^2 m_\mu^2}{m_e^4}\right]+\mathcal{O}\left(\frac{m_e^2}{m_\mu^2}\frac{m_\mu^2}{m_\tau^2}\right)$$
$$=0.00052766(17)\,. \tag{4.14}$$

The result is in agreement with the numerical evaluation [18]. The error in the result is due to the τ–lepton mass uncertainty. The leading–logarithmic term of this expansion corresponds to simply replacing $\alpha(q^2 = 0)$ by $\alpha(m_\mu^2)$ in the two–loop diagram with a τ loop. We have included the last term, with odd powers of m_e and m_μ, even though it is not relevant numerically. It illustrates typical contributions of the eikonal expansion, the only source of terms non–analytical in masses squared.

With (3.42) and (4.9) to (4.14) the complete three–loop QED contribution to a_μ is now known analytically, either in form of a series expansion or exact. The mass dependent terms may be summarized as follows:

$$\begin{aligned} A_2^{(6)}(m_\mu/m_e) &= 22.868\,380\,02(20),\\ A_2^{(6)}(m_\mu/m_\tau) &= 0.000\,360\,51(21),\\ A_3^{(6)}{}_{\text{vap}}(m_\mu/m_e, m_\mu/m_\tau) &= 0.000\,527\,66(17). \end{aligned} \tag{4.15}$$

As already mentioned above, the $A_2^{(6)}(m_\mu/m_e)$ contribution is surprisingly large and predominantly from light–by–light scattering via an electron loop. The importance of this term was discovered in [29], improved by numerical calculation in [2] and calculated analytically in [19]. Adding up the relevant terms we have

$$C_3 = 24.050\,509\,65\,(46)$$

or

218   4 Electromagnetic and Weak Radiative Corrections

$$a_\mu^{(6)\ \mathrm{QED}} = 24.050\,509\,65\,(46) \left(\frac{\alpha}{\pi}\right)^3 \simeq 30141.902(1) \times 10^{-11} \quad (4.16)$$

as a result for the complete 3–loop QED contribution to a_μ. We have combined the first two errors of (4.15) in quadrature and the last linearly, as the latter depends on the same errors in the mass ratios.

4.1.4 Four–Loop QED Contribution

The calculation of the four–loop contribution to a_μ is a formidable task, as there are more than one thousand diagrams to be calculated. Since the individual diagrams are much more complicated than the three–loop ones, only a few have been calculated analytically so far [30]. In most cases one has to resort to numerical calculations. This approach has been developed and perfected over the past 25 years by Kinoshita and his collaborators [1, 2, 31, 32, 33, 34, 35] with the very recent recalculations and improvements [36, 37, 38, 39]. This $O(\alpha^4)$ contribution is sizable, about 6 standard deviations at current experimental accuracy, and a precise knowledge of this term is absolutely crucial for the comparison between theory and experiment.

As a first term we mention the mass independent term $A_1^{(8)}$, where 891 diagrams (see Fig. 4.4) contribute, which represents the leading four–loop contribution to the electron anomaly a_e. As a result of the enduring heroic effort by Kinoshita a final answer has been obtained recently by Kinoshita

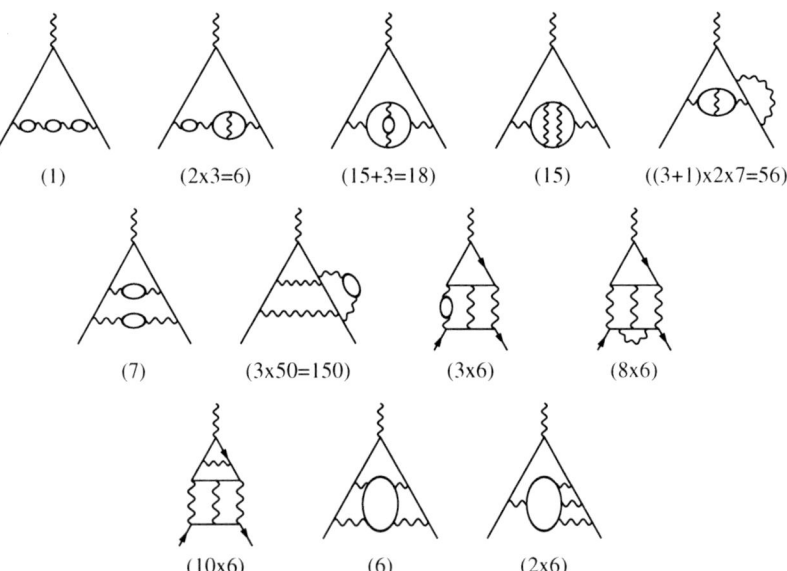

Fig. 4.4. Some typical eight order contributions to a_ℓ. In brackets the number of diagrams of a given type

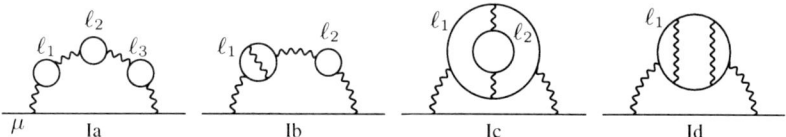

Fig. 4.5. Typical diagrams of subgroups Ia (7 diagrams), Ib (18 diagrams), Ic (9 diagrams) and Id (15 diagrams). The lepton lines represent fermions propagating in an external magnetic field. ℓ_i denote VP insertions

and collaborators [38, 39] who find[4]

$$A_1^{(8)} = -1.9144(35) \qquad (4.17)$$

where the error is due to the Monte Carlo integration.

Again the by far largest contribution to a_μ is due to $A_2^{(8)}(m_\mu/m_e)$, which collects the effects by the light internal electron loops. Here 469 diagrams contribute which may be divided into four gauge invariant (g-i) groups:

Group I: 49 diagrams obtained from the 1–loop muon vertex by inserting 1–, 2– and 3–loop lepton VP subdiagrams, i.e. the internal photon line of Fig. 4.1 is replaced by the full propagator at 3–loops. The group is subdivided into four g-i subclasses I(a), I(b), I(c) and I(d) as shown in Fig. 4.5.

---

[4]This challenging project has been initiated in the early 1980s by Kinoshita and Lindquist and lead to a first result in 1990 [32, 33]. As the subsequent ones, this result was obtained by numerical integration of the appropriately prepared Feynman integrals using the Monte Carlo integration routine **VEGAS** [40]. Since then a number of improved preliminary results have been published, which are collected in the following tabular form

$A_1^{(8)}$	year	Ref.
−1.434 (138)	1983–1990	[32, 33],
−1.557 (70)	1995	[41],
−1.4092 (384)	1997	[42],
−1.5098 (384)	2001	[43],
−1.7366 (60)	1999	[44],
−1.7260 (50)	2004	[45, 37],
−1.7283 (35)	2005	[38],
−1.9144 (35)	2007	[39],

which illustrates the stability and continuous progress of the project. Such evaluations take typically three to six month of intense runs on high performance computers. To a large extend progress was driven by the growing computing power which became available.

Results for this group have been obtained by numerical and analytic methods [30, 36]. The numerical result [36]

$$A^{(8)}_{2\,I} = 16.720\,359\,(20)\,,$$

has been obtained by using simple integral representations[5].

Group II: 90 diagrams generated from the 2–loop muon vertex by inserting 1–loop and/or 2–loop lepton VP subdiagrams as shown in Fig. 4.6. Again results for this group have been obtained by numerical and analytic methods [30, 36]. The result here is [36]

$$A^{(8)}_{2\,II} = -16.674\,591\,(68)\,.$$

Group III: 150 diagrams generated from the 3–loop muon vertex Fig. 4.3 by inserting one 1–loop electron VP subdiagrams in each internal photon line in all possible ways. Examples are depicted in Fig. 4.7. This group has been calculated numerically only, with the result [36]

$$A^{(8)}_{2\,III} = 10.793\,43\,(414)\,.$$

Group IV: 180 diagrams with muon vertex containing LbL subgraphs decorated with additional radiative corrections. This group is subdivided into g-i subsets IV(a), IV(b), IV(c) and IV(d) as illustrated in Fig. 4.8.

---

[5]Subgroup Ia has the integral representation

$$A^{(8)}_{2\,Ia} = \int_0^1 dx\,(1-x)\left(\int_0^1 dt\,\frac{\rho_2(t)}{1+[4/(1-t^2)](1-x)/x^2}\right)^3$$

where $\rho_2(t)$ is given by (3.147). Carrying out the t integral one obtains

$$A^{(8)}_{2\,Ia} = \int_0^1 dx\,(1-x)\left[-\frac{8}{9}+\frac{a^2}{3}+\left(\frac{a}{2}-\frac{a^3}{6}\right)\ln\frac{a+1}{a-1}\right]^3$$

with $a = 2/(1-x)$. In this case also the last integration may be carried out analytically [46, 47]. Similarly, subgroup Ib has the representation

$$A^{(8)}_{2\,Ib} = 2\int_0^1 dx\,(1-x)\left(\int_0^1 dt_1\,\frac{\rho_2(t_1)}{1+[4/(1-t_1^2)](1-x)/x^2}\right)$$

$$\times\left(\int_0^1 dt_2\,\frac{\rho_4(t_2)}{1+[4/(1-t_2^2)](1-x)/x^2}\right)$$

with ρ_2 given by (3.147) and ρ_4 by (3.151), respectively.

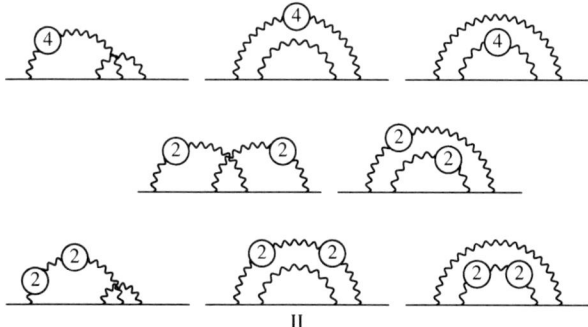

Fig. 4.6. Typical diagrams of group II (90 diagrams). The lepton lines as in Fig. 4.5. 2 and 4, respectively, indicate second (1–loop subdiagrams) and fourth (2–loop subdiagrams) order lepton–loops

Fig. 4.7. Typical diagrams of group III (150 diagrams). The lepton lines as in Fig. 4.5

The result of this calculation, which is at the limit of present possibilities, was obtained by two independent methods in [36] and reads

$$A_{2\,IV}^{(8)} = 121.8431\,(59)\,.$$

Adding up the results from the different groups the new value for $A_2^{(8)}(m_\mu/m_e)$ reads

$$A_2^{(8)}(m_\mu/m_e) = 132.6823(72)[127.50(41)]\,, \qquad (4.18)$$

in brackets the old value which was presented in [35]. In order to get some impression about the techniques and difficultieswhich have to be mastered

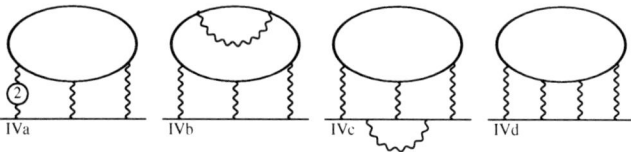

Fig. 4.8. Some typical diagrams of subgroups IVa (54 diagrams), IVb (60 diagrams), IVc (48 diagrams) and IVd (18 diagrams). The lepton lines as in Fig. 4.5

we recommend the reader to study more carefully the original work like the recent articles [36, 38, 39].

There is also a small contribution from the term $A_3^{(8)}$, which depends on 3 masses, and which arises from 102 diagrams containing two or three closed loops of VP and/or LbL type. There are contributions from the classes I (30 diagrams), II (36 diagrams) and IV (36 diagrams) defined above and the results found in [36] read

$$\begin{aligned} A_{3I}^{(8)}(m_\mu/m_e, m_\mu/m_\tau) &= 0.007\,630\,(01) \\ A_{3II}^{(8)}(m_\mu/m_e, m_\mu/m_\tau) &= -0.053\,818\,(37) \\ A_{3IV}^{(8)}(m_\mu/m_e, m_\mu/m_\tau) &= 0.083\,782\,(75) \end{aligned} \qquad (4.19)$$

which sums up to the value

$$A_3^{(8)}(m_\mu/m_e, m_\mu/m_\tau) = 0.037\,594\,(83) \,. \qquad (4.20)$$

A rough estimate of the τ–loops contribution is also given in [36] with the result

$$A_2^{(8)}(m_\mu/m_\tau) = 0.005(3) \,. \qquad (4.21)$$

In summary: all mass dependent as well as the mass independent $O(\alpha^4)$ QED contributions to a_μ have been recalculated by different methods by Kinoshita's group [36, 38, 39]. There is also some progress in analytic calculations [48]. Collecting the $A^{(8)}$ terms discussed above we obtain

$$C_4 = 130.8105(85)$$

or

$$a_\mu^{(8)\ \mathrm{QED}} = 130.810\,5\,(85)\left(\frac{\alpha}{\pi}\right)^4 \simeq 380.807(25) \times 10^{-11} \qquad (4.22)$$

as a result for the complete 4–loop QED contribution to a_μ.

4.1.5 Five–Loop QED Contribution

Here the number of diagrams (see Fig. 4.9) is 9080, a very discouraging number even for Kinoshita [37]. This contribution originally was evaluated using renormalization group (RG) arguments in [2, 49]. The new estimate by Kinoshita and Nio [37, 50] is[6]

$$A_2^{(10)}(m_\mu/m_e) = 663(20),$$

and was subsequently cross–checked by Kataev [51] using renormalization group arguments. As mentioned earlier, a bound for the size of the universal part has also been estimated [52] which is taken into account as

---

[6]The first estimate $A_2^{(10)}(m_\mu/m_e) \sim 930(170)$ has been given by Karshenboim [49].

4.1 $g-2$ in Quantum Electrodynamics   223

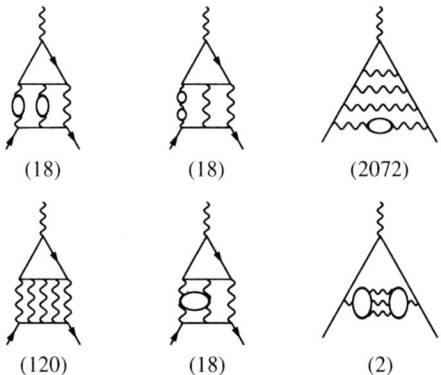

Fig. 4.9. Some typical tenth order contributions to a_ℓ including fermion loops. In brackets the number of diagrams of the given type

$$A_1^{(10)} = 0.0(3.8) \,. \qquad (4.23)$$

Thus we arrive at

$$C_5 \sim 663(20)(3.8)$$

or

$$a_\mu^{(10)\text{ QED}} \sim 663(20)(3.8) \left(\frac{\alpha}{\pi}\right)^5 \simeq 4.483(135)(26) \times 10^{-11} \qquad (4.24)$$

as an estimate of the 5–loop QED contribution.

In Table 4.2 we summarize the results of the QED calculations. The expansion coefficients C_i which multiply $(\alpha/\pi)^i$ grow rapidly with the order. Nevertheless, because of the smallness of the expansion parameter α/π, the convergence of the perturbative expansion of a_μ^{QED} is good. We conclude that the perturbative truncation error looks to be well under control at the present level of accuracy.

The universal QED terms have been summarized in (3.43) and adding up the mass dependent QED terms of the 3 flavors (e, μ, τ) we finally obtain

$$a_\mu^{\text{QED}} = 116\,584\,718.113(.082)(.014)(.025)(.137)[.162] \times 10^{-11} \qquad (4.25)$$

Table 4.2. Summary of QED contributions to a_μ

	C_i		$a_\mu^{(2i)\text{ QED}} \times 10^{11}$
C_1	0.5	$a^{(2)}$	116140973.301(81)
C_2	0.765\,857\,410\,(26)	$a^{(4)}$	413217.621(14)
C_3	24.050\,509\,65\,(46)	$a^{(6)}$	30141.902(1)
C_4	130.8105(85)	$a^{(8)}$	380.807(25)
C_5	663.0(20.0)(3.8)	$a^{(10)}$	4.483(135)(26)

where the errors are due, respectively, to the uncertainties in α_{input}, in the mass ratios, the numerical error on α^4 terms and the guessed uncertainty of the α^5 contribution.

4.2 Weak Contributions

The weak interaction contribution to $g-2$ attracted attention of theoreticians long time before it started to play a relevant role in the comparison with the experimental result. Actually the "weak contribution sensitivity" was reached only with the recent BNL experiment. With the emergence of the SM [53] and establishing its renormalizability [54] for the first time it was possible to make real predictions for a_μ beyond QED. Before, in non–renormalizable low energy effective theories, corresponding attempts were not convincing, since, as we discussed earlier only in a renormalizable theory a_μ is a finite unambiguously predictable quantity and hence an unambiguous monitor for testing the theory. Soon after a unified electroweak theory seemed established a number of groups presented the one–loop result for a_μ in 1972 [55]. At that time, the weak term turned out to be almost two orders of magnitude smaller then the experimental accuracy of the CERN $g-2$ experiment. At present the weak term is an effect of almost three standard deviations.

Weak interaction effects are mediated by exchange of the heavy weak gauge bosons $W^\pm$, which mediate charged current (CC) processes, and Z, which mediates the neutral current (NC) processes. Beyond the electroweak $SU(2)_L \otimes U(1)_Y$ Yang-Mills gauge theory, a Higgs sector is required which allows to generate the masses of the gauge bosons W and Z, as well as the masses of the fermions, without spoiling renormalizability. Thereby the gauge symmetry is broken down $SU(2)_L \otimes U(1)_Y \to U(1)_{\text{em}}$ to the Abelian subgroup of QED, and an additional physical particle has to be taken into account the famous Higgs particle particle physicists are still hunting for.

In the SM the fermions are organized in three lepton–quark families, with the left–handed fields in $SU(2)_L$ doublets and the right–handed fields in singlets:

$$\text{1st family:} \quad \begin{pmatrix} \nu_e \\ e^- \end{pmatrix}_L, \begin{pmatrix} u \\ \tilde{d} \end{pmatrix}_L, \quad \nu_{e_R}, e_R^-, u_R, d_R$$

$$\text{2nd family:} \quad \begin{pmatrix} \nu_\mu \\ \mu^- \end{pmatrix}_L, \begin{pmatrix} c \\ \tilde{s} \end{pmatrix}_L, \quad \nu_{\mu_R}, \mu_R^-, c_R, s_R$$

$$\text{3rd family:} \quad \begin{pmatrix} \nu_\tau \\ \tau^- \end{pmatrix}_L, \begin{pmatrix} t \\ \tilde{b} \end{pmatrix}_L, \quad \nu_{\tau_R}, \tau_R^-, t_R, b_R \ .$$

The Abelian subgroup $U(1)_Y$ is associated with the weak hypercharge, related to the charge and the 3rd component of weak isospin by the Gell-Mann–Nishijima relation $Y = 2(Q - T_3)$[7]. Denoting by $\nu_\ell = (\nu_e, \nu_\mu, \nu_\tau)$, $\ell = (e, \mu, \tau)$, $q_u = (u, c, t)$ and $q_d = (d, s, b)$ the four horizontal vectors in "family space" of fermion fields with identical electroweak quantum numbers, the charged current (CC) has the form

$$J_\mu^+ = J_{\mu 1} - i J_{\mu 2} = \bar{\nu}_\ell \gamma_\mu (1 - \gamma_5) U_{\mathrm{MNS}} \, \ell + \bar{q}_u \gamma_\mu (1 - \gamma_5) U_{\mathrm{CKM}} \, q_d \tag{4.26}$$

and exhibits quark family flavor changing, through mixing by the unitary 3×3 *Cabibbo-Kobayashi-Maskawa* matrix U_{CKM} as well as neutrino flavor mixing by the corresponding *Maki-Nakagawa-Sakata* matrix U_{MNS}. The $SU(2)_L$ currents have strict V–A (V = vector $[\gamma_\mu]$, A = axial–vector $[\gamma_\mu \gamma_5]$) form, which in particular implies that the CC is maximally parity (P) violating (Lee and Yang 1957). The mixing matrices exhibit a CP violating phase, which also implies the existence of a tiny electrical dipole moment. In a local QFT a non–vanishing EDM is possible only if CP is violated, as we noted earlier. For the magnetic moments CP has no special impact and the CP violating effects are too small to play any role. For our purpose the 3×3 family mixing matrices may be taken to be unit matrices. The neutral current (NC) is strictly flavor conserving

$$J_\mu^Z = J_{\mu 3} - 2 \sin^2 \Theta_W j_\mu^{em} = \sum_f \bar{\psi}_f \gamma_\mu (v_f - a_f \gamma_5) \psi_f \tag{4.27}$$

with

$$j_\mu^{em} = \sum_f Q_f \bar{\psi}_f \gamma_\mu \psi_f \tag{4.28}$$

the P conserving electromagnetic current. The weak mixing parameter $\sin^2 \Theta_W$ is responsible for the $\gamma - Z$ mixing. The sums extend over the individual

---

[7] $SU(2)_L \otimes U(1)_Y$ quantum numbers of fermions read

	Doublets				Singlets			
	$(\nu_\ell)_L$	$(\ell^-)_L$	$(u,c,t)_L$	$(\tilde{d},\tilde{s},\tilde{b})_L$	$(\nu_\ell)_R$	$(\ell^-)_R$	$(u,c,t)_R$	$(d,s,b)_R$
Q	0	-1	$2/3$	$-1/3$	0	-1	$2/3$	$-1/3$
T_3	$1/2$	$-1/2$	$1/2$	$-1/2$	0	0	0	0
Y	-1	-1	$1/3$	$1/3$	0	-2	$4/3$	$-2/3$

Quarks in addition carry $SU(3)_c$ color. The *color factor* N_{cf} is 3 for quarks and 1 for leptons, which are color singlets. Note that in the SM all matter fields are in the fundamental ($SU(2)_L$–doublets, $SU(3)_c$–triplets[antitriplets]) or trivial (singlet) representations. The simplest ones possible.

fermion flavors f (and color). In our convention the NC vector and axial–vector neutral current coefficients are given by

$$v_f = T_{3f} - 2Q_f \sin^2 \Theta_W , \quad a_f = T_{3f} \tag{4.29}$$

where T_{3f} is the weak isospin ($\pm\frac{1}{2}$) of the fermion f. The matter field Lagrangian thus takes the form

$$\mathcal{L}_{\mathrm{matter}} = \sum_f \bar{\psi}_f i\gamma^\mu \partial_\mu \psi_f + \frac{g}{2\sqrt{2}} (J_\mu^+ W^{\mu-} + \mathrm{h.c.}) + \frac{g}{2\cos\Theta_W} J_\mu^Z Z^\mu + e j_\mu^{em} A^\mu \tag{4.30}$$

where g is the $SU(2)_L$ gauge coupling constant and $e = g\sin\Theta_W$ is the charge of the positron (unification condition).

We should mention that before symmetry breaking the theory has the two gauge couplings g and g' as free parameters, after the breaking we have in addition the *vacuum expectation value* (VEV) of the Higgs field v, thus three parameters in total, if we disregard the fermion masses and their mixing parameters for the moment. The most precisely known parameters are the fine structure constant α (electromagnetic coupling strength), the Fermi constant G_μ (weak interaction strength) and the Z mass M_Z. Apart from the unification relation

$$\alpha = \frac{e^2}{4\pi} , \quad e = g\sin\Theta_W , \quad \tan\Theta_W = g'/g$$

we have the mass generation by the Higgs mechanism which yields

$$M_W = \frac{gv}{2} , \quad M_Z = \frac{gv}{2\cos\Theta_W} ,$$

while lowest order CC Fermi decay defines the Fermi or muon decay constant

$$G_\mu = \frac{g^2}{4\sqrt{2}M_W^2} = \frac{1}{\sqrt{2}v^2} .$$

The neutral to charged current ratio, called ρ–parameter, follows from

$$G_{\mathrm{NC}} = \frac{g^2}{4\sqrt{2}M_Z^2 \cos^2\Theta_W} = \frac{\rho}{\sqrt{2}v^2} ,$$

with $\rho_0 = 1$ at the tree level. These relations are subject to radiative corrections. Given α, G_μ and M_Z as input parameters, all further parameters like M_W, $\sin^2\Theta_W$, g, etc are dependent parameters. Typically when calculating versions of the weak mixing parameter $\sin^2\Theta_i$ in terms of the input parameters one obtains

$$\sin^2\Theta_i \cos^2\Theta_i = \frac{\pi\alpha}{\sqrt{2}\,G_\mu\, M_Z^2} \frac{1}{1-\Delta r_i} , \tag{4.31}$$

where
$$\Delta r_i = \Delta r_i(\alpha, G_\mu, M_Z, m_H, m_{f\neq t}, m_t)$$

includes the higher order corrections which can be calculated in the SM or in alternative models. For example,

$$\rho = 1 + \Delta\rho \ , \quad \Delta\rho = \frac{\sqrt{2} G_\mu}{16\pi^2} 3 |m_t^2 - m_b^2|$$

with a large correction proportional to m_t^2 due to the heavy top [56]. In the SM today the Higgs mass m_H is the only relevant unknown parameter and by confronting the calculated with the experimentally determined value of $\sin^2 \Theta_i$ one obtains important indirect constraints on the Higgs mass. Δr_i depends on the definition of $\sin^2 \Theta_i$. The various definitions coincide at tree level and hence only differ by quantum effects. From the weak gauge boson masses, the electroweak gauge couplings and the neutral current couplings of the charged fermions we obtain

$$\sin^2 \Theta_W = 1 - \frac{M_W^2}{M_Z^2}$$

$$\sin^2 \Theta_g = e^2/g^2 = \frac{\pi\alpha}{\sqrt{2} G_\mu M_W^2}$$

$$\sin^2 \Theta_f = \frac{1}{4|Q_f|}\left(1 - \frac{v_f}{a_f}\right), \ f \neq \nu ,$$

for the most important cases and the general form of Δr_i ($i = W, g, f$) reads

$$\Delta r_i = \Delta\alpha - f_i(\sin^2 \Theta_i) \Delta\rho + \Delta r_{i\,\text{rem}}$$

with $f_W(\sin^2 \Theta_W) = \cos^2 \Theta_W / \sin^2 \Theta_W$; $f_g(\sin^2 \Theta_g) = f_f(\sin^2 \Theta_f) = 1$ and a universal term $\Delta\alpha$ which affects the predictions of M_W via $\sin^2 \Theta_W$, etc. For M_W we have [57]

$$M_W^2 = \frac{\rho M_Z^2}{2}\left(1 + \sqrt{1 - \frac{4A_0^2}{\rho M_Z^2}\left(\frac{1}{1-\Delta\alpha} + \Delta r_{\text{rem}}\right)}\right) , \quad (4.32)$$

with

$$A_0 = \sqrt{\pi\alpha/\sqrt{2}G_\mu} = 37.2802(3) \text{ GeV} .$$

The leading dependence on the Higgs mass m_H is logarithmic with

$$\Delta r_W^{\text{Higgs}} = \frac{11}{3}\left(\ln\frac{m_H^2}{M_W^2} - \frac{5}{6}\right), \ \Delta r_f^{\text{Higgs}} = \frac{1 + 9\sin^2 \Theta_f}{3\cos^2 \Theta_f}\left(\ln\frac{m_H^2}{M_W^2} - \frac{5}{6}\right)$$

assuming $m_H \gg M_W$ (see e.g. [58] for more details).

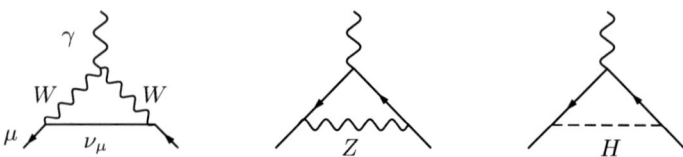

Fig. 4.10. The leading weak contributions to a_ℓ; diagrams in the physical unitary gauge

4.2.1 Weak One–Loop Effects

The relevant diagrams are shown in the following Fig. 4.10 in the unitary gauge. For the Feynman rules of the SM we refer to SM textbooks or to my TASI lecture notes [58] for a short overview. In spite of the fact that the unitary gauge is not renormalizable, the relevant gauge invariant S–matrix element, may be calculated directly in the unitary gauge. The advantage is that in this gauge only physical particles are present and diagrams exhibiting Higgs ghosts and Faddeev-Popov ghosts are absent. What is most interesting is the occurrence of the first diagram of Fig. 4.10 which exhibits a non–Abelian triple gauge vertex and the corresponding contribution provides a test of the Yang-Mills structure involved. It is of course not surprising that the photon couples to the charged W boson the way it is dictated by gauge invariance. The gauge boson contributions are given by

$$a_\mu^{(2)\,\text{EW}}(W) = \frac{\sqrt{2}G_\mu m_\mu^2}{16\pi^2}\,\frac{10}{3} \simeq +388.70(0)\times 10^{-11}$$

$$a_\mu^{(2)\,\text{EW}}(Z) = \frac{\sqrt{2}G_\mu m_\mu^2}{16\pi^2}\,\frac{(-1+4\sin^2\Theta_W)^2-5}{3} \simeq -193.88(2)\times 10^{-11} \quad (4.33)$$

while the diagram with the Higgs exchange yields[8]

$$a_\mu^{(2)\,\text{EW}}(H) = \frac{\sqrt{2}G_\mu m_\mu^2}{4\pi^2}\int_0^1 dy\,\frac{(2-y)y^2}{y^2+(1-y)(m_H/m_\mu)^2}$$

$$\simeq \frac{\sqrt{2}G_\mu m_\mu^2}{4\pi^2}\begin{cases}\frac{m_\mu^2}{m_H^2}\ln\frac{m_\mu^2}{m_H^2} & \text{for}\quad m_H \gg m_\mu \\ \frac{3}{2} & \text{for}\quad m_H \ll m_\mu\end{cases}$$

$$\leq 5\times 10^{-14}\quad \text{for}\quad m_H \geq 114\text{ GeV}\,, \quad (4.34)$$

---

[8] The exact analytic result for the Higgs reads

$$a_\mu^{(2)\,\text{EW}}(H) = \frac{\sqrt{2}G_\mu m_\mu^2}{4\pi^2}\left\{\xi(1-\xi)\ln(-\xi) + \xi^{-2}(1-\xi)(1-\xi^3)\ln(1-\xi) + \xi^{-1}(1-\xi)^2 + \frac{3}{2}\right\}$$

$$\simeq \frac{\sqrt{2}G_\mu m_\mu^2}{4\pi^2}\left\{z^{-1}(\ln z - \frac{7}{6}) + z^{-2}(3\ln z - \frac{13}{4}) + z^{-3}(9\ln z - \frac{201}{20}) + O(z^{-4}\ln z)\right\}$$

in which $z = m_H^2/m_\mu^2$, and $\xi = (\sqrt{1-y}-1)/(\sqrt{1-y}+1)$ with $y = 4/z$.

in view of the LEP bound (3.33). Employing the SM parameters given in (3.30) and (3.31) we obtain

$$a_\mu^{(2)\,\text{EW}} = (194.82 \pm 0.02) \times 10^{-11} \qquad (4.35)$$

The error comes from the uncertainty in $\sin^2\Theta_W$ given above.

4.2.2 Weak Two–Loop Effects

Part of the electroweak two–loop corrections were calculated first in 1992 by Kukhto, Kuraev, Schiller and Silagadze [59] with an unexpected result, the corrections turned out to be enhanced by very large logarithms $\ln M_Z/m_f$, which mainly come from fermion triangular–loops like in Fig. 4.11a. In QED loops with three photons attached do not contribute due to Furry's theorem and the $\gamma\gamma\gamma$–amplitude vanishes. In presence of weak interactions, because of parity violation, contributions from the two orientations of the closed fermion loops do not cancel such that the $\gamma\gamma Z$, γZZ and γWW amplitudes do not vanish. In fact for the γWW triangle charge conservation only allows one orientation of the fermion loop.

Diagrams $a)$ and $b)$, with an internal photon, appear enhanced by a large logarithm. In fact the lepton loops contributing to the $\gamma\gamma Z$ vertex lead to corrections

$$a_\mu^{(4)\,\text{EW}}([f]) \simeq \frac{\sqrt{2} G_\mu\, m_\mu^2}{16\pi^2}\, \frac{\alpha}{\pi}\, 2T_{3f} N_{cf} Q_f^2 \left[3\ln\frac{M_Z^2}{m_{f'}^2} + C_f \right]$$

in which $m_{f'} = m_\mu$ if $m_f \leq m_\mu$ and $m_{f'} = m_f$ if $m_f > m_\mu$ and

$$C_f = \begin{cases} 5/2 & \text{for } m_f < m_\mu \\ 11/6 - 8/9\,\pi^2 & \text{for } m_f = m_\mu \\ -6 & \text{for } m_f > m_\mu \end{cases}.$$

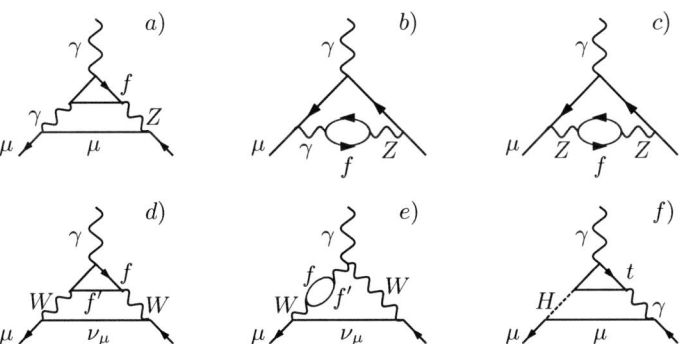

Fig. 4.11. Some of the relevant electroweak two–loop diagrams in the unitary gauge, $f = (\nu_e, \nu_\mu, \nu_\tau,)\, e, \mu, \tau, u, c, t, d, s, b$ with weak doublet partners $f' = (e, \mu, \tau,)\, \nu_e, \nu_\mu, \nu_\tau, d, s, b, u, c, t$ of course the neutrinos (in brackets) do not couple directly to the photon and hence are absent in the triangular subgraphs

230   4 Electromagnetic and Weak Radiative Corrections

For an individual fermion f the contribution is proportional to $N_{cf}Q_f^2 a_f$. In [59] only lepton loops were taken into account, and it is well known that the triangular subdiagram has an Adler-Bell-Jackiw (ABJ) or VVA anomaly [60], which cancels if all fermions are included. The *anomaly cancellation* is mandatory in a renormalizable theory and it forces the fermions in the SM to come in families of leptons and quarks [61]. The latter compensate the anomaly of the former. The cancellation condition of the SM reads

$$\sum_f N_{cf} Q_f^2 a_f = 0, \qquad (4.36)$$

and such a cancellation is expected also for the leading short distance logarithms proportional to $\ln M_Z$ and in fact this has been checked to happen on the level of the QPM for the 1st and 2nd fermion family [62, 63]. Assuming dressed constituent quarks masses $M_u, M_d > m_\mu$, the QPM result for the first family reads [63]

$$a_\mu^{(4)\,\mathrm{EW}}([e,u,d])_{\mathrm{QPM}} \simeq -\frac{\sqrt{2} G_\mu m_\mu^2}{16\pi^2} \frac{\alpha}{\pi} \left[\ln \frac{M_u^8}{m_\mu^6 M_d^2} + \frac{17}{2} \right] \simeq -4.00 \times 10^{-11}, \qquad (4.37)$$

while for the second family, with $M_s, M_c > m_\mu$, we have

$$a_\mu^{(4)\,\mathrm{EW}}([\mu,c,s])_{\mathrm{QPM}} \simeq -\frac{\sqrt{2} G_\mu m_\mu^2}{16\pi^2} \frac{\alpha}{\pi} \left[\ln \frac{M_c^8}{m_\mu^6 M_s^2} + \frac{47}{6} - \frac{8\pi^2}{9} \right] \simeq -4.65 \times 10^{-11}. \qquad (4.38)$$

For the numerical evaluation we had to insert some quark masses and we resorted to the not very well defined *constituent quark masses* used in [63]:

$$M_u = M_d = 300 \text{ MeV}, \quad M_s = 500 \text{ MeV}, \quad M_c = 1.5 \text{ GeV} \quad \text{and} \quad M_b = 4.5 \text{ GeV}. \qquad (4.39)$$

It should be noted that such large effective light quark masses violate basic Ward-Takahashi identities of low energy QCD. The latter requires values like (3.36) for the so called *current quark masses* to properly account for the pattern of chiral symmetry breaking[9]. The ambiguity in the choice of the quark

---

[9] Adopting the values (3.36) one would have to replace the masses satisfying $m_q < m_\mu$ ($q = u,d,s$) by m_μ ($SU(3)$ chiral limit), such that [62]

$$a_\mu^{(4)\,\mathrm{EW}}([e,u,d])_{\mathrm{QPM}} \simeq 0$$

and

$$a_\mu^{(4)\,\mathrm{EW}}([\mu,c,s])_{\mathrm{QPM}} \simeq -\frac{\sqrt{2} G_\mu m_\mu^2}{16\pi^2} \frac{\alpha}{\pi} \left[4 \ln \frac{m_c^2}{m_\mu^2} + \frac{32}{3} - \frac{8\pi^2}{9} \right] \simeq -5.87 \times 10^{-11}.$$

However, this free current quarks result cannot be a reasonable approximation, as it completely ignores the non–perturbative QCD effects.

masses reflects the fact that we are not in the perturbative regime. If one uses the above constituent quark masses to calculate the hadronic photon VP one does not get an answer which is close to what is obtained non–perturbatively from the dispersion integral of e^+e^-–data [64].

Concerning the third family, D'Hoker in [65] pointed out that a super–heavy fermion like the top, which usually is expected to decouple, generates a large log, because the heavy fermion does not participate in the cancellation of the large logs, while it still participates in the cancellation of the mass independent ABJ anomaly (see also [66]). The origin of the effect is the large weak isospin breaking in the top–bottom doublet, which is manifest in the large mass splitting $m_t \gg M_Z \gg m_b$. Consequently, one has to expect that the large logs from the leptons cancel against the ones from the quarks, with only partial cancellation in the 3rd family ($[\tau, t, b]$).

It should be stressed that results from individual fermions are gauge dependent and only sums of contributions for complete fermion families are physically meaningful. Nevertheless, we will give at intermediate steps partial result either in the Feynman gauge or in the unitary gauge.

The leading contributions Fig. 4.11a were investigated first by Peris, Perrottet and de Rafael [62], by evaluating the hadronic effects in a low energy effective approach. The full set of diagrams of Fig. 4.11 was calculated by Czarnecki, Krause and Marciano [63], using the QPM. The results were later refined and extended in the leading log approximation by renormalization group methods at the two– as well as at the three–loop level by Degrassi and Giudice in [67]. Thereby also smaller effects, like the ones from diagram b), were included. The latter does not give a large effect because the $\gamma - Z$ mixing propagator is of type VV with coupling strength $Q_f v_f Q_\mu v_\mu$ which is suppressed like $(1-4\sin^2\Theta_W) \sim 0.1$ for quarks and like $(1-4\sin^2\Theta_W)^2 \sim 0.01$ for leptons. Diagrams c) to e) have an additional heavy propagator and thus yield sub–leading terms only. In the enhanced contributions proportional to the large logs $\ln M_Z/m_f$ or $(m_t/M_W)^2$ the exact $\sin^2\Theta_W$ dependence has been worked out. Results may be summarized as follows:

Summary of Perturbative Leading Log Results

Two loop corrections to a_μ^{weak} naturally divide into leading logs (LL), i.e. terms enhanced by a factor of $\ln(M_Z/m_f)$ where m_f is a fermion mass scale much smaller than M_Z, and everything else, which we call non–leading logs (NLL). The 2–loop leading logs are[10] [62, 63, 67, 68]

---

[10] The LL contributions may be grouped into

$$a_\mu^{(4)\,\text{EW}}(W)_{\text{LL}} = -\frac{\sqrt{2}G_\mu m_\mu^2}{16\pi^2}\frac{\alpha}{\pi}\left[\frac{40}{3}\right]\ln\frac{M_Z}{m_\mu}$$

232   4 Electromagnetic and Weak Radiative Corrections

$$
\begin{aligned}
a_{\mu\,\mathrm{LL}}^{(4)\,\mathrm{EW}} = &-\frac{\sqrt{2}G_\mu m_\mu^2}{16\pi^2}\frac{\alpha}{\pi}\left\{\left[\frac{215}{9}+\frac{31}{9}(1-4s_W^2)^2\right]\ln\frac{m_Z}{m_\mu}\right. \\
&\left.-\sum_{f\in F}N_f Q_f\left[12\,T_f^3\,Q_f-\frac{8}{9}\left(T_f^3-2Q_f s_W^2\right)(1-4s_W^2)\right]\ln\frac{M_Z}{m_f}\right\},
\end{aligned}
$$
(4.40)

in the notation introduced above. Electron and muon loops as well as non–fermionic loops produce the $\ln(M_Z/m_\mu)$ terms in this expression (the first line) while the sum runs over $F=\tau, u, d, s, c, b$. The logarithm $\ln(M_Z/m_f)$ in the sum implies that the fermion mass m_f is larger than m_μ. For the light quarks, such as u and d, whose current masses are very small, m_f has a meaning of some effective hadronic mass scale.

The issue about how to treat the light quarks appropriately was reconsidered and discussed somewhat controversial in [68, 69, 70]. Corresponding problems and results will be considered next.

Hadronic Effects in Quark Triangle Graphs

As leptons and quarks can be treated only family–wise we have to think about how to include the quarks and hadrons. Here the hadronic corrections involved by virtue of asymptotic freedom of QCD are calculable in pQCD if a heavy mass sets the scale from where the integrals get their dominant contribution. Since all the weak contributions involve at least one heavy scale, it seems justified to work with the QPM in a first step. In doing so we will be confronted again with the question about the meaning of the quark masses to be used in case of the light quarks. As already mentioned, the crucial constraint is the ABJ anomaly cancellation[11]. The nature of the ABJ triangle anomaly

$$a_\mu^{(4)\,\mathrm{EW}}(Z,\text{no }f\text{–loops})_{\mathrm{LL}} = -\frac{\sqrt{2}G_\mu m_\mu^2}{16\pi^2}\frac{\alpha}{\pi}\left[\frac{13}{9}(g_A^\mu)^2-\frac{23}{9}(g_V^\mu)^2\right]\ln\frac{M_Z}{m_\mu}$$

$$a_\mu^{(4)\,\mathrm{EW}}(Z,\mu\text{–loop})_{\mathrm{LL}} = -\frac{\sqrt{2}G_\mu m_\mu^2}{16\pi^2}\frac{\alpha}{\pi}N_\mu\left[-6\,(g_A^\mu)^2-\frac{4}{9}(g_V^\mu)^2\right]\ln\frac{M_Z}{m_\mu}$$

where the first term comes from the triangular loop (only VVA, VVV vanishing by Furry's theorem) (diagram a) of Fig. 4.11), the second from the $\gamma-Z$ mixing propagator muon loop (only VV can contribute) (diagram e) of Fig. 4.11).

$$a_\mu^{(4)\,\mathrm{EW}}(Z,f\text{–loops})_{\mathrm{LL}} = -\frac{\sqrt{2}G_\mu m_\mu^2}{16\pi^2}\frac{\alpha}{\pi}\sum_f N_f Q_f\left[-6\,g_A^\mu g_A^f Q_f+\frac{4}{9}g_V^\mu g_V^f\right]\ln\frac{M_Z}{m_{f'}}$$

in which $m_{f'}\equiv\max[m_f, m_\mu]$.

[11] Renormalizability, gauge invariance and current conservation is intimately related. Axial anomalies showing up in the weak interaction currents for individual fermions must cancel in order not to spoil gauge invariance and hence renormalizability.

is controlled by the Adler-Bardeen non–renormalization theorem [71], which says that the one–loop anomaly is exact to all orders, by the Wess-Zumino integrability condition and the Wess-Zumino effective action [72] (see below), by Witten's algebraic/geometrical interpretation, which requires the axial current to be normalized to an integer [73]. Phenomenologically, it plays a key role in the prediction of $\pi^0 \to \gamma\gamma$, and in the solution of the η' mass problem. Last but not least, renormalizability of the electroweak Standard Model requires the anomaly cancellation which dictates the lepton–quark family structure.

Digression on the Anomaly

The axial anomaly is a quantum phenomenon which doesn't get renormalized by higher order effects. In QED the axial current anomaly is given by

$$\partial_\mu j_5^\mu(x) = \frac{e^2}{8\pi^2} \tilde{F}_{\mu\nu}(x) F^{\mu\nu}(x) \neq 0 \tag{4.41}$$

where $F^{\mu\nu} = \partial^\mu A^\nu - \partial^\nu A^\mu$ is the electromagnetic field strength tensor and $\tilde{F}_{\mu\nu} = \frac{1}{2}\varepsilon_{\mu\nu\rho\sigma} F^{\rho\sigma}$ its dual pseudotensor (parity odd). The pseudoscalar density is a divergence of a gauge dependent pseudovector

$$\tilde{F}_{\mu\nu} F^{\mu\nu} = \partial^\mu K_\mu \ ; \quad K_\mu = 2\epsilon_{\mu\rho\nu\sigma} A^\rho \partial^\nu A^\sigma \ .$$

In general, in perturbation theory the axial anomaly shows up in closed fermion loops with an odd number of axial–vector couplings if a non–vanishing γ_5–odd trace of γ–matrices like[12]

$$\mathrm{Tr}\ (\gamma^\mu \gamma^\nu \gamma^\rho \gamma^\sigma \gamma_5) = 4i\varepsilon^{\mu\nu\rho\sigma} \tag{4.42}$$

(in $d = 4$ dimensions) is involved and if the corresponding Feynman integral is not ultraviolet convergent such that it requires regularization. The basic diagram exhibiting the axial anomaly is the linearly divergent triangle diagram Fig. 4.12 which leads to the amplitude (1st diagram)

$$\tilde{T}_{ijk}^{\mu\nu\lambda}(p_1, p_2) = (-1) \, \mathrm{i}^5 \, \mathrm{Tr}\ (T_j T_i T_k) \frac{g^2}{(2\pi)^4} \int \mathrm{d}^4 k$$

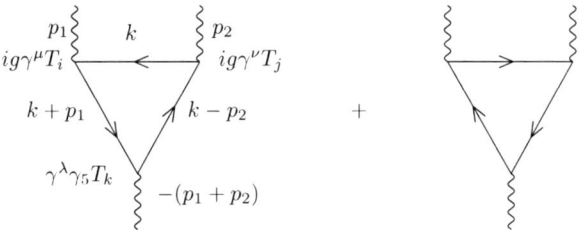

Fig. 4.12. Fermion triangle diagrams exhibiting the axial anomaly

---
[12] Notice that $\mathrm{Tr}\ \left(\prod_{i=1}^n \gamma^{\mu_i} \gamma_5\right) = 0$ for $n < 4$ and for all $n =$ odd.

$$\times \, \mathrm{Tr}\left(\frac{1}{\slashed{k} - \slashed{p}_2 + \mathrm{i}\epsilon} \gamma^\nu \frac{1}{\slashed{k} + \mathrm{i}\epsilon} \gamma^\mu \frac{1}{\slashed{k} + \slashed{p}_1 + \mathrm{i}\epsilon} \gamma^\lambda \gamma_5 \right).$$

If we include the bose symmetric contribution (2nd diagram)

$$T^{\mu\nu\lambda}_{ijk}(p_1, p_2) = \tilde{T}^{\mu\nu\lambda}_{ijk}(p_1, p_2) + \tilde{T}^{\nu\mu\lambda}_{jik}(p_2, p_1)$$

and **impose** vector current conservation

$$p_{1\mu} T^{\mu\nu\lambda}_{ijk}(p_1, p_2) = p_{2\nu} T^{\mu\nu\lambda}_{ijk}(p_1, p_2) = 0$$

we obtain the unambiguous regularization independent result

$$-(p_1 + p_2)_\lambda \, T^{\mu\nu\lambda}_{ijk}(p_1, p_2) = \mathrm{i}\, \frac{g^2}{16\pi^2} D_{ijk} \, 4\, \varepsilon^{\mu\nu\rho\sigma} p_{1\rho} p_{2\sigma} \neq 0$$

with $D_{ijk} = Tr\left(\{T_i, T_j\} T_k\right)$.

This result is independent of the masses of the fermion lines and is not changed by higher order corrections. Therefore the result is **exact** beyond perturbation theory! All anomalous fermion loops may be traced back to the basic triangular fermion loop, and in fact all other possible anomalous matrix-elements of the axial current are summarized in the general form of the anomaly equation

$$\partial_\lambda j_{5k}^\lambda(x) = \frac{g^2}{16\pi^2} D_{ijk} \, \tilde{G}_i^{\mu\nu}(x) G_{j\mu\nu}(x) \tag{4.43}$$

where $G_{i\mu\nu}(x)$ is the non-Abelian field strength tensor and $\tilde{G}_i^{\mu\nu}$ its dual pseudotensor. Equation (4.43) is the non-Abelian generalization of (4.41) in the Abelian case. As a result the condition for the absence of an anomaly reads

$$D_{ijk} = Tr\left(\{T_i, T_j\} T_k\right) = 0 \ \forall \ (ijk).$$

In fact the contributions to the anomaly being independent of the mass may be represented in terms of fixed helicity fields, and opposite helicities contribute with opposite signs

$$D_{ijk} \equiv \mathrm{Tr}\left(\{T_{Li}, T_{Lj}\} T_{Lk}\right) - \mathrm{Tr}\left(\{T_{Ri}, T_{Rj}\} T_{Rk}\right) \tag{4.44}$$

which tells us that left-handed and right-handed fields give independent contributions to the anomaly. Only theories which are democratic with respect to helicities in the axial anomaly coefficient are anomaly free. Since $SU(2)$ has only real representations $R^* \sim R$ (in particular $2 \sim 2^*$) it cannot produce any anomaly. In contrast $SU(3)$ is not anomaly safe, because the fundamental representations 3 and the complex conjugate 3^* are inequivalent. However, as quarks in the triplet representation 3 and antiquarks in the anti-triplet representation 3^* enter symmetrically in QCD (a pure vector theory), $SU(3)_c$ cannot give rise to anomalies. Only the Abelian hypercharge group $U(1)_Y$ produces anomalies, which must cancel as required by the above condition.

End of the Digression

Due to the fact that perturbative QCD breaks down at low energies the handling of the quark loops or the related hadronic fluctuations pose a particular problem as the anomaly cancellation originally works on the level of quarks. Here another important theorem comes into play, however, namely 't Hooft's anomaly matching condition [74], which states that the anomaly on the level of the hadrons must be the same as the one on the level of the quarks, as a consequence of the anomaly non–renormalization theorem. An improved treatment of the hadronic contributions using an effective field theory approach has been elaborated in [69].

Structure of Contributions from Quark Triangles

Following [68], in order to discuss the contribution from VVA triangle fermions loops one has to consider the $Z^*\gamma\gamma^*$ amplitude

$$T_{\nu\lambda} = \mathrm{i} \int \mathrm{d}^4 x\, \mathrm{e}^{\mathrm{i}qx} \langle 0|T\{j_\nu(x)\, j_{5\lambda}(0)\}|\gamma(k)\rangle \qquad (4.45)$$

which by the LSZ reduction formula is equivalent to

$$T_{\nu\lambda} = e\, \varepsilon^\mu(k)\, T_{\mu\nu\lambda}\,,$$

$$T_{\mu\nu\lambda} = -\int \mathrm{d}^4 x\, \mathrm{d}^4 y\, \mathrm{e}^{\mathrm{i}(qx-ky)} \langle 0|T\{j_\mu(x)\, j_\nu(y)\, j_{5\lambda}(0)\}|0\rangle$$

in which $\varepsilon_\mu(k)$ is the polarization vector for the external photon. At small k up to quadratic terms one may write the covariant decomposition

$$T_{\nu\lambda} = -\frac{\mathrm{i}\,e}{4\pi^2}\left[w_T(q^2)\,(-q^2 \tilde{f}_{\nu\lambda} + q_\nu q^\alpha \tilde{f}_{\alpha\lambda} - q_\lambda q^\alpha \tilde{f}_{\alpha\nu}) + w_L(q^2)\, q_\lambda q^\alpha \tilde{f}_{\alpha\nu}\right] \qquad (4.46)$$

$$\tilde{f}_{\mu\nu} = \frac{1}{2}\varepsilon_{\mu\nu\alpha\beta} f^{\alpha\beta}\,, \quad f_{\mu\nu} = k_\mu \varepsilon_\nu - k_\nu \varepsilon_\mu$$

in terms of a transversal amplitude w_T and a longitudinal one w_L, with respect the the axial current index λ[13].

The contribution of a fermion f via the $Z^*\gamma\gamma^*$ amplitude to the muon anomaly $a_\mu^{(4)\,\mathrm{EW}}([f])_{\mathrm{AVV}}$, in the unitary gauge, where the Z propagator is $\mathrm{i}\,(-g_{\mu\nu} + q_\mu q_\nu/M_Z^2)/(q^2 - M_Z^2)$, is given by

$$\Delta a_\mu^{(4)\,\mathrm{EW}}([f])_{\mathrm{AVV}} = \frac{\sqrt{2} G_\mu\, m_\mu^2}{8\pi^4}\, \frac{\alpha}{\pi}\, \mathrm{i} \int \mathrm{d}^4 q\, \frac{1}{q^2 + 2qp}\left[\frac{1}{3}\left(1 + \frac{2(qp)^2}{q^2 m_\mu^2}\right)\right.$$

$$\left.\times\left(w_L - \frac{M_Z^2}{M_Z^2 - q^2}\, w_T\right) + \frac{M_Z^2}{M_Z^2 - q^2}\, w_T\right] \qquad (4.47)$$

---

[13] The second rank tensor $-\mathrm{i} f_{\mu\nu}$ corresponds to the external electromagnetic field strength tensor $F_{\mu\nu}$ with $\partial_\mu \to -\mathrm{i}k_\mu$ and $A_\nu \to \varepsilon_\nu$.

in terms of the two scalar amplitudes $w_{L,T}(q^2)$. p is the momentum of the external muon. For leading estimates it is sufficient to set $p = 0$ except in the phase space where it would produce an IR singularity, then the result takes the much simpler form

$$\Delta a_\mu^{(4)\,\text{EW}}([f])_{\text{VVA}} \simeq \frac{\sqrt{2} G_\mu m_\mu^2}{16\pi^2} \frac{\alpha}{\pi} \int_{m_\mu^2}^{\Lambda^2} dQ^2 \left(w_L(Q^2) + \frac{M_Z^2}{M_Z^2 + Q^2} w_T(Q^2) \right),$$
(4.48)

where $Q^2 = -q^2$ and Λ is a cutoff to be taken to ∞ at the end, after summing over a family. For a perturbative fermion loop to leading order [75]

$$w_L^{1-\text{loop}}(Q^2) \;=\; 2 w_T^{1-\text{loop}}(Q^2) = \sum_f 4 T_f N_{cf} Q_f^2 \int_0^1 \frac{dx\, x\, (1-x)}{x\,(1-x)\, Q^2 + m_f^2}$$

$$\stackrel{m_f^2 \ll Q^2}{=} \sum_f 4 T_f N_{cf} Q_f^2 \left[\frac{1}{Q^2} - \frac{2 m_f^2}{Q^4} \ln \frac{Q^2}{m_f^2} + O(\frac{1}{Q^6}) \right].$$

Vainshtein [76] has shown that in the chiral limit the relation

$$w_T(Q^2)_{\text{pQCD}}\big|_{m=0} = \frac{1}{2} w_L(Q^2)\big|_{m=0}$$
(4.49)

is valid actually to all orders of perturbative QCD in the kinematical limit relevant for the $g-2$ contribution. Thus the non–renormalization theorem valid beyond pQCD for the anomalous amplitude w_L (considering the quarks $q = u, d, s, c, b, t$ only):

$$w_L(Q^2)\big|_{m=0} = w_L^{1-\text{loop}}(Q^2)\big|_{m=0} = \sum_q (2 T_q Q_q^2) \frac{2 N_c}{Q^2}$$
(4.50)

carries over to the perturbative part of the transversal amplitude. Thus in the chiral limit the perturbative QPM result for w_T is exact in pQCD. This may be somewhat puzzling, since in low energy effective QCD, which encodes the non–perturbative strong interaction effects, this kind of term seems to be absent. The non–renormalization theorem has been proven independently in [77] and was extended to the full off–shell triangle amplitude to 2–loops in [78].

One knows that there are non–perturbative corrections to Vainshtein's relation (4.49) but no ones of perturbative origin. A simple heuristic proof of Vainshtein's theorem proceeds by first looking at the imaginary part of (4.45) and the covariant decomposition (4.46). In accordance with the Cutkosky rules (see footnote[28] on p. 83 in Chap. 2) the imaginary part of an amplitude is always more convergent than the amplitude itself. The imaginary part of the one–loop result is finite and one does not need a regularization to calculate it unambiguously. In particular, it allows us to use anti–commuting γ_5 to move it from the axial vertex $\gamma_\lambda \gamma_5$ to the vector vertex γ_ν.

In the limit $m_f = 0$, this involves anti–commuting γ_5 with an even number of γ–matrices, no matter how many gluons are attached to the quark line joining the two vertices. As a result Im $T_{\nu\lambda}$ must be symmetric under $\nu \leftrightarrow \lambda$, $q \leftrightarrow -q$:

$$\mathrm{Im}\left[w_T(q^2)\left(-q^2 \tilde{f}_{\nu\lambda} + q_\nu q^\alpha \tilde{f}_{\alpha\lambda} - q_\lambda q^\alpha \tilde{f}_{\alpha\nu}\right) + w_L(q^2)\, q_\lambda q^\alpha \tilde{f}_{\alpha\nu}\right]$$
$$\propto q_\nu q^\alpha \tilde{f}_{\alpha\lambda} + q_\lambda q^\alpha \tilde{f}_{\alpha\nu}$$

which, on the r.h.s., requires that $q^2 = 0$, to get rid of the antisymmetric term proportional to $\tilde{f}_{\nu\lambda}$, and that w_T is proportional to w_L: $w_L = c\, w_T$; the symmetry follows when $c = 2$. Thus the absence of an antisymmetric part is possible only if

$$2\mathrm{Im}\, w_T(q^2) = \mathrm{Im}\, w_L(q^2) = \mathrm{constant}\, \delta(q^2) \,, \tag{4.51}$$

where the constant is fixed to be $2\pi \cdot 2T_{3f} N_{cf} Q_f^2$ by the exact form of w_L. Both w_L and w_T are analytic functions which fall off sufficiently fast at large q^2 such that they satisfy convergent DRs

$$w_{T,L}(q^2) = \frac{1}{\pi}\int_0^\infty ds\, \frac{\mathrm{Im}\, w_{T,L}(s)}{s - q^2}$$

which together with (4.51) implies (4.49). While w_L as given by (4.50) is exact beyond perturbation theory, according to the Adler-Bardeen non–renormalization theorem and by the topological nature of the anomaly [73], as a consequence of Vainshtein's non–renormalization theorem for w_T we have

$$w_T(q^2) = \frac{2T_{3f} N_{cf} Q_f^2}{Q^2} + \text{non} - \text{perturbative corrections} \,. \tag{4.52}$$

Coming back to the calculation of (4.48), we observe that the contributions from w_L for individual fermions is logarithmically divergent, but it completely drops for a complete family due to the vanishing anomaly cancellation coefficient. The contribution from w_T is convergent for individual fermions due to the damping by the Z propagator. In fact it is the leading $1/Q^2$ term of the w_T amplitude which produces the $\ln \frac{M_Z}{m}$ terms. However, the coefficient is the same as the one for the anomalous term and thus for each complete family also the $\ln M_Z$ terms must drop out. Due to the non–renormalization theorem (4.49) the perturbative leading $1/Q^2$ term of w_T has to carry over to a low energy effective approach of QCD (see below).

Results for Contributions from Fermion Loops

For the third family the calculation is perturbative and thus straight forward with the result [62, 63, 69]

$$a_\mu^{(4)\,\mathrm{EW}}([\tau,b,t]) = -\frac{\sqrt{2}G_\mu\,m_\mu^2}{16\pi^2}\frac{\alpha}{\pi}\left[\frac{8}{3}\ln\frac{m_t^2}{M_Z^2} - \frac{2}{9}\frac{M_Z^2}{m_t^2}\left(\ln\frac{m_t^2}{M_Z^2}+\frac{5}{3}\right)\right.$$
$$\left. + \ln\frac{M_Z^2}{m_b^2} + 3\ln\frac{M_Z^2}{m_\tau^2} - \frac{8}{3} + \cdots\right]$$
$$\simeq -\frac{\sqrt{2}G_\mu\,m_\mu^2}{16\pi^2}\frac{\alpha}{\pi}\times 30.3(3) \simeq -8.21(10)\times 10^{-11}\,. \quad (4.53)$$

Small terms of order m_μ^2/m_τ^2, m_b^2/M_Z^2, M_Z^4/m_t^4 and smaller mass ratios have been neglected.

While the QPM results presented above, indeed confirmed the complete cancellation of the $\ln M_Z$ terms for the 1st and 2nd family, in the third family the corresponding terms $\ln M_Z/m_\tau$ and $\ln M_Z/m_b$ remain unbalanced by a corresponding top contribution.

Since in the perturbative regime QCD corrections are of $O(\alpha_s(\mu^2)/\pi)$, where μ is in the range from m_f to M_Z, pQCD is applicable for c, b and t quarks, only (see Fig. 3.3). For the lighter quarks u, d and s, however, the QPM estimate certainly is not appropriate because strong interaction corrections are expected to contribute beyond perturbation theory and assuming that non–perturbative effects just lead to a dressing of the quark masses into constituent quarks masses certainly is an over simplification of reality. Most importantly, pQCD does not account for the fact that the chiral symmetry is spontaneously broken the mechanism responsible for the emergence of the pions as quasi Goldstone bosons. The failure of the QPM we have illustrated in the discussion following p. 153 for the much simpler case of the hadronic vacuum polarization, already. We thus have to think about other means to take into account properly the low energy hadronic effects, if possible.

Digression on the Chiral Structure of Low Energy Effective QCD

Fortunately, a firm low energy effective theory of QCD exists and is very well developed: chiral perturbation theory (CHPT) [79], an expansion for low momenta p and in the light current quark masses as chiral symmetry breaking parameters. CHPT is based on the chiral flavor structure $SU(3)_L \otimes SU(3)_R$ of the low lying hadron spectrum (u,d,s quark bound states). The $SU(3)_V$ vector currents $j_k^\mu = \sum_{ij}\bar{\psi}_i(T_k)_{ij}\gamma^\mu\psi_j$ as well as the $SU(3)_A$ axial currents $j_{5k}^\mu = \sum_{ij}\bar{\psi}_i(T_k)_{ij}\gamma^\mu\gamma_5\psi_j$ [14] are partially conserved in the $SU(3)$ sector of the (u,d,s) quark flavors, and strictly conserved in the chiral limit of vanishing quark masses $m_u, m_d, m_s \to 0$, modulo the axial anomaly in the axial singlet current. The partial conservation of the chiral currents[15] derives from

---

[14] T_k ($k=1,\ldots,8$) are the generators of the global $SU(3)$ transformations and $i,j = u,d,s$ flavor indices.

[15] Especially in the $SU(2)$ isospin subspace, the small mass splitting $|m_1 - m_2| \ll m_1 + m_2$ motivates the terminology: conserved vector current (CVC) and partially conserved axial vector current (PCAC) (see Sect. 4.2.2 below).

$\partial_\mu(\bar\psi_1\gamma^\mu\psi_2) = \mathrm{i}(m_1 - m_2)\,\bar\psi_1\psi_2$ (CVC in the isospin limit $m_u = m_d$) and $\partial_\mu(\bar\psi_1\gamma^\mu\gamma_5\psi_2) = \mathrm{i}(m_1 + m_2)\,\bar\psi_1\gamma_5\psi_2$ (PCAC) and the setup of a perturbative scheme is based on the phenomenologically observed smallness of the current quark masses (3.36).

The chiral expansion is an expansion in $\hbar$

$$\mathcal{L}_{\mathrm{eff}} = \mathcal{L}_2 + \hbar\mathcal{L}_4 + \hbar^2\mathcal{L}_6 + \cdots \tag{4.54}$$

which is equivalent to an expansion in powers of derivatives and quark masses. In standard chiral counting one power of quark mass counts as two powers of derivatives, or momentum p in momentum space. In chiral $SU(3)$ there exists an octet of massless pseudoscalar particles (π, K, η), the Goldstone bosons in the chiral limit. The leading term of the expansion is the non–linear σ–model, where the pseudoscalars are encoded in a unitary 3×3 matrix field

$$U(\phi) = \exp\left(-\mathrm{i}\sqrt{2}\,\frac{\phi(x)}{F}\right) \tag{4.55}$$

with (T_i the SU(3) generators)

$$\phi(x) = \sum_i T_i\phi_i = \begin{pmatrix} \frac{\pi^0}{\sqrt{2}} + \frac{\eta}{\sqrt{6}} & \pi^+ & K^+ \\ \pi^- & \frac{-\pi^0}{\sqrt{2}} + \frac{\eta}{\sqrt{6}} & K^0 \\ K^- & \overline{K}^0 & -2\frac{\eta}{\sqrt{6}} \end{pmatrix} + \frac{1}{\sqrt{3}}\begin{pmatrix} \eta' & & \\ & \eta' & \\ & & \eta' \end{pmatrix} \tag{4.56}$$

where the second term is the diagonal singlet contribution by the η' meson. The latter is not a Goldstone boson, however it is of leading order in $1/N_c$. The leading order Lagrangian at $O(p^2)$ is then given by

$$\mathcal{L}_2 = \frac{F^2}{4}\operatorname{Tr}\left\{D^\mu U D_\mu U^\dagger + M^2\,(U + U^\dagger)\right\} \tag{4.57}$$

where, in absence of external fields, the covariant derivative $D_\mu U = \partial_\mu U$ coincides with the normal derivative. Furthermore, $M^2 = 2B\hat{m}$, where B is proportional to the quark condensate $\langle 0|\bar{u}u|0\rangle$ and $\hat{m} = \frac{1}{2}(m_u + m_d)$. In the chiral limit of exact $SU(3)_R \otimes SU(3)_L$ symmetry we have

$$\langle 0|\bar{u}u|0\rangle = \langle 0|\bar{d}d|0\rangle = \langle 0|\bar{s}s|0\rangle\;.$$

The parameters M and F are the leading order versions of the pion mass and the pion decay constant, respectively:

$$m_\pi^2 = M^2\left[1 + O(\hat{m})\right],\qquad F_\pi = F\left[1 + O(\hat{m})\right]\;.$$

The low energy effective currents again are nonlinear in the pion fields and in CHPT again appear expanded in the derivatives of U and the quark masses. For vector and axial–vector current one obtains

$$V_\mu^i = \frac{\mathrm{i}F^2}{4}\langle\sigma^i(U^\dagger D_\mu U + U D_\mu U^\dagger)\rangle + O(p^3) = \left[\varepsilon^{ijk}\phi^j\partial_\mu\phi^k + O(\phi^3)\right] + O(p^3)\;,$$

$$A^i_\mu = \frac{iF^2}{4}\langle\sigma^i(U^\dagger D_\mu U - U D_\mu U^\dagger)\rangle + O(p^3) = \left[-F\partial_\mu\phi^i + O(\phi^3)\right] + O(p^3) ,$$

which implies the conserved vector current (CVC) and the partially conserved axial vector current (PCAC) relations. Despite the fact that this Lagrangian is non–renormalizable, one can use it to calculate matrix elements like in standard perturbation theory. However, unlike in renormalizable theories where only terms already present in the original bare Lagrangian get reshuffled by renormalization, in non–renormalizable theories order by order in the expansion new vertices of increasing dimensions and associated new free couplings called *low energy constants* show up and limit the predictive power of the effective theory.

At physical quark masses the value of the condensate is estimated to be $\langle m_q \bar{q}q\rangle \sim -(0.098 \text{ GeV})^4$ for $q = u, d$. The key relation to identify the quark condensates in terms of physical quantities is the Gell-Mann, Oakes and Renner (GOR) [80] relation. In the chiral limit the mass operators $\bar{q}_R u_L$, or $\bar{q}_L u_R$ transform under $(3^*, 3)$ of the chiral group $SU(3)_L \otimes SU(3)_R$. Hence the quark condensates would have to vanish identically in case of an exact chirally symmetric world. In fact the symmetry is spontaneously broken and the vacuum of the real world is not chirally symmetric, and the quark condensates do not have to vanish. In order to determine the quark condensates, consider the charged axial currents and the related pseudoscalar density

$$A_\mu = \bar{d}\gamma_\mu\gamma_5 u$$
$$P = \bar{d}\, i\gamma_5 u$$

and the OPE of the product

$$A_\mu(x)\, P^+(y) = \sum_i C^i_\mu(x-y)\, \mathcal{O}^i(\frac{x+y}{2}) .$$

In QCD we may inspect the short distance expansion and study its consequences. One observation is that taking the VEV only the scalar operators contribute and one obtains the exact relation

$$\langle 0|A_\mu(x)\, P^+(y)|0\rangle = \frac{(x-y)_\mu}{2\pi^2\,(x-y)^4}\, \langle 0|\bar{u}u + \bar{d}d|0\rangle .$$

The spectral representation (see (3.123)) for the two–point function on the l.h.s. is of the form $p_\mu\, \rho(p^2)$ and current conservation requires $p^2\, \rho(p^2) = 0$ such that only the Goldstone modes, the massless pions, contribute, such that with

$$\langle 0|A_\mu(0)|\pi^+\rangle = i\, F_\pi\, p_\mu$$
$$\langle 0|P^+(0)|\pi^+\rangle = g_\pi$$

we get

$$F_\pi g_\pi = -\langle 0|\bar{u}u + \bar{d}d|0\rangle \ .$$

For nonvanishing quark masses the PCAC relation $\partial^\mu A_\mu = (m_u + m_d) P$ then implies the exact relation

$$F_\pi m_{\pi^+}^2 = (m_u + m_d) g_\pi$$

and the famous GOR relation

$$F_\pi^2 m_{\pi^+}^2 = -(m_u + m_d)\langle 0|\bar{u}u + \bar{d}d|0\rangle \qquad (4.58)$$

follows from the last two relations. Note that the quark condensates must be negative! They are a measure for the asymmetry of the vacuum in the chiral limit, and thus are true order parameters. If both F_π and $\langle 0|\bar{u}u + \bar{d}d|0\rangle$ have finite limits as $m_q \to 0$ the pion mass square must go to zero linear with the quark masses

$$m_{\pi^+}^2 = B(m_u + m_d) \ ; \quad B \equiv -\frac{1}{F_\pi^2}\langle 0|\bar{u}u + \bar{d}d|0\rangle \ ; \quad B > 0 \ .$$

The deviation from the chiral limit is controlled by CHPT. The quark masses as well as the quark condensates depend on the renormalization scale μ, however, the product $\langle 0|m_q \bar{q}q|0\rangle$ is RG invariant as is inferred by the GOR relation.

End of the Digression

In [62] the light quark contribution to Fig. 4.11a were evaluated using the low energy effective form of QCD which is CHPT. To lowest order in the chiral expansion, the hadronic $Z\gamma\gamma$ interaction is dominated by the pseudoscalar meson (the quasi Goldstone bosons) exchange. The corresponding effective couplings are given by

$$\mathcal{L}^{(2)} = -\frac{e}{2\sin\Theta_W \cos\Theta_W} F_\pi \partial_\mu \left(\pi^0 + \frac{1}{\sqrt{3}}\eta_8 - \frac{1}{\sqrt{6}}\eta_0\right) Z^\mu \ , \qquad (4.59)$$

which is the relevant part of the $O(p^2)$ chiral effective Lagrangian, and the effective $O(p^4)$ coupling

$$\mathcal{L}_{\mathrm{WZW}} = \frac{\alpha}{\pi}\frac{N_c}{12 F_\pi}\left(\pi^0 + \frac{1}{\sqrt{3}}\eta_8 + 2\sqrt{\frac{2}{3}}\eta_0\right)\tilde{F}_{\mu\nu}F^{\mu\nu} \ , \qquad (4.60)$$

which is the Wess-Zumino-Witten Lagrangian. The latter reproduces the ABJ anomaly on the level of the hadrons. π^0 is the neutral pion field, F_π the pion decay constant ($F_\pi = 92.4$ MeV). The pseudoscalars η_8, η_0 are mixing into the physical states η, η'. The $[u, d, s]$ contribution with long distance (L.D.) part

242   4 Electromagnetic and Weak Radiative Corrections

($E < \mu$) evaluated in CHPT and a short distance (S.D.) part ($E > \mu$) to be evaluated in the QPM. The cut–off for matching L.D. and S.D. part typically is $M_\Lambda = m_P \sim 1$ GeV to $M_\Lambda = M_\tau \sim 2$ GeV. The corresponding diagrams are shown in Fig. 4.13, which together with its crossed version in the unitary gauge and in the chiral limit, up to terms suppressed by m_μ^2/M_Λ^2, yields[16]

$$a_\mu^{(4)\,\mathrm{EW}}([u,d,s]; p < M_\Lambda)_{\mathrm{CHPT}} = \frac{\sqrt{2} G_\mu\, m_\mu^2}{16\pi^2} \frac{\alpha}{\pi} \times \left[\frac{4}{3}\ln\frac{M_\Lambda^2}{m_\mu^2} + \frac{2}{3}\right]$$

$$\simeq 2.10 \times 10^{-11}\,,$$

$$a_\mu^{(4)\,\mathrm{EW}}([u,d,s]; p > M_\Lambda)_{\mathrm{QPM}} = \frac{\sqrt{2} G_\mu\, m_\mu^2}{16\pi^2} \frac{\alpha}{\pi} \left[2\ln\frac{M_Z^2}{M_\Lambda^2}\right]$$

$$\simeq 4.45 \times 10^{-11}\,.$$

Note that the last diagram of Fig. 4.13 in fact takes into account the leading term of (4.52) which is protected by Vainshtein's relation (4.49).

Above a divergent term has been dropped, which cancels against corresponding terms from the complementary contributions from e, μ and c fermion–loops. Including the finite contributions from e, μ and c:

$$a_\mu^{(4)\,\mathrm{EW}}([e,\mu,c])_{QPM} = \frac{\sqrt{2} G_\mu\, m_\mu^2}{16\pi^2} \frac{\alpha}{\pi} \left[-6\ln\frac{M_Z^2}{m_\mu^2} + 4\ln\frac{M_Z^2}{M_c^2} - \frac{37}{3} + \frac{8}{9}\pi^2\right]$$

$$\simeq -\frac{\sqrt{2} G_\mu\, m_\mu^2}{16\pi^2} \frac{\alpha}{\pi} \times 50.37 \simeq -13.64 \times 10^{-11}$$

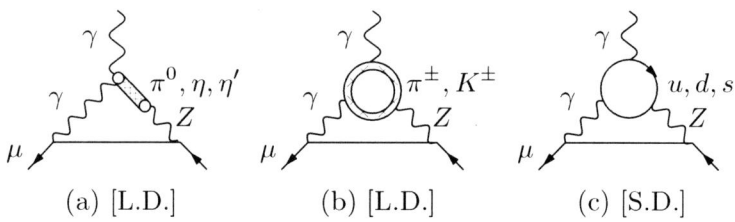

(a) [L.D.]   (b) [L.D.]   (c) [S.D.]

Fig. 4.13. The two leading CHPT diagrams (L.D.) and the QPM diagram (S.D.). The charged pion loop is sub–leading and will be discarded

---

[16] The simplest way to implement the lower cut–off M_Λ to the low energy effective field theory (EFT) is to write in (4.48)

$$\frac{1}{M_Z^2 + Q^2} = \underbrace{\frac{1}{M_\Lambda^2 + Q^2}}_{EFT} + \underbrace{\frac{1}{M_Z^2 + Q^2} - \frac{1}{M_\Lambda^2 + Q^2}}_{QPM}$$

by using the QPM for the second term. In the first term M_Z is replaced by M_Λ, in the second term constant terms drop out in the difference as the quark masses in any case have values far below the cut–offs.

the complete answer for the 1st plus 2nd family reads [62]

$$a_\mu^{(4)\,\mathrm{EW}}\left(\begin{bmatrix} e,u,d \\ \mu,c,s \end{bmatrix}\right)_{\mathrm{CHPT}} = \frac{\sqrt{2}G_\mu m_\mu^2}{16\pi^2}\frac{\alpha}{\pi}\left[-\frac{14}{3}\ln\frac{M_A^2}{m_\mu^2} + 4\ln\frac{M_A^2}{M_c^2} - \frac{35}{3} + \frac{8}{9}\pi^2\right]$$

$$\simeq -\frac{\sqrt{2}G_\mu m_\mu^2}{16\pi^2}\frac{\alpha}{\pi}\times 26.2(5) \simeq -7.09(13)\times 10^{-11}\,. \quad (4.61)$$

In (4.61) the error comes from varying the cut–off M_A between 1 GeV and 2 GeV. Below 1 GeV CHPT can be trusted above 2 GeV we can trust pQCD. Fortunately the result is not very sensitive to the choice of the cut–off[17].

On the other hand results depend quite strongly on the quark masses utilized. This result was refined by a more elaborate analysis in which sub-leading terms were calculated using the operator product expansion (OPE).

Digression on the Operator Product Expansion

The operator product expansion (Wilson short distance expansion) [81] is a formal expansion of the product of two local field operators $A(x)\,B(y)$ in powers of the distance $(x-y)\to 0$ in terms of singular coefficient functions and regular composite operators:

$$A(x)\,B(y) \simeq \sum_i C_i(x-y)\,\mathcal{O}_i(\frac{x+y}{2})$$

where the operators $\mathcal{O}_i(\frac{x+y}{2})$ represent a complete system of local operators of increasing dimensions. The coefficients may be calculated formally by normal perturbation theory by looking at the Green functions

$$\langle 0|TA(x)\,B(y)\,X|0\rangle = \sum_{i=0}^{N} C_i(x-y)\,\langle 0|T\mathcal{O}_i(\frac{x+y}{2})\,X|0\rangle + R_N(x,y)$$

constructed such that

$$R_N \to 0 \text{ as } (x-y)^{a_N} \,; \, (x-y)^2 < 0$$
$$a_N < a_{N+1}\,\forall\,N$$

(asymptotic expansion). By X we denoted any product of fields suitable to define a physical state $|X\rangle$ via the LSZ reduction formula (see Table 2.1).

---

[17] If no cut–off is applied to the validity of the effective theory as in [62] one gets -8.58×10^{-11}, in which case an unphysical residual $\ln M_Z$ dependence persists. The QPM result taking the rather arbitrary constituent quark masses (4.39) is -8.65×10^{-11}. The QPM result taking current quark masses (3.36) is -5.87×10^{-11}. In [68] the leading logarithmic estimate is -6.72×10^{-11} (Equations (26) plus (28) of [68]), while a refined estimate yields -6.65×10^{-11} (Equations (60) plus (65) of [68]) fairly close to our estimate (4.61).

The OPE is a very important tool in particular in the intrinsically non–perturbative strong interaction dynamics, which is perturbative at short distances only, by virtue of asymptotic freedom. It serves to separate soft non–perturbative low energy effects from hard perturbative high energy effects in case a hadronic process involves a highly energetic subprocess. Typically, the short distance singular coefficient functions are often accessible to pQCD while the soft effects are factored out into a non–perturbative matrix elements of appropriate composite operators. The latter in many cases may be determined by experiment or by non–perturbative methods like QCD on a lattice. One of the most prominent examples of the application of the OPE is deep inelastic electron–nucleon scattering (DIS), which uncovered the quark structure of hadrons at short wave lengths. The factorization into coefficients and matrix elements in the OPE is renormalization scheme dependent and in particular depends on the renormalization scale μ. The factorization into hard and soft physics requires the condition $m_f \ll \mu \ll Q$, which we will assume to be satisfied in the following. For a more comprehensive elaboration of the subject I recommend Shifman's lectures [82].

At the heart of the OPE is the following basic problem: Local products of quantum fields in general are singular, for two scalar fields in scalar φ^4–theory for example

$$T\{\varphi(x)\,\varphi(y)\,X\}|_{\lim x \to y} \sim$$

creates a loop which in general in UV singular, the obtained composite field $\varphi^2(x=y)$ is defined after subtraction of an UV singular term only, i.e. it requires renormalization. In fact a series of new divergences shows up: all superficially divergent sub–diagrams, which contain the generated vertex:

The dots represent derivatives in configuration space or multiplication of the line with the corresponding momentum in momentum space. The dashed circles enclose a renormalization part which corresponds to a constant, and graphically contracts into a point. The superficial divergence of the corresponding sub–diagrams γ_i in $d = 4$ dimensions is given by dim $\gamma_i = 4 - N_i - L_i + \dim \varphi^2$; dim $\varphi^2 = 2$, where N_i is the number of φ–lines and L_i the number of derivatives on φ–lines. The subtraction factors multiply

Green functions or matrix elements with insertions of operators of increasing dimensions. The Wilson expansion isolates the subtraction terms related to sub–diagrams $\tilde{\gamma}_i$ which translate into γ_i by identifying $x = y$:

The first factor of each term represents the coefficient $C_i(x - y)$ the second the operator matrix element $\langle 0|T\mathcal{O}_i(\frac{x+y}{2})|X\rangle$.

For a product of two currents the procedure is similar. The object of interest in our case is

$$T\{j_\nu(x)\, j_{5\lambda}(0)\} = T\{:\bar{\psi}(x)\gamma_\nu\,\psi(x)::\bar{\psi}(0)\gamma_\lambda\gamma_5\psi(0):\}$$

where the Wick ordering $:\cdots:$ is the prescription that only fields from different vertices are to be contracted (see p. 45). A contraction of two free Fermi fields under the T–product represents a Dirac propagator

$$T\{\psi_{\alpha c i}(x)\,\bar{\psi}_{\beta c' j}(y)\}_{\text{free}} = \mathrm{i}\, S_{F\alpha\beta}(x - y; m_i)\,\delta_{cc'}\delta_{ij} - :\bar{\psi}_{\beta c' j}(y)\,\psi_{\alpha c i}(x):$$

for a free field. In our example the currents are diagonal in color and flavor and we hence suppress color and flavor indices. We thus obtain in the case of free fields

$$T\{j_\nu(x)\, j_{5\lambda}(0)\}_{\text{free}} = T\{:\bar{\psi}_\alpha(x)(\gamma_\nu)_{\alpha\beta}\,\psi_\beta(x)::\bar{\psi}_{\alpha'}(0)(\gamma_\lambda\gamma_5)_{\alpha'\beta'}\psi_{\beta'}(0):\}$$
$$= (-1)\,\mathrm{i}\, S_{F\beta'\alpha}(-x; m_f)\,(\gamma_\nu)_{\alpha\beta}\,\mathrm{i}\,S_{F\beta\alpha'}(x; m_f)\,(\gamma_\lambda\gamma_5)_{\alpha'\beta'}$$
$$+ \mathrm{i}\,S_{F\beta\alpha'}(x; m_f):\bar{\psi}_\alpha(x)(\gamma_\nu)_{\alpha\beta}\,(\gamma_\lambda\gamma_5)_{\alpha'\beta'}\psi_{\beta'}(0):$$
$$+ \mathrm{i}\,S_{F\beta'\alpha}(-x; m_f):\bar{\psi}_{\alpha'}(0)(\gamma_\lambda\gamma_5)_{\alpha'\beta'}(\gamma_\nu)_{\alpha\beta}\,\psi_\beta(x):$$
$$+ :j_\nu(x)\, j_{5\lambda}(0):$$

$$= (-1)\;\text{⬭}\; + \;\text{⬭}\; + \;\text{⬭}\; + \;\text{⬭}$$
(4.62)

The first term in fact is zero. A two point correlator of VA–type vanishes identically[18], however, for VV– or AA–type of products of currents such a

---

[18] In momentum space the γ_5–odd trace yields terms proportional to $\varepsilon_{\nu\lambda\alpha\beta}$ where the two indices α and β have to be contracted with momenta or with $g^{\alpha\beta}$, yielding a vanishing result. In a propagator there is only one momentum p available, but $p^\alpha p^\beta$ is symmetric and contracts to zero with the anti–symmetric ε–tensor.

term in general is present. For the second and third term we may proceed as follows: the Dirac propagators have the form

$$S_{F\alpha\beta}(x - y; m_i) = (i\gamma^\mu \partial_\mu + m_i)_{\alpha\beta}\, \Delta_F(x - y, m_i)$$

where $\Delta_F(x-y, m_i)$ is the scalar Feynman propagator (see 2.2) $i/(p^2 - m_i^2 + i\varepsilon)$ in momentum space, and the Dirac algebra may be easily worked out by using the Chisholm identity

$$\gamma^\nu \gamma^\alpha \gamma^\lambda = (g^{\nu\alpha}g^{\lambda\beta} + g^{\lambda\alpha}g^{\nu\beta} - g^{\nu\lambda}g^{\alpha\beta})\gamma_\beta + i\varepsilon^{\nu\alpha\lambda\beta}\gamma_5\gamma_\beta \ .$$

The two terms correspond to the symmetric and the antisymmetric part. In the chiral limit then only terms exhibiting one γ matrix are left which enter bilocal vector or axial vector currents of the form

$$\begin{aligned} J_\beta^V(x,0) &\equiv\, :\bar\psi(x)\gamma_\beta\psi(0): \\ J_\beta^A(x,0) &\equiv\, :\bar\psi(x)\gamma_\beta\gamma_5\psi(0): \ . \end{aligned} \qquad (4.63)$$

In the presence of interactions and a set of other fields X characterizing a state $|X\rangle$ we graphically may write

$$T\{j_\nu(x)\, j_{5\lambda}(0)\, X\} =$$

$$(4.64)$$

The Wilson OPE is obtained know by expanding the bilocal current $:\bar\psi(x)\cdots\psi(0):$ in x. In the free field case these Wick monomials are regular for $x \to 0$ as the singular term, the first term of (4.62), has been split off. It is therefore possible to perform a Taylor series expansion in x

$$:\bar\psi(x)\cdots\psi(0): = \sum_{n=0}^{\infty} \frac{1}{n!} x^{\mu_1}\cdots x^{\mu_n} :\bar\psi(0)\, \overleftarrow{\partial}_{\mu_1}\cdots\overleftarrow{\partial}_{\mu_n}\cdots\psi(0):$$

and

$$:\bar\psi(0)\cdots\psi(x): = \sum_{n=0}^{\infty} \frac{1}{n!} x^{\mu_1}\cdots x^{\mu_n} :\bar\psi(0)\cdots\overrightarrow{\partial}_{\mu_1}\cdots\overrightarrow{\partial}_{\mu_n}\psi(0):$$

The bilocal operators thus take the form

$$J_\mu^X(x,0) = \sum_{n=0}^{\infty} \frac{1}{n!} x^{\mu_1}\cdots x^{\mu_n}\, \mathcal{O}^X_{\mu_1\cdots\mu_n;\mu}(0) \ .$$

In momentum space factors x^μ are represented by a derivative with respect to momentum $-\mathrm{i}\frac{\partial}{\partial p_\mu}$. In gauge theories, like QED and QCD, of course derivatives in x–space have to be replaced by covariant derivatives in order to keep track of gauge invariance. In general it is not too difficult to guess the form of the possible leading, sub–leading etc. operators from the tensor structure and the other symmetries. For the second term above, as an example, diagrammatically we have

$$\text{(diagrams)} \quad (4.65)$$

where the 1st coefficient diagram in leading order is the VVA triangle diagram, the 2nd coefficient diagram in leading order is a Compton scattering like tree diagram. The second line shows the leading perturbative terms in case the "final state" X is a photon γ. The other terms of (4.64) may be worked out along the same lines.

We now turn back to the application of the OPE in calculating hadronic effects in the weak contributions to $g - 2$. For this purpose the state $|X\rangle$ is the external one–photon state $|\gamma(k)\rangle$ in the classical limit, where it describes an external magnetic field. The first term of (4.64) in this case does not contribute. The diagrammatic representation of the OPE allows us an easy transition from configuration to momentum space.

End of the Digression

Non–perturbative Effects via the OPE

For the purpose of the anomalous magnetic moment (see (4.45)) one need consider only two currents

$$\hat{T}_{\nu\lambda} = \mathrm{i} \int \mathrm{d}^4 x \, \mathrm{e}^{\mathrm{i}qx} \, T\{j_\nu(x)\, j_{5\lambda}(0)\} = \sum_i c^i_{\nu\lambda\alpha_1\ldots\alpha_i}(q)\, \mathcal{O}_i^{\alpha_1\ldots\alpha_i}$$

where the operators $\mathcal{O}$ are local operators constructed from the light fields, the photon, light quarks and gluon fields. The Wilson coefficients c^i encode the short distance properties while the operator matrix elements describe the non–perturbative long range strong interaction features. The matrix element of our concern is

248   4 Electromagnetic and Weak Radiative Corrections

$$T_{\nu\lambda} = \langle 0|\hat{T}_{\nu\lambda}|\gamma(k)\rangle = \sum_i c^i_{\nu\lambda\alpha_1\ldots\alpha_i}(q)\,\langle 0|\mathcal{O}_i^{\alpha_1\ldots\alpha_i}|\gamma(k)\rangle \qquad (4.66)$$

in the classical limit $k \to 0$, where the leading contribution becomes linear in $\tilde{f}_{\alpha\beta}$ the dual of $f_{\alpha\beta} = k_\alpha \varepsilon_\beta - k_\beta \varepsilon_\alpha$. Hence, only those operators contribute which have the structure of an antisymmetric tensor

$$\langle 0|\mathcal{O}_i^{\alpha\beta}|\gamma(k)\rangle = -\mathrm{i}\,\frac{1}{4\pi^2}\kappa_i \tilde{f}^{\alpha\beta} \qquad (4.67)$$

with constants κ_i which depend on the renormalization scale μ. The operators contributing to $T_{\nu\lambda}$ in the OPE, in view of the tensor structure (4.46), are of the form

$$T_{\nu\lambda} = \sum_i \left\{ c^i_T(q^2)\,(-q^2 \mathcal{O}^i_{\nu\lambda} + q_\nu q^\alpha \mathcal{O}^i_{\alpha\lambda} - q_\lambda q^\alpha \mathcal{O}^i_{\alpha\nu}) + c^i_L(q^2)\, q_\lambda q^\alpha \mathcal{O}^i_{\alpha\nu} \right\}$$

such that (4.68)

$$w_{T,L}(q^2) = \sum_i c^i_{T,L}(q^2, \mu^2)\,\kappa_i(\mu^2)\,. \qquad (4.69)$$

The OPE is an expansion for large $Q^2 = -q^2$ and the relevance of the terms are determined by the dimension of the operator, the low dimensional ones being the most relevant, unless they are nullified or suppressed by small coefficients due to exact or approximate symmetries, like chiral symmetry. Note that the functions we expand are analytic in the q^2–plane and an asymptotic expansion for large Q^2 is a formal power series in $1/Q^2$ up to logarithms. Therefore operators of odd dimension must give contributions proportional to the mass m_f of the light fermion field from which the operator is constructed. In the chiral limit the operators must be of *even* dimension and antisymmetric.

In the following we include the factors T_{3f} at the $Z^\lambda j_{5\lambda}(0)$ vertex (axial current coefficient) and Q_f at the $A^\nu j_\nu(x)$ vertex (vector current coefficient) as well as the color multiplicity factor N_{cf} where appropriate. A further factor Q_f (coupling to the external photon) comes in via the matrix elements κ_i of fermion operators $\bar{f}\cdots f$. In case of helicity flip operators $\bar{f}_R \cdots f_L$ or $\bar{f}_L \cdots f_R$ the corresponding κ_i will be proportional to m_f.

The first non–vanishing term of the OPE is the 1st term on the r.h.s. of (4.65), which requires a parity odd operator linear in the photon field. In fact, the leading operator has dimension $d_\mathcal{O} = 2$ given by the parity odd dual electromagnetic field strength tensor

$$\mathcal{O}_F^{\alpha\beta} = \frac{1}{4\pi^2}\tilde{F}^{\alpha\beta} = \frac{1}{4\pi^2}\varepsilon^{\alpha\beta\rho\sigma}\partial_\rho A_\sigma\,.$$

The normalization is chosen such that $\kappa_F = 1$ and hence $w^F_{L,T} = c^F_{L,T}$. The corresponding coefficient for this leading term is given by the perturbative one–loop triangle diagram and yields

$$c_L^F[f] = 2c_T^F[f] = \frac{4T_{3f}N_{cf}Q_f^2}{Q^2}\left[1 - \frac{2m_f^2}{Q^2}\ln\frac{Q^2}{\mu^2} + O(\frac{m_f^4}{Q^4})\right] \quad (4.70)$$

where the leading $1/Q^2$ term cancels family-wise due to quark–lepton duality. In the chiral limit we know that this is the only contribution to w_L.

Next higher term is the 2nd term on the r.h.s. of (4.65). The $d_\mathcal{O} = 3$ operators which can contribute to the amplitudes under consideration are given by

$$\mathcal{O}_f^{\alpha\beta} = -i\bar{f}\sigma^{\alpha\beta}\gamma_5 f \equiv \frac{1}{2}\varepsilon^{\alpha\beta\rho\sigma}\bar{f}\sigma^{\rho\sigma}f \ .$$

These helicity flip operators only may contribute if chiral symmetry is broken and the corresponding coefficients must be of the form $c^f \propto m_f/Q^4$. These coefficients are determined by tree level diagrams of Compton scattering type and again contribute equally to both amplitudes

$$c_L^f[f] = 2c_T^f[f] = \frac{8T_{3f}Q_f m_f}{Q^4}\ .$$

Misusing the spirit of the OPE for the moment and neglecting the soft strong interaction effects, we may calculate the soft photon quark matrix element in the QPM from the one–loop diagram shown in (4.65) (last diagram) which is UV divergent and in the $\overline{\text{MS}}$ scheme yields

$$\kappa_f = -Q_f\, N_f\, m_f \, \ln\frac{\mu^2}{m_f^2} \ .$$

Inserting this in

$$\Delta^{(d_\mathcal{O}=3)}w_L = 2\Delta^{(d_\mathcal{O}=3)}w_T = \frac{8}{Q^4}\sum_f T_{3f}\, Q_f\, m_f\, \kappa_f$$

one recovers precisely the $1/Q^4$ term of (4.70). So far we have reproduced the known perturbative result. Nevertheless the calculation illustrates the use of the OPE. While the leading $1/Q^2$ term is not modified by soft gluon interactions, i.e. $\kappa_F = 1$ is exact as the state $|\gamma\rangle$ represents a physical on–shell photon, undressed from possible self–energy corrections, the physical κ_f cannot be obtained from pQCD. So far it is an unknown constant. Here again, the spontaneous breakdown of the chiral symmetry and the existence of, in the chiral limit, non–vanishing quark condensates $\langle\bar\psi\psi\rangle_0 \neq 0$ plays a central role. Now, unlike in perturbation theory, κ_f need not be proportional to m_f. In fact it is proportional to $\langle\bar\psi\psi\rangle_0$. As the condensate is of dimensionality 3, another quantity must enter carrying dimension of a mass and which is finite in the chiral limit. In the u,d quark sector this is either the pion decay constant F_0 or the ρ mass M_{ρ^0}. As it is given by the matrix element (4.67) (see also the last graph of Fig. 4.65) κ_f must be proportional to $N_{cf}Q_f$ such that

$$\kappa_f = N_{cf} Q_f \frac{\langle \bar{\psi}_f \psi_f \rangle_0}{F_0^2}$$

and hence [69, 76]

$$\Delta^{(d_\mathcal{O}=3)} w_L = 2 \Delta^{(d_\mathcal{O}=3)} w_T = \frac{8}{Q^4} \sum_f N_{cf} T_{3f} Q_f^2 \, m_f \, \frac{\langle \bar{\psi}_f \psi_f \rangle_0}{F_0^2} \; . \quad (4.71)$$

An overall constant, in fact is not yet fixed, however, it was chosen such that it reproduces the expansion of non–perturbative modification of w_L as a pion propagator beyond the chiral limit:

$$w_L = \frac{2}{Q^2 + m_\pi^2} = \frac{2}{Q^2} - \frac{2 m_\pi^2}{Q^4} + \cdots$$

as we will see below.

All operators of $d_\mathcal{O} = 4$ may be reduced via the equation of motion to $d_\mathcal{O} = 3$ operators carrying a factor of mass in front:

$$\bar{f} \left(D^\alpha \gamma^\beta - D^\beta \gamma^\alpha \right) \gamma_5 f = -m_f \, \bar{f} \sigma^{\alpha\beta} \gamma_5 f \; .$$

They thus do not yield new type of corrections and will not be considered further, as they are suppressed by the light quark masses as m_f^2/Q^4.

Similarly the dimension $d_\mathcal{O} = 5$ operators

$$\bar{f} f \tilde{F}^{\alpha\beta}, \; \bar{f} \gamma_5 f \tilde{F}^{\alpha\beta}, \; \cdots$$

which are contributing to the $1/Q^6$ coefficient, require a factor m_f and thus again are suppressed by nearby chiral symmetry.

More important are the dimension $d_\mathcal{O} = 6$ operators, which yield $1/Q^6$ terms and give non–vanishing contributions in the chiral limit. Here again the specific low energy structure of QCD comes into play, namely the spontaneous symmetry breaking of the chiral symmetry (in the symmetry limit). The latter is characterized by the existence of an *orderparameter*[19], which in QCD are the color singlet quark condensates $\langle \bar{\psi}_q \psi_q \rangle$ of the light quarks $q = u, d, s$, where we have implicitly summed over color. The point is that the condensates are non–vanishing in the chiral limit $m_q = 0$, typically they take values $\langle \bar{\psi}_q \psi_q \rangle \simeq -(240 \text{ MeV})^3$. Note that in pQCD chiral symmetry (in the symmetry limit) remains unbroken, $\langle \bar{\psi}_q \psi_q \rangle$ vanishes identically. Higher order color singlet contributions are possible which include hard gluon exchange represented by the Feynman diagrams of Fig. 4.14. They are of the type as represented by the last diagram of (4.64). The operators responsible derive from : $j_\nu(x) j_{5\lambda}(0)$: corrected by second order QCD (two quark gluon interaction vertices as given in Fig. 2.9 in Sect. 2.8) with the gluon and two quark pairs contacted, like

---

[19] Spontaneous symmetry breaking is best known from ferromagnets, where rotational invariance is spontaneously broken, leading to spontaneous magnetization $\langle S_z \rangle = M \neq 0$ in a frame where M is directed along the z–axis.

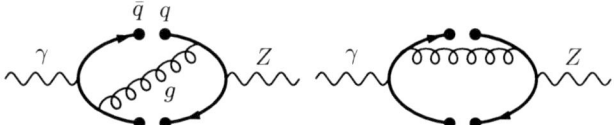

Fig. 4.14. Non–perturbative quark condensate contributions due to spontaneous breaking of chiral symmetry. The scalars $\bar{q}q$ couple to the vacuum $\langle \bar{q}q \rangle \neq 0$. Two other diagrams are obtained by attaching the gluons to the quark lines by other permutations

$$:\bar{\psi}(x)\gamma_\nu \psi(x)\, \bar{\psi}(0)\gamma_\lambda\gamma_5\psi(0): \;::\underbrace{\bar{\psi}_a \gamma^\alpha (T_i)_{aa'} \psi_{a'} G^i_\alpha (z_1)}::\underbrace{\bar{\psi}_b \gamma^\beta (T_j)_{bb'} \psi_{b'} G^j_\beta (z_2)}:$$

where T_i are the $SU(3)$ generators satisfying

$$\sum_i (T_i)_{aa'}(T_i)_{bb'} = \frac{1}{2}\left(\delta_{ab'}\delta_{a'b} - \frac{1}{N_c}\delta_{aa'}\delta_{bb'}\right).$$

The terms have been worked out in detail in [69] and are of the form

$$\hat{T}_{\nu\lambda}(q) = i\,[q_\beta \varepsilon_{\nu\lambda\rho\alpha} q^\rho - q_\alpha \varepsilon_{\nu\lambda\rho\beta} q^\rho]\left(-2\pi^2 \frac{\alpha_s}{\pi}\right)\frac{\mathcal{O}^{\alpha\beta}(0)}{Q^6} + \cdots$$

with

$$\mathcal{O}^{\alpha\beta} = \left[\frac{2}{3}\left(\bar{u}\sigma^{\alpha\beta}u\right)(\bar{u}u) + \frac{1}{3}\left(\bar{d}\sigma^{\alpha\beta}d\right)(\bar{d}d) + \frac{1}{3}\left(\bar{s}\sigma^{\alpha\beta}s\right)(\bar{s}s)\right](0)\,.$$

These terms yield the leading non–perturbative (NP) contributions and persist in the chiral limit. They only contribute to the transversal amplitude, and using estimates presented in [83] one obtains

$$w_T(Q^2)_{\mathrm{NP}} \simeq -\frac{16}{9}\pi^2\,\frac{2}{F_0^2}\,\frac{\alpha_s}{\pi}\,\frac{\langle\bar{\psi}\psi\rangle^2}{Q^6} \qquad (4.72)$$

for large enough Q^2, the ρ mass being the typical scale. This NP contribution breaks the degeneracy $w_T(Q^2) = \frac{1}{2} w_L(Q^2)$ which is valid in perturbation theory[20]. Taking into account the quark condensates together with explicit chiral symmetry breaking, according to (4.71), also a term

$$\Delta w_T(Q^2)_{\mathrm{NP}} = \frac{1}{2}\Delta w_L(Q^2)_{\mathrm{NP}} \simeq \frac{4}{9}\,\frac{3}{2F_0^2}\,\frac{(4m_u - m_d - m_s)\langle\bar{\psi}\psi\rangle}{Q^4}\,, \qquad (4.73)$$

---

[20] The OPE only provides information on w_T for Q^2 large. At low Q^2 we only know that $w_T(0) = 128\pi^2\, C_{22}^W$ where C_{22}^W is one of the unknown CHPT constants in the $O(p^6)$ parity odd part of the chiral Lagrangian [84].

yields an NP contribution, but this time to both w_T and w_L.

The consequences of the OPE for the light quarks u, d and s in the chiral limit may be summarized as follows [68]:

$$w_L[u,d]_{m_{u,d}=0} = -3\, w_L[s]_{m_s=0} = \frac{2}{Q^2}\;, \qquad (4.74)$$

$$w_T[u,d]_{m_{u,d}=0} = -3\, w_T[s]_{m_s=0} = \frac{1}{Q^2} - \frac{32\pi\alpha_s}{9\,Q^6}\frac{\langle\bar\psi\psi\rangle_0^2}{F_\pi^2} + O(Q^{-8})\;.$$

The condensates are fixed essentially by the Gell-Mann-Oakes-Renner (GOR) relations (4.58)

$$(m_u + m_d)\,\langle\bar\psi\psi\rangle_0 = -F_0^2 m_\pi^2$$
$$m_s\langle\bar\psi\psi\rangle_0 \simeq -F_0^2 M_K^2\;.$$

and the last term of (4.74) numerically estimates to

$$w_T(Q^2)_{\rm NP} \sim -\alpha_s\,(0.772~{\rm GeV})^4/Q^6\;,$$

i.e. the scale is close to the ρ mass. Our estimates are rough leading order estimates in the sense of CHPT. The index $_0$ denotes quantities in the chiral limit. Except from the masses of the pseudoscalars, which vanish in the chiral limit, we do not distinguish between quantities like the pseudoscalar decay constants F_0, F_π and F_K. Similarly, we assume the light quark condensates $\langle\bar\psi\psi\rangle_0$ to be approximately equal for u, d and s quarks. Furthermore, we use $m_\eta^2 \simeq \frac{4}{3}m_K^2$ and $M_{\eta'}^2 \simeq M_0^2$ with $M_0 \simeq 950$ MeV (for CHPT refinements we refer to [79]). Also isospin symmetry will be assumed where appropriate.

In fact the non–perturbative refinements of the leading π^0, η, η' exchange contributions in w_L requires the inclusion of vector–meson exchanges which contribute to w_T. More precisely, for the transversal function the intermediate states have to be 1^+ mesons with isospin 1 and 0 or 1^- mesons with isospin 1. The lightest ones are ρ, ω and a_1. They are massive also in the chiral limit.

In principle, the incorporation of vector–mesons, like the ρ, in accordance with the basic symmetries is possible using the Resonance Lagrangian Approach (RLA) [85, 86], an extended form of CHPT. The more recent analyses are based on quark–hadron duality, as it holds in the large N_c limit of QCD [87, 88], for modeling the hadronic amplitudes [89]. The infinite series of narrow vector states known to show up in the large N_c limit is then approximated by a suitable lowest meson dominance, i.e. amplitudes are assumed to be saturated by known low lying physical states of appropriate quantum numbers. This approach was adopted in an analysis by the Marseille group [69][21].

---

[21] In this analysis, the leading $1/Q^2$ term of w_T in (4.74) got lost, which produces a fake $\ln M_Z$ term in the leading hadronic contribution. This was rectified in [76, 68] and confirmed by the authors of [69] in [77]. The $1/Q^6$ correction was estimated using "large N_c limit of QCD" type of arguments and taking into account the three

An analysis which takes into account the complete structure (4.74) was finalized in [68]. In the narrow width approximation one may write

$$\mathrm{Im}\, w_T = \pi \sum_i g_i \, \delta(s - m_i^2)$$

where the weight factors g_i satisfy

$$\sum_i g_i = 1\,, \quad \sum_i g_i m_i^2 = 0$$

in order to reproduce (4.74) in the chiral limit. Beyond the chiral limit the corrections (4.73) should be implemented by modifying the second constraint to match the coefficient of the second terms in the OPE.

While for the leptons we have

$$w_L[\ell] = -\frac{2}{Q^2}\,, \quad (\ell = e, \mu, \tau)$$

tha hadronic amplitudes get modified by strong interaction effects as mentioned: a sufficient number of states with appropriate weight factors has to be included in order to be able to satisfy the S.D. constraints, obtained via the OPE. Since the Z does not have fixed parity both vector and axial vector states couple (see Fig. 4.13a). For the 1st family π^0, $\rho(770)$ and $a_1(1260)$ are taken into account[22]

---

lowest lying hadrons with appropriate quantum numbers as poles: the ρ, ρ' and a_1, yields

$$\Delta a_\mu|_T^{\mathrm{HA\ 3-poles}} = \frac{G_\mu}{\sqrt{2}} \frac{m_\mu^2}{8\pi^2} \frac{\alpha}{\pi} \times (0.04 \pm 0.02) \simeq (0.011 \pm 0.005) \times 10^{-11}\,.$$

Thus, these interesting NP corrections at the present level of precision turn out to be completely negligible. However also the longitudinal amplitude is modified by mass effects. While for the first family quarks the effects are very small, for the strange quark the contribution turns out to be relevant. The estimate here yields

$$\Delta a_\mu|_L = \frac{\sqrt{2} G_\mu\, m_\mu^2}{16\pi^2} \frac{\alpha}{\pi} \times (4.57 \pm 1.17 \pm 1.37) \simeq (1.2 \pm 0.3 \pm 0.4) \times 10^{-11}\,.$$

Still the effect is small, however one has to estimate such possible effects in order to reduce as much as possible the hadronic uncertainties.

[22] It should be noted that the "pole" in $w_L[\ell] = 2/q^2$ has nothing to do with a massless one–particle exchange, it is just a kinematic singularity which follows from the tensor decomposition (4.46). Therefore the hadronic counterpart $w_L[u,d] = -2/(q^2 - m_\pi^2 + i\varepsilon)$ is not just a chiral symmetry breaking shift of the Goldstone pole, which is the result of the spontaneous chiral symmetry breaking. What matters is that in physical quantities the residue of the "pole" must be checked in order to know, whether there is a true pole or not. The pion–pole in $w_L[u,d]$ certainly has a different origin than the spurious one of $w_L[\ell]$.

$$w_L[u,d] = \frac{2}{Q^2 + m_\pi^2} \simeq 2\left(\frac{1}{Q^2} - \frac{m_\pi^2}{Q^4} + \cdots\right)$$

$$w_T[u,d] = \frac{1}{M_{a_1}^2 - M_\rho^2}\left[\frac{M_{a_1}^2 - m_\pi^2}{Q^2 + M_\rho^2} - \frac{M_\rho^2 - m_\pi^2}{Q^2 + M_{a_1}^2}\right] \simeq \left(\frac{1}{Q^2} - \frac{m_\pi^2}{Q^4} + \cdots\right),$$

for the 2nd family $\eta'(960)$, $\eta(550)$, $\phi(1020)$ and $f_1(1420)$ are included

$$w_L[s] = -\frac{2}{3}\left[\frac{2}{Q^2 + M_{\eta'}^2} - \frac{1}{Q^2 + m_\eta^2}\right] \simeq -\frac{2}{3}\left(\frac{1}{Q^2} - \frac{\tilde{M}_\eta^2}{Q^4} + \cdots\right)$$

$$w_T[s] = -\frac{1}{3}\frac{1}{M_{f_1}^2 - M_\phi^2}\left[\frac{M_{f_1}^2 - m_\eta^2}{Q^2 + M_\phi^2} - \frac{M_\phi^2 - m_\eta^2}{Q^2 + M_{f_1}^2}\right] \simeq -\frac{1}{3}\left(\frac{1}{Q^2} - \frac{m_\eta^2}{Q^4} + \cdots\right).$$

with $\tilde{M}_\eta^2 = 2M_{\eta'}^2 - m_\eta^2$. The expansion shows how it fits to what we got from the OPE. Numerically the differences are not crucial, however, and we adopt the specific forms given above.

While the contributions to a_μ from the heavier states may be calculated using the simplified integral (4.48), for the leading π^0 contribution we have to use (4.47), which also works for $m_\pi \sim m_\mu$. In terms of the integrals $I_1(z) \doteq \int_0^1 dx\,(1+x)\,\ln(x + (1-x)^2 z)$ and $I_2(z) \doteq \int_0^1 dx\,(1-x)^2\,\ln(x + (1-x)^2 z)$, the results obtained for the 1st family reads [68]

$$a_\mu^{(4)\,\text{EW}}([e,u,d]) \simeq -\frac{\sqrt{2}G_\mu\, m_\mu^2}{16\pi^2}\frac{\alpha}{\pi}\left\{\frac{4}{3}I_1(m_\mu^2/m_\pi^2) + 2\ln\frac{m_\pi^2}{m_\mu^2} + \frac{8}{3}\right.$$

$$+ 4\frac{m_\pi^2}{m_\mu^2}\left[I_2(m_\mu^2/m_\pi^2) - \frac{1}{3}\ln\frac{m_\pi^2}{m_\mu^2} + \frac{2}{9}\right]$$

$$\left. + \ln\frac{M_\rho^2}{m_\mu^2} - \frac{M_\rho^2}{M_{a_1}^2 - M_\rho^2}\ln\frac{M_{a_1}^2}{M_\rho^2} + \frac{2}{3}\right\}$$

$$\simeq -\frac{\sqrt{2}G_\mu\, m_\mu^2}{16\pi^2}\frac{\alpha}{\pi} \times 7.46(74) = -2.02(20) \times 10^{-11}\,. \quad (4.75)$$

This may be compared with the QPM result (4.37), which is about a factor two larger and again illustrates the problem of perturbative calculation in the light quark sector. For the 2nd family adding the μ and the perturbative charm contribution one obtains

$$a_\mu^{(4)\,\text{EW}}([\mu,c,s]) \simeq -\frac{\sqrt{2}G_\mu\, m_\mu^2}{16\pi^2}\frac{\alpha}{\pi}\left[\frac{2}{3}\ln\frac{M_\phi^2}{M_{\eta'}^2} - \frac{2}{3}\ln\frac{M_{\eta'}^2}{m_\eta^2}\right.$$

$$\left. + \frac{1}{3}\frac{M_\phi^2 - m_\eta^2}{M_{f_1}^2 - M_\phi^2}\ln\frac{M_{f_1}^2}{M_\phi^2} + 4\ln\frac{M_c^2}{M_\phi^2} + 3\ln\frac{M_\phi^2}{m_\mu^2} - \frac{8\pi^2}{9} + \frac{56}{9}\right]$$

$$\simeq -\frac{\sqrt{2}G_\mu\, m_\mu^2}{16\pi^2}\frac{\alpha}{\pi} \times 17.1(1.1) \simeq -4.63(30) \times 10^{-11}\,, \quad (4.76)$$

which yields a result close to the one obtained in the QPM (4.38). Here the QPM works better because the non–perturbative light s-quark contribution is suppressed by a factor four relative to the c due to the different charge.

Note that this large N_c QCD (LNC) inspired result

$$a_\mu^{(4)\,\mathrm{EW}}\left(\begin{bmatrix} e,u,d \\ \mu,c,s \end{bmatrix}\right)_{\mathrm{LNC}} \simeq -\frac{\sqrt{2}G_\mu\, m_\mu^2}{16\pi^2}\frac{\alpha}{\pi} \times 24.56 \simeq -6.65 \times 10^{-11}, \tag{4.77}$$

obtained here for the 1st plus 2nd family, is close to the very simple estimate (4.61) based on separating L.D. and S.D. by a cut–off in the range 1 to 2 GeV.

Perturbative Residual Fermion–loop Effects

So far unaccounted are sub–leading contributions which come from diagrams $c), d), e)$ and $f)$. They have been calculated in [63] with the result

$$a_{\mu\,\mathrm{NLL}}^{(4)\,\mathrm{EW}} = -\frac{\sqrt{2}G_\mu\, m_\mu^2}{16\pi^2}\frac{\alpha}{\pi}\left\{\frac{1}{2s_W^2}\left[\frac{5}{8}\frac{m_t^2}{M_W^2} + \ln\frac{m_t^2}{M_W^2} + \frac{7}{3}\right] + \Delta C^{tH}\right\}$$
$$\simeq -4.15(11) \times 10^{-11} - (1.1^{-0.1}_{+1.4}) \times 10^{-11} \tag{4.78}$$

where ΔC^{tH} is the coefficient from diagram f)

$$\Delta C^{tH} = \begin{cases} \frac{16}{9}\ln\frac{m_t^2}{m_H^2} + \frac{104}{27} & m_H \ll m_t \\ \frac{32}{3}\left(1 - \frac{1}{\sqrt{3}}\mathrm{Cl}_2(\pi/3)\right) & m_H = m_t \\ \frac{m_t^2}{m_H^2}\left(8 + \frac{8}{9}\pi^2 + \frac{8}{3}\left(\ln\frac{m_H^2}{m_t^2} - 1\right)^2\right) & m_H \gg m_t \end{cases}$$

with typical values $\Delta C^{tH} = (5.84, 4.14, 5.66)$ contributing to (4.78) by $(-1.58, -1.12, -1.53) \times 10^{-11}$, respectively, for $m_H = (100, m_t, 300)$ GeV. The first term in (4.78) is for $\Delta C^{tH} = 0$, the second is the ΔC^{tH} contribution for $m_H = m_t$ with uncertainty corresponding to the range $m_H = 100$ GeV to $m_H = 300$ GeV.

Results for the Bosonic Contributions

Full electroweak bosonic corrections have been calculated in [90]. At the two–loop level there are 1678 diagrams (fermion loops included) in the linear 't Hooft gauge, and the many mass scales involved complicate the exact calculation considerably. However, the heavy masses M_W, M_Z and m_H, which appear in the corresponding propagators, reveal these particles to be essentially static, and one may perform asymptotic expansions in $(m_\mu/M_V)^2$ and

$(M_V/m_H)^2$, such that the calculation simplifies considerably. A further approximation is possible taking advantage of the smallness of the NC vector couplings, which are suppressed like $(1 - 4\sin^2\Theta_W) \sim 0.1$ for quarks and $(1 - 4\sin^2\Theta_W)^2 \sim 0.01$ for leptons, i.e. in view of the experimental value $\sin^2\Theta_W \sim 0.23$ we may take $\sin^2\Theta_W = 1/4$ as a good approximation. This remarkable calculation was performed by Czarnecki, Krause and Marciano in 1995 [90]. Altogether, they find for the two–loop electroweak corrections

$$a_\mu^{(4)\,\mathrm{EW}}(\mathrm{bosonic}) = \frac{\sqrt{2}G_\mu m_\mu^2}{16\pi^2}\frac{\alpha}{\pi}\left(\sum_{i=-1}^{2}\left[a_{2i}s_W^{2i} + \frac{M_W^2}{m_H^2}b_{2i}s_W^{2i}\right] + O(s_W^6)\right)$$
$$\simeq -21.4^{+4.3}_{-1.0} \times 10^{-11} \qquad (4.79)$$

for $M_W = 80.392$ ($\sin^2\Theta_W = 1 - M_W^2/M_Z^2$) and $m_H = 250$ GeV ranging between $m_H = 100$ GeV and $m_H = 500$ GeV. The expansion coefficients are given by

$$a_{-2} = \tfrac{19}{36} - \tfrac{99}{8}S_2 - \tfrac{1}{24}\ln\tfrac{m_H^2}{M_W^2}$$

$$a_0 = -\tfrac{859}{18} + 11\tfrac{\pi}{\sqrt{3}} + \tfrac{20}{9}\pi^2 + \tfrac{393}{8}S_2 - \tfrac{65}{9}\ln\tfrac{M_W^2}{m_\mu^2} + \tfrac{31}{72}\ln\tfrac{m_H^2}{M_W^2}$$

$$a_2 = \tfrac{165169}{1080} - \tfrac{385}{6}\tfrac{\pi}{\sqrt{3}} - \tfrac{29}{6}\pi^2 + \tfrac{33}{8}S_2 + \tfrac{92}{9}\ln\tfrac{M_W^2}{m_\mu^2} - \tfrac{133}{72}\ln\tfrac{m_H^2}{M_W^2}$$

$$a_4 = -\tfrac{195965}{864} + \tfrac{265}{3}\tfrac{\pi}{\sqrt{3}} + \tfrac{163}{18}\pi^2 + \tfrac{223}{12}S_2 - \tfrac{184}{9}\ln\tfrac{M_W^2}{m_\mu^2} - \tfrac{5}{8}\ln\tfrac{m_H^2}{M_W^2}$$

$$b_{-2} = \tfrac{155}{192} + \tfrac{3}{8}\pi^2 - \tfrac{9}{8}S_2 + \tfrac{3}{2}\ln^2\tfrac{m_H^2}{M_W^2} - \tfrac{21}{16}\ln\tfrac{m_H^2}{M_W^2}$$

$$b_0 = \tfrac{433}{36} + \tfrac{5}{24}\pi^2 - \tfrac{51}{8}S_2 + \tfrac{3}{8}\ln^2\tfrac{m_H^2}{M_W^2} + \tfrac{9}{4}\ln\tfrac{m_H^2}{M_W^2}$$

$$b_2 = -\tfrac{431}{144} + \tfrac{3}{8}\pi^2 + \tfrac{315}{8}S_2 + \tfrac{3}{2}\ln^2\tfrac{m_H^2}{M_W^2} - \tfrac{11}{8}\ln\tfrac{m_H^2}{M_W^2}$$

$$b_4 = \tfrac{433}{216} + \tfrac{13}{24}\pi^2 + \tfrac{349}{24}S_2 + \tfrac{21}{8}\ln^2\tfrac{m_H^2}{M_W^2} - \tfrac{49}{12}\ln\tfrac{m_H^2}{M_W^2}$$

and

$$S_2 \equiv \frac{4}{9\sqrt{3}}\mathrm{Cl}_2\left(\frac{\pi}{3}\right) = 0.2604341...$$

The on mass–shell renormalization prescription has been used. Part of the two–loop bosonic corrections have been absorbed into the lowest order result, by expressing the one–loop contributions in (4.33) in terms of the muon decay constant G_μ[23]. For the lower Higgs masses the heavy Higgs mass expansion is not accurate and an exact calculation has been performed by Heinemeyer,

---

[23] In [91] using asymptotic expansions and setting $m_H \sim M_W$ and $\sin^2\Theta_W \sim 0$ an approximate form for the bosonic corrections

$$a_\mu^{(4)\,\mathrm{EW}}(\mathrm{bosonic}) = -\frac{\sqrt{2}G_\mu m_\mu^2}{16\pi^2}\frac{\alpha}{\pi}\left[\frac{65}{9}\ln\frac{M_W^2}{m_\mu^2} + O(\sin^2\Theta_W \ln\frac{M_W^2}{m_\mu^2})\right]$$

Stöckinger and Weiglein [92] and by Gribouk and Czarnecki [93]. The result has the form

$$a_\mu^{(4)\,\text{EW}}(\text{bosonic}) = \frac{\sqrt{2} G_\mu\, m_\mu^2}{16\pi^2}\,\frac{\alpha}{\pi}\left(c_L^{\text{bos},2L}\ln\frac{m_\mu^2}{M_W^2} + c_0^{\text{bos},2L}\right), \quad (4.80)$$

where the coefficient of the large logarithm $\ln\frac{m_\mu^2}{M_W^2} \sim -13.27$ is given by the simple expression

$$c_L^{\text{bos},2L} = \frac{1}{18}[107 + 23\,(1-4s_W^2)^2] \sim 5.96\,.$$

In contrast to the leading term the Higgs mass dependent function $c_0^{\text{bos},2L}$ in its exact analytic form is rather unwieldy and therefore has not been published. It has been calculated numerically first in [92]. The result was confirmed in [93] which also presents a number of semi–analytic intermediate results which give more insight into the calculation. In the range of interest, $m_H = 50$ GeV to $m_H = 500$ GeV, say, on may expand the result as a function of the unknown Higgs mass in terms of Tschebycheff polynomials defined on the interval $[-1, 1]$, for example. With

$$x = (2m_H - 550\text{ GeV})/(450\text{ GeV})$$

and the polynomials

$$t_1 = 1\,, \quad t_2 = x\,, \quad t_{i+2} = 2xt_{i+1} - t_i\,, \quad i = 1,\cdots, 4$$

we may approximate (4.80) in the given range by

$$a_\mu^{(4)\,\text{EW}}(\text{bosonic}) \simeq \sum_{i=1}^{6} a_i\, t_i(x) \times 10^{-10} \quad (4.81)$$

with the coefficients $a_1 = 80.0483$, $a_2 = 8.4526$, $a_3 = -3.3912$, $a_4 = 1.4024$, $a_5 = -0.5420$ and $a_6 = 0.2227$. The result is plotted in Fig. 4.15 and may be summarized in the form

$$a_\mu^{(4)\,\text{EW}}(\text{bosonic}) = (-21.56^{+1.49}_{-1.05}) \times 10^{-11} \quad (4.82)$$

was given, which is not too far from the full result (4.79) which for $\sin^2\Theta_W = 0.224$ and $m_H = 250$ GeV may be cast into the effective form

$$a_\mu^{(4)\,\text{EW}}(\text{bosonic}) \simeq -\frac{\sqrt{2} G_\mu\, m_\mu^2}{16\pi^2}\,\frac{\alpha}{\pi}\left[5.96\ln\frac{M_W^2}{m_\mu^2} - 0.19\right]$$

$$\simeq -\frac{\sqrt{2} G_\mu\, m_\mu^2}{16\pi^2}\,\frac{\alpha}{\pi} \times 78.9 \simeq -21.1 \times 10^{-11}\,.$$

258   4 Electromagnetic and Weak Radiative Corrections

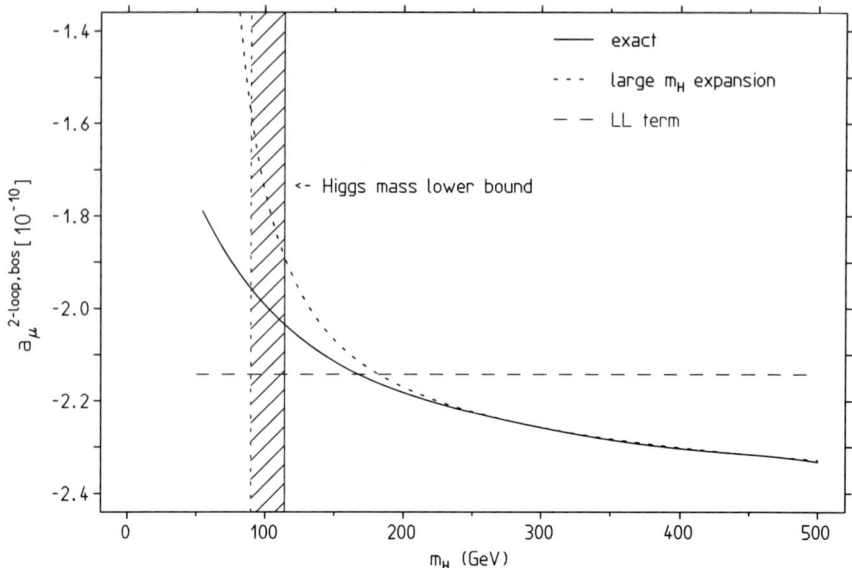

Fig. 4.15. Exact result for the bosonic correction vs. the asymptotic expansion (4.79) minus a correction 0.88×10^{-11} and the LL approximation (first term of (4.80))

where the central value is obtained for $m_H = m_t$ and the validity range given holds for $m_H = 100$ GeV to $m_H = 300$ GeV. The exact result exhibits a much more moderate Higgs mass dependence at lower Higgs masses and the uncertainty caused by the unknown Higgs mass is reduced substantially.

Summary of the Results for the Weak Contributions

The various weak contributions are collected in Table 4.3 and add up to the total weak 2–loop contribution

$$a_\mu^{(4)\,\text{EW}} \simeq (-41.76^{+1.11}_{-1.39}[m_H, m_t] \pm 1.0[\text{had}]) \times 10^{-11} \ . \qquad (4.83)$$

The high value -40.65 for low $m_H = 100$ GeV, the central value is for $m_H = m_t$ GeV and the minimum -43.15 for a high $m_H = 300$ GeV (see (3.33)).

Table 4.3. Summary of weak 2–loop effects. Fermion triangle loops: 1st, 2nd and 3rd family LO, fermion loops NLL and bosonic loops (with equation numbers)

[eud] LO (4.75)	[μsc] LO (4.76)	[τbt] LO (4.53)	NLL (4.78)	bosonic (4.82)
-2.02 ± 0.2	-4.63 ± 0.3	-8.21 ± 0.1	$-5.3^{+0.1}_{-1.4}$	$-21.6^{+1.5}_{-1.0}$

Three–loop effects have been estimated by RG methods first in [67] and confirmed in [68] with the result

$$a^{(6)\,\mathrm{EW}}_{\mu\,\mathrm{LL}} \simeq (0.4 \pm 0.2) \times 10^{-11} \tag{4.84}$$

where the error stands for uncalculated 3–loop contributions.

By adding up (4.35), (4.83) and (4.84) we find the result

$$a^{\mathrm{EW}}_\mu = (153.5 \pm 1.0[\mathrm{had}]\,^{+1.1}_{-1.4}[\mathrm{m_H, m_t, 3-loop}]) \times 10^{-11}\,. \tag{4.85}$$

based on [92, 68, 93][24].

References

1. T. Kinoshita, B. Nizic, Y. Okamoto, Phys. Rev. D **41** (1990) 593
2. T. Kinoshita, W. J. Marciano, In: *Quantum Electrodynamics*, ed by T. Kinoshita (World Scientific, Singapore 1990) pp. 419–478
3. G. W. Bennett et al. [Muon (g-2) Collaboration], Phys. Rev. Lett. **92** (2004) 161802
4. B. L. Roberts Nucl. Phys. B (Proc. Suppl.) **131** (2004) 157;
   R. M. Carey et al., Proposal of the BNL Experiment E969, 2004 (www.bnl.gov/ henp/docs/pac0904/P969.pdf);
   J-PARC Letter of Intent L17, B. L. Roberts contact person
5. B. E. Lautrup, E. de Rafael, Nucl. Phys. B **70** (1974) 317
6. A. Petermann, Helv. Phys. Acta **30** (1957) 407; Nucl. Phys. **5** (1958) 677
7. C. M. Sommerfield, Phys. Rev. **107** (1957) 328; Ann. Phys. (NY) **5** (1958) 26
8. B. E. Lautrup, E. De Rafael, Nuovo Cim. A **64** (1969) 322
9. H. Suura, E. Wichmann, Phys. Rev. **105** (1957) 1930;
   A. Petermann, Phys. Rev. **105** (1957) 1931
10. H. H. Elend, Phys. Lett. **20** (1966) 682; Erratum-ibid. **21** (1966) 720
11. M. Passera, J. Phys. G **31** (2005) R75; Phys. Rev. D **75** (2007) 013002
12. G. Li, R. Mendel, M. A. Samuel, Phys. Rev. D **47** (1993) 1723
13. J. A. Mignaco, E. Remiddi, Nuovo Cim. A **60** (1969) 519; R. Barbieri, E. Remiddi, Phys. Lett. B **49** (1974) 468; Nucl. Phys. B **90** (1975) 233; R. Barbieri, M. Caffo, E. Remiddi, Phys. Lett. B **57** (1975) 460; M. J. Levine, E. Remiddi, R. Roskies, Phys. Rev. D **20** (1979) 2068; S. Laporta, E. Remiddi, Phys. Lett. B **265** (1991) 182; S. Laporta, Phys. Rev. D **47** (1993) 4793; Phys. Lett. B **343** (1995) 421; S. Laporta, E. Remiddi, Phys. Lett. B **356** (1995) 390

---

[24]The result is essentially the same as

$$a^{\mathrm{EW}}_\mu = (154 \pm 1[\mathrm{had}] \pm 2[\mathrm{m_H, m_t, 3-loop}]) \times 10^{-11}$$

of Czarnecki, Marciano and Vainshtein [68], which also agrees numerically with the one

$$a^{\mathrm{EW}}_\mu = (152 \pm 1[\mathrm{had}]) \times 10^{-11}$$

obtained by Knecht, Peris, Perrottet and de Rafael [69].

14. S. Laporta, E. Remiddi, Phys. Lett. B **379** (1996) 283
15. T. Kinoshita, Phys. Rev. Lett. **75** (1995) 4728
16. T. Kinoshita, Nuovo Cim. B **51** (1967) 140
17. B. E. Lautrup, E. De Rafael, Phys. Rev. **174** (1968) 1835;
    B. E. Lautrup, M. A. Samuel, Phys. Lett. B **72** (1977) 114
18. S. Laporta, Nuovo Cim. A **106** (1993) 675
19. S. Laporta, E. Remiddi, Phys. Lett. B **301** (1993) 440
20. J. H. Kühn, A. I. Onishchenko, A. A. Pivovarov, O. L. Veretin, Phys. Rev. D **68** (2003) 033018
21. M. A. Samuel, G. Li, Phys. Rev. D **44** (1991) 3935 [Errata-ibid. D **46** (1992) 4782; D **48** (1993) 1879]
22. A. Czarnecki, M. Skrzypek, Phys. Lett. B **449** (1999) 354
23. S. Friot, D. Greynat, E. De Rafael, Phys. Lett. B **628** (2005) 73
24. B. E. Lautrup, M. A. Samuel, Phys. Lett. B **72** (1977) 114
25. T. Kinoshita, Phys. Rev. Lett. **61** (1988) 2898
26. M. A. Samuel, Phys. Rev. D **45** (1992) 2168
27. R. Barbieri, E. Remiddi, Nucl. Phys. B **90** (1975) 233
28. R. Barbieri, E. Remiddi, Phys. Lett. B **57** (1975) 273
29. J. Aldins, T. Kinoshita, S. J. Brodsky, A. J. Dufner, Phys. Rev. Lett. **23** (1969) 441; Phys. Rev. D **1** (1970) 2378
30. M. Caffo, S. Turrini, E. Remiddi, Phys. Rev. D **30** (1984) 483;
    E. Remiddi, S. P. Sorella, Lett. Nuovo Cim. **44** (1985) 231;
    D. J. Broadhurst, A. L. Kataev, O. V. Tarasov, Phys. Lett. B **298** (1993) 445;
    S. Laporta, Phys. Lett. B **312** (1993) 495;
    P. A. Baikov, D. J. Broadhurst: Three–loop QED Vacuum Polarization and the Four–loop Muon Anomalous Magnetic Moment. In: *New Computing Techniques in Physics Research IV. International Workshop on Software Engineering and Artificial Intelligence for High Energy, Nuclear Physics*, ed by B. Denby, D. Perret-Gallix (World Scientific, Singapore 1995) pp. 167–172; hep-ph/9504398
31. T. Kinoshita, W. B. Lindquist, Phys. Rev. Lett. **47** (1981) 1573
32. T. Kinoshita, W. B. Lindquist, Phys. Rev. D **27** (1983) 867; Phys. Rev. D **27** (1983) 877; Phys. Rev. D **27** (1983) 886; Phys. Rev. D **39** (1989) 2407; Phys. Rev. D **42** (1990) 636
33. T. Kinoshita, In: *Quantum Electrodynamics*, ed by T. Kinoshita (World Scientific, Singapore 1990) pp. 218–321
34. T. Kinoshita, Phys. Rev. D **47** (1993) 5013
35. V. W. Hughes, T. Kinoshita, Rev. Mod. Phys. **71** (1999) S133
36. T. Kinoshita, M. Nio, Phys. Rev. Lett. **90** (2003) 021803; Phys. Rev. D **70** (2004) 113001
37. T. Kinoshita, Nucl. Phys. B (Proc. Suppl.) **144** (2005) 206
38. T. Kinoshita, M. Nio, Phys. Rev. D **73** (2006) 013003
39. T. Aoyama, M. Hayakawa, T. Kinoshita, M. Nio, arXiv:0706.3496 [hep-ph]
40. G. P. Lepage, J. Comput. Phys. **27** (1978) 192
41. T. Kinoshita, IEEE Trans. Instrum. Meas. **44** (1995) 498
42. T. Kinoshita, IEEE Trans. Instrum. Meas. **46** (1997) 108
43. V. W. Hughes, T. Kinoshita, Rev. Mod. Phys. **71** (1999) S133;
    T. Kinoshita, IEEE Trans. Instrum. Meas. **50** (2001) 568

44. T. Kinoshita, *Anomalous Magnetic Moment of the Electron and the Fine Structure Constant* in *An Isolated Atomic Particle at Rest in Free Space: Tribute to Hans Dehmelt, Nobel Laureate*, eds E. N. Fortson, E. M. Henley, (Narosa Publishing House, New Delhi, India, 2005)
45. T. Kinoshita, *Recent Developments of the Theory of Muon and Electron g-2*, In: *"In Memory of Vernon Willard Hughes"*, ed E. W. Hughes, F. Iachello (World Scientific, Singapore 2004) pp. 58–77
46. M. A. Samuel, Lett. Nuovo Cim. **21** (1978) 227
47. M. L. Laursen, M. A. Samuel, J. Math. Phys. **22** (1981) 1114
48. S. Laporta, P. Mastrolia, E. Remiddi, Nucl. Phys. B **688** (2004) 165; P. Mastrolia, E. Remiddi, Nucl. Phys. B (Proc. Suppl.) **89** (2000) 76
49. S. G. Karshenboim, Phys. Atom. Nucl. **56** (1993) 857 [Yad. Fiz. **56N6** (1993) 252]
50. T. Kinoshita, M. Nio, Phys. Rev. D **73** (2006) 053007
51. A. L. Kataev, Nucl. Phys. Proc. Suppl. **155** (2006) 369; hep-ph/0602098; Phys. Rev. D **74** (2006) 073011
52. P. J. Mohr, B. N. Taylor, Rev. Mod. Phys. **77** (2005) 1
53. S. L. Glashow, Nucl. Phys. B **22** (1961) 579; S. Weinberg, Phys. Rev. Lett. **19** (1967) 1264; A. Salam, Weak and electromagnetic interactions. In: *Elementary Particle Theory*, ed by N. Svartholm, (Amquist and Wiksells, Stockholm 1969) pp. 367–377
54. G. 't Hooft, Nucl. Phys. B **33** (1971) 173; **35** (1971) 167; G. 't Hooft, M. Veltman, Nucl. Phys. B **50** (1972) 318
55. R. Jackiw, S. Weinberg, Phys. Rev. D **5** (1972) 2396;
    I. Bars, M. Yoshimura, Phys. Rev. D **6** (1972) 374;
    G. Altarelli, N. Cabibbo, L. Maiani, Phys. Lett. B **40** (1972) 415;
    W. A. Bardeen, R. Gastmans, B. Lautrup, Nucl. Phys. B **46** (1972) 319;
    K. Fujikawa, B. W. Lee, A. I. Sanda, Phys. Rev. D **6** (1972) 2923
56. M. Veltman, Nucl. Phys. B **123** (1977) 89; M. S. Chanowitz et al., Phys. Lett. **78B** (1978) 1; M. Consoli, S. Lo Presti, L. Maiani, Nucl. Phys. B **223** (1983) 474; J. Fleischer, F. Jegerlehner, Nucl. Phys. B **228** (1983) 1
57. M. Consoli, W. Hollik, F. Jegerlehner, Phys. Lett. B **227** (1989) 167
58. F. Jegerlehner, Renormalizing the Standard Model. In: *Testing the Standard Model*, ed by M. Cvetič, P. Langacker (World Scientific, Singapore 1991) pp. 476–590; see http://www-com.physik.hu-berlin.de/~fjeger/books.html
59. T. V. Kukhto, E. A. Kuraev, A. Schiller, Z. K. Silagadze, Nucl. Phys. B **371** (1992) 567
60. S. L. Adler, Phys. Rev. **177** (1969) 2426;
    J. S. Bell, R. Jackiw, Nuovo Cim. **60A** (1969) 47;
    W. A. Bardeen, Phys. Rev. **184** (1969) 1848
61. C. Bouchiat, J. Iliopoulos, P. Meyer, Phys. Lett. **38B** (1972) 519;
    D. Gross, R. Jackiw, Phys. Rev. D **6** (1972) 477;
    C. P. Korthals Altes, M. Perrottet, Phys. Lett. **39B** (1972) 546
62. S. Peris, M. Perrottet, E. de Rafael, Phys. Lett. **B355** (1995) 523
63. A. Czarnecki, B. Krause, W. Marciano, Phys. Rev. D **52** (1995) R2619
64. F. Jegerlehner, Nucl. Phys. (Proc. Suppl.) C **51** (1996) 131
65. E. D'Hoker, Phys. Rev. Lett. **69** (1992) 1316
66. T. Sterling, M. J. G. Veltman, Nucl. Phys. B **189** (1981) 557
67. G. Degrassi, G. F. Giudice, Phys. Rev. **58D** (1998) 053007

68. A. Czarnecki, W. J. Marciano, A. Vainshtein, Phys. Rev. D **67** (2003) 073006
69. M. Knecht, S. Peris, M. Perrottet, E. de Rafael, JHEP **0211** (2002) 003; E. de Rafael, *The muon g − 2 revisited,* hep-ph/0208251
70. A. Czarnecki, W. J. Marciano and A. Vainshtein, Acta Phys. Polon. B **34** (2003) 5669
71. S. L. Adler, W. A. Bardeen, Phys. Rev. **182** (1969) 1517
72. J. Wess and B. Zumino, Phys. Lett. B **37** (1971) 95
73. E. Witten, Nucl. Phys. B **223** (1983) 422
74. G. 't Hooft, In: *Recent Developments in Gauge Theories*, Proceedings of the Summer-Institute, Cargese, France, 1979, ed by G. 't Hooft et al., NATO Advanced Study Institute Series B: Physics Vol. 59 (Plenum Press, New York 1980)
75. L. Rosenberg, Phys. Rev. **129** (1963) 2786
76. A. Vainshtein, Phys. Lett. B **569** (2003) 187
77. M. Knecht, S. Peris, M. Perrottet, E. de Rafael, JHEP **0403** (2004) 035.
78. F. Jegerlehner, O. V. Tarasov, Phys. Lett. B **639** (2006) 299
79. J. Gasser, H. Leutwyler, Annals Phys. **158** (1984) 142; Nucl. Phys. B **250** (1985) 465
80. M. Gell-Mann, R. J. Oakes, B. Renner, Phys. Rev. **175** (1968) 2195
81. K. G. Wilson, Phys. Rev. **179** (1969) 1499;
    K. G. Wilson, J. B. Kogut, Phys. Rept. **12** (1974) 75;
    V. A. Novikov, M. A. Shifman, A. I. Vainshtein, V. I. Zakharov, Nucl. Phys. B **249**, 445 (1985) [Yad. Fiz. **41**, 1063 (1985)]
82. M. A. Shifman, World Sci. Lect. Notes Phys. **62** (1999) 1 (Chap. 1,2)
83. S. Peris, M. Perrottet, E. de Rafael, JHEP **9805** (1998) 011;
    M. Knecht, S. Peris, M. Perrottet, E. de Rafael, Phys. Rev. Lett. **83** (1999) 5230; M. Knecht, A. Nyffeler, Eur. Phys. J. C **21** (2001) 659
84. J. Bijnens, L. Girlanda, P. Talavera, Eur. Phys. J. C **23** (2002) 539
85. G. Ecker, J. Gasser, A. Pich, E. de Rafael, Nucl. Phys. B **321** (1989) 311
86. G. Ecker, J. Gasser, H. Leutwyler, A. Pich, E. de Rafael, Phys. Lett. B **223** (1989) 425
87. G. 't Hooft, Nucl. Phys. B **72** (1974) 461; ibid. **75** (1974) 461
88. A. V. Manohar, Hadrons in the $1/N$ Expansion, In: *At the frontier of Particle Physics*, ed M. Shifman, (World Scientific, Singapore 2001) Vol. 1, pp. 507–568
89. E. de Rafael, Phys. Lett. B **322** (1994) 239
90. A. Czarnecki, B. Krause, W. J. Marciano, Phys. Rev. Lett. **76** (1996) 3267
91. V. A. Smirnov, Mod. Phys. Lett. A **10** (1995) 1485
92. S. Heinemeyer, D. Stöckinger, G. Weiglein, Nucl. Phys. B **699** (2004) 103
93. T. Gribouk, A. Czarnecki, Phys. Rev. D **72** (2005) 053016

5
Hadronic Effects

The basic problems we are confronted with when we have to include the non–perturbative hadronic contributions to $g-2$, we have outlined in Sect. 3.2.1 pp. 153ff. and in Sect. 4.2.2 pp. 232ff, already. We will distinguish three types of contributions, which will be analyzed in different subsections below:

i) The most sizable hadronic effect is the $O(\alpha^2)$ vacuum polarization insertion in the internal photon line of the leading one–loop muon vertex diagram Fig. 5.1. The hadronic "blob" can be calculated with help of the method discussed in Sect. 3.7.1. While perturbation theory fails and ab initio non–perturbative calculations are not yet available, it may be obtained via a DR from the measured cross–section $e^+e^- \to$ hadrons via (3.135) and (3.134).

ii) An order of magnitude smaller but still of relevance are the hadronic VP insertions contributing at order $O(\alpha^3)$. They are represented by diagrams exhibiting one additional VP insertion, leptonic or hadronic, in the photon line or by diagrams with an additional virtual photon attached in all possible ways in Fig. 5.1. As long as hadronic effects enter via photon vacuum polarization only, they can be safely evaluated in terms of

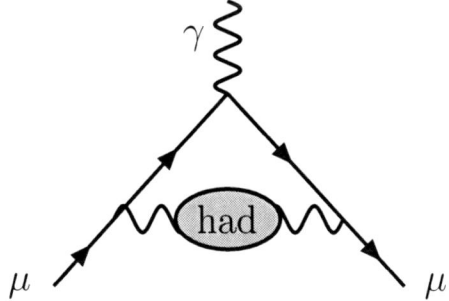

Fig. 5.1. Leading hadronic contribution to $g-2$

Fig. 5.2. Leading hadronic light–by–light scattering contribution to $g-2$

experimental data via the basic DR (3.135). The errors of the data here appear suppressed by one power in α relative to the leading hadronic contribution and therefore do not play a critical role.

iii) More involved and problematic is the hadronic light–by–light contribution, represented by the diagram Fig. 5.2, and entering at $O(\alpha^3)$. Here, a low energy effective field theory (EFT) approach beyond CHPT is needed and some model assumptions are unavoidable. Unfortunately such effective model predictions depend in a relevant way on model assumptions, as we will see. What saves the day at present is the fact that the size of the effect is only about twice the size of the uncertainty of the leading hadronic VP contribution. Therefore, a rough estimate only cannot spoil the otherwise reliable prediction. For the future it remains a real challenge for theory since further progress in $g-2$ precision physics depends on progress in putting this calculation on a theoretically saver basis.

Since the different types of contributions are confronted with different kinds of problems, which require a detailed discussion in each case, we will consider them in turn in the following subsections.

5.1 Vacuum Polarization Effects and e^+e^- Data

Fortunately vacuum polarization effects may be handled via dispersion relations together with available $e^+e^- \to$ hadrons data (see p. 13 for remarks on the early history). The tools which we need to overcome the main difficulties we have developed in Sect. 3.7.1 and at the end of Sect. 3.8. For the evaluation of the leading order contribution the main problem is the handling of the experimental e^+e^-–annihilation data and in particular of their systematic errors. The latter turn out to be the limiting factor for the precision of the theoretical prediction of a_μ.

To leading order in α the hadronic "blob" in Fig. 5.1 has to be identified with the photon self–energy function $\Pi_\gamma'^{\mathrm{had}}(s)$ which we relate to the cross–section $e^+e^- \to$ hadrons by means of the DR (3.155) based on the correspondence:

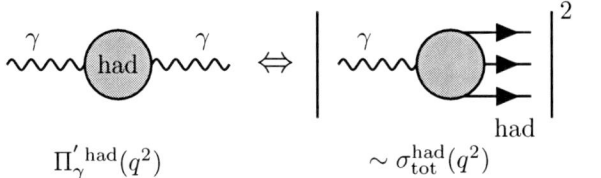

$\Pi_\gamma^{'\,\mathrm{had}}(q^2)$ $\sim \sigma_{\mathrm{tot}}^{\mathrm{had}}(q^2)$

The interrelationship is based on unitarity (optical theorem) and causality (analyticity), as elaborated before. Remember that $\Pi_\gamma^{'\,\mathrm{had}}(q^2)$ is a one particle irreducible (1PI) object, represented by diagrams which cannot be cut into two disconnected parts by cutting a single photon line. At low energies the imaginary part is related to intermediate hadronic states like $\pi^0\gamma, \rho, \omega, \phi, \cdots$, $\pi\pi, 3\pi, 4\pi, \cdots, \pi\pi\gamma,,\cdots, KK, \cdots \pi\pi Z, \cdots, \pi\pi H, \cdots$ (at least one hadron plus any strong, electromagnetic or weak interaction contribution), which in the DR correspond to the states produced in e^+e^-–annihilation via a virtual photon (at energies sufficiently below the point where $\gamma-Z$ interference comes into play).

At low energies, near flavor thresholds and in domains exhibiting resonances $\sigma_{\mathrm{tot}}^{\mathrm{had}}(q^2)$ cannot be calculated from first principles, because at present we lack appropriate non–perturbative methods to perform calculations in the time–like region[1].

Fortunately, the cross–sections required are available in form of existing experimental data. Since the leading hadronic contribution is rather large, an elaborate handling of the experimental data is mandatory because the experimental errors are substantial and of course limit the precision of the "theoretical" prediction of a_μ. Like the deep inelastic electron–nucleon scattering experiments, the e^+e^-–annihilation experiments played an eminent role in establishing QCD as the underlying theory of the strong interaction and have a long history. Touschek initiated the construction of an e^+e^- storage ring accelerator in the early 1960's at Frascati near Rome. Improved e^+e^- storage ring facilities and first cross–section measurements followed at Orsay, Novosibirsk and Frascati. The observed rise in the total hadronic cross–section at these times looked very puzzling, as actually a drop as $1/E^2$ was expected at high energies from unitarity arguments. The CEA experiment [1], however, which operated at slightly higher energy, left no room for doubts that the cross–section was far higher than theoretical expectations. For the first time,

---

[1] Note that in place of the representation (3.154) which requires the hadronic cross–section, per se a time–like quantity, there is the alternative representation (3.157) in terms of the space–like Adler-function, which at low energies is accessible to non–perturbative lattice QCD simulations. At higher energies pQCD is applicable (see Sect. 2.8). Progress in lattice QCD together with perturbation theory will allow us to calculate a_μ^{had} from the QCD Lagrangian one day in future.

QCD[2], which predicted a cross–section enhanced by the color multiplicity factor 3, was clearly favored by experiment and as we know in the sequel revolutioned strong interaction physics [3, 4, 5]. SLAC and DESY, reaching higher energies, followed and unexpectedly new states were discovered at SLAC, the τ lepton, the charm quark c and the bottom quark b. The highest energies so far were reached with LEP at CERN going up to 200 GeV. Important for the evaluation of the hadronic contributions to $g-2$ are recent and ongoing hadronic cross–section measurements at Novosibirsk, Frascati and Beijing which provided much more accurate e^+e^- data. Table 5.1 gives a more complete overview of the history of e^+e^- machines and experiments and the maximum center of mass energy they reached. Unfortunately, some of the energy ranges have been covered only by old experiments with typically 20% systematic errors. For a precise evaluation of the hadronic effects we need to combine data sets from many experiments of very different quality and performed in different energy intervals. The key problem here is the proper handling of the systematic errors, which are of different origin and depend on the experiment (machine and detector) as well as on theory input like radiative corrections. The statistical errors commonly are assumed to be Gaussian and hence may be added in quadrature. A problem here may be the low statistics of many of the older experiments which may not always justify this treatment. In the

Table 5.1. Chronology of e^+e^- facilities

Year	Accelerator	$E_{\max}$ (GeV)	Experiments	Laboratory
1961–1962	AdA	0.250		LNF Frascati (Italy)
1965–1973	ACO	0.6–1.1	DM1	Orsay (France)
1967–1970	VEPP-2	1.02–1.4	'spark chamber'	Novosibirsk (Russia)
1967–1993	ADONE	3.0	BCF,$\gamma\gamma$, $\gamma\gamma$2, MEA, $\mu\pi$, FENICE	LNF Frascati (Italy)
1971–1973	CEA	4,5		Cambridge (USA)
1972–1990	SPEAR	2.4–8	MARK I, CB, MARK 2	SLAC Stanford (USA)
1974–1992	DORIS	~11	ARGUS, CB, DASP 2, LENA, PLUTO	DESY Hamburg (D)
1975–1984	DCI	3.7	DM1,DM2,M3N,B$\overline{\text{B}}$	Orsay (France)
1975–2000	VEPP-2M	0.4–1.4	OLYA, CMD, CMD-2, ND,SND	Novosibirsk (Russia)
1978–1986	PETRA	12–47	PLUTO, CELLO, JADE, MARK-J, TASSO	DESY Hamburg (D)
1979–1985	VEPP-4	~11	MD1	Novosibirsk (Russia)
1979-	CESR	9–12	CLEO, CUSB	Cornell (USA)
1980–1990	PEP	~29	MAC, MARK-2	SLAC Stanford (USA)
1987–1995	TRISTAN	50–64	AMY, TOPAZ, VENUS	KEK Tsukuba (Japan)
1989	SLC	90 GeV	SLD	SLAC Stanford (USA)
1989–2001	BEPC	2.0–4.8	BES, BES-II	IHEP Beijing (China)
1989–2000	LEP I/II	110/210	ALEPH, DELPHI, L3, OPAL	CERN Geneva (CH)
1999–2007	DAΦNE	Φ factory	KLOE	LNF Frascati (Italy)
1999-	PEP-II	B factory	BaBar	SLAC Stanford (USA)
1999-	KEKB	B factory	Belle	KEK Tsukuba (Japan)

---

[2]Apart from its role in explaining Bjorken scaling in deep inelastic ep–scattering [2].

low energy region particularly important for $g-2$, however, data have improved dramatically in recent years (CMD-2, SND/Novosibirsk, BES/Beijing, CLEO/Cornell, KLOE/Frascati, BaBar/SLAC) and the statistical errors are a minor problem now.

The main uncertainty, related to the systematic errors of the experimental data, is evaluated via a certain common sense type error handling, which often cannot be justified unambiguously. This "freedom" of choice has lead to a large number of estimates by different groups which mainly differ by individual taste and the level of effort which is made in the analysis of the data. Issues here are: the completeness of the data utilized, interpolation and modeling procedures, e.g. direct integration of the data by applying the trapezoidal rule versus fitting the data to some smooth functional form before integration, separation of energy ranges where data or theory (pQCD and/or hadronic models) are considered to be more reliable, combining the data before or after integration etc.

A reliable combination of the data requires to know more or less precisely what experiments have actually measured and what they have published. As mentioned earlier hadronic cross–section data are represented usually by the cross–section ratio

$$R_\gamma^{\mathrm{had}}(s) \equiv \frac{\sigma(e^+e^- \to \gamma^* \to \mathrm{hadrons})}{\sigma(e^+e^- \to \gamma^* \to \mu^+\mu^-)} \quad (5.1)$$

which measures the hadronic cross–section in units of the leptonic point–cross–section. One of the key questions here is: what is the precise definition of $R(s)$ as a "measured" quantity? In theory we would consider (5.1), which also may be written in terms of lowest order cross–sections, with respect to QED effects. In short notation

$$R_\gamma^{\mathrm{had}}(s) \equiv \frac{\sigma_{\mathrm{had}}(s)}{\sigma_{\mu\mu}(s)} = \frac{\sigma_{\mathrm{had}}^0(s)}{\sigma_{\mu\mu}^0(s)}$$

which reveals $R(s)$ defined in this way as an *undressed R(s)* quantity, since in the ratio common effects, like dressing by VP effects (iterated VP insertions), normalization[3] (luminosity measurement) and the like cancel from the ratio automatically. While the dressed[4] physical cross–sections $\sigma_{\mathrm{had}}(s)$ and $\sigma_{\mu\mu}(s)$ are proportional to the square of the effective running fine structure constant $\alpha(s)$ see (3.115 and Fig. 3.13) the "bare" or "undressed" ones $\sigma_{\mathrm{had}}^0(s)$ and $\sigma_{\mu\mu}^0(s)$ are proportional to the square of the classical fine structure constant α determined at zero momentum transfer. The ratio obviously is insensitive to

---

[3] Note that the *initial state radiation* (ISR) bremsstrahlung only cancels if the same cuts are applied to hadro–production and to $\mu^+\mu^-$ pair production, a condition, which usually is not satisfied. We should keep in mind that experimentally it is not possible to distinguish an initial state photon from a final state photon.

[4] The terminus "*dressed*" refers to the inclusion of higher order effects which are always included in measured quantities.

dressing by vacuum polarization. For the leading diagram Fig. 5.1 "dressed" would mean that the full photon propagator is inserted, "undressed" means that just the 1PI photon self–energy is inserted.

In principle, one could attempt to treat self–energy insertions in terms of the full photon propagator according to (3.145), however, vertices cannot be resumed in a similar way such that working consistently with full propagators and full vertices as building blocks, known as the "skeleton expansion", is technically not feasible. One should avoid as much as possible treating part of the contributions in a different way than others. One has to remind that many fundamental properties of a QFT like gauge invariance, unitarity or locality, only can be controlled systematically order by order in perturbation theory. We therefore advocate to stick as much as possible to an order by order approach for what concerns the expansion in the electromagnetic coupling α, i.e. we will use (3.145) only in expanded form which allows a systematic order by order treatment in α.

It turns out that at the level of accuracy we are aiming at, the quantity $R(s)$ we need is not really the ratio (5.1). We have seen that some unwanted effects cancel but others do not. In particular all kinds of electromagnetic radiation effects do not cancel in the ratio. This is obvious if we consider the low energy region, particularly important for the a_μ^{had} evaluation, where $\pi^+\pi^-$–production dominates and according to (5.1) should be compared with $\mu^+\mu^-$–production. Neither the final state radiation (FSR) bremsstrahlung contributions nor the phase spaces are commensurate and drop out, and the $\mu^+\mu^-$–production phase space in the threshold region of $\pi^+\pi^-$–pair production is certainly in the wrong place here. What we need is the hadronic contribution to Im $\Pi'_\gamma(s)$, which is what enters in the DR for $\Pi'_\gamma(s)$. Thus, what one has to extract from the measurements for the use in the DR is

$$R_\gamma^{\text{had}}(s) = 12\pi \text{Im} \Pi'_\gamma{}^{\text{had}}(s) \tag{5.2}$$

as accurately as possible, where $\Pi'_\gamma{}^{\text{had}}(s)$ is the hadronic component of the 1PI photon self–energy.

In fact the high energy asymptotic form of $\sigma_{\mu\mu}(s)$ is the quantity appropriate for the normalization:

$$R_\gamma^{\text{had}}(s) = \sigma(e^+e^- \to \text{hadrons})/\frac{4\pi\alpha(s)^2}{3s} . \tag{5.3}$$

At first, the cross–section here must have been corrected for bremsstrahlung effects, because the latter are process and detector dependent and are of higher order in α. The detector dependence is due to finite detector resolution and other so called *cuts* which we have discussed in Sect. 2.6.6. Cuts are unavoidable as real detectors by construction have some blind zones, e.g. the beam tube, and detection thresholds where events get lost. This requires *acceptance* and *efficiency* corrections. As a matter of fact a total cross–section can be obtained only by extrapolations and theory or some modeling assumptions may be required to extract the quantity of interest.

5.1 Vacuum Polarization Effects and e^+e^- Data   269

There are two types of total cross–section "measurements". At low energies, in practice up to 1.4 to 2.1 GeV, one has to identify individual final states, because there is no typical characteristic "stamp", which allows the experimenter to identify a hadronic versus a non–hadronic event. One has to identify individual states by mass, charge, multiplicity, the number of final state particles. At high energies the primary quark pair produced hadronizes into two or more bunches, called jets, of hadrons of multiplicity increasing with energy. With increasing energy one passes more and more multi–hadron thresholds, like the ones of the n pion channels: $\pi^+\pi^-$, $\pi^+\pi^-\pi^0$, $\pi^+\pi^-\pi^+\pi^-$, $\pi^+\pi^-\pi^0\pi^0$ and so on, and the energy available distributes preferably into many–particle states if the corresponding phase space is available (see Fig. 5.3). The non–perturbative nature of the strong interaction is clearly manifest here since a perturbative order by order hierarchy is obviously absent on the level of the hadrons produced. In contrast, created lepton pairs can be easily identified in a detector as a two–body state and other non–hadronic states are down in the rate at least by one order in α. Therefore, at high enough energy one may easily separate leptons from hadrons because they have clearly distinguishable signatures, in the first place the multiplicity. This allows for an *inclusive* measurement of the total cross–section, all hadronic states count and there is no need for identification of individual channels. Such measurements are

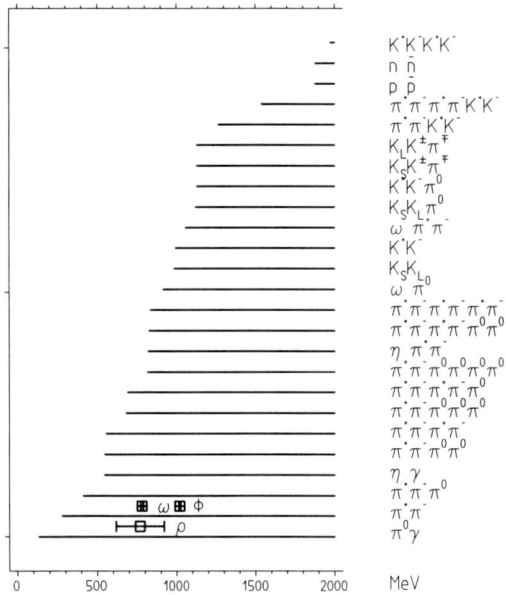

Fig. 5.3. Thresholds for exclusive multi particle channels below 2 GeV

available[5] above about 1.4 GeV (MEA, $\gamma\gamma 2$). Above 2.1 GeV inclusive measurements are standard. The amazing fact is that at the level of the inclusive cross–section, for high enough energies when the effective strong coupling constant α_s is small enough (see Fig. 3.3), perturbative QCD starts to work well away from threshold regions, where resonances show up, in the sense

$$\sigma(e^+e^- \to \text{hadrons})(s) = \sum_{X_h} \sigma(e^+e^- \to X_h)(s) \simeq \sum_q \sigma(e^+e^- \to q\bar{q})(s) \,,$$

where the sums go over all states which are possible by conservation laws and phase space. The sum over quarks q is subject to the constraint $4m_q^2 \ll s$. The quark–pair production cross–section is calculable in pQCD. Here the asymptotic freedom of QCD (see p. 129) comes into play in a way similar to what is familiar from deep inelastic ep–scattering and Bjorken scaling.

At low energies an inclusive measurement of the total hadronic cross–sections is not possible and pQCD completely fails. Experimentally, it becomes a highly non–trivial task to separate muon–pairs from pion–pairs, neutral pions from photons, $\pi^+\pi^-\pi^0$ from $\pi^+\pi^-\gamma$ etc. Here only *exclusive* measurements are possible, each channel has to be identified individually and the cross–section is obtained by adding up all possible channels. Many channels, e.g. those with π^0's are not easy to identify and often one uses isospin relations or other kind of theory input to estimate the total cross–section.

Experimentally, what is determined is of the form (see (2.105))

$$R^{\text{had exp}}(s) = \frac{N_{\text{had}} (1 + \delta_{\text{RC}})}{N_{\text{norm}} \varepsilon} \frac{\sigma_{\text{norm}}(s)}{\sigma_{\mu\mu,0}(s)} \,,$$

where N_{had} is the number of observed hadronic events, N_{norm} is the number of observed normalizing events, ε is the detector efficiency–acceptance product of hadronic events while δ_{RC} are radiative corrections to hadron production. $\sigma_{\text{norm}}(s)$ is the physical cross–section for normalizing events (including all radiative corrections integrated over the acceptance used for the luminosity measurement) and $\sigma_{\mu\mu,0}(s) = 4\pi\alpha^2/3s$ is the normalization. In particular this shows that a precise measurement of R requires precise knowledge of the relevant radiative corrections.

For the normalization mostly the Bhabha scattering process is utilized [or $\mu\mu$ itself in some cases]. In general, it is important to be aware of the fact that the effective fine structure constant $\alpha(\mu)$ enters radiative correction calculations with different scales μ in "had" and "norm" and thus must be taken into account carefully[6]. Care also is needed concerning the ISR corrections because cuts for the Bhabha process $(e^+e^- \to e^+e^-)$ typically are different from the

---

[5]Identifying the many channel (see Fig. 5.3) is difficult in particular when neutrals are involved. There is plenty of problems both with missing events or double counting states.

[6]Bhabha scattering $e^+(p_+) \ e^-(p_-) \to e^+(p'_+) \ e^-(p'_-)$ has two tree level diagrams Fig. 5.4 the t– and the s–channel. With the positive c.m. energy square

5.1 Vacuum Polarization Effects and e^+e^- Data

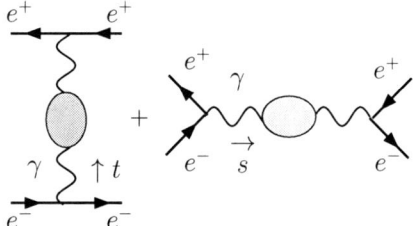

Fig. 5.4. VP dressed tree level Bhabha scattering in QED

ones applied to $e^+e^- \to$ hadrons. Usually, experiments have included corresponding uncertainties in their systematic errors, if they not have explicitly accounted for all appropriate radiative corrections.

The most important contribution for calculating a_μ^{had} comes from the low energy region below about 1 GeV. In Fig. 5.5 we show a compilation of the measurements of the square of the pion form factor $|F_\pi(s)|^2 = 4 R_{\pi\pi}(s)/\beta_\pi^3$ with $\beta_\pi = (1 - 4m_\pi^2/s)^{1/2}$ the pion velocity.

A collection of e^+e^-–data above 1 GeV is shown in Fig. 5.6 [6], an up–to–date version of earlier compilations [7, 8, 9, 10, 11, 12, 13] by different groups. For detailed references and comments on the data we refer to [7] and the more recent experimental papers by MD-1 [14], BES [15], CMD-2 [16, 17], KLOE [18], SND [19] and BaBar [20]. A list of experiments and references is given in [13], where the data available are collected.

A lot of effort went into the perturbative QCD calculation of $R(s)$. The leading term is given by the QPM result

$$R(s)^{\text{QPM}} \simeq N_c \sum_q Q_q^2 \,, \qquad (5.4)$$

$s = (p^+ + p^-)^2$ and the negative momentum transfer square $t = (p_- - p'_-)^2 = -\frac{1}{2}(s - 4m_e^2)(1 - \cos\theta)$, θ the e^- scattering angle, there are two very different scales involved. The VP dressed lowest order cross–section is

$$\frac{d\sigma}{d\cos\Theta} = \frac{s}{48\pi} \sum_{ik} |A_{ik}|^2$$

in terms of the tree level helicity amplitudes A_{ik}, $i,k =$ L,R denoting left– and right–handed electrons. The dressed transition amplitudes, in the approximation of vanishing electron mass, read

$$|A_{\text{LL,RR}}|^2 = \frac{3}{8}(1+\cos\theta)^2 \left| \frac{e^2(s)}{s} + \frac{e^2(t)}{t} \right|^2$$

$$|A_{\text{LR,RL}}|^2 = \frac{3}{8}(1-\cos\theta)^2 \left| \frac{e^2(s)}{s} + \frac{e^2(t)}{t} \right|^2 \,.$$

Preferably one uses small angle Bhabha scattering (small $|t|$) as a normalizing process which is dominated by the t–channel $\sim 1/t$, however, detecting electrons and positrons along the beam axis often has its technical limitations.

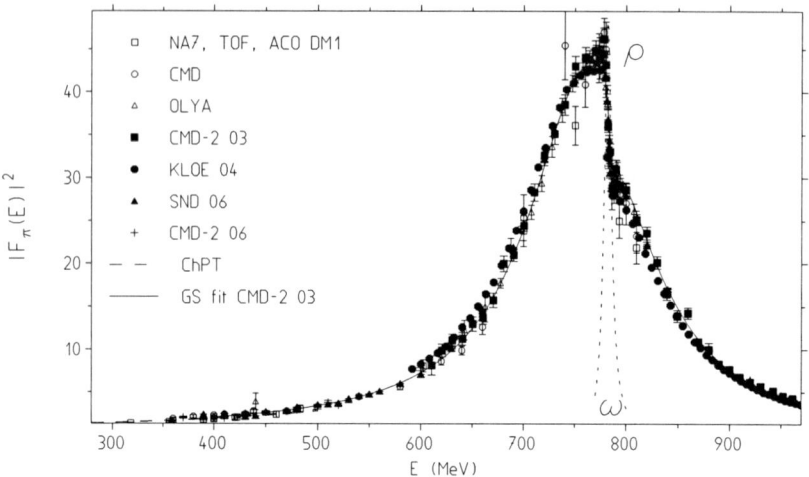

Fig. 5.5. The dominating low energy tail is given by the channel $e^+e^- \to \pi^+\pi^-$ which forms the ρ–resonance. The $\rho - \omega$ mixing caused by isospin breaking ($m_u - m_d \neq 0$) is distorting the ideal Breit-Wigner resonance shape of the ρ

where the sum extends over quarks q with $4m_q^2 \ll s$. Thus depending on the number of quark thresholds passed $R = 2$, $10/3$ and $11/3$ for $N_q = 3, 4$ and 5, respectively. In Fig. 5.6 one may nicely observe the jumps in R when a new threshold is passed. The higher order corrections are very important for a precise calculation of the contributions from the perturbative regions. Fortunately they are moderate sufficiently far above the thresholds. In pQCD the $\overline{MS}$ scheme (see Sect. 2.6.5) is generally adopted and normal order by order calculations are always improved by RG resummations. Corrections are known to $O(\alpha_s^3)$ [4, 21, 22, 23]. The last term was first obtained by Gorishnii, Kataev, Larin and Surguladze, Samuel [22] in the massless limit, and then extended to include the mass effect by Chetyrkin, Kühn et al. [23]. The state of the art was implemented recently in the program RHAD by Harlander and Steinhauser [24]. Away from the resonance regions the agreement between theory and experiment looks fairly convincing, however, one has to keep in mind that the systematic errors, which vary widely between a few % up to 20% are not shown in the plot. Typically, the theory result is much more accurate than the experimental one, in regions where it applies. This is possible, because, the QCD parameters α_s and the charm and bottom quark masses relevant here are known from plenty of all kinds of experiments rather accurately now. Nevertheless, it is not obvious that applying pQCD in place of the data, as frequently done, is not missing some non–perturbative contributions. The non–perturbative quark condensate terms ($1/Q^4$ power corrections) which enter the OPE are not a real problem in our context as they are small at energies where pQCD applies [25]. There are other kinds of NP

5.1 Vacuum Polarization Effects and e^+e^- Data

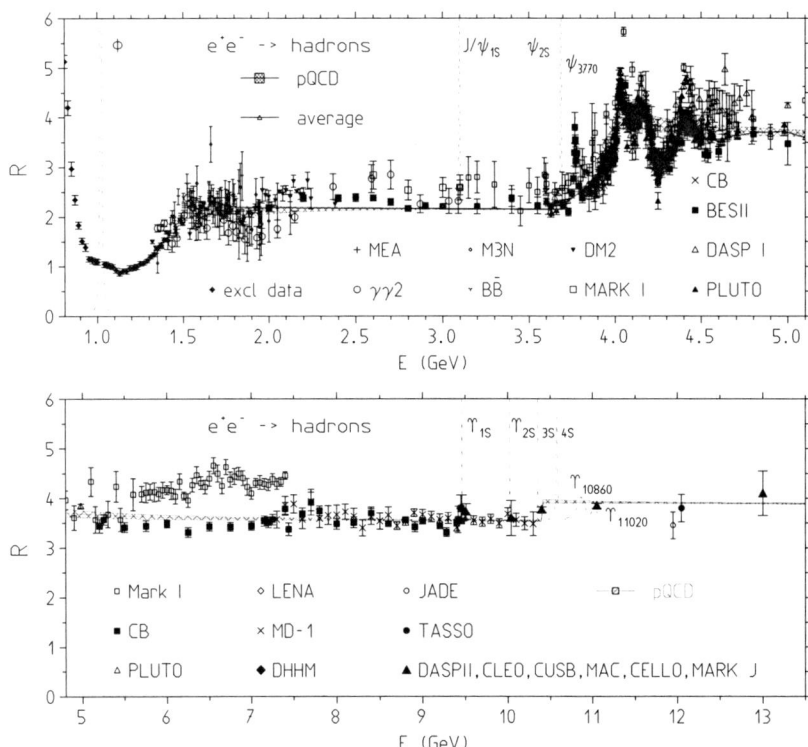

Fig. 5.6. Experimental results for $R_\gamma^{\mathrm{had}}(s)$ in the range 1 GeV $< E = \sqrt{s} <$ 13 GeV, obtained at the e^+e^- storage rings listed in Table 5.1. The perturbative quark–antiquark pair–production cross–section is also displayed (pQCD). Parameters: $\alpha_s(M_Z) = 0.118 \pm 0.003$, $M_c = 1.6 \pm 0.15$ GeV, $M_b = 4.75 \pm 0.2$ GeV and $\mu \in (\frac{\sqrt{s}}{2}, 2\sqrt{s})$

phenomena like bound states, resonances, instantons and in particular the hadronization of the quarks. In applying pQCD to describe real physical cross–sections of hadro–production one needs a "rule" which bridges the asymptotic freedom regime with the confinement regime, since the hadronization of the colored partons produced in the hard kicks into color singlet hadrons eludes a quantitative understanding. The rule is referred to as *quark hadron duality*[7] [27, 28], which states that for large s the average non–perturbative hadron cross–section equals the perturbative quark cross–section:

$$\overline{\sigma(e^+e^- \to \mathrm{hadrons})}(s) \simeq \sum_q \sigma(e^+e^- \to q\bar{q})(s) , \qquad (5.5)$$

---

[7] Quark–hadron duality was first observed phenomenologically for the structure function in deep inelastic electron–proton scattering [26].

where the averaging extends from threshold up to the given s value which must lie far enough above a threshold (global duality). Approximately, such duality relations then would hold for energy intervals which start just below the last threshold passed up to s. Qualitatively, such a behavior is visible in the data, however, for precise reliable predictions it has not yet been possible to quantify the accuracy of the duality conjecture. A quantitative check would require much more precise cross–section measurements than the ones available today. Ideally, one should attempt to reach the accuracy of pQCD predictions. In addition, in dispersion integrals the cross–sections are weighted by different s–dependent kernels, while the duality statement is claimed to hold for weight unity. One procedure definitely is contradicting duality reasonings: to "take pQCD plus resonances" or to "take pQCD where $R(s)$ is smooth and data in the complementary ranges". Also adjusting the normalization of experimental data to conform with pQCD within energy intervals (assuming local duality) has no solid foundation.

In view of the problematic quality of the data in some regions a "theory–driven" approach replacing data by pQCD results in smaller or larger intervals [29, 30, 31] may well be adequate to reduce the hadronic uncertainties. However, the uncertainty of the pQCD results evaluated by varying just the QCD parameters $\alpha_s(\mu)$, the quark masses $m_q(\mu)$ and the renormalization scale μ, conventionally, in a range $\mu \in (\frac{\sqrt{s}}{2}, 2\sqrt{s})$, generally does not account for possible non–perturbative uncertainties, related to the hadronization process. Thus the problem of the theory driven approach is a reliable error estimate, and not the shift in the central value, which may well be shifted in the right direction. In the following we generally present a conservative approach of the evaluation of the hadronic effects, taking the data and directly integrating them in all regions where pQCD cannot be trusted in the sense as advocated in [24].

The following *data integration procedure* has been used for the evaluation of the dispersion integral:

1. Take data as they are and apply the trapezoidal rule (connecting data points by straight lines) for integration.
2. To combine results from different experiments: i) integrate data for individual experiments and combine the results, ii) combine data from different experiments before integration and integrate the combined "integrand". Check consistency of the two possible procedures to estimate the reliability of the results.
3. Error analysis: 1) statistical errors are added in quadrature, 2) systematic errors are added linearly for different experiments, 3) combined results are obtained by taking weighted averages. 4) all errors are added in quadrature for "independent" data sets. We assume this to be allowed in particular for different energy regions and/or different accelerators.
4. The ρ–resonance region is integrated using the Gounaris-Sakurai (GS) parametrization of the pion form factor [32]. Other pronounced resonances

have been parametrized by Breit-Wigner shapes with parameters taken from the Particle Data Tables [33].

5.1.1 Integrating the Experimental Data and Estimating the Error

Here we briefly elaborate on procedures and problems related to the integration of the function $R(s)$ given in terms of experimental data sets with statistical and systematic errors. Obviously one needs some interpolation procedure between the data points. The simplest is to use the trapezoidal rule in which data points are joined by straight lines. This procedure is problematic if data points are sparse in relation to the functional shape of $R(s)$. Note that in pQCD $R(s)$ is close to piecewise constant away from thresholds and resonances (where pQCD fails) and the trapezoidal rule should work reliably. For resonances the trapezoidal rule is not very suitable and therefor one uses Breit-Wigner type parametrizations in terms of resonance parameters given in the particle data table. Here it is important to check which type of BW parametrization has been used to determine the resonance parameters (see [7] for a detailed discussion). Some analyses use other smoothing procedures, by fitting the data to some guessed functional form (see e.g. [34, 35]).

While statistical errors commonly are added in quadrature (Gaussian error propagation), the systematic errors of an experiment have to be added linearly, because they encode overall errors like normalization or acceptance errors. Usually the experiments give systematic errors as a relative systematic uncertainty and the systematic error to be added linearly is given by the central value times the relative uncertainty. For data from different experiments the combination of the systematic errors is more problematic. If one would add systematic errors linearly everywhere, the error would be obviously overestimated since one would not take into account the fact that independent experiments have been performed. However, often experiments use common simulation techniques for acceptance and luminosity determinations and the same state–of–the–art calculations for radiative corrections such that correlations between different experiments cannot be excluded. Since we are interested in the integral over the data only, a natural procedure seems to be the following: for a given energy range (scan region) we integrate the data points for each individual experiment and then take a weighted mean, based on the quadratically combined statistical and systematic error, of the experiments which have been performed in this energy range. By doing so we have assumed that different experiments have independent systematic errors, which of course often is only partially true[8]. The problem with this method is that there exist regions where data are sparse yet the cross–section varies

---

[8] If there are known common errors, like the normalization errors for experiments performed at the same facility, one has to add the common error after averaging. In some cases we correct for possible common errors by scaling up the systematic error appropriately.

rapidly, like in the ρ–resonance region. The applicability of the trapezoidal rule is then not reliable, but taking other models for the extrapolation introduces another source of systematic errors. It was noticed some time ago in [36] that fitting data to some function by minimizing χ^2 may lead to misleading results. Fortunately, the problem may be circumvented by the appropriate definition of the χ^2 to be minimized (see below).

In order to start from a better defined integrand we do better to combine all available data points into a single dataset. If we would take just the collection of points as if they were from *one* experiment we not only would get a too pessimistic error estimate but a serious problem could be that scarcely distributed precise data points do not get the appropriate weight relative to densely spaced data point with larger errors. What seems to be more adequate is to take for each point of the combined set the weighted average of the given point and the linearly interpolated points of the other experiments:

$$\bar{R} = \frac{1}{w} \sum_i w_i R_i$$

with total error $\delta_{tot} = 1/\sqrt{w}$, where $w = \sum_i w_i$ and $w_i = 1/\delta_{i\,tot}^2$. By $\delta_{i\,tot} = \sqrt{\delta_{i\,sta}^2 + \delta_{i\,sys}^2}$ we denote the combined error of the individual measurements. In addition, to each point a statistical and a systematic error is assigned by taking weighted averages of the squared errors:

$$\delta_{sta} = \sqrt{\frac{1}{w} \sum_i w_i\, \delta_{i\,sta}^2}\;, \quad \delta_{sys} = \sqrt{\frac{1}{w} \sum_i w_i\, \delta_{i\,sys}^2}\;.$$

There is of course an ambiguity in separating the well–defined combined error into a statistical and a systematic one. We may also calculate separately the total error and the statistical one and obtain a systematic error $\delta_{sys} = \sqrt{\delta_{tot}^2 - \delta_{sta}^2}$. Both procedures give very similar results. We also calculate $\chi^2 = \sum_i w_i (R_i - \bar{R})^2$ and compare it with $N-1$, where N is the number of experiments. Whenever $S = \sqrt{\chi^2/(N-1)} > 1$, we scale the errors by the factor S, unless there are plausible arguments which allow one to discard inconsistent data points.

In order to extract the maximum of information, weighted averages of different experiments at a given energy are calculated. The solution of the averaging problem may be found by minimizing χ^2 as defined by

$$\chi^2 = \sum_{n=1}^{N_{\exp}} \sum_{i,j=1}^{N_n} (R_i^n - \bar{R}_i)\,(C_{ij}^n)^{-1}\,(R_j^n - \bar{R}_j)$$

where R_i^n is the R measurement of the nth experiment at energy $\sqrt{s_i}$, $N_{\exp}$ the number of experiments, C_{ij}^n is the covariance matrix between the ith and jth data point of the nth experiment, and $\bar{R}$ is the average to be determined. The covariance matrix is given by

$$C_{ij}^n = \begin{cases} (\delta_{i\,\text{sta}}^n)^2 + (\delta_{i\,\text{sys}}^n)^2 & \text{for } j = i \\ \delta_{i\,\text{sys}}^n \cdot \delta_{j\,\text{sys}}^n & \text{for } j \neq i \end{cases}, \quad i,j = 1,\cdots,N_n$$

where $\delta_{i\,\text{sta}}^n$ and $\delta_{i\,\text{sys}}^n$ denote the statistical and systematic error, respectively, of R_i^n. The minimum condition $\frac{\mathrm{d}\chi^2}{\mathrm{d}\bar{R}_i} = 0$, for all i yields the system of linear equations

$$\sum_{n=1}^{N_{\text{exp}}} \sum_{j=1}^{N_n} (C_{ij}^n)^{-1} (R_j^n - \bar{R}_j) = 0, \quad i = 1,\cdots,N_n$$

The inverse covariance matrix $\bar{C}_{ij}^{-1}$ between the calculated averages $\bar{R}_i$ and $\bar{R}_j$ is the sum over the inverse covariances of every experiment

$$\bar{C}_{ij}^{-1} = \sum_{n=1}^{N_{\text{exp}}} (C_{ij}^n)^{-1} .$$

This procedure, if taken literally, would yield reliable fits only if the errors would be small enough, which would require in particular sufficiently high statistics. In fact, many of the older experiments suffer from low statistics and uncertain normalization and the fits obtained in this manner are biased towards too low values (compare [34] with [35], for example). The correct χ^2 minimization requires to replace the experimental covariance matrices C_{ij}^n by the ones of the fit result $\bar{C}_{ij}$ [36]. This is possible by iteration with the experimental covariance as a start value.

5.1.2 The Cross–Section $e^+e^- \to$ Hadrons

The total cross–section for hadron production in e^+e^-–annihilation (a typical s-channel process) may be written in the form

$$\sigma_{\text{had}}(s) = \frac{\alpha}{s} \frac{4\pi}{\sqrt{1 - \frac{4m_e^2}{s}}} \left(1 + \frac{2m_e^2}{s}\right) \operatorname{Im} \Pi_\gamma^{'\text{had}}(s)$$

$$\simeq \frac{4\pi\alpha}{s} \operatorname{Im} \Pi_\gamma^{'\text{had}}(s), \quad \text{since } s \geq 4m_\pi^2 \gg m_e^2,$$

where $\Pi_\gamma^{'\text{had}}(s)$ is the hadronic part of the photon vacuum polarization with (see (3.131) and Sect. 3.7.1)

$$\operatorname{Im} \Pi_\gamma^{'\text{had}}(s) = \frac{e^2}{12\pi} R_\gamma^{\text{had}}(s) .$$

From (2.175) we easily get the lowest order quark/antiquark pair–production cross–section encoded in

$$R_\gamma^{\mathrm{pQCD}}(s) = \sum_q N_{cq} Q_q^2 \sqrt{1 - \frac{4m_q^2}{s}} \left(1 + \frac{2m_q^2}{s}\right), \qquad (5.6)$$

which however is a reasonable approximation to hadro–production only at high energies away from thresholds and resonances (see below) and to the extent that quark–hadron duality (5.5) holds. At low energies $4m_\pi^2 < s < 9\,m_\pi^2$ $\pi\pi$–production is the dominant hadro–production process. The pion–pair production is commonly parametrized in terms of a non–perturbative amplitude, the pion form–factor $F_\pi(s)$,

$$R_\gamma^{\mathrm{had}}(s) = \frac{1}{4}\left(1 - \frac{4m_\pi^2}{s}\right)^{\frac{3}{2}} |F_\pi^{(0)}(s)|^2, \qquad s < 9\,m_\pi^2. \qquad (5.7)$$

For point–like pions we would have $F_\pi(s) = F_\pi(0) = 1$. At this point it is important to remind the reader that we have been deriving a set of relations and formulae to leading order $O(\alpha^2)$ in QED in Sect. 2.7. For a precise analysis of the hadronic effects higher order QED corrections are important as well. Furthermore, we have assumed that the center of mass energy $E = \sqrt{s}$ is small enough, typically, $E \leq 12$ GeV say, such that virtual Z exchange contributions $e^+e^- \to Z^* \to$ hadrons or $e^+e^- \to Z^* \to \mu^+\mu^-$ are sufficiently suppressed relative to virtual γ^* exchange at the precision we are aiming at. Since a_μ^{had} is rather insensitive to the high energy tail such a condition is not a problem.

In order to obtain the observed cross–section, we have to include the QED corrections, the virtual, soft and hard photon effects. The basic problems have been discussed in Sect. 2.6.6. For the important $\pi\pi$ channel, assuming scalar QED for the pions (see Fig. 2.8 for the Feynman rules) the one–loop diagrams are depicted in Fig. 5.7. In calculating the corrected

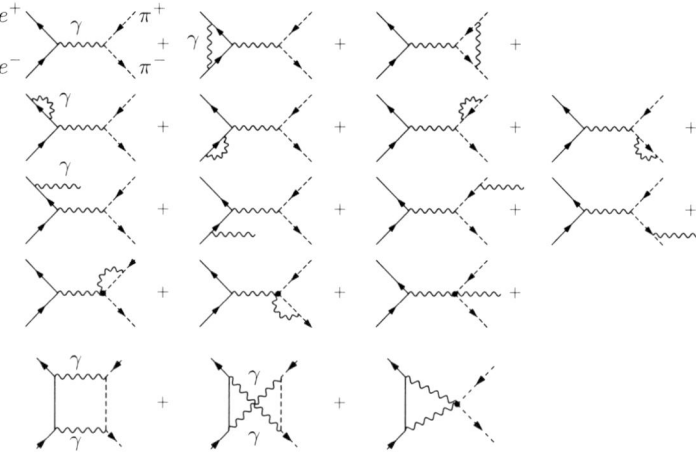

Fig. 5.7. One–loop sQED radiative corrections to pion–pair production assuming point–like pions

cross–section one starts with point–like pions and replaces the point form–factor $F_\pi^{\text{point}}(s) \equiv 1$ (strong interaction switched off) by the strong interaction dressed one with $F_\pi(s)$ a generic function of s. At least to $O(\alpha^2)$ this is possible due to the simple structure (see (5.8)) of the observed cross–section [37, 38, 39, 40].

Particularly important is the initial state radiation (ISR) which may lead to huge corrections in the shape of the cross–section. The most dramatic effects are of kinematical nature and may be used for cross–section measurements by the *radiative return* (RR) mechanism shown in Fig. 5.8: in the radiative process $e^+e^- \to \pi^+\pi^-\gamma$, photon radiation from the initial state reduces the invariant mass from s to $s' = s(1-k)$ of the produced final state, where k is the fraction of energy carried away by the photon radiated from the initial state. This may be used to measure $\sigma_{\text{had}}(s')$ at all energies $\sqrt{s'}$ lower than the fixed energy $\sqrt{s}$ at which the accelerator is running [41]. This is particularly interesting for machines running on–resonance like the ϕ– and B–factories, which typically have huge event rates as they are running on top of a peak [42, 43, 44]. The first dedicated radiative return experiment has been performed by KLOE at DAΦNE/Frascati, by measuring the $\pi^+\pi^-$ cross–section [18] (see Fig. 5.5)[9]. Results from BABAR will be discussed later.

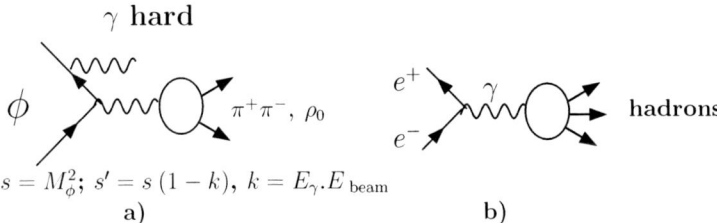

Fig. 5.8. a) Radiative return measurement of the $\pi^+\pi^-$ cross–section by KLOE at the ϕ–factory DAΦNE. At the B–factory at SLAC, using the same principle, BABAR has measured many other channels at higher energies. b) Standard measurement of σ_{had} in an energy scan, by tuning the beam energy

---

[9] The KLOE measurement is a radiative return measurement witch is a next to leading order approach. On the theory side one expects that the handling of the photon radiation requires one order in α more than the scan method for obtaining the same accuracy. At present the cross–section determined by KLOE using the PHOKARA Monte Carlo, as a function of the pion–pair invariant mass appears tilted relative to the CMD-2/SND data. KLOE data lie higher below the ρ^0 and lower above the ρ^0, with deviations at the few % level at the boundaries of the measured energy range. What is observed is that there is no deviation in the integral taken over the measured range, while there is a difference in the distribution (see comments after Table 7.3). My speculation is that this may be due to a slight misidentification of the pion–pair invariant mass by unreconstructed higher order radiation effects. In fact additional unidentified final state photon radiation tends

The "observed" cross–section at $O(\alpha^2)$ may be written in the form

$$\sigma^{\mathrm{obs}}(s) = \sigma_0(s)\left[1 + \delta_{\mathrm{ini}}(\omega) + \delta_{\mathrm{fin}}(\omega)\right]$$
$$+ \int_{4m_\pi^2}^{s-2\omega\sqrt{s}} ds'\, \sigma_0(s')\, \rho_{\mathrm{ini}}(s,s') + \sigma_0(s) \int_{4m_\pi^2}^{s-2\omega\sqrt{s}} ds'\, \rho_{\mathrm{fin}}(s,s')\,, \quad (5.8)$$

which also illustrates the unfolding problem one is confronted with in determining the cross–section of interest $\sigma_0(s)$. This "bare" cross–section, undressed from electromagnetic effects, is formally given by the point cross–section (2.255) times the absolute square of the pion form–factor which encodes the strong interaction effects

$$\sigma_0(s) = |F_\pi(s)|^2\, \sigma^{\mathrm{point}}(s) = \frac{\pi\alpha^2}{3s}\, \beta_\pi^3\, |F_\pi(s)|^2\,. \qquad (5.9)$$

ω is an IR cut parameter as introduced in Sect. 2.6.6. It drops out in the sum (5.8). The initial state corrections, in the approximation $O(\alpha m_e^2/m_\pi^2)$, are given by the following virtual+soft (V+S) and hard (H) parts:

$$\delta_{\mathrm{ini}}(\omega) = \ln\left(\frac{2\omega}{\sqrt{s}}\right) B_e(s) + \frac{\alpha}{\pi}\left[-2 + \frac{\pi^2}{3} + \frac{3}{2}L_e\right]$$

where $L_e = \ln\left(\frac{s}{m_e^2}\right)$ and $B_e(s) = \frac{2\alpha}{\pi}[L_e - 1]$. The hard ISR radiator function is given by

$$\rho_{\mathrm{ini}}(s,s') = \frac{1}{s}\left[\frac{B_e(s)}{1-z} - \frac{\alpha}{\pi}(1+z)(L_e - 1)\right]\,,$$

with $z = s'/s$. We denote by $\beta_\pi = (1 - 4m_\pi^2/s)^{1/2}$ the pion velocity. The final state corrections again we separate into a virtual+soft part and a hard part:

$$\delta_{\mathrm{fin}}(\omega) = \ln\left(\frac{2\omega}{\sqrt{s}}\right) B_\pi(s,s') + \frac{\alpha}{\pi}\Bigg\{\frac{3s - 4m_\pi^2}{s\beta_\pi} \ln\left(\frac{1+\beta_\pi}{1-\beta_\pi}\right) - 2$$
$$-\frac{1}{2}\ln\left(\frac{1-\beta_\pi^2}{4}\right) - \frac{3}{2}\ln\left(\frac{s}{m_\pi^2}\right) - \frac{1+\beta_\pi^2}{2\beta_\pi}\left[\ln\left(\frac{1+\beta_\pi}{1-\beta_\pi}\right)\right.$$
$$\left.\left[\ln\left(\frac{1+\beta_\pi}{2}\right) + \ln(\beta_\pi)\right] + \ln\left(\frac{1+\beta_\pi}{2\beta_\pi}\right)\ln\left(\frac{1-\beta_\pi}{2\beta_\pi}\right)\right.$$
$$\left. + 2\mathrm{Li}_2\left(\frac{2\beta_\pi}{1+\beta_\pi}\right) + 2\mathrm{Li}_2\left(-\frac{1-\beta_\pi}{2\beta_\pi}\right) - \frac{2}{3}\pi^2\right]\Bigg\}\,,$$

with

$$B_\pi(s,s') = \frac{2\alpha}{\pi}\frac{s'\beta_\pi(s')}{s\beta_\pi(s)}\left[\frac{1+\beta_\pi^2(s')}{2\beta_\pi(s')}\ln\left(\frac{1+\beta_\pi(s')}{1-\beta_\pi(s')}\right) - 1\right]\,.$$

to move events from a higher energy bin into a lower one. The integral is obviously less sensitive to the correct even–by–event energy determination.

5.1 Vacuum Polarization Effects and e^+e^- Data

The hard FSR radiator function reads

$$\rho_{\text{fin}}(s, s') = \frac{1}{s}\left[\frac{B_\pi(s, s')}{1-z} + \frac{2\alpha}{\pi}(1-z)\frac{\beta_\pi(s')}{\beta_\pi^3(s)}\right].$$

At the level of precision of interest also higher order corrections should be included. The $O(\alpha^2)$ corrections are partially known only and we refer to [39] and references therein for more details.

The crucial point is that the radiator functions $\rho_{\text{ini}}(s, s')$ and to some extent also $\rho_{\text{fin}}(s, s')$ are calculable in QED. Pion pair production is C-invariant and it is very important that experimental angular cuts, which always have to be applied, are symmetric such that C invariance is respected. Then, as in (5.8) for the total cross–section, at the one–loop level initial-final state (IFS) interference terms are vanishing, also for the cut cross–sections. Generally, the IFS interference derives from the box diagrams of Fig. 5.7 and the cross terms

which are obtained in calculating the transition probability $|T|^2$. Under this condition the cross–section factorizes into initial state and final state radiation as in (5.8). Still we have a problem, the FSR is not calculable from first principles [45, 46]. Such $\rho_{\text{fin}}(s, s')$ is model–dependent, however the soft photon part is well modeled by sQED[10].

One other important point should be added here. A look at Fig. 5.8b tells us that there are two factors of e in the related matrix element, the absolute square of which determines the hadronic cross–section. One from the initial $e^+e^-\gamma^*$-vertex the other from the hadronic vertex. The physical hadronic cross–section is proportional to $\alpha(s)^2$, because in the physical cross–section

---

[10] In radiative return measurements at low energy one looks at the $\pi^+\pi^-$ invariant mass distribution $\left(\frac{d\sigma}{ds'}\right)$ plus any photon. Once s' is fixed the missing energy $s - s'$ is fixed and an "automatic" unfolding is obtained. Using the pion form factor ansatz:

$$\left(\frac{d\sigma}{ds'}\right)_{\text{sym-cut}} = |F_\pi(s')|^2 \left(\frac{d\sigma}{ds'}\right)^{\text{point}}_{\text{ini, sym-cut}} + |F_\pi(s)|^2 \left(\frac{d\sigma}{ds'}\right)^{\text{point}}_{\text{fin, sym-cut}},$$

we may directly resolve for the pion form factor as

$$|F_\pi(s')|^2 = \frac{1}{\left(\frac{d\sigma}{ds'}\right)^{\text{point}}_{\text{ini, sym-cut}}} \left\{\left(\frac{d\sigma}{ds'}\right)_{\text{sym-cut}} - |F_\pi(s)|^2 \left(\frac{d\sigma}{ds'}\right)^{\text{point}}_{\text{fin, sym-cut}}\right\}.$$

This is a remarkable equation since it tells us that the inclusive pion–pair invariant mass spectrum allows us to get the pion form factor unfolded from photon radiation directly as for fixed s and a given s' the photon energy is determined. The point cross sections are assumed to be given by theory and $d\sigma/ds'$ is the observed experimental pion–pair spectral function.

the full photon propagator including all radiative corrections contributes in the measurement, as discussed in Sect. 3.7.1. In order to obtain the 1PI photon self–energy, which is our building–block for systematic order by order (in α) calculations, we have the undress the physical cross–section from multiple 1PI insertions, which make up the dressed propagator. This requires to replace the running $\alpha(s)$ by the classical α:

$$\sigma_{\text{tot}}(e^+e^- \to \text{hadrons}) \to \sigma_{\text{tot}}^{(0)}(e^+e^- \to \text{hadrons}) \left(\frac{\alpha}{\alpha(s)}\right)^2 \quad (5.10)$$

and, using (3.137) we obtain

$$\Pi'_\gamma(k^2) - \Pi'_\gamma(0) = \frac{k^2}{4\pi^2\alpha} \int_0^\infty ds \, \frac{\sigma_{\text{tot}}^{(0)}(e^+e^- \to \text{hadrons})}{(s - k^2 - i\varepsilon)} \, . \quad (5.11)$$

Note that using the physical cross section in the DR gives a nonsensical result, since in order to get the photon propagator we have to subtract in any case the external charge at the right scale. Thus while

$$\frac{k^2}{4\pi^2\alpha} \int_0^\infty ds \, \frac{\sigma_{\text{tot}}(e^+e^- \to \text{hadrons})}{(s - k^2 - i\varepsilon)}$$

is double counting the VP effects, and therefore does not yield something useful, the linearly $\alpha/\alpha(s)$–rescaled cross–section

$$\frac{k^2}{4\pi^2} \int_0^\infty ds \, \frac{1}{\alpha(s)} \frac{\sigma_{\text{tot}}(e^+e^- \to \text{hadrons})}{(s - k^2 - i\varepsilon)} \, , \quad (5.12)$$

yields the hadronic shift in the full photon propagator. Only at least once VP–subtracted physical cross–sections are useful in DRs!

5.1.3 $R(s)$ in Perturbative QCD

Due to the property of *asymptotic freedom*, which infers that the effective strong interaction constant $\alpha_s(s)$ becomes weaker the higher the energy scale $E = \sqrt{s}$, we may calculate the hadronic current correlators in perturbation theory as a power series in α_s/π. According to the general analysis presented above, the object of interest is

$$\rho(s) = \frac{1}{\pi} \text{Im} \, \Pi_\gamma(s) \, ; \quad \Pi_\gamma(q) : \;\; \text{\scriptsize \raisebox{0.5ex}{$\sim\!\!\!\bigotimes\!\!\!\sim$}} \, . \quad (5.13)$$

The QCD perturbation expansion diagrammatically is given by

5.1 Vacuum Polarization Effects and e^+e^- Data

Lines ⁓⁓⁓ show external photons, →— propagating quarks/antiquarks and ⁓⁓⁓ propagating gluons. See Fig. 2.9 for the Feynman rules of QCD. The vertices ⊗ are marking renormalization counter term insertions. They correspond to subtraction terms which render the divergent integrals finite.

In QED (the above diagrams with gluons replaced by photons) the phenomenon of vacuum polarization was discussed first by Dirac [47] and finalized at the one–loop level by Schwinger [48] and Feynman [49]. Soon later Jost and Luttinger [50] presented the first two–loop calculation.

In 0th order in the strong coupling α_s we have

$$2\text{Im} \; \text{⁓◯⁓} \; = \left| \text{⁓◁} \right|^2$$

which is proportional to the the free quark–antiquark production cross–section [5] in the so called *Quark Parton Model* describing quarks with the strong interaction turned off, which gets true in the high energy limit of QCD. As it is common practice, rather than considering the total hadronic production cross-section $\sigma_{\text{tot}}(e^+e^- \to \gamma^* \to \text{hadrons})$ itself, we again use

$$R(s) \doteq \frac{\sigma_{\text{tot}}(e^+e^- \to \gamma^* \to \text{hadrons})}{\frac{4\pi\alpha^2}{3s}} = 12\pi^2 \rho(s) \;, \quad (5.14)$$

which for sufficiently large s can be calculated in QCD perturbation theory. The result is given by [4, 21]

$$R(s)^{\text{pert}} = N_c \sum_f Q_f^2 \frac{v_f}{2} \left(3 - v_f^2\right) \Theta(s - 4m_f^2)$$
$$\times \left\{1 + a c_1(v_f) + a^2 c_2 + a^3 c_3 + \cdots \right\} \quad (5.15)$$

where $a = \alpha_s(s)/\pi$ and, assuming $4m_f^2 \ll s$, i.e. in the massless approximation

$$c_1 = 1$$
$$c_2 = C_2(R) \left\{ -\frac{3}{32} C_2(R) - \frac{3}{4} \beta_0 \zeta(3) - \frac{33}{48} N_f + \frac{123}{32} N_c \right\}$$

$$= \frac{365}{24} - \frac{11}{12}N_f - \beta_0\zeta(3) \simeq 1.9857 - 0.1153 N_f$$
$$c_3 = -6.6368 - 1.2002 N_f - 0.0052 N_f^2 - 1.2395 \left(\sum_f Q_f\right)^2 / \left(3\sum_f Q_f^2\right)$$

in the $\overline{MS}$ scheme. $N_f = \sum_{f:4m_f^2 \leq s} 1$ is the number of active flavors. The mass dependent threshold factor in front of the curly brackets is a function of the velocity $v_f = \left(1 - \frac{4m_f^2}{s}\right)^{1/2}$ and the exact mass dependence of the first correction term

$$c_1(v) = \frac{2\pi^2}{3v} - (3+v)\left(\frac{\pi^2}{6} - \frac{1}{4}\right)$$

is singular (Coulomb singularity due to soft gluon final state interaction) at threshold. The singular terms exponentiate [51]:

$$1 + x \to \frac{2x}{1 - e^{-2x}} \; ; \; x = \frac{2\pi\alpha_s}{3\beta}$$

$$\left(1 + c_1(v)\frac{\alpha_s}{\pi} + \cdots\right) \to \left(1 + c_1(v)\frac{\alpha_s}{\pi} - \frac{2\pi\alpha_s}{3v}\right)\frac{4\pi\alpha_s}{3v}\frac{1}{1 - \exp\left\{-\frac{4\pi\alpha_s}{3v}\right\}} \; .$$

Applying renormalization group improvement, the coupling α_s and the masses m_q have to be understood as running parameters

$$R\left(\frac{m_{0f}^2}{s_0}, \alpha_s(s_0)\right) = R\left(\frac{m_f^2(\mu^2)}{s}, \alpha_s(\mu^2)\right) \; ; \; \mu = \sqrt{s} \; .$$

where $\sqrt{s_0}$ is a reference energy. Mass effects are important once one approaches a threshold from the perturbatively save region sufficiently far above the thresholds where mass effects may be safely neglected. They have been calculated up to three loops by Chetyrkin, Kühn and collaborators [23] and have been implemented in the FORTRAN routine RHAD by Harlander and Steinhauser [24].

Where can we trust the perturbative result? Perturbative QCD is supposed to work best in the deep Euclidean region away from the physical region characterized by the cut in the analyticity plane Fig. 5.9. Fortunately, the physical region to a large extent is accessible to pQCD as well provided the energy scale is sufficiently large and one looks for the appropriate observable.

The imaginary part (total cross-section) corresponds to the jump of the vacuum polarization function $\Pi(q^2)$ across the cut. On the cut we have the thresholds of the physical states, with lowest lying channels: $\pi^+\pi^-$, $\pi^0\pi^+\pi^-$, $\cdots$ and resonances $\rho, \omega, \phi, J/\psi \cdots, \Upsilon \cdots, \cdots$. QCD is confining the quarks (a final proof of confinement is yet missing) in hadrons. In any case the quarks

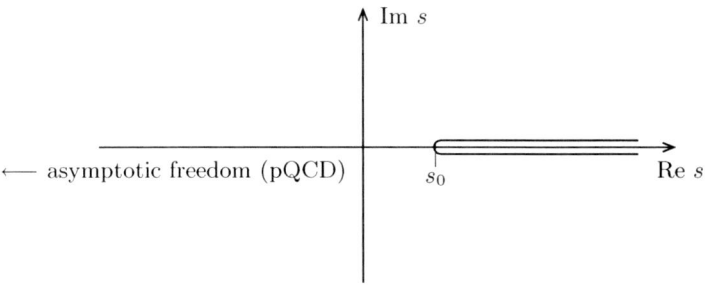

Fig. 5.9. Analyticity domain for the photon vacuum polarization function. In the complex s–plane there is a cut along the positive real axis for $s > s_0 = 4m^2$ where m is the mass of the lightest particles which can be pair–produced

hadronize (see Fig. 5.10), a highly non–perturbative phenomenon which is poorly understood in detail.

Neither the physical *thresholds* nor the *resonances* are obtained with perturbation theory! In particular, the perturbative quark–pair thresholds in (5.15) do not nearly approximate the physical thresholds for the low energy region below about 2 GeV, say. At higher energies pQCD works sufficiently far away from thresholds and resonances, i.e. in regions where $R(s)$ is a slowly varying function. This may be learned from Fig. 5.6 where the e^+e^-–data are shown together with the perturbative QCD prediction. Less problematic is the space–like (Euclidean) region $-q^2 \to \infty$, since it is away from thresholds and resonances. The best monitor for a comparison between theory and experiment has been proposed by Adler [52] long time ago: the so called *Adler–function*, up to a normalization factor, the derivative of the vacuum polarization function in the space–like region, introduced in (3.158) (see Fig. 5.13). In any case on has to ask the e^+e^-–annihilation data and to proceed in a semi–phenomenological way.

At higher energies highly energetic *partons*, quarks and/or gluons, are produced and due to asymptotic freedom perturbative QCD should somehow be applicable. As we will see this in fact manifests itself, for example, in the correct prediction of $\sigma_{\mathrm{tot}}(e^+e^- \to \gamma^* \to \text{hadrons})$ in non–resonant regions at high enough energies, in the sense of quark–hadron duality (5.5). However, the consequences of the validity of pQCD are more far–reaching. According to perturbation theory the production of hadrons in e^+e^-–annihilation proceeds via the primary creation of a quark–antiquark pair (see Figs. 5.10, 5.11) where the quarks hadronize. The elementary process tells us that in a high energy collision of positrons and electrons (in the center of mass frame) q and $\bar{q}$ are produced with high momentum in opposite directions (back–to–back).

The differential cross–section, up to a color factor the same as for $e^+e^- \to \mu^+\mu^-$, reads

$$\frac{d\sigma}{d\Omega}(e^+e^- \to q\bar{q}) = \frac{3}{4}\frac{\alpha_s^2}{s}\sum Q_f^2\,(1+\cos^2\theta)$$

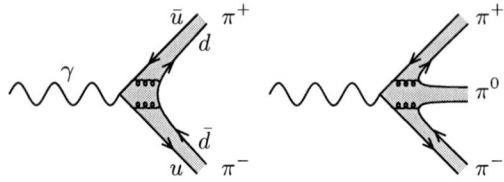

Fig. 5.10. Hadron production in low energy e^+e^-–annihilation: the primarily created quarks must hadronize. The shaded zone indicates strong interactions via gluons which confine the quarks inside hadrons

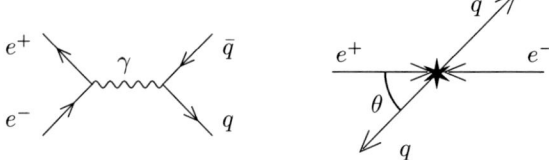

Fig. 5.11. Fermion pair production in e^+e^-–annihilation. The lowest order Feynman diagram (**left**) and the same process in the c.m. frame (**right**). The arrows represent the spacial momentum vectors and θ is the production angle of the quark relative to the electron in the c.m. frame

typical for an angular distribution of a spin 1/2 particle. Indeed, the quark and the antiquark seemingly hadronize individually in that they form *jets* [53]. Jets are bunches of hadrons which concentrate in a relatively narrow angular cone. This in spite of the fact that the quarks have unphysical charge and color, true physical states only can have integer charge and must be color singlets. Apparently, while charge and color have enough time to recombine into color singlets of integer charge, the momentum apparently has not sufficient time to distribute isotropically. The extra quarks needed to form physical states are virtual pairs created from the vacuum and carried along by the primary quarks. As a rule pQCD is applicable to the extent that "hard partons", quarks or gluons, may be interpreted as jets. Fig. 5.12 illustrates such $q\bar{q}$ (two–jet event) and $q\bar{q}g$ (three–jet event) jets. Three jet events produced with the electron positron storage ring PETRA at DESY in 1979 revealed the existence of the gluon. The higher the energy the narrower the jets, quite opposite to expectations at pre QCD times when most people believed events with increasing energy will be more and more isotropic multi–hadron states.

5.1.4 Non–Perturbative Effects, Operator Product Expansion

The non–perturbative (NP) effects are parametrized as prescribed by the operator product expansion of the electromagnetic current correlator [54]

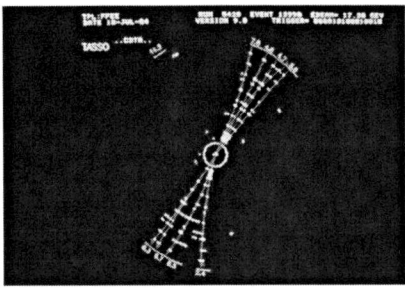 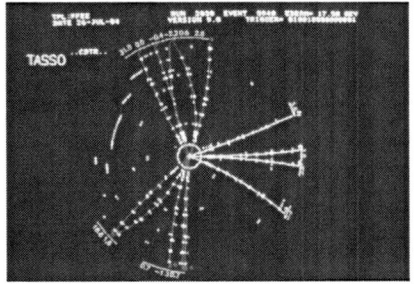

Fig. 5.12. Two and three jet event first seen by TASSO at DESY in 1979

$$\Pi_\gamma'^{NP}(Q^2) = \frac{4\pi\alpha}{3} \sum_{q=u,d,s} Q_q^2 N_{cq} \cdot \left[\frac{1}{12}\left(1 - \frac{11}{18}a\right) \frac{\langle \frac{\alpha_s}{\pi}GG\rangle}{Q^4} \right.$$
$$+ 2\left(1 + \frac{a}{3} + \left(\frac{11}{2} - \frac{3}{4}l_{q\mu}\right)a^2\right) \frac{\langle m_q \bar{q}q\rangle}{Q^4} \qquad (5.16)$$
$$+ \left(\frac{4}{27}a + \left(\frac{4}{3}\zeta_3 - \frac{257}{486} - \frac{1}{3}l_{q\mu}\right)a^2\right) \sum_{q'=u,d,s} \frac{\langle m_{q'}\bar{q}'q'\rangle}{Q^4} \Bigg]$$
$$+ \cdots$$

where $a \equiv \alpha_s(\mu^2)/\pi$ and $l_{q\mu} \equiv \ln(Q^2/\mu^2)$. $\langle\frac{\alpha_s}{\pi}GG\rangle$ and $\langle m_q\bar{q}q\rangle$ are the scale–invariantly defined condensates. Sum rule estimates of the condensates yield typically (large uncertainties) $\langle\frac{\alpha_s}{\pi}GG\rangle \sim (0.389 \text{ GeV})^4$, $\langle m_q\bar{q}q\rangle \sim -(0.098 \text{ GeV})^4$ for $q = u, d$, and $\langle m_q\bar{q}q\rangle \sim -(0.218 \text{ GeV})^4$ for $q = s$. Note that the above expansion is just a parametrization of the high energy tail of NP effects associated with the existence of non–vanishing condensates. The dilemma with the OPE in our context is that it works for large enough Q^2 only and in this form fails do describe NP physics at lower Q^2. Once it starts to be numerically relevant pQCD starts to fail because of the growth of the strong coupling constant. In $R(s)$ NP effects as parametrized by (5.16) have been shown to be small in [25, 30, 55]. Note that the quark condensate, the vacuum expectation value (VEV) $\langle \mathcal{O}_q\rangle$ of the dimension 3 operator $\mathcal{O}_q \doteq \bar{q}q$, is a well defined non–vanishing order parameter in the chiral limit of QCD. In pQCD it is vanishing to all orders. In contrast the VEV of the dimension 4 operator $\mathcal{O}_G \doteq \frac{\alpha_s}{\pi}GG$ is non–vanishing in pQCD but ill–defined at first as it diverges like Λ^4 in the UV cut–off. $\mathcal{O}_G$ contributes to the trace of the energy momentum tensor[11][56, 57, 58]

---

[11]In a QFT a symmetric energy momentum tensor $\Theta_{\mu\nu}(x)$ should exist such that the generators of the Poincaré group are represented by (see (2.5, 2.6))

$$P_\mu = \int d^3x\, \Theta_{0\mu}(x)\,, \quad M_{\mu\nu} = \int d^3x\, (x_\mu \Theta_{0\nu} - x_\nu \Theta_{0\mu})(x)\,.$$

$$\Theta^\mu_{\ \mu} = \frac{\beta(g_s)}{2g_s} GG + (1+\gamma(g_s))\{m_u \bar{u}u + m_d \bar{d}d + \cdots\} \quad (5.17)$$

where $\beta(g_s)$ and $\gamma(g_s)$ are the RG coefficients (2.277) and in the chiral limit

$$\varepsilon_{\text{vac}} = -\left\{\frac{\beta_0}{32} + O(\alpha_s)\right\}\langle\mathcal{O}_G\rangle$$

represents the vacuum energy density which is not a bona fide observable in a continuum QFT. In the Shifman-Vainshtein-Zakharov (SVZ) approach [54] it is treated to represent the soft part with respect to the renormalization scale μ, while the corresponding OPE coefficient comprises the hard physics from scales above μ.

Note that in the chiral limit $m_q \to 0$ the trace (5.17) does not vanish as expected on the classical level. Thus scale invariance (more generally conformal invariance) is broken in any QFT unless the β-function has a zero. This is another renormalization anomaly, which is a quantum effect not existing in a classical field theory. The renormalization group is another form of encoding the broken dilatation Ward identity. It's role for the description of the asymptotic behavior of the theory under dilatations (scale transformations) has been discussed in Sect. 2.6.5, where it was shown that under dilatations the effective coupling is driven into a zero of the β-function. For an asymptotically free theory like QCD we reach the scaling limit in the high energy limit. At finite energies we always have scaling violations, as they are well known from deep inelastic electron nucleon scattering. In e^+e^-–annihilation the scaling violation are responsible for the energy dependence (via the running coupling) of $R(s)$ in regions where mass effects are negligible.

As mentioned earlier the Adler–function is a good monitor to compare the pQCD as well as the NP results with experimental data. Fig. 5.13 shows that pQCD in the Euclidean region works very well for $\sqrt{Q^2} \gtrsim 2.5$ GeV [55]. The NP effects just start to be numerically significant where pQCD starts to fail. Thus, no significant NP effects can be established from this plot.

This corresponds to Noether's theorem for the Poincaré group (see (2.155)). In a strictly renormalizable massless QFT which exhibits only dimensionless couplings classically one would expect the theory to be conformally invariant. The energy momentum tensor then would also implement infinitesimal dilatations and special conformal transformations. That is, the currents

$$D_\mu(x) = x^\rho\, \Theta_{\mu\rho} \,;\quad K_{\mu\nu} = 2\,x^\rho\, x_\nu\, \Theta_{\mu\rho} - x^2 \Theta_{\mu\nu}$$

ought to be conserved, which requires the trace of the energy momentum tensor to vanish $\Theta^\mu_{\ \mu} = 0$.

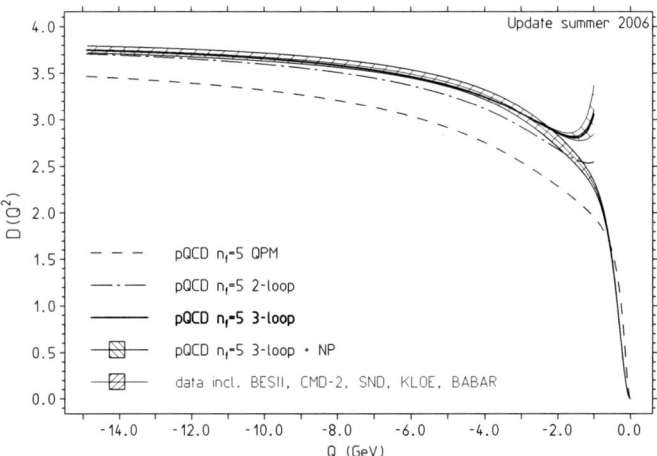

Fig. 5.13. "Experimental" Adler–function versus theory (pQCD + NP) in the low energy region (as discussed in [55]). Note that the error includes both statistical and systematic ones, in contrast to Fig. 5.6 where only statistical errors are shown

5.2 Leading Hadronic Contribution to $(g-2)$ of the Muon

We now are going to evaluate the hadronic vacuum polarization effects coming from the 5 "light" quarks $q = u, d, s, c, b$ in terms of the experimental e^+e^- data[12]. Quarks contribute to the electromagnetic current according to their charge

$$j_{\text{em}}^{\mu\,\text{had}} = \sum_c \left(\frac{2}{3}\bar{u}_c\gamma^\mu u_c - \frac{1}{3}\bar{d}_c\gamma^\mu d_c - \frac{1}{3}\bar{s}_c\gamma^\mu s_c + \frac{2}{3}\bar{c}_c\gamma^\mu c_c - \frac{1}{3}\bar{b}_c\gamma^\mu b_c + \frac{2}{3}\bar{t}_c\gamma^\mu t_c \right) \,.$$

The hadronic electromagnetic current $j_{\text{em}}^{\mu\,\text{had}}$ is a color singlet and hence includes a sum over colors indexed by c. Its contribution to the electromagnetic current correlator (3.125) defines $\Pi'^{\,\text{had}}_\gamma(s)$, which enters the calculation of the leading order hadronic contribution to a_μ^{had}, diagrammatically given by Fig. 5.1. The representation as a dispersion integral has been developed in

---

[12] The heavy top quark of mass $m_t \simeq 171.4(2.1)$ GeV we certainly may treat perturbatively, as at the scale m_t the strong interaction coupling is weak (see Fig. 3.3). Actually, the top quark t is irrelevant here since, as we know, heavy particles decouple in QED in the limit $m_t \to \infty$ and contribute like a VP τ–loop with an extra factor $N_c Q_t^2 = 4/3$, thus

$$a_\mu^{(4)}(\text{vap}, top) = \frac{4}{3}\left[\frac{1}{45}\left(\frac{m_\mu}{m_t}\right)^2 + \cdots\right]\left(\frac{\alpha}{\pi}\right)^2 \sim 5.9 \times 10^{-14}.$$

Sect. 3.8 on p. 197 (see also p. 189). Using (3.155) a_μ^{had} may be directly evaluated in terms of $R_\gamma(s)$ defined in (5.3). More precisely we may write

$$a_\mu^{\text{had}} = \left(\frac{\alpha m_\mu}{3\pi}\right)^2 \left(\int_{m_{\pi^0}^2}^{E_{\text{cut}}^2} ds\, \frac{R_\gamma^{\text{data}}(s)\,\hat{K}(s)}{s^2} + \int_{E_{\text{cut}}^2}^{\infty} ds\, \frac{R_\gamma^{\text{pQCD}}(s)\,\hat{K}(s)}{s^2} \right) \quad (5.18)$$

with a cut E_{cut} in the energy, separating the non–perturbative part to be evaluated from the data and the perturbative high energy tail to be calculated using pQCD. The kernel $K(s)$ is represented by (3.140), discarding the factor α/π. This integral can be performed analytically. Written in terms of the variable

$$x = \frac{1-\beta_\mu}{1+\beta_\mu}, \quad \beta_\mu = \sqrt{1-4m_\mu^2/s}$$

the result reads[13] [59]

$$K(s) = \frac{x^2}{2}(2-x^2) + \frac{(1+x^2)(1+x)^2}{x^2}\left(\ln(1+x) - x + \frac{x^2}{2}\right) + \frac{(1+x)}{(1-x)}x^2 \ln(x). \quad (5.20)$$

We have written the integral (5.18) in terms of the rescaled function

$$\hat{K}(s) = \frac{3s}{m_\mu^2} K(s)$$

which is slowly varying only in the range of integration. It increases monotonically from 0.63... at $\pi\pi$ threshold $s = 4m_\pi^2$ to 1 at ∞. The graph is shown in Fig. 5.14.

It should be noted that for small x the calculation of the function $K(s)$, in the form given above, is numerically instable and we instead use the asymptotic expansion (used typically for $x \leq 0.0006$)

$$K(s) = \left(\frac{1}{3} + \left(\frac{17}{12} + \left(\frac{11}{30} + \left(-\frac{1}{10} + \frac{3}{70}x\right)x\right)x\right)x\right)x + \frac{1+x}{1-x}x^2 \ln(x).$$

Other representations of $K(s)$, like the simpler–looking form

---

[13]The representation (5.20) of $K(s)$ is valid for the muon (or electron) where we have $s > 4m_\mu^2$ in the domain of integration $s > 4m_\pi^2$, and x is real, and $0 \leq x \leq 1$. For the τ (5.20) applies for $s > 4m_\tau^2$. In the region $4m_\pi^2 < s < 4m_\tau^2$, where $0 < r = s/m_\tau^2 < 4$, we may use the form

$$K(s) = \frac{1}{2} - r + \frac{1}{2}r(r-2)\ln(r) - \left(1 - 2r + \frac{1}{2}r^2\right)\varphi/w \quad (5.19)$$

with $w = \sqrt{4/r - 1}$ and $\varphi = 2\arctan(w)$.

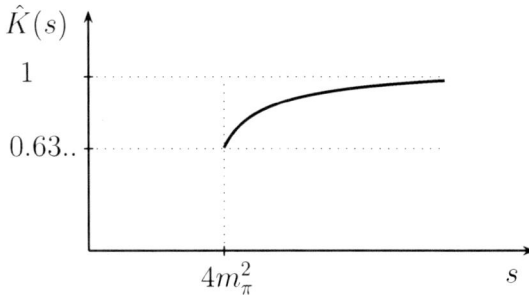

Fig. 5.14. Graph of weight function $\hat{K}(s)$ of the $g - 2$ dispersion integral

$$K(s) = \frac{1}{2} - r + \frac{1}{2}r(r-2)\ln(r) + \left(1 - 2r + \frac{1}{2}r^2\right)\ln(x)/\beta_\mu ,$$

with $r = s/m_\mu^2$, are much less suitable for numerical evaluation because of much more severe numerical cancellation.

Note the $1/s^2$–enhancement of contributions from low energies in a_μ. Thus the $g - 2$ kernel gives very high weight to the low energy range, in particular to the lowest lying resonance, the ρ^0. Thus, this $1/E^4$ magnification of the low energy region by the a_μ kernel–function together with the existence of the pronounced ρ^0 resonance in the $\pi^+\pi^-$ cross–section are responsible for the fact that pion pair production $e^+e^- \to \pi^+\pi^-$ gives the by far largest contribution to a_μ^{had} [14]. The ρ is the lowest lying vector–meson resonance

---

[14] As we need the VP–undressed hadronic cross–section in the DR, the physical form factor $F_\pi(s)$ which includes VP effects has to be corrected accordingly:

$$|F_\pi^{(0)}(s)|^2 = |F_\pi(s)|^2 \, (\alpha/\alpha(s))^2 . \tag{5.21}$$

Fig. 5.18 shows $\Delta\alpha(s) = 1 - \alpha/\alpha(s)$. In the time–like region. The resonances lead to pronounced variations of the effective charge (shown in the $\rho - \omega$, ϕ and J/ψ region).

For an order by order in α procedure of including corrections in a systematic manner, final state radiation should be subtracted as suggested in Sect. 5.1.2. The initial state radiation must and can be subtracted in any case, the final state radiation should be subtracted if possible. Note that measurements unavoidably include all virtual plus the unobserved soft photons. However, the hard virtual part for hadronic final states cannot be calculated in a model–independent manner, such that the subtraction seems not possible. It is therefore better to include as much as possible all photons in an inclusive measurement. The LKN theorem (see Sect. 2.6.6) infers that the inclusive cross section of virtual, soft plus hard real photons is $O(\alpha)$ without any logarithmic enhancement. Which also means a moderate model–dependence of the FSR correction, as a consequence of the absence of potentially large logs. We thus include the FSR (including full photon phase space) as

$$|F_\pi^{(\gamma)}(s)|^2 = |F_\pi^{(0)}(s)|^2 \left(1 + \eta(s)\frac{\alpha}{\pi}\right) \tag{5.22}$$

and shows up in $\pi^+\pi^- \to \rho^0$ at $m_\rho \sim 770$ MeV (see Fig. 5.5). This dominance of the low energy hadronic cross–section by a single simple two–body channel is good luck for a precise determination of a_μ, although a very precise determination of the $\pi^+\pi^-$ cross–section is a rather difficult task. Below about 810 MeV $\sigma_{tot}^{had}(s) \simeq \sigma_{\pi\pi}(s)$ to a good approximation but at increasing energies more and more channels open (see Fig. 5.3) and "measurements of R" get more difficult. In the light sector of $q = u, d, s$ quarks, besides the ρ there is the ω, which is mixing with the ρ, and the ϕ resonance, essentially a $\bar{s}s$ bound system. In the charm region we have the pronounced $\bar{c}c$–resonances, the $J/\psi_{1S}, \psi_{2S}, \cdots$ resonance series and in the bottom region the $\bar{b}b$–resonances $\Upsilon_{1S}, \Upsilon_{2S}, \cdots$. Many of the resonances are very narrow as indicated in Fig. 5.6.

For the evaluation of the basic integral (5.18) we take $R(s)$–data up to $\sqrt{s} = E_{cut} = 5.2$ GeV and for the Υ resonance–region between 9.46 and 13 GeV and apply perturbative QCD from 5.2 to 9.46 GeV and for the high energy tail above 13 GeV. The result obtained is [6]

$$a_\mu^{had(1)} = (692.10 \pm 5.66) \times 10^{-10} \tag{5.23}$$

and is based on a direct integration of all available e^+e^-–data. The contributions and errors from different energy regions is shown in Table 5.2. Most noticeable about this result are three features (see also Table 3.1)

– the experimental errors of the data lead to a substantial theoretical uncertainty, which is of the same size as the present experimental error of the BNL $g-2$ experiment;
– the low energy region is dominated by the $\pi\pi$–channel and the ρ–resonance contributions is dramatically enhanced: $\sim 78\%$ of the contribution and $\sim 42\%$ of error of a_μ^{had} comes from region $2m_\pi < \sqrt{s} < M_\phi$.
– the "intermediate" energy region, between 1 and 2 GeV, still gives a substantial contribution of about 15%. Unfortunately, because of the low quality of the R–data in the region, it contributes 52% of the total error, and therefore, together with the slightly more precisely known low energy contribution, is now the main source of uncertainty in the theoretical determination of a_μ .

Here we also refer to the brief summary which has been given in Sect. 3.2.1 after p. 153.

Integration of various exclusive channels yields the results of Table 5.3, which illustrates the relative weight of different channels in the region of exclusive channel measurements. Inclusive measurements are available above

---

to order $O(\alpha)$, where $\eta(s)$ (5.3) is a known correction factor in sQED (Schwinger 1989) (see p. 308 below). Here $F_\pi^{(0)}(s)$ is obtained from the measured cross section by subtracting photonic effects using the sQED with the applied experimental cuts on the real hard photons.

5.2 Leading Hadronic Contribution to $(g-2)$ of the Muon

Table 5.2. Results for $a_\mu^{\text{had}} \times 10^{10}$ from different energy ranges. Given are statistical, systematic and the total error, the relative precision in % [rel] and the contribution to the final error[2] in % [abs].

final state	range (GeV)	result	(stat)	(syst)	[tot]	rel	abs
ρ	(0.28, 0.99)	501.37	(1.89)	(2.93)	[3.49]	0.7%	37.9%
ω	(0.42, 0.81)	36.96	(0.44)	(1.00)	[1.09]	3.0%	3.7%
ϕ	(1.00, 1.04)	34.42	(0.48)	(0.79)	[0.93]	2.7%	2.7%
J/ψ		8.51	(0.40)	(0.38)	[0.55]	6.5%	1.0%
Υ		0.10	(0.00)	(0.01)	[0.01]	6.7%	0.0%
had	(0.99, 2.00)	67.89	(0.24)	(3.99)	[3.99]	5.9%	49.8%
had	(2.00, 3.10)	22.13	(0.15)	(1.22)	[1.23]	5.6%	4.7%
had	(3.10, 3.60)	4.02	(0.08)	(0.08)	[0.11]	2.8%	0.0%
had	(3.60, 9.46)	13.87	(0.10)	(0.14)	[0.17]	1.3%	0.1%
had	(9.46, 13.00)	1.30	(0.01)	(0.08)	[0.09]	6.6%	0.0%
pQCD	(13.0, ∞)	1.53	(0.00)	(0.00)	[0.00]	0.1%	0.0%
data	(0.28, 13.00)	690.57	(2.07)	(5.27)	[5.66]	0.8%	0.0%
total		692.10	(2.07)	(5.27)	[5.66]	0.8%	100.0%

1.2 GeV, however, recent progress in this problematic range comes from measurements based on the *radiative return* mechanism by BABAR [20] for the exclusive channels $e^+e^- \to \pi^+\pi^-\pi^0$, $\pi^+\pi^-\pi^+\pi^-$, $K^+K^-\pi^+\pi^-$, $2(K^+K^-)$, $3(\pi^+\pi^-)$, $2(\pi^+\pi^-\pi^0)$, $K^+K^-2(\pi^+\pi^-)$ and $\bar{p}p$. These data cover a much

Table 5.3. Contributions to a_μ^{had} and $\Delta\alpha_{\text{had}}^{(5)}(-s_0)$, $\sqrt{s_0} = 10$ GeV, from the energy region 0.318 GeV $< E <$ 2 GeV. $X^* = X(\to \pi^0\gamma)$, $_{iso}$=evaluated using isospin relations

channel X	a_μ^X	%	$\Delta\alpha^X$	%	channel X	a_μ^X	%	$\Delta\alpha^X$	%
$\pi^0\gamma$	3.32	0.52	0.26	0.42	$\omega\pi^+\pi^-[*]$	0.10	0.01	0.03	0.05
$\pi^+\pi^-$	503.43	78.15	34.34	55.25	$K^+K^-\pi^0$	0.41	0.06	0.15	0.24
$\pi^+\pi^-\pi^0$	46.54	7.22	4.63	7.45	$[K_S^0 K_L^0 \pi^0]_{iso}$	0.41	0.06	0.15	0.24
$\eta\gamma$	0.47	0.07	0.06	0.10	$K_S^0 K^\pm \pi^\mp$	1.32	0.21	0.48	0.77
$\pi^+\pi^-2\pi^0$	21.20	3.29	5.97	9.60	$[K_L^0 K^\pm \pi^\mp]_{iso}$	1.32	0.21	0.48	0.77
$2\pi^+2\pi^-$	15.41	2.39	4.54	7.30	$K^+K^-\pi^+\pi^-$	1.95	0.30	0.92	1.48
$\pi^+\pi^-3\pi^0$	1.29	0.20	0.42	0.68	$[K\bar{K}\pi\pi]_{iso}$	2.93	0.46	1.12	1.79
$2\pi^+2\pi^-\pi^0$	2.19	0.34	0.73	1.17	$K^+K^-2\pi^+2\pi^-$	0.07	0.01	0.04	0.07
$\pi^+\pi^-4\pi^0$	0.20	0.03	0.10	0.16	$p\bar{p}$	0.20	0.03	0.10	0.16
$\eta\pi^+\pi^-[*]$	0.26	0.04	0.07	0.11	$n\bar{n}$	0.33	0.05	0.17	0.28
$2\pi^+2\pi^-2\pi^0$	3.80	0.59	1.86	3.00	$2K^+2K^-$	0.02	0.00	0.02	0.02
$3\pi^+3\pi^-$	0.84	0.13	0.43	0.70	$\omega \to$ missing	0.09	0.01	0.01	0.01
$\omega\pi^0[*]$	0.82	0.13	0.17	0.28	$\phi \to$ missing	0.03	0.00	0.00	0.01
K^+K^-	22.05	3.42	3.16	5.08	sum	644.19	100.00	62.16	100.00
$K_S^0 K_L^0$	13.19	2.05	1.74	2.80	tot [sum in %]	692.10	[93.08]	73.65	[84.40]

broader energy interval and extend to much higher energies than previous experiments.

The sum of the exclusive channels from Table 5.3 is 644.19 which together with the sum of contributions from energies $E > 2$ GeV 51.46 from Table 5.2 yields a slightly higher value 695.65 than the 692.10 we get by including also the inclusive data below 2 GeV. Results are well within errors and this is a good consistence test.

There have been many independent evaluations of a_μ^{had} in the past[15] [6, 7, 8, 9, 10, 11, 76, 77, 78, 79, 80, 81, 82], and some of the more recent ones are shown in Table 3.2 and Fig. 5.16. For more detailed explanations of the differences see the comments to Fig. 7.1.

A compilation of the e^+e^-–data in the most important low energy region is shown in Fig. 5.5. The relative importance of various regions is illustrated in Fig. 5.15. The update of the results [7], including the more recent data from MD-1, BES-II, CMD-2, SND, KLOE and BABAR [14, 15, 16, 17, 18, 19, 20].

The possibility of using hadronic τ–decay data was briefly discussed in Sect. 3.2.1 on p. 156. More details are given as an Addendum 5.2.2 to this section. Taking into account the τ-data increases the contribution to a_μ^{had} by 2σ (see Table 3.2 and Fig. 5.16). As the discrepancy between isospin rotated τ–data (see Fig. 5.17), corrected for isospin violations, and the direct e^+e^-–data is not completely understood, at present only the e^+e^-–data can be used for the evaluation of a_μ^{had}.

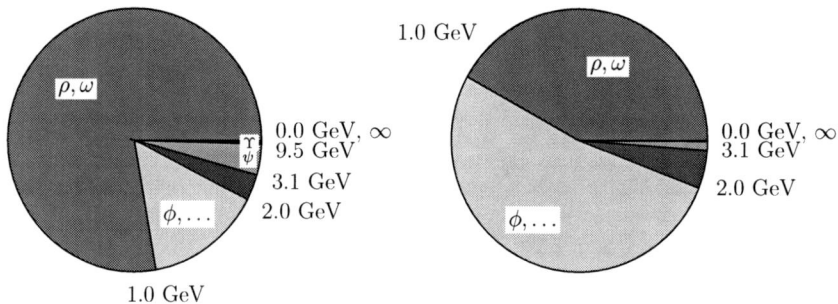

Fig. 5.15. The distribution of contributions (**left**) and errors (**right**) in % for a_μ^{had} from different energy regions. The error of a contribution i shown is $\delta_{i\,\mathrm{tot}}^2 / \sum_i \delta_{i\,\mathrm{tot}}^2$ in %. The total error combines statistical and systematic errors in quadrature

---

[15]The method how to calculate hadronic vacuum polarization effects in terms of hadronic cross–sections was developed long time ago by Cabibbo and Gatto [62]. First estimations were performed in [63, 64, 65]. As cross–section measurements made further progress much more precise estimates became possible in the mid 80's [71, 72, 73, 74, 75]. A more detailed analysis based on a complete up–to–date collection of data followed about 10 years later [7].

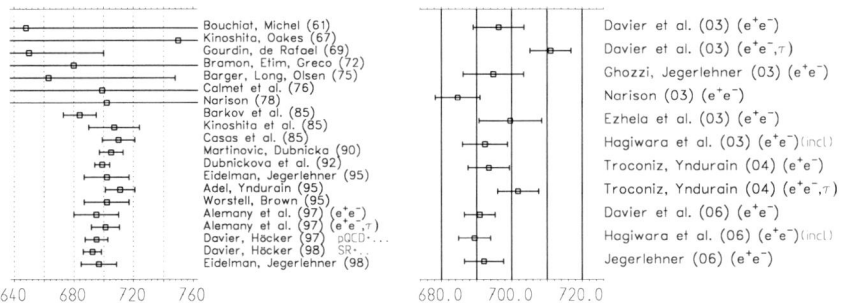

Fig. 5.16. History of evaluations before 2000 (**left**) [7, 8, 9, 10, 63, 64, 65, 66, 67, 68, 69, 70, 71, 72, 73, 74, 75, 83], and some more recent ones (**right**) [6, 76, 77, 78, 79, 80, 81, 82]; $(e^+e^-) = e^+e^-$–data based, $(e^+e^-,\tau) =$ in addition include data from τ spectral functions (see text)

5.2.1 Addendum I: The Hadronic Contribution to the Running Fine Structure Constant

By the same procedure, we have evaluated a_μ^{had}, the renormalized VP function can be calculated. The latter is identical to the shift in the fine structure constant, which encodes the charge screening:

$$\Delta\alpha(s) \equiv -\text{Re}\left[\Pi'_\gamma(s) - \Pi'_\gamma(0)\right] . \tag{5.24}$$

For the evaluation of the hadronic contribution we apply the DR (3.135). The integral to be evaluated is

$$\Delta\alpha_{\text{had}}^{(5)}(s) = -\frac{\alpha s}{3\pi}\left(\fint_{m_{\pi^0}^2}^{E_{\text{cut}}^2} ds' \frac{R_\gamma^{\text{data}}(s')}{s'(s'-s)} + \fint_{E_{\text{cut}}^2}^{\infty} ds' \frac{R_\gamma^{\text{pQCD}}(s')}{s'(s'-s)}\right) . \tag{5.25}$$

Since, in this case the kernel behaves like $1/s$ (as compared to $1/s^2$ for a_μ) data from higher energies are much more important here. The hadronic contribution due to the 5 light quarks $\Delta\alpha_{\text{had}}^{(5)}(s)$ supplemented by the leptonic contribution is presented in Fig. 5.18. A particularly important parameter for

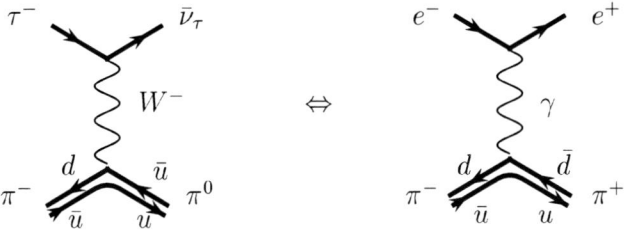

Fig. 5.17. τ–decay vs. e^+e^-–annihilation: the involved hadronic matrix–elements $\langle \text{out } \pi^+\pi^- | j_\mu^{I=1}(0) | 0 \rangle$ and $\langle \text{out } \pi^0\pi^- | J_{\bar{V}\mu}^-(0) | 0 \rangle$ are related by isospin

precision physics at the Z–resonance (LEP/SLD experiments) is the precise value of the effective fine structure constant at the Z mass scale $\sqrt{s} = M_Z = 91.19$ GeV $\alpha(M_Z^2)$. The hadronic contribution to the shift is

$$\Delta\alpha^{(5)}_{\text{hadrons}}(M_Z^2) = 0.027607 \pm 0.000225 \qquad (5.26)$$

which together with the leptonic contribution (3.117) and using (3.115) yields

$$\alpha^{-1}(M_Z^2) = 128.947 \pm 0.035 \ . \qquad (5.27)$$

With more theory input, based on the Adler–function method [6, 55, 83], we obtain (see Fig. 5.13)

$$\Delta\alpha^{(5)}_{\text{hadrons}}(M_Z^2) = 0.027593 \pm 0.000169 \qquad (5.28)$$
$$\alpha^{-1}(M_Z^2) = 128.938 \pm 0.023 \ .$$

The effective fine structure constant shown in Fig. 5.18 is very important also for removing the VP effects from the physical cross–section in order to get the undressed one which is needed in the DR (5.18).

5.2.2 Addendum II: τ Spectral Functions vs. e^+e^- Annihilation Data

In 1997 precise τ–spectral functions became available [84, 86, 87] which, to the extent that flavor $SU(2)_f$ in the light hadron sector is a good symmetry, allows

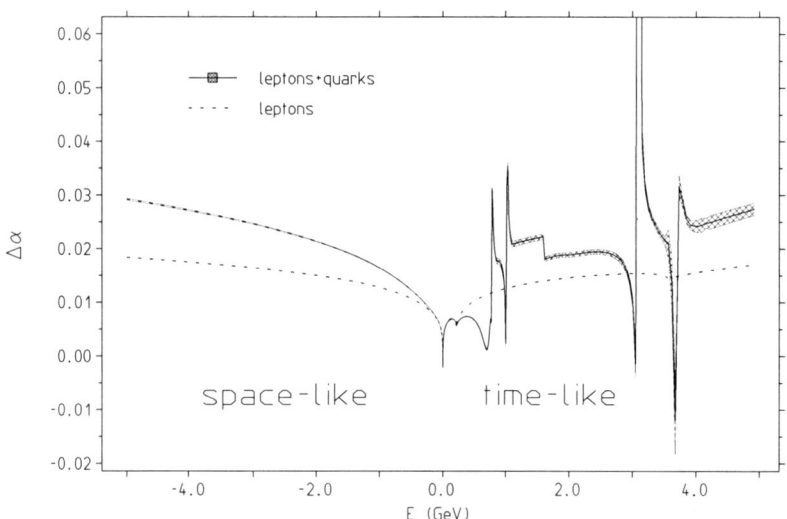

Fig. 5.18. Shift of the effective fine structure constant $\Delta\alpha$ as a function of the energy scale in the time–like region $s > 0$ ($E = \sqrt{s}$) vs. the space–like region $-s > 0$ ($E = -\sqrt{-s}$). The band indicates the uncertainties

5.2 Leading Hadronic Contribution to $(g-2)$ of the Muon

to obtain the iso–vector part of the e^+e^- cross–section [88]. The idea to use the τ spectral data to improve the evaluation of the hadronic contributions $a_\mu^{\rm had}$ was realized by Alemany, Davier and Höcker [10].

The iso–vector part of $\sigma(e^+e^- \to \text{hadrons})$ may be calculated by an isospin rotation, like $\pi^0\pi^- \to \pi^+\pi^-$, from τ–decay spectra, to the extent that the so–called conserved vector current (CVC) would be exactly conserved (which it is not, see below). In the following we will explicitly consider the dominating 2π channel only. The relation we are looking for may be derived by comparing the relevant lowest order diagrams Fig. 5.17, which for the e^+e^- case translates into

$$\sigma_{\pi\pi}^{(0)} \equiv \sigma_0(e^+e^- \to \pi^+\pi^-) = \frac{4\pi\alpha^2}{s} v_0(s) \tag{5.29}$$

and for the τ case into

$$\frac{1}{\Gamma}\frac{d\Gamma}{ds}(\tau^- \to \pi^-\pi^0\nu_\tau) = \frac{6\pi|V_{ud}|^2 S_{\rm EW}}{m_\tau^2} \frac{B(\tau^- \to \nu_\tau e^- \bar{\nu}_e)}{B(\tau^- \to \nu_\tau \pi^-\pi^0)}$$

$$\times \left(1 - \frac{s}{m_\tau^2}\right)\left(1 + \frac{2s}{m_\tau^2}\right) v_-(s) \tag{5.30}$$

where $|V_{ud}| = 0.9752 \pm 0.0007$ [33] denotes the CKM weak mixing matrix element and $S_{\rm EW} = 1.0233 \pm 0.0006$ accounts for electroweak radiative corrections [76, 89, 90, 91, 92, 93]. The spectral functions are obtained from the corresponding invariant mass distributions. The $B(i)$'s are branching ratios. SU(2) symmetry (CVC) would imply

$$v_-(s) = v_0(s) \ . \tag{5.31}$$

The spectral functions $v_i(s)$ are related to the pion form factors $F_\pi^i(s)$ by

$$v_i(s) = \frac{\beta_i^3(s)}{12\pi}|F_\pi^i(s)|^2 \ ; \quad (i = 0, -) \tag{5.32}$$

where $\beta_i(s)$ is the pion velocity. The difference in phase space of the pion pairs gives rise to the relative factor $\beta_{\pi^-\pi^0}^3/\beta_{\pi^-\pi^+}^3$.

Before a precise comparison via (5.31) is possible all kinds of isospin breaking effects have to be taken into account. As mentioned earlier, this has been investigated in [93] for the most relevant $\pi\pi$ channel. The corrected version of (5.31) (see [93] for details) may be written in the form

$$\sigma_{\pi\pi}^{(0)} = \left[\frac{K_\sigma(s)}{K_\Gamma(s)}\right] \frac{d\Gamma_{\pi\pi[\gamma]}}{ds} \times \frac{R_{\rm IB}(s)}{S_{\rm EW}} \tag{5.33}$$

with

$$K_\Gamma(s) = \frac{G_F^2 |V_{ud}|^2 m_\tau^3}{384\pi^3}\left(1 - \frac{s}{m_\tau^2}\right)^2\left(1 + 2\frac{s}{m_\tau^2}\right); \ K_\sigma(s) = \frac{\pi\alpha^2}{3s},$$

and the isospin breaking correction

$$R_{\mathrm{IB}}(s) = \frac{1}{G_{\mathrm{EM}}(s)} \frac{\beta_{\pi^-\pi^+}^3}{\beta_{\pi^-\pi^0}^3} \left|\frac{F_V(s)}{f_+(s)}\right|^2 \qquad (5.34)$$

includes the QED corrections to $\tau^- \to \nu_\tau \pi^- \pi^0$ decay with virtual plus real soft and hard (integrated over all phase space) photon radiation.

However, photon radiation by hadrons is poorly understood theoretically. The commonly accepted recipe is to treat radiative corrections of the pions by scalar QED, except for the short distance (SD) logarithm proportional the $\ln M_W/m_\pi$ which is replaced by the quark parton model result and included in S_{EW} by convention. This SD log is present only in the weak charged current transition $W^{+*} \to \pi^+\pi^0\,(\gamma)$, while in the charge neutral electromagnetic current transition $\gamma^* \to \pi^+\pi^-\,(\gamma)$ this kind of leading log is absent. In any case there is an uncertainty in the correction of the isospin violations by virtual and real photon radiation which is hard to quantify.

Originating from (5.32), $\beta_{\pi^-\pi^+}^3/\beta_{\pi^-\pi^0}^3$ is a phase space correction due to the $\pi^\pm - \pi^0$ mass difference. $F_V(s) = F_\pi^0(s)$ is the NC vector current form factor, which exhibits besides the $I = 1$ part an $I = 0$ contribution. The latter $\rho - \omega$ mixing term is due to the SU(2) breaking ($m_d - m_u$ mass difference). Finally, $f_+(s) = F_\pi^-$ is the CC $I = 1$ vector form factor. One of the leading isospin breaking effects is the $\rho-\omega$ mixing correction included in $|F_V(s)|^2$. The form–factor corrections, in principle, also should include the electromagnetic shifts in the masses and the widths of the ρ's[16]. Up to this last mentioned effect, discussed in [77], which was considered to be small, all the corrections were applied in [76] but were not able to eliminate the observed discrepancy between $v_-(s)$ and $v_0(s)$. The deviation is starting at the peak of the ρ and is increasing with energy to about 10–20%.

5.2.3 Digression: Exercises on the Low Energy Contribution

One important question we may ask here is to what extent are we able to understand and model the low energy hadronic piece theoretically? This excursion is manly thought to shed light on what has a chance to work and what not in modeling low–energy hadronic effects. It is a kind of preparation for the discussion of the hadronic light–by–light scattering. As a starting point for understanding strong interaction physics at the muon mass scale one could attempt to use chiral perturbation theory, the low energy effective description of QCD, where quarks and gluons are replaced by hadrons, primarily the pions, the quasi Goldstone bosons of spontaneous chiral symmetry breaking. One would then calculate $\pi^\pm$–loops as shown in Fig. 5.19, and as discussed earlier in Sect. 2.7.

---

[16]Because of the strong resonance enhancement, especially in the ρ region, a small isospin breaking shift in mass and width between ρ^0 and $\rho^\pm$, typically $\Delta m_\rho = m_{\rho^\pm} - m_{\rho^0} \sim 2.5$ MeV and $\Delta \Gamma_\rho = \Gamma_{\rho^\pm} - \Gamma_{\rho^0} \sim 1.5$ MeV and similar for the higher resonances $\rho', \rho'', \cdots$ and the mixing of these states, causes a large effect in the tails by the kinematical shift this implies.

5.2 Leading Hadronic Contribution to $(g-2)$ of the Muon

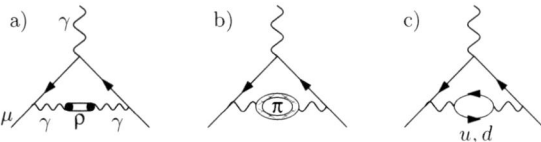

Fig. 5.19. Low energy effective graphs a) and b) and high energy graph c)

The charged spin 0 pions $\pi^\pm$ are assumed to couple to photons via minimal coupling, assuming the pions to be point–like as a leading approximation (see Sect. 2.7). However, the result given in Table 5.4 is underestimating the effect by about a factor 3. The main parameter for the size of the contribution is the mass and the coefficient $N_{ci}Q_i^2$, for color and charge of a particle species i (see (2.174)). If we would treat the quarks like leptons, switching off strong interactions and hence using the quark parton model (which is a good approximation only at sufficiently high energies) we would get for the sum of u and d quarks the result given in square brackets which is similar in size to the contribution from an electron, about a factor 100 too large! The large difference between the $\pi^\pm$ result and the (u,d) doublet result illustrates the dilemma with naive perturbative approaches. The huge contribution on the quark level was obtained using the current quark masses $m_u \sim 3\text{MeV}, m_d \sim 8\text{MeV}$, which appear in the QCD Lagrangian as chiral symmetry breaking parameters. Strong interactions lead to dressed quarks with effective "constituent quark masses", a concept which is not very well–defined e.g. if we choose $m_u \sim m_d \sim 300$ MeV (about 1/3 of the proton mass) one now gets a result which, this time, is a factor of two too small. In any case it is much closer to reality. This illustrates how sensitive these perturbative results are to the precise choice of the values of the quark masses. The failure of these trials is that one main non–perturbative effect is missing, namely, the ρ^0–resonance: a neutral spin 1 vector–meson, produced in $e^+e^- \to \rho^0 \to \pi^+\pi^-$. Spin 1 vector–mesons can be incorporated in the framework of CHPT (see p. 238) which leads to the Resonance Lagrangian Approach [94, 95]. The result obtained by integrating the corresponding non–relativistic Breit-Wigner ρ^0 resonance in the range (280,810) MeV gives a remarkably good result if we compare it with what we get using experimental data (see the first entry in Table 5.4). This also shows that adding up the ρ–exchange and the $\pi^\pm$–loop as independent effects would lead to a wrong answer. This is not so surprising since working with pions and vector–mesons as independent fields necessarily at some point produces

Table 5.4. Low energy effective estimates of the leading vacuum polarization effects $a_\mu^{(4)}(\text{vap})$. For comparison: 5.8420×10^{-8} for μ–loop, 5.9041×10^{-6} for e–loop

data [280,810] MeV	ρ^0–exchange	$\pi^\pm$–loop	$[u,d]$ loops
4.2666×10^{-8}	4.2099×10^{-8}	1.4154×10^{-8}	$2.2511 \times 10^{-8} [4.4925 \times 10^{-6}]$

a double counting problem, because the ρ may be understood to some extent as a $\pi^+\pi^-$ resonance. A much more reasonable approach would be to apply the low energy effective theory up to an energy scale Λ (L.D. part) and pQCD above the same cut off Λ (S.D. part). For more educated estimations of a_μ^{had} in low energy effective theory see [96] (see also [97]). We have been discussing the various possibilities in order to get some feeling about the reliability of such estimates, because in higher orders in general we will not be able to resort to experimental data to estimate the non–perturbative effect.

Fortunately, firm theoretical predictions are not only possible for the perturbative high energy tail. Also the low energy tail is strongly constrained, by the low energy effective CHPT briefly introduced on p. 238 in Sect. 4.2.2. The quantity of interest here is the vector form factor, defined by the hadronic pion pair production matrix element

$$\langle \text{out } \pi^+(p_+)\pi^-(p_-)|V_\mu(0)|0\rangle = -i(p_+ - p_-)_\mu F_V(s), \tag{5.35}$$

where $V_\mu(x)$ is the isovector current and $s = (p_+ + p_-)^2$. $F_V(s)$ has been calculated in CHPT in [98, 99] (one–loop), [100] (two–loop numerical) and [101] (two–loop analytical). The last reference gives a compact analytical result

$$F_V(s) = 1 + \frac{1}{6}\langle r^2\rangle_V^\pi s + c_V^\pi s^2 + f_V^U\left(\frac{s}{m_\pi^2}\right), \tag{5.36}$$

and a fit to the space–like NA7 data [102] with the expression (5.36) leaving $\langle r^2\rangle_V^\pi$ and c_V^π as free parameters, and including the theoretical error, leads to

$$\langle r^2\rangle_V^\pi = 0.431 \pm 0.020 \pm 0.016 \text{ fm}^2$$
$$c_V^\pi = 3.2 \pm 0.5 \pm 0.9 \text{ GeV}^{-4} \tag{5.37}$$

where the first and second errors indicate the statistical and theoretical uncertainties, respectively. The central value of c_V^π is rather close to the value obtained by resonance saturation, $c_V^\pi = 4.1 \text{ GeV}^{-4}$ [100]. Since experimental $\pi\pi$ production data below 300 MeV are poor or inexistent and the key integral (5.18) exhibits a $1/E^4$ enhancement of the low energy tail, (5.36) provides an important and firm parametrization of the low energy region and allows for a save evaluation of the contribution to a_μ^{had} as has been shown in [7].

The crucial point here is that the threshold behavior is severely constrained by the chiral structure of QCD via the rather precise data for the pion form factor in the space–like region. The space–like fit provides a good description of the data in the time–like region. Pure chiral perturbation theory is able to make predictions only for the low energy tail of the form factor.

The electromagnetic form factor of the pion $F_\pi(s)$ usually is defined in an idealized world of strong interactions with two quark flavors (u and d) only, and electroweak interactions switched off. $F_\pi(s)$ has an iso–vector part $I = 1$ as well as an iso–scalar part $I = 0$. The latter is due to isospin breaking by the mass difference of the u and d quarks: $m_u - m_s \neq 0$, which leads to $\rho - \omega$ mixing:

5.2 Leading Hadronic Contribution to $(g-2)$ of the Muon

$$|\rho\rangle = |\rho_0\rangle - \varepsilon |\omega_0\rangle \, , \, |\omega\rangle = |\omega_0\rangle + \varepsilon |\rho_0\rangle \, ,$$

where $|\omega_0\rangle$ and $|\rho_0\rangle$ are the pure isoscalar and isovector states, respectively, and ε is the $\rho - \omega$ mixing parameter. Then, in the energy region close to the $\rho(770)$– and $\omega(782)$–meson masses, the form factor can be written as

$$\begin{aligned} F_\pi(s) &\simeq \left[\frac{F_\rho}{s - M_\rho^2} + \varepsilon \frac{F_\omega}{s - M_\omega^2}\right] \left[\frac{F_\rho}{-M_\rho^2} + \varepsilon \frac{F_\omega}{-M_\omega^2}\right]^{-1} \\ &\approx -\frac{M_\rho^2}{s - M_\rho^2} \left[1 + \varepsilon \frac{F_\omega(M_\omega^2 - M_\rho^2)s}{F_\rho M_\omega^2(s - M_\omega^2)}\right] \, , \end{aligned} \quad (5.38)$$

where we only keep the terms linear in ε. The quantities M_ω and M_ρ are complex and contain the corresponding widths.

The mixing is responsible for the typical distortion of the ρ–resonance (see Fig. 5.5), which originally would be a pure isospin $I = 1$ Breit-Wigner type resonance. The pion form factor (5.38) is the basic ansatz for the Gounaris-Sakurai formula [32] which is often used to represent experimental data by a phenomenological fit (see e.g. [16]).

However, theory in this case can do much more by exploiting systematically *analyticity*, *unitarity* and the properties of the *chiral limit*. A key point is that the phase of the pion form factor is determined by the $\pi\pi$–scattering phase shifts [103]. Known experimental $\pi\pi$–scattering data [104] together with progress in theory (combining two–loop CHPT and dispersion theory) lead to much more precise pion scattering lengths a_0^0 and a_0^2 [105, 106]. As a consequence, combining space–like data, $\pi\pi$–scattering phase shifts and time–like data one obtains severe theoretical constraints on the pion form factor $F_\pi(s)$ for $s \leq 2M_K$ [107, 108]. A similar approach has been used previously in [73, 81, 109].

To be more specific, the corresponding electromagnetic vector current form factor $F_\pi(s)$ has the following properties:

1) $F_\pi(s)$ is an analytic function of s in the whole complex s–plane, except for a cut on the positive real axis for $4m_\pi^2 \leq s < \infty$. If we approach the cut from above $s \to s + i\varepsilon$, $\varepsilon > 0$, $\varepsilon \to 0$ the form factor remains complex and is characterized by two real functions, the modulus and the phase

$$F_\pi(s) = |F_\pi(s)| \, e^{i\delta(s)} \, ; \quad (5.39)$$

2) analyticity relates $\mathrm{Re} F_\pi(s)$ and $\mathrm{Im} F_\pi(s)$ by a DR, which may be expressed as a relation between modulus and phase $\delta(s) = \arctan(\mathrm{Im} F_\pi(s)/\mathrm{Re} F_\pi(s))$, known as the Omnès representation [103]

$$F_\pi(s) = G_1(s) \, P(s) \, , \quad G_1(s) = \exp\left\{\frac{s}{\pi} \int_{4m_\pi^2}^\infty ds' \, \frac{\delta(s')}{s'(s' - s)}\right\} , \quad (5.40)$$

where $P(s)$ is a polynomial, which determines the behavior at infinity, or, equivalently, the number and position of the zeros;

3) charge conservation $F_\pi(0) = 1$, which fixes $P(0) = 1$;
4) $F_\pi(s)$ is real below the 2 pion threshold ($-\infty < s < 4m_\pi^2$), which implies that $P(s)$ must be a polynomial with real coefficients;
5) the inelastic threshold is $s_{\text{in}} = 16m_\pi^2$;
6) finally, we have to take into account the isospin breaking by another factor which accounts for the $I = 0$ contribution:

$$P(s) \to G_\omega(s) \cdot G_2(s) , \tag{5.41}$$

where $G_\omega(s)$ accounts for the ω–pole contribution due to $\rho - \omega$–mixing with mixing amplitude ε:

$$G_\omega(s) = 1 + \varepsilon \frac{s}{s_\omega - s} + \ldots \qquad s_\omega = (M_\omega - \frac{1}{2}i\Gamma_\omega)^2 . \tag{5.42}$$

In order to get it real below the physical thresholds we use an energy dependent width

$$\Gamma_\omega \to \Gamma_\omega(s) = \sum_X \Gamma(\omega \to X, s) = \frac{s}{M_\omega^2} \Gamma_\omega \left\{ \sum_X Br(\omega \to X) \frac{F_X(s)}{F_X(M_\omega^2)} \right\}, \tag{5.43}$$

where $Br(V \to X)$ denotes the branching fraction for the channel $X = 3\pi, \pi^0\gamma, 2\pi$ and $F_X(s)$ is the phase space function for the corresponding channel normalized such that $F_X(s) \to $ const for $s \to \infty$ [111].

The representation (5.40) tells us that once we know the phase on the cut and the location of the zeros of $G_2(s)$ the form factor is calculable in the entire s-plane. In the elastic region $s \le s_{\text{in}}$ Watson's theorem[17], exploiting unitarity, relates the phase of the form factor to the P wave phase shift of the $\pi\pi$ scattering amplitude with the same quantum numbers, $I = 1, J = 1$:

---

[17] The pion isovector form factor is defined by the matrix element (5.35). The $\pi^+\pi^-$ state in this matrix element, in order not to vanish, must be in a $I = 1, J = 1$ (P wave) state, J the angular momentum. If we look at the charge density j_0, time-reversal (T) invariance tells us that

$$\langle \text{out } \pi^+\pi^- | j_0(0) | 0 \rangle = \langle \text{in } \pi^+\pi^- | j_0(0) | 0 \rangle^* , \tag{5.44}$$

as for fixed J only "in" and "out" get interchanged. The complex conjugation follows from the fact that T must be implemented by an anti–unitary transformation. Now, with S the unitary scattering operator, which transforms in and out scattering states according to $|X \text{ out}\rangle = S^+ |X \text{ in}\rangle$ (X the label of the state) we have (using (5.44))

$$\langle \text{out } \pi^+\pi^- | j_0(0) | 0 \rangle = \langle \text{in } \pi^+\pi^- | S j_0(0) | 0 \rangle = e^{2i\delta_{\pi\pi}} \langle \text{out } \pi^+\pi^- | j_0(0) | 0 \rangle^*$$

which implies $F_\pi(s) = e^{2i\delta_{\pi\pi}} F_\pi^*(s)$. As two pions below the inelastic thresholds may scatter elastically only, by unitarity the S–matrix must be a pure phase in this case. The factor 2 is a convention, $\delta_{\pi\pi}(s)$ is the $\pi\pi$–scattering phase shift.

5.2 Leading Hadronic Contribution to $(g-2)$ of the Muon

$$\left.\begin{array}{r}\delta(s) = \delta_1^1(s) \\ \eta_1(s) \equiv 1\end{array}\right\} \quad \text{for} \quad s \leq s_\text{in} = 16 m_\pi^2 \,, \tag{5.45}$$

where $\eta_1 = |F_\pi(s)|$ is the elasticity parameter. However, it is an experimental fact that the inelasticity is negligible until the quasi two–body channels $\omega\pi$, $a_1\pi$, are open, thus in practice one can take (5.45) as an excellent approximation up to about 1 GeV (while $\sqrt{s_\text{in}} \simeq 0.56$ GeV). Actually, the phase difference (5.45) satisfies the bound [110]

$$\sin^2(\delta(s) - \delta_1^1(s)) \leq \frac{1}{2}[1 - \sqrt{1-r^2(s)}] \,, \quad r(s) = \frac{\sigma_\text{non-}2\pi^{I=1}}{\sigma_{e^+e^- \to \pi^+\pi^-}} \,, \tag{5.46}$$

and $\eta_1 \leq (1-r)/(1+r)$, provided $r < 1$, which holds true below 1.13 GeV (below 1 GeV $r < 0.143 \pm 0.024$, or $\delta - \delta_1^1 \lesssim 6°$, strongly decreasing towards lower energies).

The $\pi\pi$ scattering phase shift is due to elastic re–scattering of the pions in the final state (*final state interaction*) as illustrated by Fig. 5.20 The $\pi\pi$ scattering phase shift has been studied recently in the framework of the Roy equations, also exploiting chiral symmetry [105]. As a result it turns out that $\delta_1^1(s)$ is constrained to a remarkable degree of accuracy up to about $E_0 = 0.8$ GeV (matching point). The behavior of $\delta_1^1(s)$ in the region below the matching point is controlled by three parameters: two S–wave scattering lengths a_0^0, a_0^2 and by the boundary value $\phi \equiv \delta_1^1(E_0)$. One may treat ϕ as a free parameter and rely on the very accurate predictions for a_0^0, a_0^2 from chiral perturbation theory. This information may be used to improve the accuracy of the pion form factor and thus to reduce the uncertainty of the hadronic contribution to the muon $g-2$.

The remaining function $G_2(s)$ represents the smooth background that contains the curvature generated by the remaining singularities. The 4π channel opens at $s = 16\, m_\pi^2$ but phase space strongly suppresses the strength of the corresponding branch point singularity of the form $(1 - s_\text{in}/s)^{9/2}$ – a significant inelasticity only manifests itself for $s > s_\text{in} = (M_\omega + m_\pi)^2$. The conformal mapping

$$z = \frac{\sqrt{s_\text{in} - s_1} - \sqrt{s_\text{in} - s}}{\sqrt{s_\text{in} - s_1} + \sqrt{s_\text{in} - s}} \tag{5.47}$$

maps the plane cut along $s > s_\text{in}$ onto the unit disk in the z–plane. It contains a free parameter s_1 - the value of s which gets maps into the origin. $G_2(s)$ may be approximated by a polynomial in z:

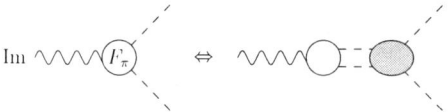

Fig. 5.20. Final state interaction due to $\pi\pi \to \pi\pi$ scattering

$$G_2(s) = 1 + \sum_{i=1}^{n_P} c_i \left(z^i - z_0^i\right), \qquad (5.48)$$

where z_0 is the image of $s = 0$. The shift of z by $z \to z - z_0$ is required to preserves the charge normalization condition $G_2(0) = 1$. The form of the branch point singularity $(1 - s_{\text{in}}/s)^{9/2}$ imposes four constraints on the polynomial; a non–trivial contribution from $G_2(s)$ thus requires a polynomial of fifth order at least. An important issue is the need for a normalization point at the upper end of the energy range under consideration $(M_\rho \cdots 2M_K)$. In fact, the present dispersion in the $\pi\pi$–data (see Fig. 5.5) makes it difficult to fully exploit this approach as it seems not possible to get a convincing simultaneous fit to the different data sets. Details have been worked out in [107, 108].

5.3 Higher Order Contributions

At order $O(\alpha^3)$ there are several classes of hadronic contributions with typical diagrams shown in Fig. 5.21. They have been estimated first in [69]. Classes (a) to (c) involve leading hadronic VP insertions and may be treated using DRs together with experimental e^+e^-–annihilation data. Class (d) involves leading QED corrections of the charged hadrons and related problems were discussed at the end of Sect. 5.2 on p. 291, already. The last class (e) is a new class of non–perturbative contributions, the *hadronic light–by–light scattering* which is constrained by experimental data only for one exceptional line of phase space. The evaluation of this contribution is particularly difficult and it will be discussed in the next section.

The $O(\alpha^3)$ hadronic contributions from classes (a), (b) and (c) may be evaluated without particular problems as described in the following.

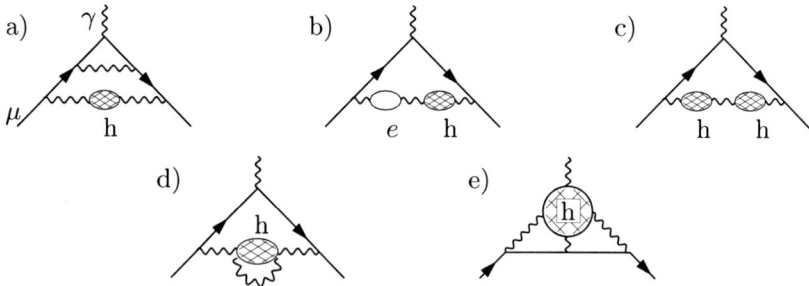

Fig. 5.21. Hadronic higher order contributions: a)–c) involving LO vacuum polarization, d) involving HO vacuum polarization and e) involving light-by-light scattering

5.3 Higher Order Contributions

At the three–loop level all diagrams of Fig. 4.3 which involve closed muon–loops are contributing to the hadronic corrections when at least one muon–loop is replaced by a quark–loop dressed by strong interactions mediated by virtual gluons.

Class (a) consists of a subset of 12 diagrams of Fig. 4.3: diagrams 7) to 18) plus 2 diagrams obtained from diagram 22) by replacing one muon–loop by a hadronic "bubble", and yields a contribution of the type

$$a_\mu^{(6)\,\mathrm{had}_{[(a)]}} = \left(\frac{\alpha}{\pi}\right)^3 \frac{2}{3} \int_{4m_\pi^2}^\infty \frac{ds}{s} R(s)\, K^{[(a)]}\left(s/m_\mu^2\right) \tag{5.49}$$

where $K^{[(a)]}(s/m_\mu^2)$ is a QED function which was obtained analytically by Barbieri and Remiddi [112]. The kernel function is the contribution to a_μ of the 14 two–loop diagrams obtained from diagrams 1) to 7) of Fig. 4.2 by replacing one of the two photons by a "heavy photon" of mass $\sqrt{s}$. The convolution (5.49) then provides the insertion of a photon self–energy part into the photon line represented by the "heavy photon" according to the method outlined in Sect. 3.8. Explicitly, the kernel is given by

$$\begin{aligned}
K^{[(a)]}(b) = & -\frac{139}{144} + \frac{115}{72} b + \left(\frac{19}{12} - \frac{7}{36} b + \frac{23}{144} b^2 + \frac{1}{b-4}\right) \ln b \\
& + \left(-\frac{4}{3} + \frac{127}{36} b - \frac{115}{72} b^2 + \frac{23}{144} b^3\right) \frac{\ln y}{\sqrt{b(b-4)}} \\
& + \left(\frac{9}{4} + \frac{5}{24} b - \frac{1}{2} b^2 - \frac{2}{b}\right) \zeta(2) + \frac{5}{96} b^2 \ln^2 b \\
& + \left(-\frac{1}{2} b + \frac{17}{24} b^2 - \frac{7}{48} b^3\right) \frac{\ln y}{\sqrt{b(b-4)}} \ln b \\
& + \left(\frac{19}{24} + \frac{53}{48} b - \frac{29}{96} b^2 - \frac{1}{3b} + \frac{2}{b-4}\right) \ln^2 y \\
& + \left(-2b + \frac{17}{6} b^2 - \frac{7}{12} b^3\right) \frac{D_p(b)}{\sqrt{b(b-4)}} \\
& + \left(\frac{13}{3} - \frac{7}{6} b + \frac{1}{4} b^2 - \frac{1}{6} b^3 - \frac{4}{b-4}\right) \frac{D_m(b)}{\sqrt{b(b-4)}} \\
& + \left(\frac{1}{2} - \frac{7}{6} b + \frac{1}{2} b^2\right) T(b) \tag{5.50}
\end{aligned}$$

where

$$y = \frac{\sqrt{b} - \sqrt{b-4}}{\sqrt{b} + \sqrt{b-4}}$$

and

$$D_p(b) = \text{Li}_2(y) + \ln y \ln(1-y) - \frac{1}{4}\ln^2 y - \zeta(2),$$

$$D_m(b) = \text{Li}_2(-y) + \frac{1}{4}\ln^2 y + \frac{1}{2}\zeta(2),$$

$$T(b) = -6\,\text{Li}_3(y) - 3\,\text{Li}_3(-y) + \ln^2 y \ln(1-y)$$
$$+ \frac{1}{2}\left(\ln^2 y + 6\,\zeta(2)\right)\ln(1+y) + 2\ln y\,(\text{Li}_2(-y) + 2\text{Li}_2(y))\,.$$

Again $\text{Li}_2(y) = -\int_0^y \frac{dt}{t}\ln(1-y)$ is the dilogarithm and $\text{Li}_3(y) = \int_0^y \frac{dt}{t}\text{Li}_2(t)$ the trilogarithm defined earlier in (3.38). Limiting cases are

$$K^{[(a)]}(0) = \frac{197}{144} + \frac{1}{2}\zeta(2) - 3\zeta(2)\ln 2 + \frac{3}{4}\zeta(3)$$

$$K^{[(a)]}_\infty(b) \stackrel{b\to\infty}{=} -\frac{1}{b}\left(\frac{23}{36}\ln b + 2\zeta(2) - \frac{223}{54}\right).$$

For the subclass which corresponds to the leading hadronic VP graph Fig. 5.1 decorated in all possible ways with an additional virtual photon the result reads

$$\Delta K^{[(a)]}(b) = \frac{35}{36} + \frac{8}{9}b + \left(\frac{4}{3} - \frac{1}{9}b - \frac{5}{18}b^2\right)\ln b$$
$$+ \left(-\frac{4}{3} + \frac{19}{9}b + \frac{4}{9}b^2 - \frac{15}{8}b^3\right)\frac{\ln y}{\sqrt{b(b-4)}}$$
$$+ \left(1 + \frac{1}{3}b - \frac{1}{6}b^2 - \frac{2}{b}\right)\zeta(2) + \left(\frac{1}{2} + \frac{1}{6}b - \frac{1}{12}b^2 - \frac{1}{3b}\right)\ln^2 y$$
$$+ \left(\frac{16}{3} - \frac{4}{3}b - \frac{4}{3}b^2 + \frac{1}{3}b^3\right)\frac{D_m(b)}{\sqrt{b(b-4)}} \tag{5.51}$$

Krause [113] has given an expansion up to fourth order which reads

$$K^{[(a)]}(s/m^2) = \frac{m^2}{s}\left\{\left[\frac{223}{54} - 2\zeta(2) - \frac{23}{36}\ln\frac{s}{m^2}\right]\right. \tag{5.52}$$
$$+ \frac{m^2}{s}\left[\frac{8785}{1152} - \frac{37}{8}\zeta(2) - \frac{367}{216}\ln\frac{s}{m^2} + \frac{19}{144}\ln^2\frac{s}{m^2}\right]$$
$$+ \frac{m^4}{s^2}\left[\frac{13072841}{432000} - \frac{883}{40}\zeta(2) - \frac{10079}{3600}\ln\frac{s}{m^2} + \frac{141}{80}\ln^2\frac{s}{m^2}\right]$$
$$\left. + \frac{m^6}{s^3}\left[\frac{2034703}{16000} - \frac{3903}{40}\zeta(2) - \frac{6517}{1800}\ln\frac{s}{m^2} + \frac{961}{80}\ln^2\frac{s}{m^2}\right]\right\}.$$

Here m is the mass of the external lepton $m = m_\mu$ in our case. The expanded approximation is more practical for the evaluation of the dispersion integral, because it is numerically more stable in general.

5.3 Higher Order Contributions

Class (b) consists of 2 diagrams only, obtained from diagram 22) of Fig. 4.3, and one may write this contribution in the form

$$a_\mu^{(6)\,\mathrm{had}_{[(b)]}} = \left(\frac{\alpha}{\pi}\right)^3 \frac{2}{3} \int_{4m_\pi^2}^\infty \frac{ds}{s} R(s)\, K^{[(b)]}(s/m_\mu^2) \tag{5.53}$$

with

$$K^{[(b)]}(s/m_\mu^2) = \int_0^1 dx\, \frac{x^2(1-x)}{x^2+(1-x)s/m_\mu^2}\left[-\hat{\Pi}_\gamma^{\prime\,e}\left(-\frac{x^2}{1-x}\frac{m_\mu^2}{m_e^2}\right)\right]$$

where we have set $\Pi' = \frac{\alpha}{\pi}\hat{\Pi}'$. Using (2.174) with $z = -\frac{x^2}{1-x}\frac{m_\mu^2}{m_e^2}$,

$$\hat{\Pi}_\gamma^{\prime\,e}(z) = -2\int_0^1 dy\, y(1-y)\ln(1-z\,y(1-y)) = \frac{8}{9} - \frac{\beta^2}{3} + \left(\frac{1}{2} - \frac{\beta^2}{6}\right)\beta\ln\frac{\beta-1}{\beta+1}$$

with $\beta = \sqrt{1 + 4\frac{1-x}{x^2}\frac{m_e^2}{m_\mu^2}}$.

Here the kernel function is the contribution to a_μ of the 2 two–loop diagrams obtained from diagrams 8) of Fig. 4.2 by replacing one of the two photons by a "heavy photon" of mass $\sqrt{s}$.

In diagram b) $m_f^2/m^2 = (m_e/m_\mu)^2$ is very small and one may expand β in terms of this small parameter. The x–integration afterwards may be performed analytically. Up to terms of order $\mathcal{O}(\frac{m_f^2}{m^2})$ the result reads [113]

$$K^{[(b)]}(s) = -\left(\frac{5}{9} + \frac{1}{3}\ln\frac{m_f^2}{m^2}\right) \times \left\{\frac{1}{2} - (x_1 + x_2)\right. \tag{5.54}$$

$$+ \frac{1}{x_1 - x_2}\left[x_1^2(x_1-1)\ln\left(\frac{-x_1}{1-x_1}\right) - x_2^2(x_2-1)\ln\left(\frac{-x_2}{1-x_2}\right)\right]\bigg\}$$

$$-\frac{5}{12} + \frac{1}{3}(x_1+x_2) + \frac{1}{3(x_1-x_2)}\left\{x_1^2(1-x_1)\left[\mathrm{Li}_2\left(\frac{1}{x_1}\right) - \frac{1}{2}\ln^2\left(\frac{-x_1}{1-x_1}\right)\right]\right.$$

$$\left.-x_2^2(1-x_2)\left[\mathrm{Li}_2\left(\frac{1}{x_2}\right) - \frac{1}{2}\ln^2\left(\frac{-x_2}{1-x_2}\right)\right]\right\},$$

with $x_{1,2} = \frac{1}{2}(b \pm \sqrt{b^2 - 4b})$ and $b = s/m^2$. The expansion to fifth order is given by

$$K^{[(b)]}(s) = \frac{m^2}{s}\left\{\left(-\frac{1}{18} + \frac{1}{9}\ln\frac{m^2}{m_f^2}\right)\right.$$

$$+ \frac{m^2}{s}\left(-\frac{55}{48} + \frac{\pi^2}{18} + \frac{5}{9}\ln\frac{s}{m_f^2} + \frac{5}{36}\ln\frac{m^2}{m_f^2} - \frac{1}{6}\ln^2\frac{s}{m_f^2} + \frac{1}{6}\ln^2\frac{m^2}{m_f^2}\right)$$

$$+ \frac{m^4}{s^2}\left(-\frac{11299}{1800} + \frac{\pi^2}{3} + \frac{10}{3}\ln\frac{s}{m_f^2} - \frac{1}{10}\ln\frac{m^2}{m_f^2} - \ln^2\frac{s}{m_f^2} + \ln^2\frac{m^2}{m_f^2}\right)$$

$$-\frac{m^6}{s^3}\left(\frac{6419}{225} - \frac{14}{9}\pi^2 + \frac{76}{45}\ln\frac{m^2}{m_f^2} - \frac{14}{3}\ln^2\frac{m^2}{m_f^2} - \frac{140}{9}\ln\frac{s}{m_f^2} + \frac{14}{3}\ln^2\frac{s}{m_f^2}\right)$$

$$-\frac{m^8}{s^4}\left(\frac{53350}{441} - \frac{20}{3}\pi^2 + \frac{592}{63}\ln\frac{m^2}{m_f^2} - 20\ln^2\frac{m^2}{m_f^2} - \frac{200}{3}\ln\frac{s}{m_f^2} + 20\ln^2\frac{s}{m_f^2}\right)\Big\}$$

$$+\frac{m_f^2}{m^2}\left[\frac{m^2}{s} - \frac{2\,m^4}{3\,s^2} - \frac{m^6}{s^3}\left(-2\ln\frac{s}{m^2} + \frac{25}{6}\right) - \frac{m^8}{s^4}\left(-12\ln\frac{s}{m^2} + \frac{97}{5}\right)\right.$$

$$\left. - \frac{m^{10}}{s^5}\left(-56\ln\frac{s}{m^2} + \frac{416}{5}\right)\right]\,. \tag{5.55}$$

Class (c) includes the double hadronic VP insertion, which is given by

$$a_\mu^{(6)\,\mathrm{had}_{[(c)]}} = \left(\frac{\alpha}{\pi}\right)^3 \frac{1}{9}\int_{4m_\pi^2}^{\infty}\frac{ds\,ds'}{s\,s'} R(s)\,R(s')\,K^{[(c)]}(s,s') \tag{5.56}$$

where

$$K^{[(c)]}(s,s') = \int_0^1 dx\, \frac{x^4\,(1-x)}{[x^2 + (1-x)\,s/m_\mu^2][x^2 + (1-x)\,s'/m_\mu^2]}\,.$$

This integral may be performed analytically. Setting $b = s/m^2$ and $c = s'/m^2$ one obtains for $b \neq c$

$$K^{[(c)]}(s,s') = \frac{1}{2} - b - c - \frac{(2-b)\,b^2\,\ln(b)}{2\,(b-c)} - \frac{b^2\,(2 - 4b + b^2)\,\ln\!\left(\frac{b+\sqrt{-(4-b)\,b}}{b-\sqrt{-(4-b)\,b}}\right)}{2\,(b-c)\,\sqrt{-(4-b)\,b}}$$

$$- \frac{(-2+c)\,c^2\,\ln(c)}{2\,(b-c)} + \frac{c^2\,(2 - 4c + c^2)\,\ln\!\left(\frac{c+\sqrt{-(4-c)\,c}}{c-\sqrt{-(4-c)\,c}}\right)}{2\,(b-c)\,\sqrt{-(4-c)\,c}}\,, \tag{5.57}$$

and for $b = c$

$$K^{[(c)]}(s,s') = \frac{1}{2} - 2c + \frac{c}{2}\left(-2 + c - 4\ln(c) + 3c\ln(c)\right) + \frac{c\,(-2 + 4c - c^2)}{2(-4+c)}$$

$$+ \frac{c\,(12 - 42c + 22c^2 - 3c^3)\,\ln\!\left(\frac{c+\sqrt{(-4+c)\,c}}{c-\sqrt{(-4+c)\,c}}\right)}{2\,(-4+c)\,\sqrt{(-4+c)\,c}}\,. \tag{5.58}$$

Class (d) exhibits 3 diagrams (diagrams 19 to 21) of Fig. 4.3 and corresponds to the leading hadronic contribution with $R(s)$ corrected for final state radiation. We thus may write this correction by replacing

$$R(s) \to R(s)\,\eta(s)\,\frac{\alpha}{\pi} \tag{5.59}$$

in the basic integral (5.18). This correction is particularly important for the dominating two pion channel for which $\eta(s)$ may be calculated in scalar QED (treating the pions as point–like particles) [114, 115] and the result reads

$$\eta(s) = \frac{1+\beta_\pi^2}{\beta_\pi}\left\{4\mathrm{Li}_2\left(\frac{1-\beta_\pi}{1+\beta_\pi}\right) + 2\mathrm{Li}_2\left(-\frac{1-\beta_\pi}{1+\beta_\pi}\right)\right.$$
$$\left. -3\log\left(\frac{2}{1+\beta_\pi}\right)\log\left(\frac{1+\beta_\pi}{1-\beta_\pi}\right) - 2\log(\beta_\pi)\log\left(\frac{1+\beta_\pi}{1-\beta_\pi}\right)\right\}$$
$$-3\log\left(\frac{4}{1-\beta_\pi^2}\right) - 4\log(\beta_\pi)$$
$$+\frac{1}{\beta_\pi^3}\left[\frac{5}{4}(1+\beta_\pi^2)^2 - 2\right]\log\left(\frac{1+\beta_\pi}{1-\beta_\pi}\right) + \frac{3}{2}\frac{1+\beta_\pi^2}{\beta_\pi^2} \qquad (5.60)$$

and provides a good measure for the dependence of the observables on the pion mass. Neglecting the pion mass is obviously equivalent to taking the high energy limit

$$\eta(s \to \infty) = 3 \;.$$

In Fig. 5.22 the correction $\eta(s)$ is plotted as a function of the center of mass energy. It can be realized that for energies below 1 GeV the pion mass leads to a considerable enhancement of the FSR corrections. Regarding the desired precision, ignoring the pion mass would therefore lead to wrong results. Close to threshold for pion pair production ($s \simeq 4m_\pi^2$) the Coulomb forces between the two final state pions play an important role. In this limit the factor $\eta(s)$ becomes singular $[\eta(s) \to \pi^2/2\beta_\pi]$ which means that the $O(\alpha)$ result for the FS correction cannot be trusted anymore. Since these singularities are known to all orders of perturbation theory one can resum these contributions, which leads to an exponentiation [114]:

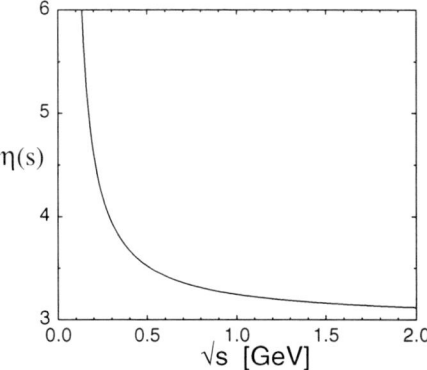

Fig. 5.22. The FSR correction factor $\eta(s)$ as a function of the center of mass energy $\sqrt{s}$

Table 5.5. Higher order contributions from diagrams a)–c) (in units 10^{-11})

$a_\mu^{(2a)}$	$a_\mu^{(2b)}$	$a_\mu^{(2c)}$	$a_\mu^{had(2)}$	Ref.
−199(4)	107(3)	2.3(0.6)	−90(5)	[72]
−211(5)	107(2)	2.7(0.1)	−101(6)	[113]
−209(4)	106(2)	2.7(1.0)	−100(5)	[10]
−207.3(1.9)	106.0(0.9)	3.4(0.1)	−98(1)	[80]
−207.5(2.0)	104.2(0.9)	3.0(0.1)	−100.3(2.2)	[6]

$$R^{(\gamma)}(s) = R(s)\left(1 + \eta(s)\frac{\alpha}{\pi} - \frac{\pi\alpha}{2\beta_\pi}\right)\frac{\pi\alpha}{\beta_\pi} \times \left[1 - \exp\left(-\frac{\pi\alpha}{\beta_\pi}\right)\right]^{-1} \quad (5.61)$$

Above a center of mass energy of $\sqrt{s} = 0.3$ GeV the exponentiated correction to the Born cross–section deviates from the non–exponentiated correction less than 1 %. The corresponding $O(\alpha)$ sQED contribution to the anomalous magnetic moment of the muon is

$$\delta^\gamma a_\mu^{had} = (38.6 \pm 1.0) \times 10^{-11}, \quad (5.62)$$

where we added a guesstimated error which of course is not the true model error, the latter remaining unknown[18]. In the inclusive region above typically 2 GeV, the FRS corrections are well represented by the inclusive photon emission from quarks. However, since in inclusive measurements experiments commonly do not subtract FSR, the latter is included already in the data and no additional contribution has to be taken into account. In more recent analyses this contribution is usually included in the leading hadronic contribution (5.23) as the $\pi^+\pi^-\gamma$ channel (see Table 5.3).

Results obtained by different groups, for so far unaccounted higher order vacuum polarization effects, are collected in Table 5.5. We will adopt the estimate

$$a_\mu^{had(2)} = (-100.3 \pm 2.2) \times 10^{-11} \quad (5.63)$$

obtained with the compilation [6].

5.4 Hadronic Light–by–Light Scattering

In perturbation theory hadronic light–by–light scattering diagrams are like leptonic ones with leptons replaced by quarks which, however, exhibit strong interactions via gluons, which at low energies lead to a breakdown of perturbation theory.

---

[18] One could expect that due to $\gamma - \rho^0$ mixing (VMD type models [116], see below) the sQED contribution gets substantially reduced. However, due to the low scales $\sim m_\mu, m_\pi$ involved, here, in relation to M_ρ the photons essentially behave classically in this case. Also, the bulk of the VP contribution at these low scales comes from

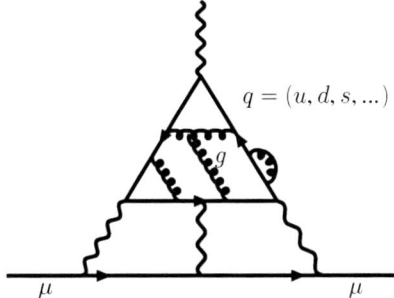

Nevertheless, it is instructive to ask what quark–loop contributions would look like, if strong interactions would be weak or turned off. Quark loops, of course, play a role in estimating the S.D. effects above a certain energy scale. In fact, we may check which energy scales contribute relevant to the LbL integrals in case of a muon loop and cutting off high energy contributions by a cut–off Λ. Typically, one obtains

Λ [GeV]	0.5	0.7	1.0	2.0
$a_\mu \times 10^{10}$	24	26	38	45

which illustrates that even for the muon the LbL contributions are rather sensitive to contributions from unexpectedly high scales. Only when the cut-off exceeds about 2 GeV the correct result $a_\mu^{(6)}(\text{lbl}, \mu) \simeq 46.50 \times 10^{-10}$ is well approximated. A constituent quark loop would yield the results summarized in Table 5.6[19]. For the light quarks the numerical results are certainly more trustable while for the heavier quarks, like the c, the asymptotic expansion (4.11) becomes more reliable (see [72]; results taken from TABLE I of [72])[20].

Table 5.6. CQM estimates of $a_\mu^{(6)}(\text{lbl}, q) \times 10^{11}$

0.3 GeV lepton	[ud]	s	c	[uds]	[udsc]	method
79.0	49.7	1.1	2.1	50.8	52.9	[72] numerical
81.0	51.0	1.2	2.2	52.1	54.4	(4.11)

the neutral ρ^0–exchange Fig. 5.19, which does not directly produce FSR, the latter thus being due to the dissociated charged $\pi^+\pi^-$ intermediate state as assumed in sQED. In fact the main contribution comes from very low energies (Fig. 5.22).

[19] The constituent quark model (CQM) result quoted in [117], including u, d, s and c quarks, reads $a_\mu^{(6)}(\text{lbl}, u, d, s, c)_{\text{CQM}} \simeq 62(3) \times 10^{-11}$.

[20] In the free quark model (parton model) with current quark masses given in (3.36) one would get $a_\mu^{(6)}(\text{lbl}, u+d) = 8229.34 \times 10^{-11}$ and $a_\mu^{(6)}(\text{lbl}, s) = 17.22 \times 10^{-11}$ by adapting color, charge and mass in (4.9) and (3.48), respectively. Apart from the

312   5 Hadronic Effects

Certainly, quark loops are far from accounting for the bulk of the hadronic LbL–effects. Actually, it is the spontaneous breakdown of the nearby chiral symmetry of QCD, an intrinsically non–perturbative phenomenon, which shapes the leading hadronic effects to be evaluated. While the non–perturbative effects which show up in the hadronic vacuum polarization may be reliably evaluated in terms of measured hadronic cross–sections $\sigma_{\text{tot}}(e^+e^- \to \gamma^* \to$ hadrons), which allows us to obtain the full photon propagator $\langle 0|T\{A^\mu(x_1)A^\nu(x_2)\}|0\rangle$, for the light–by–light scattering Green function $\langle 0|T\{A^\mu(x_1)A^\nu(x_2)A^\rho(x_3)A^\sigma(x_4)\}|0\rangle$ we have little direct experimental information when photons are off–shell. In the contribution to $g-2$ we need the light–by–light scattering amplitude with one photon real ($k^2 = 0$), or more precisely, its first derivative $\partial/\partial k^\mu$ evaluated at $k^\mu = 0$, equivalent to $E_\gamma \to 0$. But, the other three momenta are off–shell and to be integrated over the full phase space of the two remaining independent four–vectors. Unfortunately, the object in question cannot be calculated from first principles at present. Perturbation theory fails, chiral perturbation theory is limited to the low energy tail only and for lattice QCD there is a long way to go until such objects can be calculated with the required precision. One thus has to resort to models which are inspired by known properties of QCD as well as known phenomenological facts. One fact we already know from the hadronic VP discussion, the ρ meson is expected to play an important role in the game. It looks natural to apply a vector–meson dominance (VMD) like model. Electromagnetic interactions of pions treated as point–particles would be described by scalar QED, as a first step in the sense of a low energy expansion. Note that in photon–hadron interactions the photon mixes with hadronic vector–mesons like the ρ^0. The naive VMD model attempts to take into account this hadronic dressing by replacing the photon propagator as

$$\frac{\mathrm{i}\,g^{\mu\nu}}{q^2} + \cdots \to \frac{\mathrm{i}\,g^{\mu\nu}}{q^2} + \cdots - \frac{\mathrm{i}\left(g^{\mu\nu} - \frac{q^\mu q^\nu}{q^2}\right)}{q^2 - m_\rho^2} = \frac{\mathrm{i}\,g^{\mu\nu}}{q^2}\frac{m_\rho^2}{m_\rho^2 - q^2} + \cdots, \quad (5.64)$$

where the ellipses stand for the gauge terms. Of course real photons $q^2 \to 0$ in any case remain undressed and the dressing would go away for $m_\rho^2 \to \infty$. The main effect is that it provides a damping at high energies with the ρ mass as an effective cut–off (physical version of a Pauli-Villars cut–off). However, the naive VMD model does not respect chiral symmetry properties.

A way to incorporate vector–mesons $\rho, \omega, \phi, \ldots$ in accordance with the basic symmetries of QCD is the Resonance Lagrangian Approach (RLA) [94, 95], an extended version of CHPT (see p. 238) which also implements VMD in a consistent manner. Alternative versions of the RLA are the Hidden Local

---

fact that pQCD makes no sense here, one should note that results are very sensitive to the precise definition of the quark masses. Also note that the chiral limit $m_q \to 0$ of (4.9) [with $m_e \to m_q$ ($q = u, d, s$)] is IR singular. This also demonstrates the IR sensitivity of the LbL scattering contribution.

Gauge Symmetry[21] (HLS) [118] or massive Yang-Mills [119] models and the Extended Nambu-Jona-Lasinio (ENJL) [120] model. They are basically equivalent [95, 119, 121] in the context of our application.

A new quality of the problem encountered here is the fact that the integrand depends on 3 invariants q_1^2, q_2^2, q_3^2, where $q_3 = -(q_1+q_2)$. In contrast the hadronic VP correlator, or the VVA triangle with an external zero momentum vertex, only depends on a single invariant q^2. In the latter case, the invariant amplitudes (form factors) may be separated into a low energy part $q^2 \leq \Lambda^2$ (soft) where the low energy effective description applies and a high energy part $q^2 > \Lambda^2$ (hard) where pQCD works. In multi–scale problems, however, there are mixed soft–hard regions (see Fig. 5.23), where no answer is available in general, unless we have data to constrain the amplitudes in such regions. In our case, only the soft region $q_1^2, q_2^2, q_3^2 \leq \Lambda^2$ and the hard region $q_1^2, q_2^2, q_3^2 > \Lambda^2$ are under control of either the low energy EFT and of pQCD, respectively. In the other domains operator product expansions and/or soft versus hard factorization "theorems" à la Brodsky-Farrar [122] may be applied.

Another problem of the RLA is that the low energy effective theory is non–renormalizable and thus has unphysical UV behavior, while QCD is renormalizable and has the correct UV behavior (but unphysical IR behavior). As a consequence of the mismatch of the functional dependence on the cut–off, one cannot match the two pieces in a satisfactory manner and one obtains a cut–off dependent prediction. Unfortunately, the cut–off dependence of the sum is not small even if one varies the cut–off only within "reasonable" boundaries around about 1 or 2 GeV, say. Of course the resulting uncertainty just reflects the model dependence and so to say parametrizes our ignorance. An estimate of the real model dependence is difficult as long as we are not knowing the true solution of the problem. In CHPT and its extensions, the low energy constants parametrizing the effective Lagrangian are accounting for the appropriate S.D. behavior, usually. Some groups however prefer an alternative approach based

---

[21] In this approach the vector part $SU(2)_V$ of the global chiral group $SU(2)_L \otimes SU(2)_R$, realized as a non–linear σ model for the pions (see (4.57)), is promoted to a local symmetry and the ρ–mesons become the corresponding gauge vector bosons, as they do in the massive YM approach. Together with the electromagnetic $U(1)_Q$ local group one obtains the symmetry pattern: $[SU(2)_L \otimes SU(2)_R/SU(2)_V]_{\text{global}} \otimes [SU(2)_V]_{\text{hidden}} \otimes U(1)_Q$, where the local group is broken by the Higgs mechanism to $U(1)_{\text{em}}$, with $Q_{\text{em}} = Q + T_3^{\text{hidden}}$, essentially as in the electroweak SM. Unlike in the massive Yang-Mills (YM) ansatz the gauge bosons here are considered as collective fields ($V^\mu = \bar{q}\gamma^\mu q$ etc.) as in the ENJL model. The generalization to $SU(3)$ is obvious. Similar to the pseudoscalar field $\phi(x)$ (4.56), the $SU(3)$ gauge bosons conveniently may be written as a 3×3 matrix field

$$V_\mu(x) = \sum_i T_i V_{\mu i} = \begin{pmatrix} \frac{\rho^0}{\sqrt{2}} + \frac{\omega_8}{\sqrt{6}} & \rho^+ & K^{*+} \\ \rho^- & \frac{-\rho^0}{\sqrt{2}} + \frac{\omega_8}{\sqrt{6}} & K^{*0} \\ K^{*-} & \overline{K}^{*0} & -2\frac{\omega_8}{\sqrt{6}} \end{pmatrix}_\mu .$$

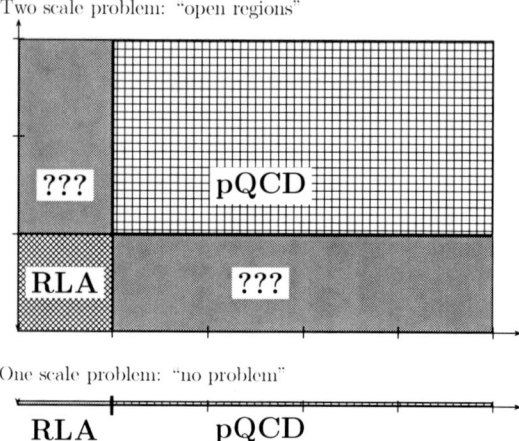

Fig. 5.23. Multi–scale strong interaction problems. For two and more scales some regions are neither modeled by low energy effective nor by perturbative QCD

on the fact that the weakly coupled large–N_c QCD, i.e. $SU(N_c)$ for $N_c \to \infty$ under the constraint $\alpha_s N_c$=constant, is theoretically better known than true QCD with $N_c = 3$. It is thus tempting to approximate QCD as an expansion in $1/N_c$ [123, 124, 125].

Of course, also applying a large–N_c expansion one has to respect the low energy properties of QCD as encoded by CHPT. In CHPT the effective Lagrangian has an overall factor N_c, while the U matrix, exhibiting the pseudoscalar fields, is N_c independent. Each additional meson field has a $1/F_\pi \propto 1/\sqrt{N_c}$. In the context of CHPT the $1/N_c$ expansion thus is equivalent to a semiclassical expansion. The chiral Lagrangian can be used at tree level, and loop effects are suppressed by powers of $1/N_c$.

5.4.1 Calculating the Hadronic LbL Contribution

Let us start now with a setup of what one has to calculate actually. The hadronic light–by–light scattering contribution to the electromagnetic vertex is represented by the diagram Fig. 5.24. According to the diagram, a complete discussion of the hadronic light–by–light contributions involves the full rank–four hadronic vacuum polarization tensor

$$\Pi_{\mu\nu\lambda\rho}(q_1, q_2, q_3) = \int d^4x_1 \, d^4x_2 \, d^4x_3 \, e^{i(q_1 x_1 + q_2 x_2 + q_3 x_3)}$$
$$\times \langle 0 | T\{j_\mu(x_1) j_\nu(x_2) j_\lambda(x_3) j_\rho(0)\} | 0 \rangle . \quad (5.65)$$

Momentum k of the external photon is incoming, while the q_i's of the virtual photons are outgoing from the hadronic "blob". Here $j_\mu(x)$ denotes the light quark part of the electromagnetic current

5.4 Hadronic Light–by–Light Scattering

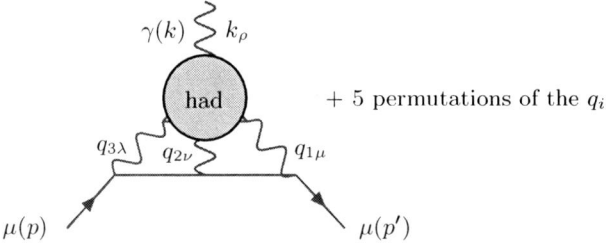

Fig. 5.24. Setup for the calculation of the hadronic contribution of the light-by-light scattering to the muon electromagnetic vertex

$$j_\mu(x) = \frac{2}{3}(\bar{u}\gamma_\mu u)(x) - \frac{1}{3}(\bar{d}\gamma_\mu d)(x) - \frac{1}{3}(\bar{s}\gamma_\mu s)(x) \equiv \bar{q}\hat{Q}\gamma_\mu q(x) \,. \quad (5.66)$$

It includes a summation over color of the color and flavor diagonal quark bilinears. Since the electromagnetic current $j_\mu(x)$ is conserved, the tensor $\Pi_{\mu\nu\lambda\rho}(q_1, q_2, q_3)$ satisfies the Ward-Takahashi identities

$$\{q_1^\mu; q_2^\nu; q_3^\lambda; k^\rho\}\Pi_{\mu\nu\lambda\rho}(q_1, q_2, q_3) = 0\,, \quad (5.67)$$

with $k = (q_1 + q_2 + q_3)$. This implies

$$\Pi_{\mu\nu\lambda\rho}(q_1, q_2, k - q_1 - q_2) = -k^\sigma (\partial/\partial k^\rho)\, \Pi_{\mu\nu\lambda\sigma}(q_1, q_2, k - q_1 - q_2)\,, \quad (5.68)$$

and thus tells us that the object of interest is linear in k when we go to the static limit $k^\mu \to 0$ in which the anomalous magnetic moment is defined.

Up to one-loop the electromagnetic $\bar{\ell}\ell\gamma$-vertex has been discussed in Sect. 2.6.3, its general structure in Sect. 3.3. Here we adopt the notation of Knecht and Nyffeler [126] ($q \to k$, $p_1 \to p$ and $p_2 \to p'$). From the diagram we easily read off the contribution of $\Pi_{\mu\nu\lambda\sigma}(q_1, q_2, q_3)$ to the electromagnetic vertex which is given by

$$\langle \mu^-(p')|(ie)j_\rho(0)|\mu^-(p)\rangle = (-ie)\,\bar{u}(p')\,\Pi_\rho(p',p)\,u(p)$$
$$= \int \frac{d^4 q_1}{(2\pi)^4} \frac{d^4 q_2}{(2\pi)^4} \frac{(-i)^3}{q_1^2 q_2^2 (q_1+q_2-k)^2} \frac{i}{(p'-q_1)^2 - m^2} \frac{i}{(p-q_1-q_2)^2 - m^2}$$
$$\times (-ie)^3\, \bar{u}(p')\,\gamma^\mu(\slashed{p}'-\slashed{q}_1+m)\,\gamma^\nu\,(\slashed{p}-\slashed{q}_1-\slashed{q}_2+m)\,\gamma^\lambda\,u(p)$$
$$\times (ie)^4 \Pi_{\mu\nu\lambda\rho}(q_1, q_2, k-q_1-q_2)\,, \quad (5.69)$$

with $k_\mu = (p' - p)_\mu$. For the contribution to the form factors

$$\bar{u}(p')\,\Pi_\rho(p',p)\,u(p) = \bar{u}(p')\left[\gamma_\rho F_E(k^2) + i\frac{\sigma_{\rho\tau}k^\tau}{2m_\mu}F_M(k^2)\right]u(p)\,, \quad (5.70)$$

(5.68) implies that $\Pi_\rho(p',p) = k^\sigma \Pi_{\rho\sigma}(p',p)$ with

$$\bar{u}(p')\,\Pi_{\rho\sigma}(p',p)\,u(p) = -\mathrm{i}e^6 \times$$

$$\int \frac{\mathrm{d}^4 q_1}{(2\pi)^4}\frac{\mathrm{d}^4 q_2}{(2\pi)^4} \frac{1}{q_1^2\,q_2^2\,(q_1+q_2-k)^2}\frac{1}{(p'-q_1)^2-m^2}\frac{1}{(p-q_1-q_2)^2-m^2}$$

$$\times\ \bar{u}(p')\gamma^\mu\,(\slashed{p}'-\slashed{q}_1+m)\,\gamma^\nu\,(\slashed{p}-\slashed{q}_1-\slashed{q}_2+m)\,\gamma^\lambda\,u(p)$$

$$\times\ \frac{\partial}{\partial k^\rho}\,\Pi_{\mu\nu\lambda\sigma}(q_1,q_2,k-q_1-q_2)\,.$$

The WT–identity takes the form $k^\rho k^\sigma \bar{u}(p')\Pi_{\rho\sigma}(p',p)\,u(p) = 0$, which implies $\delta^{\mathrm{lbl}}F_{\mathrm{E}}(0) = 0$ and, in the terminology introduced at the end of Sect. 3.5, we have $V_\rho(p) = \Pi_\rho(p',p)|_{k=0} = 0$ and $T_{\rho\sigma}(p) = \Pi_{\rho\sigma}(p',p)|_{k=0}$. Thus, using the projection technique outlined in Sect. 3.5, the hadronic light–by–light contribution to the muon anomalous magnetic moment is equal to

$$F_{\mathrm{M}}(0) = \frac{1}{48m}\,\mathrm{Tr}\,\{(\slashed{p}+m)[\gamma^\rho,\gamma^\sigma](\slashed{p}+m)\Pi_{\rho\sigma}(p,p)\}\,. \quad (5.71)$$

This is what we actually need to calculate. The integral to be performed is 8 dimensional. Thereof 3 integrations can be done analytically. In general, one has to deal with a 5 dimensional non–trivial integration over 3 angles and 2 moduli.

As mentioned before, the hadronic tensor $\Pi_{\mu\nu\lambda\sigma}(q_1,q_2,k-q_1-q_2)$ we have to deal with, is a problematic object, because it has an unexpectedly complex structure as we will see, in no way comparable with the leptonic counterpart. The general covariant decomposition involves 138 Lorentz structures of which 32 can contribute to $g-2$. Fortunately, this tensor is dominated by the pseudoscalar exchanges $\pi^0, \eta, \eta', \ldots$ (see Fig. 3.6), described by the WZW effective Lagrangian (4.60). This fact rises hope that a half–way reliable estimate should be possible. Generally, the perturbative QCD expansion only is useful to evaluate the short distance (S.D.) tail, while the dominant long distance (L.D.) part must be evaluated using some low energy effective model which includes the pseudoscalar Goldstone bosons as well as the vector mesons as shown in Fig. 5.25.

Note that, in spite of the fact that in pQCD our hadronic tensor $\Pi_{\mu\nu\lambda\sigma}(q_1,q_2,k-q_1-q_2)$ only involves parity conserving vector interactions (γ^μ–type), in full QCD the parity violating axial–vector interactions ($\gamma^\mu\gamma_5$–type) are ruling the game. Thereby the existence of the ABJ anomaly related via PCAC to the pseudoscalar states plays the key role. This connection may be illustrated as in Fig. 5.26[22].

5.4.2 Sketch on Hadronic Models

One way to "derive" the low energy structure of QCD starting from the QCD Lagrangian is to integrate out the S.D. part of the gluonic degrees

---

[22] Formally, a $\gamma_5^2 = 1$ appears inserted at one of the vertices and one of the γ_5's then anticommuted to one of the other vertices. The "quark–loop picture" is not kind of resummed pQCD, which does not know pions, rather an ENJL type diagram.

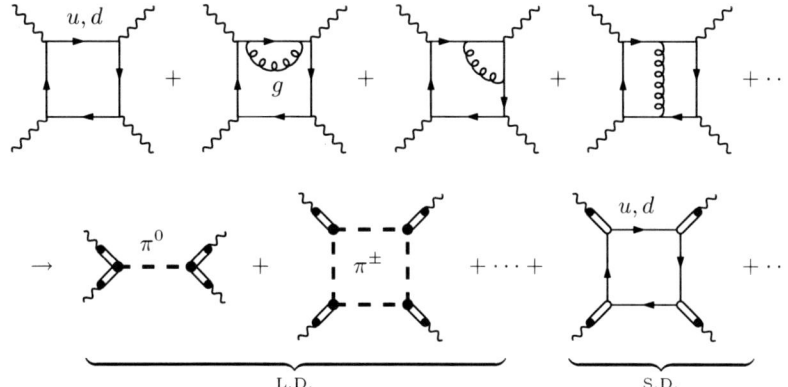

Fig. 5.25. Hadronic light–by–light scattering is dominated by π^0–exchange in the odd parity channel, pion loops etc. at long distances (L.D.) and quark loops including hard gluonic corrections at short distances (S.D.). The photons in the effective theory couple to hadrons via $\gamma - \rho^0$ mixing

of freedom, which implies effective four quark interactions and a model very similar to the Nambu-Jona-Lasinio (NJL) model [127] (compare also the linear σ-model [128]), however, with nucleons replaced by *constituent quarks*. Practically, this is done via the regulator replacement,

$$\frac{1}{Q^2} \to \int_0^{1/\Lambda^2} d\tau \, e^{-\tau Q^2} \;, \qquad (5.72)$$

in the gluon propagator and an expansion in $1/\Lambda^2$. In the leading $1/N_c$ limit this leads to the Lagrangian

$$\mathcal{L}_{\text{ENJL}} = \mathcal{L}_{\text{QCD}}^\Lambda + 2 g_S \sum_{i,j} \left(\bar{q}_R^i q_L^j\right)\left(\bar{q}_L^j q_R^i\right)$$

$$- g_V \sum_{i,j} \left[\left(\bar{q}_L^i \gamma^\mu q_L^j\right)\left(\bar{q}_L^j \gamma_\mu q_L^i\right) + (L \to R)\right] , \qquad (5.73)$$

defining the so called extended Nambu-Jona-Lasinio (ENJL) model (see [129] for a comprehensive review). Summation over colors between brackets in 5.73 is understood, i, j are flavor indices, $q_{R,L} \equiv (1/2)(1 \pm \gamma_5) q$ are the chiral

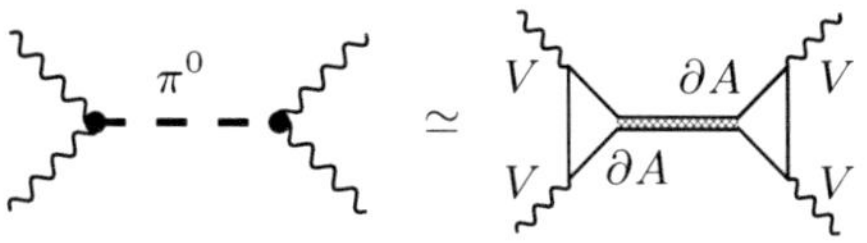

Fig. 5.26. Hadronic degrees of freedom (effective theories) versus quark gluon picture (QCD); example π^0 exchange

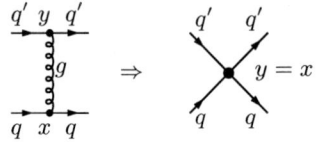

quark fields and

$$g_V \equiv \frac{8\pi^2 G_V(\Lambda)}{N_c \Lambda^2} \quad , \quad g_S \equiv \frac{4\pi^2 G_S(\Lambda)}{N_c \Lambda^2} \quad (5.74)$$

are Fermi type coupling parameters. The couplings $G_S(\Lambda)$ and $G_V(\Lambda)$ are dimensionless and $\mathcal{O}(1)$ in the $1/N_c$ expansion and to leading order the constraint

$$G_S = 4G_V = \frac{\alpha_s}{\pi} N_c \, , \text{ i.e. } \alpha_s = O(1/N_c) \quad (5.75)$$

should be satisfied at scales where pQCD applies. The ENJL model exhibits the same symmetry pattern, the spontaneously broken chiral symmetry which is inferring the existence of non–vanishing quark condensates ($\langle \bar{u}u \rangle, \langle \bar{d}d \rangle, \langle \bar{s}s \rangle \neq 0$) and of the Goldstone modes, the pions ($\pi^0, \pi^\pm$), the η and the kaons in the $SU(3)$ (u, d, s) quark sector. The Lagrangian $\mathcal{L}_{\text{QCD}}^\Lambda$ includes only low frequency (less than Λ) modes of quark and gluon fields.

In the ENJL model quarks get dressed constituent quarks in place of the much lighter current quarks which appear in the QCD Lagrangian. The constituent quark masses are obtained as a solution of the gap equation Fig. 5.27[23] and typically take values (4.39) for $\Lambda \simeq 1.16$ GeV, depending *on the cut–off* (phenomenological adjustment).
Constituent quark–antiquark pair correlators $\langle (\bar{q}\Gamma_i q')(x) (\bar{q}\Gamma_j q')(y) \rangle$[24] via iterated four–fermion interactions as illustrated in Fig. 5.28 form meson propagators such that one obtains the Fig. 5.29 type of ENJL diagrams which implies a VMD like dressing between mesons, quarks and the virtual photons. It should be clear that the ENJL model does not allow to make predictions from first principles, since although it is "derived" from QCD by "integrating

---

[23]The quark propagator in the ENJL model to leading order in $1/N_c$ is obtained by Schwinger-Dyson resummation according to Fig. 5.27. There is no wave function renormalization to this order in $1/N_c$ and the mass can be self–consistently determined from the Schwinger-Dyson equation. To leading order in N_c, this leads to the condition

$$M_i = m_i - g_S \langle \bar{q}q \rangle_i \; ; \; \langle \bar{q}q \rangle_i \equiv \langle 0 | :\bar{q}_i q_i : | 0 \rangle \, ,$$

$$\langle \bar{q}q \rangle_i = -4 N_c M_i \int_\Lambda \frac{d^4 p}{(2\pi)^4} \frac{i}{p^2 - M_i^2} \, . \quad (5.76)$$

Here i denotes the quark flavor. The constituent quark mass M_i is independent of the momentum and only a function of G_S, Λ and the current mass m_i.

[24]The Γ_i's denotes a 4×4 matrix in spinor space (see (2.21)) times a 2×2 matrix in isospin space (Pauli matrices), which specifies the channel: spin, parity, isospin, charge etc.

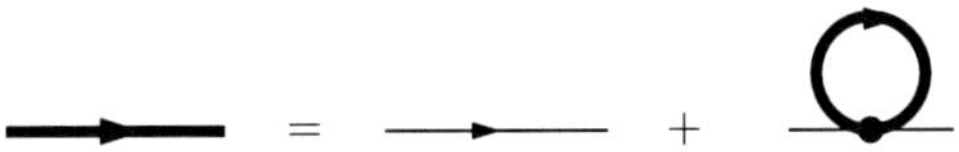

Fig. 5.27. Schwinger–Dyson equation for the inverse quark propagator (see Sect. 2.6.2 (2.179)), which at zero momentum leads to the gap equation 5.76. Free lines without endpoints denote inverse propagators; thick line: dressed or constituent quark; thin line: free current quark

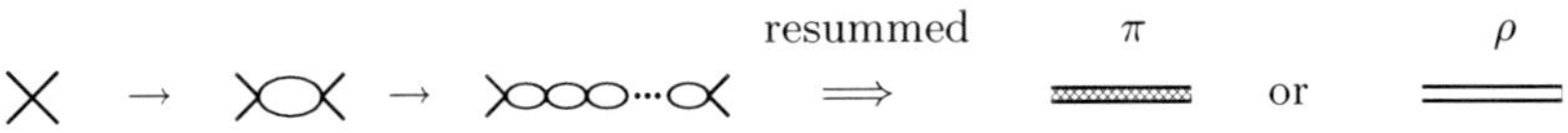

Fig. 5.28. ENJL meson propagators

out the gluons" in the functional integral such a derivation is not possible on a quantitative level, because the non–perturbative aspects are not under control with presently available methods. What emerges is a particular structure of an effective theory, sharing the correct low energy properties of QCD, with effective couplings and masses of particles to be taken from phenomenology.

In fact, in order to work with the model one has to go one step further and introduce the collective fields describing the hadrons, like the pseudo–scalars and the vector–mesons and this leads back to the RLA or HLS type of approaches where the meson fields are put in by hand from the very beginning, just using the symmetries and the symmetry breaking patterns to constrain the effective Lagrangian. However, this does not fix the Lagrangian completely. For example, a special feature of the HLS Lagrangian [118] is the absence of a $\rho^0\rho^0\pi^+\pi^-$ term, which is present in the extended chiral Lagrangian as well as in the VMD ansatz.

The spectrum of states, which eventually should be taken into account, together with the quantum numbers are given in the following Table 5.7. Nonet symmetry would correspond to states

$$\psi_8 = \frac{1}{\sqrt{6}}(u\bar{u} + d\bar{d} - 2s\bar{s})$$

$$\psi_1 = \frac{1}{\sqrt{3}}(u\bar{u} + d\bar{d} + s\bar{s}) ,$$

where ψ_1 is the ideal flavor singlet state. This symmetry is broken and the physical states are mixed through a rotation

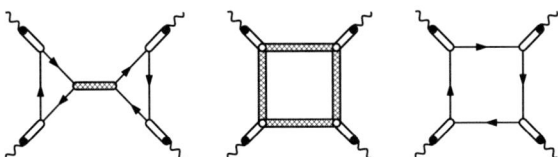

Fig. 5.29. ENJL model graphs: π^0–exchange, pion–loop and quark–loop dressed by $\rho - \gamma$ transitions

Table 5.7. Low lying mesons (hadrons) in the quark model [130]. States with a question mark are not yet (fully) established experimentally

$n^{2s+1}\ell_J$	J^{PC}	$I=1$ $u\bar{d}, d\bar{u}, \frac{1}{\sqrt{2}}(d\bar{d}-u\bar{u})$	$I=\frac{1}{2}$ $u\bar{s}, s\bar{u}, d\bar{s}, s\bar{d}$	$I=0$ f'	$I=0$ f	θ [°]
$1\,^1S_0$	0^{-+}	π^+, π^-, π^0	$K^+, K^-, K^0, \bar{K}^0$	η	$\eta'(958)$	-24.6
$1\,^3S_1$	1^{--}	$\rho(770)$	$K^*(892)$	$\phi(1020)$	$\omega(782)$	36.0
$1\,^3P_0$	0^{++}	$a_0(1450)$	$K_0^*(1430)$	$f_0(1710)$	$f_0(1370)$	
$1\,^3S_1$	1^{++}	$a_1(1260)$	K_1???	$f_1(1420)$	$f_1(1285)$	??52.0

$$f' = \psi_8 \cos\theta - \psi_1 \sin\theta$$
$$f = \psi_8 \sin\theta + \psi_1 \cos\theta$$

and the mixing angle has to be determined by experiment. For $\tan\theta = 1/\sqrt{2} \simeq 35.3°$ the state f' would be a pure $s\bar{s}$ state. This is realized to good accuracy for $\omega - \phi$ mixing where ϕ is almost pure $s\bar{s}$.

At low energies, the interaction of a neutral pion with photons is described by the Wess-Zumino-Witten Lagrangian (4.60). Since this is a non-renormalizable interaction, employing it in loop calculations generally results in ultraviolet divergences.

A simple and commonly adopted option is to introduce a form factor into the $\pi^0\gamma\gamma$ interaction vertex, which tames the contributions of highly virtual photons. This results in the following $\pi^0\gamma\gamma$ interaction vertex:

$$V^{\mu\nu}_{\pi^0\gamma\gamma}(q_1, q_2) = \frac{\alpha N_c}{3\pi F_\pi} F_{\pi^0\gamma^*\gamma^*}(m_\pi^2, q_1^2, q_2^2) \, \mathrm{i}\, \epsilon^{\mu\nu\alpha\beta} q_{1\alpha} q_{2\beta}, \quad (5.77)$$

with $F_{\pi^0\gamma\gamma}(m_\pi^2, 0, 0) = 1$ and where $q_{1,2}$ denote the momenta of the two outgoing photons.

The part of the RLA Lagrangian relevant for us here includes the terms containing the neutral vector–meson $\rho^0(770)$, and the charged axial–vector mesons $a_1^\pm(1260)$ and $\pi^\pm$, as well as the photon:

$$\mathcal{L}^{\mathrm{HLS}}_{\mathrm{int}} = -eg_\rho A^\mu \rho^0_\mu - \mathrm{i}\, g_{\rho\pi\pi} \rho^0_\mu \left(\pi^+ \overleftrightarrow{\partial}^\mu \pi^-\right) - \mathrm{i}\, g_{\gamma\pi\pi} A_\mu \left(\pi^+ \overleftrightarrow{\partial}^\mu \pi^-\right)$$
$$+ (1-a)\, e^2 A^\mu A_\mu \pi^+ \pi^- + 2eg_{\rho\pi\pi} A^\mu \rho^0_\mu \pi^+ \pi^- - e\frac{g_\rho}{F_\pi} A^\mu \left(V^+_{a_1\mu}\pi^- - V^-_{a_1\mu}\pi^+\right)$$
$$+ \cdots \quad (5.78)$$

where masses and couplings are related by

$$M_\rho^2 = ag_V^2 F_\pi^2\,, \quad g_\rho = ag_V F_\pi^2\,,$$
$$g_{\rho\pi\pi} = \tfrac{1}{2} ag_V\,, \quad g_{\gamma\pi\pi} = \left(1 - \tfrac{a}{2}\right) e\,.$$

The parameter a is not fixed by the symmetry itself. A good choice is $a = 2$ which conforms with the phenomenological facts **i)** universality of the ρ coupling $g_{\rho\pi\pi} = g_V$, **ii)** $g_{\gamma\pi\pi} = 0$, which is the ρ meson dominance of the pion

5.4 Hadronic Light–by–Light Scattering

form factor, and **iii)** the KSRF relation [131] $M_\rho^2 = 2g_{\rho\pi\pi}^2 F_\pi^2$. The corresponding Feynman rules Fig. 5.30 supplement the sQED ones Fig. 2.8. Also included we have the WZW term (4.60) and the vector–boson propagators read

$$\mathrm{i}\Delta_V^{\mu\nu}(q, M_V) = \frac{-\mathrm{i}}{(q^2 - M_V^2)} \left\{ g^{\mu\nu} - \frac{q^\mu q^\nu}{q^2} \right\} \qquad (5.79)$$

where M_V is the mass.

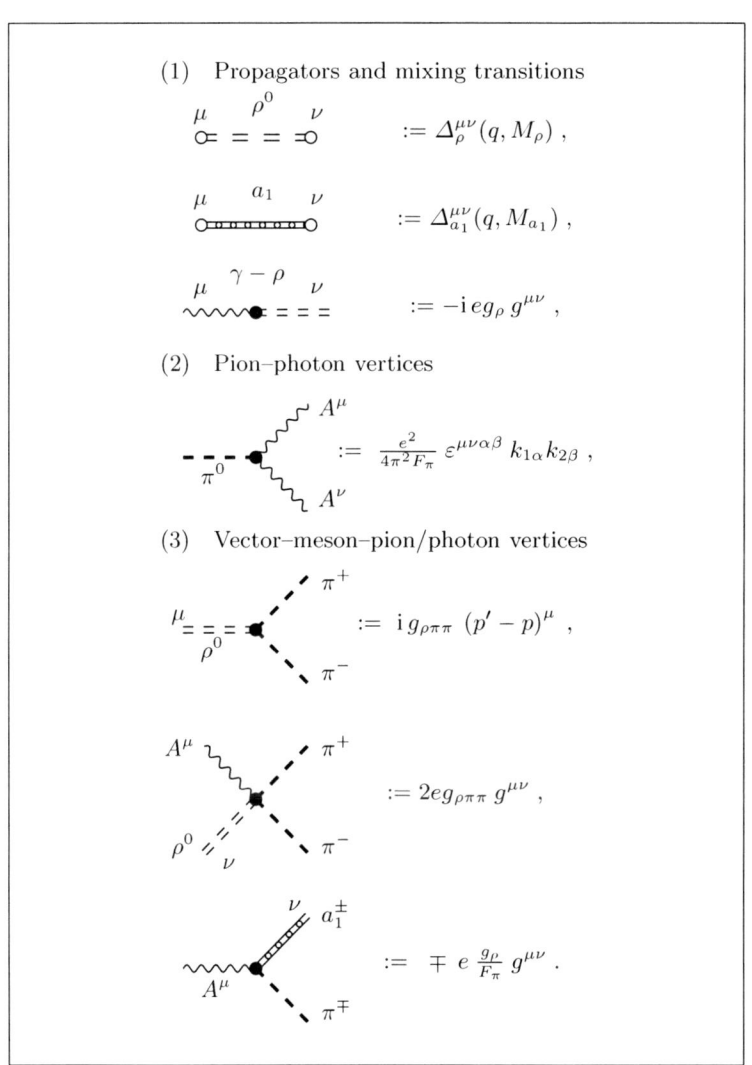

Fig. 5.30. Feynman rules for RLA; all momenta are incoming with $\gamma(k)$, $\pi^+(p')$ and $\pi^-(p)$. These rules supplement the sQED rules Fig. 2.8

Table 5.8. Orders with respect to $1/N_c$ and chiral expansion of typical leading contributions shown in Fig. 5.31

Diagram	$1/N_c$ expansion	p expansion	type
Fig. 5.31(a)	N_c	p^6	π^0, η, η' exchange
Fig. 5.31(a)	N_c	p^8	a_1, ρ, ω exchange
Fig. 5.31(b)	1	p^4	meson loops ($\pi^\pm$, $K^\pm$)
Fig. 5.31(c)	N_c	p^8	quark loops

As mentioned at the beginning of this section already, for the models presented so far one is confronted with the problem that one has to complement a non–renormalizable effective theory with renormalizable perturbative QCD above a certain cut–off. This generally results in a substantial cut–off dependence of the results. In order to avoid this matching problem, the most recent estimations attempt to resort to quark–hadron duality for matching L.D. and S.D. physics. This duality can be proven to hold in the large–N_c limit of QCD and this may be exploited in an $1/N_c$ expansion approach to QCD. However, once more the $N_c \to \infty$ limit, in which the hadrons turn out to be an infinite series of vector resonances, is not under complete quantitative control [123, 124]. Hence, a further approximation must be made by replacing the infinite series of narrow resonances by a few low lying states which are identified with existing hadronic states. As a result one obtains a modeling of the hadronic amplitudes, the simplest one being the lowest meson dominance (LMD) or minimal hadronic ansatz (MHA) approximation to large–N_c QCD [132]. An examples of this type of ansatz has been discussed on p. 253 in Sect. 4.2.2. For a detailed discussion the reader should consult the articles [132, 133].

The various hadronic LbL contributions in the effective theory are shown in Fig. 5.31 and the corresponding $1/N_c$ and chiral $O(p)$ counting is given in Table 5.8.

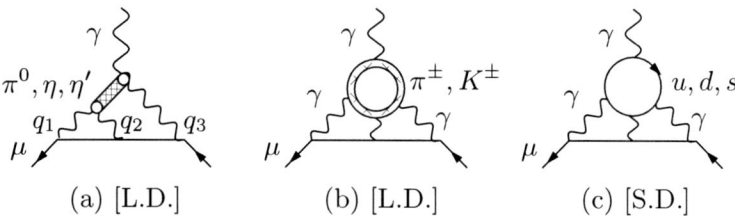

(a) [L.D.]   (b) [L.D.]   (c) [S.D.]

Fig. 5.31. Hadronic light–by–light scattering diagrams in a low energy effective model description. Diagrams (**a**) and (**b**) represent the long distance [L.D.] contributions at momenta $p \leq \Lambda$, diagram (**c**) involving a quark loop which yields the leading short distance [S.D.] part at momenta $p \geq \Lambda$ with $\Lambda \sim 1$ to 2 GeV an UV cut–off. Internal photon lines are dressed by $\rho - \gamma$ mixing

Based on effective hadronic models, major efforts in estimating a_μ^{LbL} were made by Hayakawa, Kinoshita and Sanda (HKS 1995) [117], Bijnens, Pallante and Prades (BPP 1995) [134] and Hayakawa and Kinoshita (HK 1998) [135]. In 2001 Knecht and Nyffeler (KN 2001) [126, 136] discovered a sign mistake in the π^0, η, η' exchange contribution (see also [137, 138]), which changed the central value by $+167 \times 10^{-11}$! More recently Melnikov and Vainshtein (MV 2004) [139] found additional inconsistencies in previous calculations, this time in the short distance constraints (QCD/OPE) used in matching the high energy behavior of the effective models used for the π^0, η, η' exchange contribution. Knecht and Nyffeler restrict their analysis to pion–pole approximation. At least one vector state (V) has to be included in addition to the leading one in order to be able to match the correct high energy behavior. The resulting "LMD+V" parametrization has been worked out for the calculation of the LbL π^0–pole contribution in [126] and was used later in [139] with modified parameter h_2 (see below) at the internal vertex and with a constant pion–pole form factor at the external vertex. Explicit forms of form factors will be considered later.

Before we are going to summarize and discuss the results, in the following, we are presenting a discussion of the pion–pole term mainly, for the reader interested in more details of such calculations. Needless to say that only the original literature can provide full details of these difficult calculations.

5.4.3 Pion–pole Contribution

Here we discuss the dominating hadronic contributions which are due to neutral pion–exchange diagrams Fig. 5.32

The key object here is the $\pi^0\gamma\gamma$ form factor $\mathcal{F}_{\pi^0\gamma^*\gamma^*}(m_\pi^2, q_1^2, q_2^2)$ which is defined by the matrix element

$$\mathrm{i}\int \mathrm{d}^4x\, \mathrm{e}^{\mathrm{i}q\cdot x}\langle 0|T\{j_\mu(x)j_\nu(0)\}|\pi^0(p)\rangle =$$
$$\varepsilon_{\mu\nu\alpha\beta}\, q^\alpha p^\beta\, \mathcal{F}_{\pi^0\gamma^*\gamma^*}(m_\pi^2, q^2, (p-q)^2)\ . \quad (5.80)$$

It is Bose symmetric $\mathcal{F}_{\pi^{0*}\gamma^*\gamma^*}(s, q_1^2, q_2^2) = \mathcal{F}_{\pi^{0*}\gamma^*\gamma^*}(s, q_2^2, q_1^2)$ of course, as the two photons are indistinguishable. This holds for off–shell pions as well. An

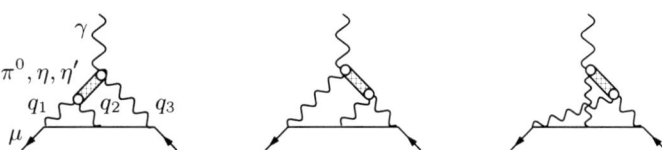

Fig. 5.32. Leading hadronic light–by–light scattering diagrams. Internal photons lines are dressed by $\rho - \gamma$ mixing

important point we should notice is that in the Feynman integral corresponding to one of the diagrams of Fig. 5.32 the pion is **not** necessarily near the pole, although pole–dominance might be expected to give a reasonable approximation. For clarity we therefore define the form factor not by the matrix element (5.80), but by the vertex function

$$i \int d^4x \, e^{iq \cdot x} \langle 0|T\{j_\mu(x) \, j_\nu(0) \, \tilde{\varphi}_{\pi^0}(p)\}|0\rangle =$$

$$\varepsilon_{\mu\nu\alpha\beta} \, q^\alpha p^\beta \, \mathcal{F}_{\pi^{0*}\gamma^*\gamma^*}(p^2, q^2, (p-q)^2) \times \frac{i}{p^2 - m_\pi^2} \,, \quad (5.81)$$

with $\tilde{\varphi}(p) = \int d^4y \, e^{ipx} \varphi(y)$ the Fourier transformed π^0–field.

The π^0–exchange contributions to $\Pi_{\mu\nu\lambda\rho}(q_1, q_2, q_3)$, according to Fig. 5.32 takes the form

$$i \, \Pi^{(\pi^0)}_{\mu\nu\lambda\rho}(q_1, q_2, q_3) =$$

$$\frac{\mathcal{F}_{\pi^{0*}\gamma^*\gamma^*}(q_3'^2, q_1^2, q_2^2) \, \mathcal{F}_{\pi^{0*}\gamma^*\gamma^*}(q_3'^2, q_3^2, k^2)}{q_3'^2 - m_\pi^2} \, \varepsilon_{\mu\nu\alpha\beta} \, q_1^\alpha q_2^\beta \, \varepsilon_{\lambda\rho\sigma\tau} \, q_3^\sigma q_3'^\tau$$

$$+ \frac{\mathcal{F}_{\pi^{0*}\gamma^*\gamma^*}(q_1'^2, q_2^2, q_3^2) \, \mathcal{F}_{\pi^{0*}\gamma^*\gamma^*}(q_1'^2, q_1^2, k^2)}{q_1'^2 - m_\pi^2} \, \varepsilon_{\mu\rho\alpha\beta} \, q_1^\alpha q_1'^\beta \, \varepsilon_{\nu\lambda\sigma\tau} \, q_2^\sigma q_3^\tau$$

$$+ \frac{\mathcal{F}_{\pi^{0*}\gamma^*\gamma^*}(q_2'^2, q_1^2, q_3^2) \, \mathcal{F}_{\pi^{0*}\gamma^*\gamma^*}(q_2'^2, q_2^2, k^2)}{q_2'^2 - m_\pi^2} \, \varepsilon_{\mu\lambda\alpha\beta} \, q_1^\alpha q_3^\beta \, \varepsilon_{\nu\rho\sigma\tau} \, q_2^\sigma q_2'^\tau$$

with $q_i' = q_i + k$. To compute $a_\mu^{\text{LbL};\pi^0} \equiv F_M(0)|_{\text{LbL};\pi^0}$, we need

$$i \, \frac{\partial}{\partial k^\rho} \Pi^{(\pi^0)}_{\mu\nu\lambda\sigma}(q_1, q_2, k - q_1 - q_2) =$$

$$\frac{\mathcal{F}_{\pi^{0*}\gamma^*\gamma^*}(q_3^2, q_1^2, q_2^2) \, \mathcal{F}_{\pi^{0*}\gamma^*\gamma}(q_3^2, q_3^2, 0)}{q_3^2 - m_\pi^2} \, \varepsilon_{\mu\nu\alpha\beta} \, q_1^\alpha q_2^\beta \, \varepsilon_{\lambda\sigma\rho\tau} \, q_3^\tau$$

$$+ \frac{\mathcal{F}_{\pi^{0*}\gamma^*\gamma^*}(q_1^2, q_2^2, q_3^2) \, \mathcal{F}_{\pi^{0*}\gamma^*\gamma}(q_1^2, q_1^2, 0)}{q_1^2 - m_\pi^2} \, \varepsilon_{\mu\sigma\tau\rho} \, q_1^\tau \, \varepsilon_{\nu\lambda\alpha\beta} \, q_1^\alpha q_2^\beta$$

$$+ \frac{\mathcal{F}_{\pi^{0*}\gamma^*\gamma^*}(q_2^2, q_1^2, q_3^2) \, \mathcal{F}_{\pi^{0*}\gamma^*\gamma}(q_2^2, q_2^2, 0)}{q_2^2 - m_\pi^2} \, \varepsilon_{\mu\lambda\alpha\beta} \, q_1^\alpha q_2^\beta \, \varepsilon_{\nu\sigma\rho\tau} \, q_2^\tau + O(k) \,.$$

Here, we may set $k^\mu = 0$ such that $q_3 = -(q_1 + q_2)$. Inserting this last expression into (5.71) and computing the corresponding Dirac traces, one obtains [126]

$$a_\mu^{\text{LbL};\pi^0} = -e^6 \int \frac{d^4q_1}{(2\pi)^4} \frac{d^4q_2}{(2\pi)^4} \frac{1}{q_1^2 q_2^2 (q_1+q_2)^2 [(p+q_1)^2 - m^2][(p-q_2)^2 - m^2]}$$

$$\times \left[\frac{\mathcal{F}_{\pi^{0*}\gamma^*\gamma^*}(q_2^2, q_1^2, q_3^2) \, \mathcal{F}_{\pi^{0*}\gamma^*\gamma}(q_2^2, q_2^2, 0)}{q_2^2 - m_\pi^2} \, T_1(q_1, q_2; p) \right.$$

$$+ \frac{\mathcal{F}_{\pi^{0*}\gamma^*\gamma^*}(q_3^2, q_1^2, q_2^2) \, \mathcal{F}_{\pi^{0*}\gamma^*\gamma}(q_3^2, q_3^2, 0)}{q_3^2 - m_\pi^2} \, T_2(q_1, q_2; p) \Bigg], \quad (5.82)$$

with

$$T_1(q_1, q_2; p) = \frac{16}{3} (p \cdot q_1)(p \cdot q_2)(q_1 \cdot q_2) - \frac{16}{3} (p \cdot q_2)^2 q_1^2$$
$$- \frac{8}{3}(p \cdot q_1)(q_1 \cdot q_2) q_2^2 + 8(p \cdot q_2) q_1^2 q_2^2 - \frac{16}{3}(p \cdot q_2)(q_1 \cdot q_2)^2$$
$$+ \frac{16}{3} m^2 q_1^2 q_2^2 - \frac{16}{3} m^2 (q_1 \cdot q_2)^2,$$

$$T_2(q_1, q_2; p) = \frac{16}{3}(p \cdot q_1)(p \cdot q_2)(q_1 \cdot q_2) - \frac{16}{3}(p \cdot q_1)^2 q_2^2$$
$$+ \frac{8}{3}(p \cdot q_1)(q_1 \cdot q_2) q_2^2 + \frac{8}{3}(p \cdot q_1) q_1^2 q_2^2$$
$$+ \frac{8}{3} m^2 q_1^2 q_2^2 - \frac{8}{3} m^2 (q_1 \cdot q_2)^2.$$

Two of the three diagrams give equal contributions and T_2 has been symmetrized with respect to the exchange $q_1 \leftrightarrow -q_2$. At this stage everything is known besides the $\pi^0 \gamma \gamma$ off-shell form factors.

5.4.4 The $\pi^0 \gamma \gamma$ Transition Form Factor

Above we have formally reduced the problem of calculating the π^0–exchange contribution diagrams Fig. 5.32 to the problem of calculating the integral (5.82). The non–perturbative aspect is now confined in the form–factor function $\mathcal{F}_{\pi^{0*}\gamma^*\gamma^*}(s, s_1, s_2)$, which is largely unknown. For the time being we have to use one of the hadronic models introduced above together with pQCD as a constraint on the high energy asymptotic behavior. Fortunately some experimental data are also available. The constant $\mathcal{F}_{\pi^0\gamma\gamma}(m_\pi^2, 0, 0)$ is well determined by the $\pi^0 \to \gamma\gamma$ decay rate. The invariant matrix element reads

$$\mathcal{M}\left[\pi^0(q) \to \gamma(p_1, \lambda_1) \, \gamma(p_2, \lambda_2)\right] =$$
$$e^2 \, \varepsilon^{\mu*}(p_1, \lambda_1) \, \varepsilon^{\nu*}(p_2, \lambda_2) \, \varepsilon_{\mu\nu\alpha\beta} \, p_1^\alpha p_2^\beta \, \mathcal{F}_{\pi^{0*}\gamma^*\gamma^*}(q^2, p_1^2, p_2^2). \quad (5.83)$$

The on–shell transition amplitude in the chiral limit follows from the WZW–Lagrangian (4.60) and is given by

$$M_{\pi^0\gamma\gamma} = e^2 \, \mathcal{F}_{\pi^0\gamma\gamma}(0, 0, 0) = \frac{e^2 N_c}{12\pi^2 F_\pi} = \frac{\alpha}{\pi F_\pi} \approx 0.025 \text{ GeV}^{-1},$$

and with $F_\pi \sim 92.4$ MeV and quark color number $N_c = 3$, rather accurately predicts the experimental result

$$|M_{\pi^0\gamma\gamma}^{\text{exp}}| = \sqrt{64\pi \Gamma_{\pi^0\gamma\gamma}/m_\pi^3} = 0.025 \pm 0.001 \text{ GeV}^{-1}.$$

Additional experimental information is available for $\mathcal{F}_{\pi^0\gamma^*\gamma}(m_\pi^2, -Q^2, 0)$ coming from experiments $e^+e^- \to e^+e^-\pi^0$ (see Fig. 5.33) where the electron (positron) gets tagged, i.e., selected according to appropriate kinematical criteria, such that $Q^2 = -(p_b - p_t)^2 = 2E_bE_t(1 - \cos\Theta_t)$ is large. p_b is the beam electron (positron) four–momentum, p_t the one of the tagged electron (positron) and Θ_t is the angle between $\boldsymbol{p}_t$ and $\boldsymbol{p}_b$. The differential cross–section

$$\frac{d\sigma}{dQ^2}(e^+e^- \to e^+e^-\pi^0) ,$$

is then strongly peaked towards zero momentum transfer of the untagged positron (electron) which allows experiments to extract the form factor.

Note that the production of an on–shell pion at large $-q_1^2 = Q^2$ is only possible if the real photon is highly energetic, i.e. $q_2^0 = |\boldsymbol{q}_2|$ large. This is different from the $g - 2$ kinematical situation at the external photon vertex, where the external photon has zero four–momentum. By four–momentum conservation thus only $\mathcal{F}_{\pi^{0*}\gamma^*\gamma}(-Q^2, -Q^2, 0)$ and **not** $\mathcal{F}_{\pi^{0*}\gamma^*\gamma}(m_\pi^2, -Q^2, 0)$ can enter at the **external** vertex. However, for a "far off–shell pion" the effective theory breaks down altogether. Indeed, $\mathcal{F}_{\pi^{0*}\gamma^*\gamma}(-Q^2, -Q^2, 0)$ is not an observable quantity away from the pion–pole and in particular for large $Q^2 \gg m_\pi^2$.

For the **internal** vertex both photons are virtual, and luckily, experimental data on $\mathcal{F}_{\pi^0\gamma^*\gamma}(m_\pi^2, -Q^2, 0)$ is available from CELLO [140] and CLEO [141]. This is **one** of the "question marks region" of Fig. 5.23 which is actually controlled by experimental data. Experiments fairly well confirm the Brodsky-Lepage [142] evaluation of the large Q^2 behavior

$$\lim_{Q^2 \to \infty} \mathcal{F}_{\pi^0\gamma^*\gamma}(m_\pi^2, -Q^2, 0) \sim \frac{2F_\pi}{Q^2} . \tag{5.84}$$

In this approach the transition form factor is represented as a convolution of a hard scattering amplitude (HSA) and the soft non–perturbative meson wave

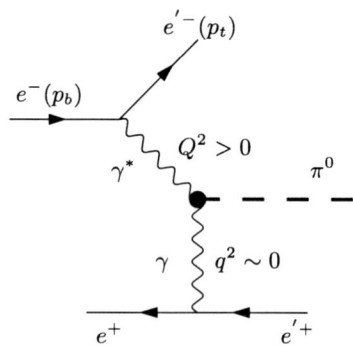

Fig. 5.33. Measurement of the π^0 form factor $\mathcal{F}_{\pi^0\gamma^*\gamma}(m_\pi^2, -Q^2, 0)$ at high space–like Q^2

function and the asymptotic behavior follows from a pQCD calculation of the HSA. Together with the constraint from π^0 decay

$$\lim_{Q^2 \to 0} \mathcal{F}_{\pi^0 \gamma^* \gamma}(m_\pi^2, -Q^2, 0) = \frac{1}{4\pi^2 F_\pi} \qquad (5.85)$$

an interpolating formula

$$\mathcal{F}_{\pi^0 \gamma^* \gamma}(m_\pi^2, -Q^2, 0) \simeq \frac{1}{4\pi^2 F_\pi} \frac{1}{1 + (Q^2/8\pi^2 F_\pi^2)} \qquad (5.86)$$

was proposed, which in fact gives an acceptable fit to the data shown in Fig. 5.34. Refinements of form factor calculations/models were discussed and compared with the data in [141] (see also [143, 144, 145]).

It is important to note here that the L.D. term $\mathcal{F}_{\pi^0 \gamma \gamma}(m_\pi^2, 0, 0)$, which is unambiguously determined by the anomaly, gets screened at large Q^2, in spite of the fact that in the chiral limit

$$\mathcal{F}_{\pi^0{}^* \gamma^* \gamma^*}(q_3^2, q_1^2, q_2^2)|_{m_q=0} = \mathcal{F}_{\pi^0 \gamma \gamma}(0,0,0)|_{m_q=0} = \frac{1}{4\pi^2 F_0} \;. \qquad (5.87)$$

This behavior is in common with the one of the quark loops when $m_q \neq 0$, as we will discuss next. A seemingly plausible approximation which helps to

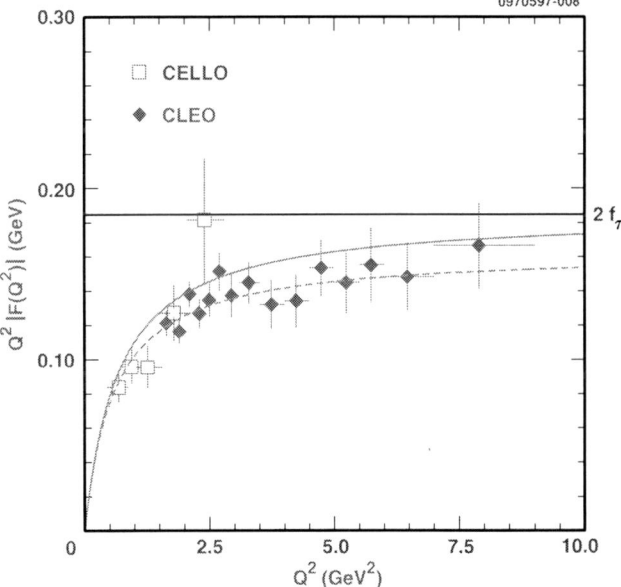

Fig. 5.34. $\mathcal{F}_{\pi^0 \gamma^* \gamma}(m_\pi^2, -Q^2, 0)$ data from CLEO and CELLO. Shown is the Brodsky-Lepage prediction (5.86) (*solid curve*) and the phenomenological fit by CLEO (*dashed curve*) Reprinted with permission from [141]. Copyright (2007) by the American Physical Society

simplify the calculation is to assume *pion–pole dominance* in the sense that one takes the form factor on the pion mass shell and uses $\mathcal{F}_{\pi^0 \gamma^* \gamma^*}$ everywhere. This pole approximation apparently has been used by all authors (HKS,BPP,KN) in the past, but has been criticized recently (MV). The point is that the form factor sitting at the external photon vertex in the pole approximation [read $\mathcal{F}_{\pi^0 \gamma^* \gamma}(m_\pi^2, -Q^2, 0)$] for $-Q^2 \neq m_\pi^2$ violates four-momentum conservation $k^\mu = 0$. The latter requires $\mathcal{F}_{\pi^{0*} \gamma^* \gamma}(-Q^2, -Q^2, 0)$ as discussed before. In the chiral limit the only consistent choice for the form factor in the pole approximation is $\mathcal{F}_{\pi^0 \gamma \gamma}(0,0,0)$ which is a constant given by (5.85); this model is advocated by Melnikov and Vainshtein, and leads to a substantially larger contribution, due to the lack of damping of the high energy modes. But, what we really need is $\mathcal{F}_{\pi^{0*} \gamma^* \gamma}(-Q^2, -Q^2, 0)$ and the question is how it behaves at high energies. Definitely, no experimental information is available here.

After the presentation of the experimental constraints we turn to the theoretical models. Let us consider first the behavior of $\mathcal{F}_{\pi^{0*} \gamma^* \gamma^*}$ in the CQM, where it is given by a quark triangular loop (see (2.143) and [135])[25]

$$F^{\text{CQM}}_{\pi^{0*} \gamma^* \gamma^*}(q^2, p_1^2, p_2^2) = 2 m_q^2 \, C_0(m_q, m_q, m_q; q^2, p_1^2, p_2^2)$$

$$\equiv \int [d\alpha] \, \frac{2 m_q^2}{m_q^2 - \alpha_2 \alpha_3 p_1^2 - \alpha_3 \alpha_1 p_2^2 - \alpha_1 \alpha_2 q^2} \,, \quad (5.88)$$

where $[d\alpha] = d\alpha_1 d\alpha_2 d\alpha_3 \, \delta(1 - \alpha_1 - \alpha_2 - \alpha_3)$ and m_q is a quark mass ($q = u, d, s$). For $p_1^2 = p_2^2 = q^2 = 0$ we obtain $F^{\text{CQM}}_{\pi^{0*} \gamma^* \gamma^*}(0,0,0) = 1$. Note the symmetry of C_0 under permutations of the arguments (p_1^2, p_2^2, q^2). C_0 is a known function in terms of logs and dilogs for arbitrary values of the arguments. For our purpose it is sufficient to calculate it at one of the square momenta set to zero. One finds

$$F^{\text{CQM}}_{\pi^{0*} \gamma^* \gamma^*}(0, p_1^2, p_2^2) = \frac{-m_q^2}{p_1^2 - p_2^2} \left\{ \ln^2 \frac{\sqrt{4 m_q^2 - p_1^2} - \sqrt{-p_1^2}}{\sqrt{4 m_q^2 - p_1^2} + \sqrt{-p_1^2}} - (p_1^2 \to p_2^2) \right\}.$$

For large p_1^2 at $p_2^2 \sim 0$, $q^2 \sim 0$ the asymptotic behavior is given by

$$F^{\text{CQM}}_{\pi^{0*} \gamma^* \gamma^*}(0, p_1^2, 0) \sim \frac{m_q^2}{-p_1^2} \left\{ \ln^2 \left(\frac{-p_1^2}{m_q^2} \right) \right\} . \quad (5.89)$$

For large $p_1^2 \sim p_2^2$ at $q^2 \sim 0$ we have

$$F^{\text{CQM}}_{\pi^{0*} \gamma^* \gamma^*}(0, p_1^2, p_1^2) \sim 2 \, \frac{m_q^2}{-p_1^2} \left\{ \ln \left(\frac{-p_1^2}{m_q^2} \right) \right\} , \quad (5.90)$$

---

[25] We actually first consider a current quark loop which is related via PCAC to the triangle anomaly (see below). Non-perturbative strong interactions effects transmute it to a constituent quark loop ($m_o \to M_o$, the latter being non-vanishing in the chiral limit).

and the same behavior follows for $q^2 \sim p_1^2$ at $p_2^2 \sim 0$. Note that in all cases we have the same power behavior $\sim m_q^2/p_1^2$ modulo logarithms. It is important to note that in the chiral limit $F^{\text{CQM}}_{\pi^{0*}\gamma^*\gamma^*} \xrightarrow{m_q \to 0} 0$ if $(q^2, p_1^2, p_2^2) \neq (0,0,0)$. Thus our consideration seems to be not quite relevant, as it says that the chiral corrections at high energies are damped by a $1/Q^2$ behavior in all the possible directions. The dominant terms come from the chiral limit, but, surprisingly, the CQM calculation also sheds light on the leading contribution, as we are going to discuss now. Actually, the singular behavior of $F^{\text{CQM}}_{\pi^{0*}\gamma^*\gamma^*}$ under exchange of limits:

$$\lim_{m_q \to 0} F^{\text{CQM}}_{\pi^{0*}\gamma^*\gamma^*}(q^2, p_1^2, p_2^2) \equiv 0 \quad \text{for all} \quad (q^2, p_1^2, p_2^2) \neq (0,0,0)$$

$$\lim_{(q^2, p_1^2, p_2^2) \to (0,0,0)} F^{\text{CQM}}_{\pi^{0*}\gamma^*\gamma^*}(q^2, p_1^2, p_2^2) \equiv 1 \quad \text{for all} \quad m_q \neq 0 \;, \tag{5.91}$$

implies that the chiral limit is either zero **or unity**,

$$\lim_{m_q \to 0} \lim_{(q^2, p_1^2, p_2^2) \to (0,0,0)} F^{\text{CQM}}_{\pi^{0*}\gamma^*\gamma^*}(q^2, p_1^2, p_2^2) \equiv 1 \;, \tag{5.92}$$

depending on whether $(q^2, p_1^2, p_2^2) \neq (0,0,0)$ and $(q^2, p_1^2, p_2^2) = (0,0,0)$, respectively. This singular behavior is an alternative form of expressing the ABJ anomaly and the non–renormalization theorem. For the pseudoscalar vertex the latter just means that the last identity to all orders of perturbation theory yields a constant, which always may be renormalized to unity by an appropriate renormalization of the axial current. The divergence of the latter being the interpolating field of the pseudoscalar Goldstone mode involved[26]. Amazingly, the pseudoscalar vertex (at one loop, in the real world of non–vanishing quark masses) is UV finite and regularization independent; the two vector currents are trivially conserved, because of the $\varepsilon_{\mu\nu\alpha\beta} p_1^\alpha p_2^\beta$ tensor structure in (5.83), and we obtain the ABJ anomaly as a IR phenomenon and not as a UV renormalization effect as it appears if one looks at the VVA matrix element. Since the anomaly is exact to all orders and at all energy scales, it is not surprising that it may be obtained from the IR region as well. Note that with the exception of the WZW point form factor, all other models considered (see e.g. (5.109) or (5.111), below) share the property of the CQM that they yield the anomaly at $(0, 0, 0)$ while dropping for large p_i^2 like $1/p_i^2$ if $(p_1^2, p_2^2, p_3^2) \neq (0, 0, 0)$. But likely only the CQM may be a half–way reasonable model for the configuration $(-Q^2, -Q^2, 0)$ needed at the external vertex.

An alternative way to look at the problem is to use the anomalous PCAC relation (5.95) and to relate $\pi^0\gamma\gamma$ directly with the ABJ anomaly (Bell-Jackiw approach). We therefore consider the VVA three–point function

$$\mathcal{W}_{\mu\nu\rho}(q_1, q_2) = i \int d^4x_1 d^4x_2 \, e^{i(q_1 \cdot x_1 + q_2 \cdot x_2)} \langle 0 | T\{V_\mu(x_1) V_\nu(x_2) A_\rho(0)\} | 0 \rangle \;, \tag{5.93}$$

---

[26] The anomaly cancellation required by renormalizability of a gauge theory here just would mean the absence of a non–smooth chiral limit.

of the flavor and color diagonal fermion currents

$$V_\mu = \bar{\psi}\gamma_\mu\psi \ , \quad A_\mu = \bar{\psi}\gamma_\mu\gamma_5\psi \ , \qquad (5.94)$$

where $\psi(x)$ is a quark field. The vector currents are strictly conserved $\partial_\mu V^\mu(x) = 0$, while the axial vector current satisfies a PCAC relation plus the anomaly (indexed by $_0$ are bare parameters),

$$\partial_\mu A^\mu(x) = 2\,\mathrm{i}\,m_0\bar{\psi}\gamma_5\psi(x) + \frac{\alpha_0}{4\pi}\varepsilon_{\mu\nu\rho\sigma}F^{\mu\nu}F^{\rho\sigma}(x) \ . \qquad (5.95)$$

To leading order the correlator of interest is associated with the one–loop triangle diagram plus its crossed $(q_1,\mu \leftrightarrow q_2,\nu)$ partner. The covariant decomposition of $\mathcal{W}_{\mu\nu\rho}(q_1,q_2)$ into invariant functions has four terms

$$\mathcal{W}_{\mu\nu\rho}(q_1,q_2) = \frac{1}{8\pi^2}\left\{w_L\left(q_1^2,q_2^2,q_3^2\right)(q_1+q_2)_\rho\,\varepsilon_{\mu\nu\alpha\beta}\,q_1^\alpha q_2^\beta \right.$$

$$\left. + \ 3 \ \mathrm{transversal}\right\} \ . \qquad (5.96)$$

The longitudinal part is entirely fixed by the anomaly,

$$w_L\left(q_1^2,q_2^2,q_3^2\right) = -\frac{2N_c}{q_3^2} \ , \qquad (5.97)$$

which is exact to all orders of perturbation theory, the famous Adler-Bardeen non–renormalization theorem. In order to obtain the coupling to pseudoscalars we have to take the derivative as required by the PCAC relation, and using (5.97) we obtain

$$(q_1+q_2)^\rho \mathcal{W}_{\mu\nu\rho}(q_1,q_2) = \frac{1}{8\pi^2}\,\varepsilon_{\mu\nu\alpha\beta}\,q_1^\alpha q_2^\beta w_L\left(q_1^2,q_2^2,q_3^2\right)q_3^2$$

$$= -\frac{N_c}{4\pi^2}\,\varepsilon_{\mu\nu\alpha\beta}\,q_1^\alpha q_2^\beta \ . \qquad (5.98)$$

This holds to all orders and for arbitrary momenta. It should be stressed that the pole in the amplitude w_L is just a kinematical singularity stemming from the covariant decomposition of the tensor amplitude and by dimensional counting. Thus, in general, the VVA correlator does not exhibit physical one particle poles and in observables all kinematical singularities must cancel out in any case.

A crucial question is the one about the correct high energy behavior of $\mathcal{F}_{\pi^{0*}\gamma^*\gamma^*}$. It is particularly this far off–shell behavior which enters in a relevant manner in the integral (5.82). This high energy behavior has to be fixed somehow in all the evaluations and was reconsidered by Knecht and Nyffeler [132, 126] and later by Melnikov and Vainshtein [139]. The latter authors criticized all previous evaluations in this respect and came up with a new estimation of the correct asymptotic behavior. Key tool again is the

OPE in order to investigate the short distance behavior of the four–current correlator in (5.65), which may be written as [139]

$$\langle 0 | T\{j_\mu(x_1) j_\nu(x_2) j_\lambda(x_3)\} | \gamma(k) \rangle \,,$$

taking into account that the external photon is in a physical state. A look at the first of the diagrams of Fig. 5.32, and taking into account the pole–dominance picture, shows that with q_1 and q_2 as independent loop integration momenta the most important region to investigate is $q_1^2 \sim q_2^2 \gg q_3^2$, which is related to a short distance expansion of $T\{j_\mu(x_1) j_\nu(x_2)\}$ for $x_1 \to x_2$. Thus the OPE again is of the form (4.64), however, now for two electromagnetic currents $T\{j_\mu(x) j_\nu(y) X\}$ and with a "state" X the third electromagnetic current $j_\lambda(z)$ times the physical external photon state $|\gamma(k)\rangle$.

$$\text{(5.99)}$$

Note that this time the first term of (4.65) is absent due to C–invariance (Furry's theorem). As usual the result of an OPE is a product of a perturbative *hard* "short distance coefficient function" times a non–perturbative *soft* "long distance matrix element". Surprisingly, for the leading possible term here, the non–perturbative factor is just given by the ABJ anomaly diagram, which is known to by given by the perturbative one–loop result, exact to all orders. This requires of course that the leading operator in the short distance expansion must involve the divergence of the axial current, as the VVV triangle is identically zero by Furry's theorem. This is how the pseudoscalar pion comes into the game in spite of the fact that LbL scattering externally involves vector currents only. Indeed, in leading order one obtains

$$\mathrm{i} \int \mathrm{d}^4 x_1 \int \mathrm{d}^4 x_2 \, e^{\mathrm{i}\,(q_1 x_1 + q_2 x_2)} \, T\{j_\mu(x_1) j_\nu(x_2) X\} =$$

$$\int \mathrm{d}^4 z \, e^{\mathrm{i}\,(q_1 + q_2) z} \, \frac{2\mathrm{i}}{\hat{q}^2} \, \varepsilon_{\mu\nu\alpha\beta} \hat{q}^\alpha T\{j_5^\beta(z) X\} + \cdots \quad (5.100)$$

with $j_5^\mu = \bar{q} \hat{Q}^2 \gamma^\mu \gamma_5 q$ the relevant axial current and $\hat{q} = (q_1 - q_2)/2 \approx q_1 \approx -q_2$. The momentum flowing through the axial vertex is $q_1 + q_2$ and in the limit $k^\mu \to 0$ of our interest $q_1 + q_2 \to -q_3$, which is assumed to be much smaller than $\hat{q}$ ($q_1^2 - q_2^2 \sim -2 q_3 \hat{q} \sim 0$). The ellipses stand for terms suppressed by powers of $\Lambda_{\mathrm{QCD}}/\hat{q}$. It is convenient to decompose the axial current into the different possible flavor channels and write it as a linear combination of isospin $j_{5\mu}^{(3)} = \bar{q} \lambda_3 \gamma_\mu \gamma_5 q$, hypercharge $j_{5\mu}^{(8)} = \bar{q} \lambda_8 \gamma_\mu \gamma_5 q$ and the $SU(3)$ singlet

$j_{5\mu}^{(0)} = \bar{q}\lambda_0\gamma_\mu\gamma_5 q$, where $\lambda_3 = \mathrm{diag}(1,-1,0)$ and $\lambda_8 = \mathrm{diag}(1,1,-2)$ are the diagonal Gell-Mann matrices of flavor $SU(3)$ and λ_0 is the unit matrix. Then

$$j_{5\mu} = \sum_{a=3,8,0} \frac{\mathrm{Tr}\,[\lambda_a \hat{Q}^2]}{\mathrm{Tr}\,[\lambda_a^2]}\, j_{5\mu}^{(a)}\,. \tag{5.101}$$

After the perturbative large q_1, q_2 behavior has been factored out the remaining soft matrix element to be calculated is

$$T_{\lambda\beta}^{(a)} = \mathrm{i}\int \mathrm{d}^4 z\, \mathrm{e}^{\mathrm{i}\,q_3 z}\,\langle 0|T\{j_{5\beta}^{(a)}(z)j_\lambda(0)\}|\gamma(k)\rangle\,, \tag{5.102}$$

which is precisely the VVA triangle correlator (4.45) discussed earlier in Sect. 4.2.2. This matrix–element may be written as

$$T_{\lambda\beta}^{(a)} = -\frac{\mathrm{i}\,eN_c\mathrm{Tr}\,[\lambda_a\hat{Q}^2]}{4\pi^2}\times$$

$$\left\{ w_L^{(a)}(q_3^2)\, q_{3\beta}q_3^\sigma\tilde{f}_{\sigma\lambda} + w_T^{(a)}(q_3^2)\left(-q_3^2\tilde{f}_{\lambda\beta} + q_{3\lambda}q_3^\sigma\tilde{f}_{\sigma\beta} - q_{3\beta}q_3^\sigma\tilde{f}_{\sigma\lambda}\right) \right\}\,. \tag{5.103}$$

Both amplitudes, the longitudinal w_L as well as the transversal w_T, are calculable from the triangle fermion one–loop diagram. In the chiral limit they are given by [146, 147, 148]

$$w_L^{(a)}(q^2) = 2w_T^{(a)}(q^2) = -2/q^2\,. \tag{5.104}$$

At this stage of the consideration it looks like a real mystery what all this has to do with π^0–exchange, as everything looks perfectly controlled by perturbation theory[27]. The clue is that as a low energy object we may evaluate this matrix element at the same time perfectly well in terms of hadronic spectral

---

[27] In the literature frequently the "pole" of (5.97) is misleadingly identified with the pion–pole, and chiral symmetry breaking is said to transmute the "Goldstone pole"

$$1/q^2 \to 1/(q^2 - m_\pi^2)$$

to the physical pion–pole. This argumentation is certainly wrong since this pole is also present for the leptons where it is obvious that there is no physical pole. In fact the "pole" is just a kinematical singularity, in any physical amplitude it gets removed by a q^2 factor coming from the contraction of the tensor coefficients in the covariant decomposition (5.96). In the PCAC relation (5.98), which relates the divergence of the axial current to the pion, the kinematical pole is removed. This happens both for quarks and for leptons. The emergence of pions has nothing to do with the anomaly primarily. Pions are quasi Goldstone bosons of the spontaneous breakdown of the chiral symmetry of strong interactions and as such a completely non–perturbative phenomenon, with other words, whether the operator $\bar\psi\gamma_5\psi(x)$ in the PCAC relation (5.95) is the interpolating field of a composite bound state, is a matter of the non–perturbative nature of the strong interactions of the quarks.

functions by saturating it by a sum over intermediate states, using (3.120). For the positive frequency part we have

$$\langle 0|j_{5\beta}^{(a)}(z)j_\lambda(0)|\gamma(k)\rangle = \int \frac{d^4 p_n}{(2\pi)^3} \sum_n \langle 0|j_{5\beta}^{(a)}(z)|n\rangle \langle n|j_\lambda(0)|\gamma(k)\rangle \,,$$

where for $a = 3$ the lowest state contributing is the π^0, thus

$$\langle 0|j_{5\beta}^{(3)}(z)j_\lambda(0)|\gamma(k)\rangle = \int \frac{d^3 p}{(2\pi)^3 \, 2\omega(p)} \langle 0|j_{5\beta}^{(3)}(z)|\pi^0(p)\rangle \langle \pi^0(p)|j_\lambda(0)|\gamma(k)\rangle$$
$$+ \text{subleading terms}\,.$$

Here, we have the matrix elements

$$\langle 0|j_{5\beta}^{(3)}(z)|\pi^0(p)\rangle = e^{i p z}\, 2 i F_\pi p_\beta$$
$$\langle \pi^0(p)|j_\lambda(0)|\gamma(k)\rangle = -4 e g_{\pi^0 \gamma\gamma} p^\alpha \tilde{f}_{\alpha\lambda}\,, \qquad (5.105)$$

with $\tilde{f}_{\alpha\lambda} = k_\alpha \varepsilon_\lambda - k_\lambda \varepsilon_\alpha$, ε_α the external photon's polarization vector and

$$g_{\pi^0\gamma\gamma} = \frac{N_c \operatorname{Tr}[\lambda_3 \hat{Q}^2]}{16\pi^2 F_\pi}\,. \qquad (5.106)$$

Omitting subleading terms, as a result we find

$$\langle 0|j_{5\beta}^{(3)}(z)j_\lambda(0)|\gamma(k)\rangle = \int \frac{d^4 p}{(2\pi)^3}\, \Theta(p^0)\, \delta(p^2 - m_\pi^2)\, e^{i p z}\, 2 i F_\pi p_\beta \frac{N_c \operatorname{Tr}[\lambda_3 \hat{Q}^2]}{16\pi^2 F_\pi} p^\alpha \tilde{f}_{\alpha\lambda}\,,$$

and finally for the time ordered correlation

$$\langle 0|T\{j_{5\beta}^{(3)}(z)j_\lambda(0)\}|\gamma(k)\rangle = \int \frac{d^4 p}{(2\pi)^3}\, \frac{1}{\pi}\, \frac{i}{p^2 - m_\pi^2 + i\varepsilon}\, e^{i p z}\, 2 i F_\pi p_\beta \frac{N_c \operatorname{Tr}[\lambda_3 \hat{Q}^2]}{16\pi^2 F_\pi} p^\alpha \tilde{f}_{\alpha\lambda}.$$

After this discussion which allows us to understand precisely how the π^0-exchange comes into play, we briefly present some typical $\pi^0\gamma\gamma$ transition form factor, which have been used in evaluations of the hadronic LbL contribution recently. With $\gamma^*\gamma^* \to \pi^0 \to \gamma^*\gamma^*$ replacing the full amplitude, and in the pion–pole approximation, Knecht and Nyffeler [126, 149] were able to reduce the problem analytically to a 2–dimensional integral representation over the moduli of the Euclidean momenta

$$a_\mu^{\text{LbL};\pi^0} = \int_0^\infty dQ_1 \int_0^\infty dQ_2 \sum_{i=1,2} w_i(Q_1, Q_2)\, f_i(Q_1, Q_2)\,, \qquad (5.107)$$

which may be integrated numerically without problems. In the pole approximation the weight functions w_i are model independent (rational functions, square roots and logarithms). The model dependence (form factors) resides in the f_i's. The representation allows a transparent investigation of the form factor dependences. However, for the general π^0–exchange diagrams there remain three integrations to be performed numerically and the analysis gets more involved (see [117, 134, 135]).

For simplicity, we focus here on the pion–pole approximation, i.e. $q_3^2 = m_\pi^2$ in any case (the corresponding argument is suppressed in the following). In order to get an idea about different possibilities we consider the following four cases here: (see also [133, 150])

$$\mathcal{F}^{\text{WZW}}_{\pi^0\gamma^*\gamma^*}(q_1^2, q_2^2) = \frac{N_c}{12\pi^2 F_\pi}, \qquad (5.108)$$

$$\mathcal{F}^{\text{VMD}}_{\pi^0\gamma^*\gamma^*}(q_1^2, q_2^2) = \frac{N_c}{12\pi^2 F_\pi} \frac{M_V^2}{(q_1^2 - M_V^2)} \frac{M_V^2}{(q_2^2 - M_V^2)}, \qquad (5.109)$$

$$\mathcal{F}^{\text{LMD}}_{\pi^0\gamma^*\gamma^*}(q_1^2, q_2^2) = \frac{F_\pi}{3} \frac{c_V - q_1^2 - q_2^2}{(q_1^2 - M_V^2)(q_2^2 - M_V^2)}, \qquad (5.110)$$

$$\mathcal{F}^{\text{LMD+V}}_{\pi^0\gamma^*\gamma^*}(q_1^2, q_2^2) = \frac{F_\pi}{3} \frac{h_0 - h_1(q_1^2 + q_2^2)^2 - h_2 q_1^2 q_2^2 - h_5(q_1^2 + q_2^2) - q_1^2 q_2^2 (q_1^2 + q_2^2)}{(q_1^2 - M_{V_1}^2)(q_1^2 - M_{V_2}^2)(q_2^2 - M_{V_1}^2)(q_2^2 - M_{V_2}^2)},$$
$$(5.111)$$

with

$$c_V = \frac{N_c}{4\pi^2} \frac{M_V^4}{F_\pi^2}, \quad h_0 = \frac{N_c}{4\pi^2} \frac{M_{V_1}^4 M_{V_2}^4}{F_\pi^2}.$$

All satisfy the low energy constraint $\mathcal{F}_{\pi^0\gamma^*\gamma^*}(m_\pi^2, q_1^2, q_2^2) = \mathcal{F}^{\text{WZW}}_{\pi^0\gamma^*\gamma^*}(m_\pi^2, 0, 0)$. The WZW form factor is a constant and if used at both vertices leads to a divergent result. This is not so surprising as physics requires some kind of VMD mechanism as we know. The VMD form factor as well as the HLS and the ENJL model do not satisfy the large momentum asymptotics required by QCD. Using these models thus leads to cut–off dependent results, where the cut–off is to be varied between reasonable values which enlarges the model error of such estimates. Nevertheless it should be stressed that such approaches are perfectly legitimate and the uncertainties just reflect the lack of precise understanding of this kind of physics. For the large–N_c inspired form factors the proper high energy behavior can only by implemented by introducing at least two vector mesons: the $\rho(770)$ and the $\rho'(1465)$, which is denoted by LMD+V. For a recent discussion of form factors beyond the pole–approximation we refer to [133].

In the most recent estimations the LMD+V form factor by Knecht and Nyffeler is used for the internal vertex. The experimental constraints subsumed in the form (5.84) fixes h_0 and requires $h_1 = 0$. Identifying the resonances with $M_1 = M_\rho = 769$ MeV, $M_2 = M_{\rho'} = 1465$ MeV, the phenomenological constraint also fixes $h_5 = 6.93$ GeV^4. h_2 was allowed to vary in a wide range in [126] with $h_2 = 0$ as a central value. As argued in [139], another OPE argument allows to pin down the parameter h_2 with the result that $h_2 = -10$ GeV^2 is a more appropriate central value. Knecht and Nyffeler apply the above LMD+V type form factor on both ends of the pion line: for the first diagram of Fig. 5.32 thus

$$\mathcal{F}^{\text{LMD+V}}_{\pi\gamma^*\gamma^*}(m_\pi^2, q_1^2, q_2^2) \cdot \mathcal{F}^{\text{LMD+V}}_{\pi\gamma^*\gamma}(m_\pi^2, q_3^2, 0)\,,$$

where, as explained above, in fact the second factor, with the given arguments, is kinematically forbidden. In order to avoid this inconsistency Melnikov and Vainshtein propose to use

$$\mathcal{F}^{\text{LMD+V}}_{\pi\gamma^*\gamma^*}(m_\pi^2, q_1^2, q_2^2) \cdot \mathcal{F}^{\text{WZW}}_{\pi\gamma^*\gamma}(m_\pi^2, 0, 0)\,,$$

where the WZW form factor is exact in the chiral limit $m_\pi^2 \to 0$. However, the pole–dominance assumed so far may not be a good approximation and taking the diagram more literally, would require

$$\mathcal{F}_{\pi^*\gamma^*\gamma^*}(q_3^2, q_1^2, q_2^2) \cdot \mathcal{F}_{\pi^*\gamma^*\gamma}(q_3^2, q_3^2, 0)\,,$$

as the more appropriate amplitude. This, however, requires a cut–off on the pion momentum and the complement has to be evaluated in pQCD. The second factor here is expected to be qualitatively well described by the CQM form factor, which includes the WZW term, but, beyond the chiral limit, exhibits $1/q^2$ screening of the latter, similar to the Brodsky-Lepage formula.

We therefore advocate to use consistently dressed form factors as inferred from the resonance Lagrangian approach. In view of the lack of any established information on what concerns the coefficient of the $1/Q^2$ damping in the $(-Q^2, -Q^2, 0)$ channel, we **assume** that the Brodsky-Lepage behavior essentially carries over to this channel, which corresponds to $M_q = 2\pi F_\pi/\sqrt{N_c} \sim 335$ MeV in the CQM. This is a problem which has to be clarified in a future investigation.

Analytic calculations of $a_\mu^{\text{LbL;had}}$ based on simplified (non–RLA) *Effective Field Theory* also yielded instructive results: these studies are based on the $O(N_c, p^8)$ WZW–Lagrangian, the $O(p^6)$ chiral Lagrangian and assuming scalar QED for the interaction of the photon with the charged pseudoscalars. The leading diagrams are shown in Fig. 5.35. Diagrams (a) and (b) in this approach are divergent and renormalized by the effective counter term Lagrangian $\mathcal{L}^{(6)} = (\alpha^2/4\pi^2 F_0)\,\delta\chi\,\bar{\psi}\gamma_\mu\gamma_5\psi\partial^\mu\pi^0 + \cdots$ generating diagrams d) and (e). Diagram (c) is finite. The overall divergence requires a lowest order anomalous magnetic moment type diagram (f). The effective Lagrangian thus must include a term of type (3.78), with $a_\mu \to \delta a_\mu$. Strictly speaking this spoils the predictive power of the effective theory by an overall subtraction, unless the divergence is removed by some other mechanism like the VMD model again, for example. Including the pion and kaon loops of Fig. 5.31, the result may be cast into the form [136]

$$a_\mu^{\text{LbL;had}} = \left(\frac{\alpha}{\pi}\right)^3 \left\{ N_c \left(\frac{m_\mu^2}{16\pi^2 F_\pi^2} \frac{N_c}{3}\right) \left[\ln^2 \frac{\mu_0}{m_\mu} + c_1 \ln \frac{\mu_0}{m_\mu} + c_0\right] \right. \quad (5.112)$$

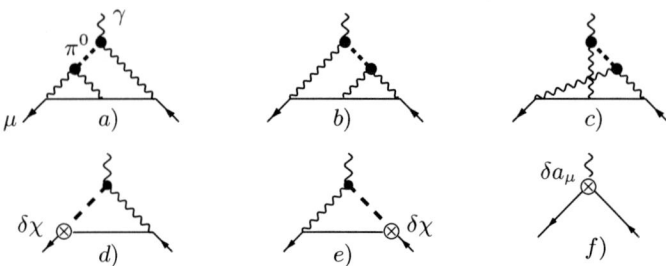

Fig. 5.35. Diagrams contributing to $a_\mu^{\text{LbL};\pi^0}$ in EFT. $\otimes$ denotes a renormalization counter term insertion. Counter terms $\delta\chi$ are needed to render the triangular subgraphs of a) and b) finite, δa_μ is needed to remove the remaining overall two–loop divergence

$$+ f\left(\frac{m_{\pi^\pm}}{m_\mu}, \frac{M_{K^\pm}}{m_\mu}\right) + O(\frac{m_\mu^2}{\mu_0^2} \times \log s, \frac{m_\mu^4}{\mu_0^4} N_c \times \log s)\right\} .$$

Since $F_\pi = O(\sqrt{N_c})$, the leading term is $O(N_c)$ (see Table 5.8) and exhibits a $\log^2$ term with universal coefficient $\mathcal{C} = (N_c^2 m_\mu^2)/(48\pi^2 F_\pi^2) \simeq 0.025$ for $N_c = 3$ [136]. The scale μ_0, originally represents the cut–off $\mu_0 = \Lambda$ or, in dimensional regularization, the $\overline{\text{MS}}$ scale or after imposing a subtraction (=renormalization) condition it is the renormalization scale. Again the VMD model (5.109) is the simplest possibility to introduce a physical cut–off $\mu_0 = M_\rho$, such that

$$a_{\mu;\text{VMD}}^{\text{LbL};\pi^0} = \left(\frac{\alpha}{\pi}\right)^3 \mathcal{C}\, X_{\pi^0} = \left(\frac{\alpha}{\pi}\right)^3 \mathcal{C}\left[\ln^2 \frac{M_\rho}{m_\mu} + c_1 \ln \frac{M_\rho}{m_\mu} + c_0\right]. \quad (5.113)$$

In this case the diagrams Fig. 5.35 exhibit three well separated scales:

$$m_\pi^2 - m_\mu^2 \ll m_\mu^2 \ll M_\rho^2 ,$$

and based on this hierarchy an expansion in $\delta \equiv (m_\pi^2 - m_\mu^2)/m_\mu^2$ and m_μ^2/M_ρ^2 is possible. The expansion in δ is especially simple and reduces to the *Taylor expansion* of the pion propagator. The expansion in m_μ^2/M_ρ^2 is a *Large Mass Expansion*. The result obtained in [137] is given by

$$X_{\pi^0} = L^2 + \left(\frac{1}{2} - \frac{\pi}{\sqrt{3}}\right) L - \frac{277}{216}$$
$$+ \frac{2\pi}{\sqrt{3}} S_2 - \frac{17\pi}{72\sqrt{3}} + \frac{57}{8} S_2 - \frac{\zeta_3}{6} - \frac{11\pi^2}{324}$$
$$+ \frac{m_\mu^2}{M_\rho^2} \left[\frac{155}{27} L^2 - \left(\frac{65}{27} + \frac{3\pi}{\sqrt{3}}\right) L - \frac{11915}{1296}\right.$$
$$\left.+ \frac{2\pi}{\sqrt{3}} S_2 + \frac{4\pi}{3\sqrt{3}} + \frac{117}{4} S_2 - \frac{1}{6}\zeta_3 + \frac{347\pi^2}{1944}\right]$$

5.4 Hadronic Light–by–Light Scattering

$$+\delta\left[\left(\frac{2}{3}-\frac{2\pi}{3\sqrt{3}}\right)L-\frac{1}{27}+\frac{5}{2}\frac{\pi}{\sqrt{3}}S_2-\frac{11\pi}{18\sqrt{3}}\right.$$
$$\left.-\frac{1}{8}S_2-\frac{2}{9}\zeta_3+\frac{53\pi^2}{648}\right]+\mathcal{O}\left(\frac{m_\mu^4}{M_\rho^4},\delta^2\right), \qquad (5.114)$$

where $L = \log(M_\rho/m_\mu)$, $\zeta_3 \simeq 1.202057$ and $S_2 = \frac{4}{9\sqrt{3}}\text{Cl}_2\left(\frac{\pi}{3}\right) \simeq 0.260434$. The numerical evaluation in terms of the known physical parameters yields

$$a_{\mu;\text{VMD}}^{\text{LbL};\pi^0} = [136 - 112 + 30] \times 10^{-11} = +54 \times 10^{-11}, \qquad (5.115)$$

and confirms the result [126]. Note that there are large cancellations between leading and subleading terms, although the leading log is sizable $\ln(M_\rho/m_\mu) \simeq 1.98$. Nyffeler obtained $a_{\mu;\text{VMD}}^{\text{LbL};\pi^0} = [123 - 103 + 34] \times 10^{-11} = +54 \times 10^{-11}$ [151] confirming this pattern, by a fit of (5.113) to the representation (5.107) for the VMD model. A similar result has been obtained in [138] by calculating $c_1 = -2/3\,\delta\chi(\mu_0) + 0.237 = -0.93^{+0.67}_{-0.83}$. The bare pion and kaon loops (with undressed photons) as expected yield a subleading correction with $f\left(\frac{m_{\pi^\pm}}{m_\mu}, \frac{M_{K^\pm}}{m_\mu}\right) = -0.038$ or $\delta a_{\mu;\text{sQED}}^{\text{LbL}}(\pi^\pm, K^\pm - \text{loops}) \simeq -48 \times 10^{-11}$.

We are now ready to summarize the results obtained by the different groups. A comparison of the different results also sheds light on the difficulties and the model dependencies in the theoretical estimations achieved so far.

5.4.5 A Summary of Results

The results of the various evaluations may be summarized as follows:

a) According to Table 5.8 the diagram Fig. 5.31(a) yields the most important contribution but requires a model for its calculation. The results for this dominating contribution are collected in Table 5.9.
b) Next in Table 5.8 are pion– and kaon–loops Fig. 5.31(b) which only yields a subleading contribution again being model dependent. Results are given in Table 5.10.

The evaluation of the bare pseudoscalar loops actually is possible in terms of the large mass expansion in m_μ/M_P. The expansion in scalar QED, relevant for the charged pion contribution, is given by [152]

$$a_\mu(\text{lbl};\pi^\pm)_{\text{sQED}} = A_{2\,\text{lbl}}^{(6)}(m_\mu/m_\pi)\left(\frac{\alpha}{\pi}\right)^3, \qquad (5.116)$$

with

$$A_{2\,\text{lbl}}^{(6)}(m_\mu/M) = \frac{m_\mu^2}{M^2}\left(\frac{1}{4}\zeta_3 - \frac{37}{96}\right)$$
$$+ \frac{m_\mu^4}{M^4}\left(\frac{1}{8}\zeta_3 + \frac{67}{6480}\zeta_2 - \frac{282319}{1944000} + \frac{67}{12960}L^2 + \frac{7553}{388800}L\right)$$

338 5 Hadronic Effects

Table 5.9. Light–by–Light: π^0, η, η'

Model for $\mathcal{F}_{\pi^{0*}\gamma^*\gamma^*}$	$a_\mu(\pi^0) \times 10^{11}$	$a_\mu(\pi^0,\eta,\eta') \times 10^{11}$
ENJL[BPP]	59(11)	85(13)
HLS [HKS,HK]	57(4)	83(6)
LMD+V[KN] ($h_2 = 0$)	58(10)	83(12)
LMD+V[KN](1) ($h_2 = -10$ GeV^2)	63(10)	88(12)
LMD+V[MV](2) ($h_2 = -10$ GeV^2)+new S.D.	77(5)	114(10)

$$+\frac{m_\mu^6}{M^6}\left(\frac{19}{216}\zeta_3 + \frac{157}{36288}\zeta_2 - \frac{767572853}{7112448000} + \frac{1943}{725760}L^2 + \frac{51103}{7620480}L\right)$$

$$+\frac{m_\mu^8}{M^8}\left(\frac{11}{160}\zeta_3 + \frac{943}{432000}\zeta_2 - \frac{3172827071}{37507050000} + \frac{8957}{6048000}L^2 + \frac{22434967}{7620480000}L\right)$$

$$+\frac{m_\mu^{10}}{M^{10}}\left(\frac{17}{300}\zeta_3 + \frac{139}{111375}\zeta_2 - \frac{999168445440307}{14377502462400000} + \frac{128437}{149688000}L^2\right.$$

$$\left.+\frac{1033765301}{691558560000}L\right) + O\left(\frac{m_\mu^{12}}{M^{12}}\right), \qquad (5.117)$$

where $L = \ln(M^2/m_\mu^2)$, M denoting the pseudoscalar meson mass m_π, $m_K, \cdots$ and $\zeta_2 = \zeta(2) = \pi^2/6$, $\zeta_3 = \zeta(3)$. The numerical evaluation of the exact sQED contribution yielded $A^{(6)}_{2\,\mathrm{lbl}}(m_\mu/m_\pi) = -0.0383(20)$ [72], more recently [152] obtains $A^{(6)}_{2\,\mathrm{lbl}}(m_\mu/m_\pi) = -0.0353$ using the heavy mass expansion approach. With our choice of parameters using (5.117) we get $a_\mu(\mathrm{lbl}; \pi^\pm)_\mathrm{sQED} = -45.3 \times 10^{-11}$.

For the dressed case $a_\mu(\mathrm{LbL})_{\mathrm{sQED+VMD}}$ an expansion in $\delta = (m_\mu - m_\pi)/m_\pi$ and $(m_\pi/M_\rho)^2$ has been given for the HLS model in [139]. For physical $m_\pi/M_\rho \sim 0.2$ this expansion is poorly convergent and therefore not of big help, as the "cut–off" M_ρ is too low.

c) Third in Table 5.8 is the quark loop Fig. 5.31(c) which only appears as a S.D. complement of the ENJL and the HLS low energy effective models. Corresponding values are included in the last column of Table 5.10.

Table 5.10. Light–by–light: $\pi^\pm, K^\pm$ & quark loops

Model $\pi^+\pi^-\gamma^*(\gamma^*)$	$a_\mu(\pi^\pm) \times 10^{11}$	$a_\mu(\pi^\pm, K^\pm) \times 10^{11}$	$a_\mu(\mathrm{quarks}) \times 10^{11}$
Point	-45.3	-49.8	62(3)
VMD	-16	-	-
ENJL[BPP]	$-18(13)$	$-19(13)$	21(3)
HLS [HKS,HK]	-4 (8)	$-4.5(8.1)$	9.7(11.1)
guesstimate [MV]	0 (10)	0 (10)	-

In the large–N_c resonance saturation approach (LMD) the S.D. behavior is incorporated as a boundary condition and no separate quark loops contributions has to be accounted for.

However, other effects which were first considered in [139] must be taken into account:

1) the constraint on the twist four $(1/q^4)$–term in the OPE requires $h_2 = -10$ GeV^2 in the Knecht-Nyffeler from factor (5.111): $\delta a_\mu \simeq +5 \pm 0$
2) the contributions from the f_1 and f'_1 isoscalar axial–vector mesons: $\delta a_\mu \simeq +10 \pm 4$ (using dressed photons)
3) for the remaining effects: scalars (f_0) + dressed $\pi^\pm, K^\pm$ loops + dressed quark loops: $\delta a_\mu \simeq -5 \pm 13$

Note that the remaining terms have been evaluated in [117, 134] only. The splitting into the different terms is model dependent and only the sum should be considered: the results read -5 ± 13 (BPP) and 5.2 ± 13.7 (HKS) and hence the contribution remains unclear[28].

Finally, including other small contributions the totals reported in the most recent estimations are shown in Table 5.11.

Note that as far as this application is concerned the ENJL and the HLS models are equivalent and in fact the HLS may be "derived" from the ENJL model by making a number of additional approximations [120]. The uncertainties quoted include the changes due to the variation of the cut–off by 0.7–8 GeV for the ENJL model and by $1-4$ GeV for the HLS model. For the LMD+V parametrization, the leading π^0–exchange contribution does not involve an explicit cut–off dependence (large–N_c duality approach).

Because of the increased accuracy of the experiments and the substantial reduction of the error on the other hadronic contributions also a reconsideration of the hadronic light–by–light contributions is needed. To what extent this is possible remains to be seen, however, some progress should be possible

Table 5.11. Summary of most recent results. The last column is my estimate based on the other results (see text)

$10^{11}\ a_\mu$	BPP	HKS	KN	MV	FJ
π^0, η, η'	85± 13	82.7± 6.4	83± 12	114± 10	88± 12
π, K loops	−19± 13	−4.5± 8.1		0± 10	−19± 13
axial vector	2.5± 1.0	1.7± 0.0		22± 5	10± 4
scalar	−6.8± 2.0	−	−	−	−7± 3
quark loops	21± 3	9.7± 11.1	−	−	21± 3
total	± 83± 32	± 89.6± 15.4	± 80± 40	± 136± 25	± 93± 34

[28] We adopt the value estimated in [133], because the sign of the scalar contribution, which dominates in the sum, has to be negative in any case (see [136]).

by taking into account various points which have been brought up in the more recent discussions.

References

1. A. Litke et al., Phys. Rev. Lett. **30** (1973) 1189; G. Tarnopolsky et al., Phys. Rev. Lett. **32** (1974) 432
2. H. D. Politzer, Phys. Rev. Lett. **30** (1973) 1346;
   D. Gross, F. Wilczek, Phys. Rev. Lett. **30** (1973) 1343
3. H. Fritzsch, M. Gell-Mann, H. Leutwyler, Phys. Lett. **47**B (1973) 365
4. T. Appelquist, H. Georgi, Phys. Rev. D **8** (1973) 4000;
   A. Zee, Phys. Rev. D **8** (1973) 4038
5. H. Fritzsch, Phys. Rev. D **10** (1974) 1624;
   H. Fritzsch, H. Leutwyler, Report CALT-68–416, 1974
6. F. Jegerlehner, Nucl. Phys. Proc. Suppl. **162** (2006) 22 [hep-ph/0608329]
7. S. Eidelman, F. Jegerlehner, Z. Phys. C **67** (1995) 585; F. Jegerlehner, Nucl. Phys. (Proc. Suppl.) C **51** (1996) 131; J. Phys. G **29** (2003) 101; Nucl. Phys. Proc. Suppl. **126** (2004) 325
8. K. Adel, F. J. Yndurain, hep-ph/9509378
9. D. H. Brown, W. A. Worstell, Phys. Rev. D **54** (1996) 3237
10. R. Alemany, M. Davier, A. Höcker, Eur. Phys. J. C **2** (1998) 123
11. A. D. Martin, J. Outhwaite, M. G. Ryskin, Phys. Lett. B **492** (2000) 69; Eur. Phys. J. C **19** (2001) 681
12. O. V. Zenin et al., hep-ph/0110176.
13. M. R. Whalley, J. Phys. G **29** (2003) A1
14. A. E. Blinov et al., Z. Phys. C **70** (1996) 31
15. J. Z. Bai et al. [BES Collaboration], Phys. Rev. Lett. **84** (2000) 594; Phys. Rev. Lett. **88** (2002) 101802
16. R. R. Akhmetshin et al. [CMD-2 Collaboration], Phys. Lett. B **578** (2004) 285
17. V. M. Aulchenko et al. [CMD-2 Collaboration], JETP Lett. **82** (2005) 743 [Pisma Zh. Eksp. Teor. Fiz. **82** (2005) 841]; R. R. Akhmetshin et al., JETP Lett. **84** (2006) 413 [Pisma Zh. Eksp. Teor. Fiz. **84** (2006) 491]; hep-ex/0610021
18. A. Aloisio et al. [KLOE Collaboration], Phys. Lett. B **606** (2005) 12
19. M. N. Achasov et al. [SND Collaboration], J. Exp. Theor. Phys. **103** (2006) 380 [Zh. Eksp. Teor. Fiz. **130** (2006) 437]
20. B. Aubert et al. [BABAR Collaboration], Phys. Rev. D **70** (2004) 072004; **71** (2005) 052001; **73** (2006) 012005; **73** (2006) 052003
21. K. G. Chetyrkin, A. L. Kataev, F. V. Tkachov, Phys. Lett. B **85** (1979) 277; M. Dine, J. Sapirstein, Phys. Rev. Lett. **43** (1979) 668; W. Celmaster, R. J. Gonsalves, Phys. Rev. Lett. **44** (1980) 560;
    K. G. Chetyrkin, Phys. Lett. B **391** (1997) 402
22. S. G. Gorishnii, A. L. Kataev, S. A. Larin, Phys. Lett. B **259** (1991) 144;
    L. R. Surguladze, M. A. Samuel, Phys. Rev. Lett. **66** (1991) 560 [Erratum-ibid. **66** (1991) 2416]
23. K. G. Chetyrkin, J. H. Kühn, Phys. Lett. B **342** (1995) 356;
    K. G. Chetyrkin, J. H. Kühn, A. Kwiatkowski, Phys. Rept. **277** (1996) 189;
    K. G. Chetyrkin, J. H. Kühn, M. Steinhauser, Phys. Lett. B **371**, 93 (1996);

Nucl. Phys. B **482**, 213 (1996); **505**, 40 (1997);
K. G. Chetyrkin, R. Harlander, J.H. Kühn, M. Steinhauser, Nucl. Phys. B **503**, 339 (1997) K. G. Chetyrkin, R. V. Harlander, J. H. Kühn, Nucl. Phys. B **586** (2000) 56 [Erratum-ibid. B **634** (2002) 413]
24. R. V. Harlander, M. Steinhauser, Comput. Phys. Commun. **153** (2003) 244
25. F. Jegerlehner, Z. Phys. C **32** (1986) 195
26. E. D. Bloom, F. J. Gilman, Phys. Rev. Lett. **25** (1970) 1140; Phys. Rev. D **4** (1971) 2901
27. E. C. Poggio, H. R. Quinn, S. Weinberg, Phys. Rev. D **13** (1976) 1958
28. M. A. Shifman, *Quark–hadron duality,* hep-ph/0009131
29. A. D. Martin, D. Zeppenfeld, Phys. Lett. B **345** (1995) 558
30. M. Davier, A. Höcker, Phys. Lett. B **419** (1998) 419; Phys. Lett. B **435** (1998) 427
31. J. H. Kühn, M. Steinhauser, Phys. Lett. B **437** (1998) 425
32. G. J. Gounaris, J. J. Sakurai, Phys. Rev. Lett. **21** (1968) 244
33. S. Eidelman et al. [Particle Data Group], Phys. Lett. B **592** (2004) 1
34. M. L. Swartz, SLAC-PUB-6710, 1994, `hep-ph/9411353` unpublished
35. M. L. Swartz, Phys. Rev. D **53** (1996) 5268
36. Nucl. Inst. Meth. A **346** (1994) 306
37. A. B. Arbuzov, V. A. Astakhov, A. V. Fedorov, G. V. Fedotovich, E. A. Kuraev, N. P. Merenkov, JHEP **9710** (1997) 006
38. V. A. Khoze, M. I. Konchatnij, N. P. Merenkov, G. Pancheri, L. Trentadue, O. N. Shekhovzova, Eur. Phys. J. C **18** (2001) 481
39. A. Hoefer, J. Gluza, F. Jegerlehner, Eur. Phys. J. C **24** (2002) 51
40. A. B. Arbuzov, G. V. Fedotovich, F. V. Ignatov, E. A. Kuraev, A. L. Sibidanov, Eur. Phys. J. C **46** (2006) 689
41. V. N. Baier, V. S. Fadin, Phys. Lett. B **27** (1968) 223
42. S. Spagnolo, Eur. Phys. J. C **6** (1999) 637; A. B. Arbuzov, E. A. Kuraev, N. P. Merenkov, L. Trentadue, JHEP **12** (1998) 009; S. Binner, J. H. Kühn, K. Melnikov, Phys. Lett. B **459** (1999) 279; M. I. Konchatnij, N. P. Merenkov, JETP Lett. **69** (1999) 811; V. A. Khoze et al., Eur. Phys. J. C **18** (2001) 481
43. M. Benayoun, S. I. Eidelman, V. N. Ivanchenko, Z. K. Silagadze, Mod. Phys. Lett. A **14** (1999) 2605.
44. H. Czyż, J. H. Kühn, Eur. Phys. J. C **18** (2001) 497; G. Rodrigo, A. Gehrmann-De Ridder, M. Guilleaume, J. H. Kühn, Eur. Phys. J. C **22** (2001) 81; G. Rodrigo, H. Czyż, J. H. Kühn, M. Szopa, Eur. Phys. J. C **24** (2002) 71; H. Czyż, A. Grzelińska, J. H. Kühn and G. Rodrigo, Eur. Phys. J. C **27** (2003) 563; Eur. Phys. J. C **33** (2004) 333; Eur. Phys. J. C **39** (2005) 411
45. J. Gluza, A. Hoefer, S. Jadach, F. Jegerlehner, Eur. Phys. J. C **28** (2003) 261
46. S. Dubinsky, A. Korchin, N. Merenkov, G. Pancheri, O. Shekhovtsova, Eur. Phys. J. C **40** (2005) 41
47. P. A. M. Dirac, Théorie du positron. In: *Septième Conseile de Physique Solvay: Structure et propriétés des noyaux atomique, October 22–29, 1933,* (Gauthier-Villars, Paris 1934), pp. 203–230;
P. A. M. Dirac, Proc. Cambridge. Phil. Soc. **30** (1934) 150
48. J. Schwinger, Phys. Rev. **75** (1949) 651
49. R. P. Feynman, Phys. Rev. **76** (1949) 749
50. R. Jost, J. M. Luttinger, Helv. Phys. Acta **23** (1950) 201

51. V. A. Novikov, L. B. Okun, M. A. Shifman, A. I. Vainshtein, M. B. Voloshin, V. I. Zakharov, Phys. Rept. **41** (1978) 1
52. S. L. Adler, Phys. Rev. D **10** (1974) 3714; A. De Rujula, H. Georgi, Phys. Rev. D **13** (1976) 1296
53. G. Sterman, S. Weinberg, Phys. Rev. Lett. **39** (1977) 1436
54. M. A. Shifman, A. I. Vainshtein, V. I. Zakharov, Nucl. Phys. B **147** (1979) 385
55. S. Eidelman, F. Jegerlehner, A. L. Kataev, O. Veretin, Phys. Lett. B **454** (1999) 369
56. R. J. Crewther, Phys. Rev. Lett. **28** (1972) 1421
57. M. S. Chanowitz, J. R. Ellis, Phys. Rev. D **7** (1973) 2490
58. S. L. Adler, J. C. Collins, A. Duncan, Phys. Rev. D **15** (1977) 1712; J. C. Collins, A. Duncan, S. D. Joglekar, Phys. Rev. D **16** (1977) 438
59. S. J. Brodsky, E. de Rafael, Phys. Rev. **168** (1968) 1620
60. H. N. Brown et al. [Muon (g-2) Collaboration], Phys. Rev. Lett. **86** (2001) 2227
61. L. M. Barkov et al., (OLYA, CMD), Nucl. Phys. B256 (1985) 365; R. R. Akhmetshin et al. [CMD-2 Collaboration], hep-ex/9904027; Phys. Lett. B **527** (2002) 161; ibid. **578** (2004) 285
62. N. Cabibbo, R. Gatto, Phys. Rev. Lett. **4** (1960) 313, Phys. Rev. **124** (1961) 1577
63. C. Bouchiat, L. Michel, J. Phys. Radium **22** (1961) 121
64. L. Durand, III., Phys. Rev. **128** (1962) 441; Erratum-ibid. **129** (1963) 2835
65. T. Kinoshita, R. J. Oakes, Phys. Lett. **25**B (1967) 143
66. M. Gourdin, E. De Rafael, Nucl. Phys. B **10** (1969) 667
67. A. Bramòn, E. Etim, M. Greco, Phys. Lett. **39**B (1972) 514
68. V. Barger, W. F. Long, M. G. Olsson, Phys. Lett. **60**B (1975) 89
69. J. Calmet, S. Narison, M. Perrottet, E. de Rafael, Phys. Lett. B **61** (1976) 283; Rev. Mod. Phys. **49** (1977) 21
70. S. Narison, J. Phys. G: Nucl. Phys. **4** (1978) 1849
71. L. M. Barkov et al., Nucl. Phys. B **256** (1985) 365
72. T. Kinoshita, B. Nizic, Y. Okamoto, Phys. Rev. Lett. **52** (1984) 717; Phys. Rev. D **31** (1985) 2108
73. J. A. Casas, C. Lopez, F. J. Ynduráin, Phys. Rev. D **32** (1985) 736
74. Ľ. Martinovič, S. Dubnička, Phys. Rev. D **42** (1990) 884
75. A. Z. Dubničková, S. Dubnička, P. Strizenec, Dubna-Report, JINR-E2-92–28, 1992
76. M. Davier, S. Eidelman, A. Höcker, Z. Zhang, Eur. Phys. J. C **27** (2003) 497; Eur. Phys. J. C **31** (2003) 503
77. S. Ghozzi, F. Jegerlehner, Phys. Lett. B **583** (2004) 222
78. S. Narison, Phys. Lett. B **568** (2003) 231
79. V. V. Ezhela, S. B. Lugovsky, O. V. Zenin, hep-ph/0312114
80. K. Hagiwara, A. D. Martin, D. Nomura, T. Teubner, Phys. Lett. B **557** (2003) 69; Phys. Rev. D **69** (2004) 093003; Phys. Lett. B **649** (2007) 173
81. J. F. De Troconiz, F. J. Yndurain, Phys. Rev. D **65** (2002) 093001; Phys. Rev. D **71** (2005) 073008
82. S. Eidelman, Proceedings of the XXXIII *International Conference on High Energy Physics,* July 27 – August 2, 2006, Moscow (Russia), World Scientific, to appear; M. Davier, Nucl. Phys. Proc. Suppl. 169 (2007) 288
83. F. Jegerlehner, In: *Radiative Corrections*, ed by J. Solà (World Scientific, Singapore 1999) pp. 75–89
84. R. Barate et al. [ALEPH Collaboration], Z. Phys. C **76** (1997) 15; Eur. Phys. J. C **4** (1998) 409;

85. S. Schael et al. [ALEPH Collaboration], Phys. Rept. **421** (2005) 191
86. K. Ackerstaff et al. [OPAL Collaboration], Eur. Phys. J. C **7** (1999) 571
87. S. Anderson et al. [CLEO Collaboration], Phys. Rev. D **61** (2000) 112002
88. Y. S. Tsai, Phys. Rev. D **4** (1971) 2821 [Erratum-ibid. D **13** (1976) 771]
89. W. J. Marciano, A. Sirlin, Phys. Rev. Lett. **61** (1988) 1815
90. E. Braaten, C. S. Li, Phys. Rev. D **42** (1990) 3888
91. R. Decker, M. Finkemeier, Nucl. Phys. B **438** (1995) 17
92. J. Erler, Rev. Mex. Fis. **50** (2004) 200 [hep-ph/0211345]
93. V. Cirigliano, G. Ecker, H. Neufeld, Phys. Lett. B **513** (2001) 361; JHEP **0208** (2002) 002
94. G. Ecker, J. Gasser, A. Pich, E. de Rafael, Nucl. Phys. B **321** (1989) 311
95. G. Ecker, J. Gasser, H. Leutwyler, A. Pich, E. de Rafael, Phys. Lett. B **223** (1989) 425
96. E. de Rafael, Phys. Lett. B **322** (1994) 239
97. B. V. Geshkenbein, V. L. Morgunov, Phys. Lett. B **352** (1995) 456
98. J. Gasser, H. Leutwyler, Ann. of Phys. (NY) **158** (1984) 142
99. J. Gasser, H. Leutwyler, Nucl. Phys. B **250** (1985) 517
100. J. Gasser, U.-G. Meißner, Nucl. Phys. B **357** (1991) 90
101. G. Colangelo, M. Finkemeier, R. Urech, Phys. Rev. D **54** (1996) 4403
102. S. R. Amendolia et al. [NA7 Collaboration], Nucl. Phys. B **277** (1986) 168
103. R. Omnès, Nuovo Cim. **8** (1958) 316
104. B. Hyams et al., Nucl. Phys. B **64** (1973) 134
105. B. Ananthanarayan, G. Colangelo, J. Gasser, H. Leutwyler, Phys. Rept. **353** (2001) 207; G. Colangelo, J. Gasser, H. Leutwyler, Nucl. Phys. B **603** (2001) 125; I. Caprini, G. Colangelo, J. Gasser, H. Leutwyler, Phys. Rev. D **68** (2003) 074006; B. Ananthanarayan, I. Caprini, G. Colangelo, J. Gasser, H. Leutwyler, Phys. Lett. B **602** (2004) 218
106. H. Leutwyler, *On the dispersion theory of $\pi\pi$ scattering*, hep-ph/0612111; $\pi\pi$ scattering, hep-ph/0612112
107. H. Leutwyler, *Electromagnetic form-factor of the pion*, hep-ph/0212324
108. G. Colangelo, Nucl. Phys. Proc. Suppl. **131** (2004) 185; ibid. **162** (2006) 256
109. B. V. Geshkenbein, Phys. Rev. D **61** (2000) 033009
110. L. Łukaszuk, Phys. Lett. B **47** (1973) 51; S. Eidelman, L. Łukaszuk, Phys. Lett. B **582** (2004) 27
111. F. M. Renard, Nucl. Phys. B **82** (1974) 1; N. N. Achasov, N. M. Budnev, A. A. Kozhevnikov, G. N. Shestakov, Sov. J. Nucl. Phys. **23** (1976) 320; N. N. Achasov, A. A. Kozhevnikov, M. S. Dubrovin, V. N. Ivanchenko, E. V. Pakhtusova, Int. J. Mod. Phys. A **7** (1992) 3187
112. R. Barbieri, E. Remiddi, Phys. Lett. B **49** (1974) 468; Nucl. Phys. B **90** (1975) 233
113. B. Krause, Phys. Lett. B **390** (1997) 392
114. J. S. Schwinger, *Particles, sources,, fields*, Vol. 3 (Addison–Wesley, Redwood City (USA) 1989), p. 99; see also: M. Drees, K. i. Hikasa, Phys. Lett. B **252** (1990) 127, where a misprint in Schwinger's formula is corrected.
115. K. Melnikov, Int. J. Mod. Phys. A **16** (2001) 4591
116. J. J. Sakurai, Annals of Phys. (NY) **11** (1960) 1; H. Joos, Acta Phys. Austriaca Suppl. **4** (1967); N. M. Kroll, T. D. Lee, B. Zumino, Phys. Rev. **157** (1967) 1376

117. M. Hayakawa, T. Kinoshita, A. I. Sanda, Phys. Rev. Lett. **75** (1995) 790; Phys. Rev. D **54** (1996) 3137
118. M. Bando, T. Kugo, S. Uehara, K. Yamawaki, T. Yanagida, Phys. Rev. Lett. **54** (1985) 1215; M. Bando, T. Kugo, K. Yamawaki, Phys. Rept. **164** (1988) 217; M. Harada, K. Yamawaki, Phys. Rept. **381** (2003) 1
119. U. G. Meissner, Phys. Rept. **161** (1988) 213; see also S. Weinberg, Phys. Rev. **166** (1968) 1568
120. A. Dhar, R. Shankar, S. R. Wadia, Phys. Rev. D **31** (1985) 3256; D. Ebert, H. Reinhardt, Phys. Lett. B **173** (1986) 453
121. J. Prades, Z. Phys. **C63**, 491 (1994); Erratum: Eur. Phys. J. **C11**, 571 (1999).
122. S. J. Brodsky, G. R. Farrar, Phys. Rev. Lett. **31** (1973) 1153; Phys. Rev. D **11** (1975) 1309
123. G. 't Hooft, Nucl. Phys. B **72** (1974) 461; ibid. **75** (1974) 461; E. Witten, Nucl. Phys. B **160** (1979) 57
124. A. V. Manohar, Hadrons in the $1/N$ Expansion, In: *At the frontier of Particle Physics*, ed M. Shifman, (World Scientific, Singapore 2001) Vol. 1, pp. 507–568
125. H. Leutwyler, Nucl. Phys. Proc. Suppl. **64** (1998) 223; R. Kaiser, H. Leutwyler, Eur. Phys. J. C **17** (2000) 623
126. M. Knecht, A. Nyffeler, Phys. Rev. D **65**, 073034 (2002)
127. Y. Nambu, G. Jona-Lasinio, Phys. Rev. **122** (1961) 345, ibid. **124** (1961) 246
128. M. Gell-Mann, M. Levy, Nuovo Cim. **16** (1960) 705
129. J. Bijnens, Phys. Rept. **265** (1996) 369
130. W. M. Yao et al. [Particle Data Group], J. Phys. G **33** (2006) 1
131. K. Kawarabayashi, M. Suzuki, Phys. Rev. Lett. **16** (1966) 255; Riazuddin, Fayyazuddin, Phys. Rev. **147** (1966) 1071
132. S. Peris, M. Perrottet, E. de Rafael, JHEP **9805** (1998) 011; M. Knecht, S. Peris, M. Perrottet, E. de Rafael, Phys. Rev. Lett. **83** (1999) 5230; M. Knecht, A. Nyffeler, Eur. Phys. J. C **21** (2001) 659
133. J. Bijnens, E. Gamiz, E. Lipartia, J. Prades, JHEP **0304** (2003) 055
134. J. Bijnens, E. Pallante, J. Prades, Phys. Rev. Lett. **75** (1995) 1447 [Erratum-ibid. **75** (1995) 3781]; Nucl. Phys. B **474** (1996) 379; [Erratum-ibid. **626** (2002) 410]
135. M. Hayakawa, T. Kinoshita, Phys. Rev. D **57** (1998) 465 [Erratum-ibid. D **66** (2002) 019902];
136. M. Knecht, A. Nyffeler, M. Perrottet, E. De Rafael, Phys. Rev. Lett. **88** (2002) 071802
137. I. Blokland, A. Czarnecki, K. Melnikov, Phys. Rev. Lett. **88** (2002) 071803
138. M. Ramsey-Musolf, M. B. Wise, Phys. Rev. Lett. **89** (2002) 041601
139. K. Melnikov, A. Vainshtein, Phys. Rev. D **70** (2004) 113006
140. H. J. Behrend et al. [CELLO Collaboration], Z. Phys. C **49** (1991) 401
141. J. Gronberg et al. [CLEO Collaboration], Phys. Rev. D **57** (1998) 33
142. G. P. Lepage, S. J. Brodsky, Phys. Rev. D **22** (1980) 2157; S. J. Brodsky, G. P. Lepage, Phys. Rev. D **24** (1981) 1808
143. A. V. Efremov, A. V. Radyushkin, Phys. Lett. B **94** (1980) 245; A. V. Radyushkin, R. Ruskov, Phys. Lett. B **374** (1996) 173; A. V. Radyushkin, Acta Phys. Polon. B **26** (1995) 2067
144. S. Ong, Phys. Rev. D **52** (1995) 3111; P. Kroll, M. Raulfs, Phys. Lett. B **387** (1996) 848; A. Khodjamirian, Eur. Phys. J. C **6** (1999) 477

145. F. del Aguila, M. K. Chase, Nucl. Phys. B **193** (1981) 517; E. Braaten, Phys. Rev. D **28** (1983) 524; E. P. Kadantseva, S. V. Mikhailov, A. V. Radyushkin, Yad. Fiz. **44** (1986) 507 [Sov. J. Nucl. Phys. **44** (1986) 326]
146. A. Vainshtein, Phys. Lett. B **569** (2003) 187
147. M. Knecht, S. Peris, M. Perrottet, E. de Rafael, JHEP **0403** (2004) 035
148. F. Jegerlehner, O. V. Tarasov, Phys. Lett. B **639** (2006) 299
149. A. Nyffeler, Nucl. Phys. B (Proc. Suppl.) **131** (2004) 162
150. J. Bijnens, F. Persson, *Effects of different form-factors in meson photon photon transitions and the muon anomalous magnetic moment*, hep-ph/0106130
151. A. Nyffeler, *Theoretical status of the muon g − 2*, hep-ph/0305135
152. J. H. Kühn, A. I. Onishchenko, A. A. Pivovarov, O. L. Veretin, Phys. Rev. D **68** (2003) 033018

6
The $g - 2$ Experiments

6.1 Overview on the Principle of the Experiment

There are a number of excellent reviews on this subject and I am following in parts the ones of Farley and Picasso [1] and of Vernon Hughes [2]. See also the more recent overviews [3, 4]. Many details on the experimental setup of the E821 experiment may be found in the dissertation written by Paley [5], which was also very helpful for me.

The principle of the BNL muon $g - 2$ experiment involves the study of the orbital and spin motion of highly polarized muons in a magnetic storage ring. This method has been applied in the last CERN experiment [6] already. The key improvements of the BLN experiment include the very high intensity of the primary proton beam from the proton storage ring AGS (Alternating Gradient Synchrotron), the injection of muons instead of pions into the storage ring, and a super–ferric storage ring magnet [7].

The muon $g-2$ experiment at Brookhaven works as illustrated in Fig. 6.1 [8, 9, 10]. Protons (mass about 1 GeV, energy 24 GeV) from the AGS hit a target and produce pions (of mass about 140 MeV). The pions are unstable and decay into muons plus a neutrino where the muons carry spin and thus a magnetic moment which is directed along the direction of the flight axis. The longitudinally polarized muons from pion decay are then injected into a uniform magnetic field B where they travel in a circle. The ring is a doughnut–shaped structure with a diameter of 14 meters. A picture of the BNL muon storage ring is shown in Fig. 6.2. In the horizontal plane of the orbit the muons execute relativistic cyclotron motion with angular frequency ω_c. By the motion of the muon magnetic moment in the homogeneous magnetic field the spin axis is changed in a particular way as described by the Larmor precession. After each circle the muon's spin axis changes by 12' (arc seconds), while the muon is traveling at the same momentum (see Fig. 3.1). The muon spin is precessing with angular frequency ω_s, which is slightly bigger than ω_c by the difference angular frequency $\omega_a = \omega_s - \omega_c$.

348   6 The $g-2$ Experiments

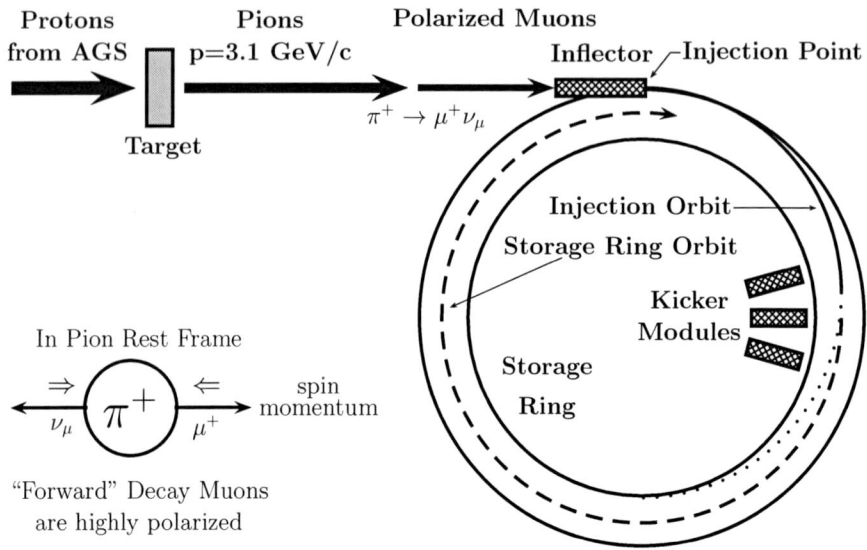

Fig. 6.1. The schematics of muon injection and storage in the $g-2$ ring

Fig. 6.2. The Brookhaven National Laboratory muon storage ring. The ring has a radius of 7.112 meters, the aperture of the beam pipe is 90 mm, the field is 1.45 Tesla and the momentum of the muon is $p_\mu = 3.094$ GeV/c. Picture taken from the Muon $g-2$ Collaboration Web Page http://www.g-2.bnl.gov/ (Courtesy of Brookhaven National Laboratory)

6.1 Overview on the Principle of the Experiment

$$\omega_c = \frac{eB}{m_\mu c \gamma}$$

$$\omega_s = \frac{eB}{m_\mu c \gamma} + \frac{e}{m_\mu c} a_\mu B$$

$$\omega_a = \frac{e}{m_\mu c} a_\mu B , \qquad (6.1)$$

where $a_\mu = (g_\mu - 2)/2$ is the muon anomaly and $\gamma = 1/\sqrt{1 - v^2/c^2}$ is the relativistic Lorentz factor, v the muon velocity[1]. In the experiment ω_a and B are measured. The muon mass m_μ is obtained from an independent experiment on muonium, which is a $(\mu^+ e^-)$ bound system. Note that if the muon would just have its Dirac magnetic moment $g = 2$ (tree level) the direction of the spin of the muon would not change at all.

In order to retain the muons in the ring an electrostatic focusing system is needed. In reality in addition to the magnetic field $\boldsymbol{B}$ an electric quadrupole field $\boldsymbol{E}$ in the plane normal to the particle orbit is applied, which changes the angular frequency according to

$$\omega_a = \frac{e}{m_\mu c} \left(a_\mu \boldsymbol{B} - \left[a_\mu - \frac{1}{\gamma^2 - 1} \right] \frac{\boldsymbol{v} \times \boldsymbol{E}}{c^2} \right) . \qquad (6.2)$$

Interestingly, one has the possibility to choose γ such that $a_\mu - 1/(\gamma^2 - 1) = 0$, in which case ω_a becomes independent of $\boldsymbol{E}$. This is the so–called *magic* γ. The muons are rather unstable and decay spontaneously after some time. When running at the magic energy the muons are highly relativistic, they travel almost at the speed of light with energies of about $E_{\text{magic}} = \gamma m_\mu c^2 \simeq 3.098$ GeV. This rather high energy is dictated by the need of a large time dilatation on the one hand and by the requirement to minimize the precession frequency shift caused by the electric quadrupole superimposed upon the uniform magnetic field. The magic γ-factor is about $\gamma = \sqrt{1 + 1/a_\mu} = 29.3$; the lifetime of a muon at rest is 2.19711 μs (micro seconds), while in the ring it is 64.435 μs (theory) [64.378 μs (experiment)]. Thus, with their lifetime being much larger than at rest, muons are circling in the ring many times before they decay into a positron plus two neutrinos: $\mu^+ \to e^+ + \nu_e + \bar{\nu}_\mu$. Since parity is violated maximally in this weak decay there is a strong correlation between the muon spin direction and the direction of emission of the positrons. The differential decay rate for the muon in the rest frame is given by (see also (2.46) and (6.55) below)

$$d\Gamma = N(E_e) \left(1 + \frac{1 - 2x_e}{3 - 2x_e} \cos\theta \right) d\Omega , \qquad (6.3)$$

in which E_e is the positron energy, x_e is E_e in units of the maximum energy $m_\mu/2$, $N(E_e)$ is a normalization factor and θ the angle between the positron momentum in the muon rest frame and the muon spin direction. The μ^+ decay

---

[1] Formulae like (6.1) presented in this first overview will be derived below.

spectrum is peaked strongly for small θ due to the non-vanishing coefficient of $\cos\theta$

$$A(E_e) \doteq \frac{1 - 2x_e}{3 - 2x_e} , \qquad (6.4)$$

which is called asymmetry factor and reflects the *parity violation*.

The positron is emitted along the spin axis of the muon as illustrated in Fig. 6.3. The decay positrons are detected by 24 lead/scintillating fiber calorimeters spread evenly around inside the muon storage ring. These counters measure the positron energy and provide the direction of the muon spin. The number of decay positrons with energy greater than E emitted at time t after muons are injected into the storage ring is

$$N(t) = N_0(E) \exp\left(\frac{-t}{\gamma\tau_\mu}\right) [1 + A(E) \sin(\omega_a t + \phi(E))] , \qquad (6.5)$$

where $N_0(E)$ is a normalization factor, τ_μ the muon life time (in the muon rest frame), and $A(E)$ is the asymmetry factor for positrons of energy greater than E. A typical example for the time structure from the BNL experiment is shown in Fig. 6.4. As we see the exponential decay law for the decaying muons is modulated by the $g - 2$ angular frequency. In this way the angular frequency ω_a is neatly determined from the time distribution of the decay positrons observed with the electromagnetic calorimeters [11, 12, 13, 14, 15].

The magnetic field is measured by *Nuclear Magnetic Resonance* (NMR) using a standard probe of H_2O [16]. This standard can be related to the magnetic moment of a free proton by

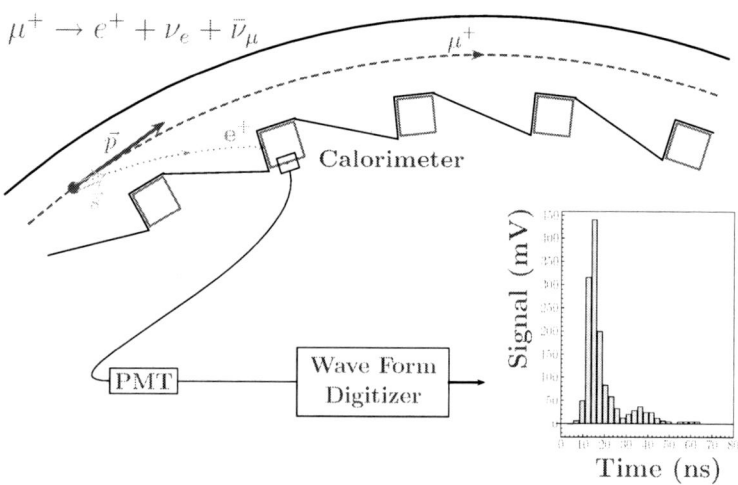

Fig. 6.3. Decay of μ^+ and detection of the emitted e^+ (PMT=Photomultiplier)

6.1 Overview on the Principle of the Experiment   351

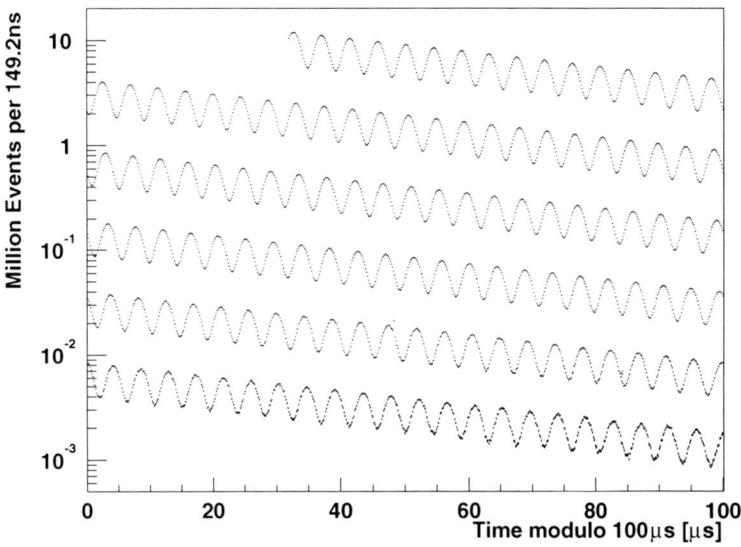

Fig. 6.4. Distribution of counts versus time for the 3.6 billion decays in the 2001 negative muon data–taking period [Courtesy of the E821 collaboration. Reprinted with permission from [7]. Copyright (2007) by the American Physical Society]

$$B = \frac{\hbar \omega_p}{2\mu_p}, \qquad (6.6)$$

where ω_p is the Larmor spin precession angular velocity of a proton in water. Using this, the frequency ω_a from (6.5), and $\mu_\mu = (1 + a_\mu)\, e\hbar/(2m_\mu c)$, one obtains

$$a_\mu = \frac{R}{\lambda - R} \qquad (6.7)$$

where

$$R = \omega_a/\omega_p \quad \text{and} \quad \lambda = \mu_\mu/\mu_p . \qquad (6.8)$$

The quantity λ appears because the value of the muon mass m_μ is needed, and also because the B field measurement involves the proton mass m_p. Measurements of the microwave spectrum of ground state muonium ($\mu^+ e^-$) [17] at LAMPF at Los Alamos, in combination with the theoretical prediction of the Muonium hyperfine splitting $\Delta\nu$ [18, 19] (and references therein), have provided the precise value

$$\frac{\mu_\mu}{\mu_p} = \lambda = 3.183\,345\,39(10)\ (30\ \text{ppb})\,, \qquad (6.9)$$

which is used by the E821 experiment to determine a_μ via (6.7). More details on the hyperfine structure of muonium will be given below in Sect. 6.6.

352   6 The $g-2$ Experiments

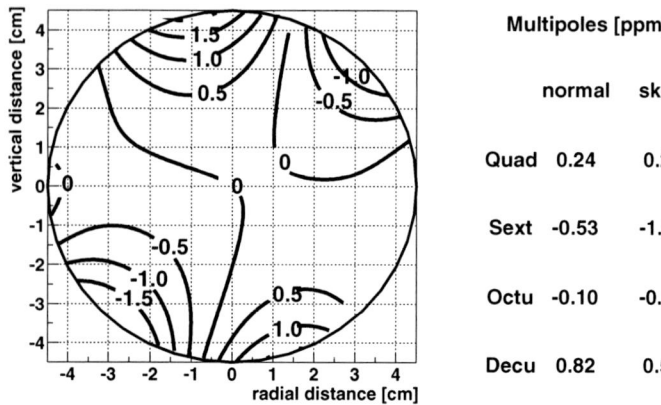

Fig. 6.5. Magnetic field profile. The contours are averaged over azimuth and interpolated using a multi–pole expansion. The circle indicates the storage aperture. The contour lines are separated by 1 ppm deviations from the central average [Courtesy of the E821 collaboration [7]]

Since the spin precession frequency can be measured very well, the precision at which $g-2$ can be measured is essentially determined by the possibility to manufacture a constant homogeneous magnetic field B and to determine its value very precisely. An example of a field map from the BNL experiment is shown in Fig. 6.5. Important but easier to achieve is the tuning to the magic energy. Possible deviations may be corrected by adjusting the effective magnetic field appropriately.

In the following we will discuss various aspects mentioned in this brief overview in more detail: beam dynamics, spin precession dynamics, some theory background about the properties of the muon. This should shed some more light on the muon spin physics as it derives from the SM. A summary of the main experimental results and two short addenda on the ground state hyperfine structure of muonium and on single electron dynamics and the electron $g-2$ will close this part on the experimental principles.

6.2 Particle Dynamics

The anomalous magnetic moment of both electrons and muons are measured by observing the motion of charged particles in a type of Penning trap, which consists of an electrical quadrupole field superimposed upon a uniform magnetic field. The configurations used in these experiments have axial symmetry. The orbital motion of charged particles in the storage ring may be discussed separately from the spin motion because the forces associated with the anomalous magnetic moment are very weak ($a_\mu \approx 1.16 \times 10^{-3}$) in comparison to the forces of the charge of the particle determining the orbital motion. The force F on a particle of charge e of velocity v in fields E and B is given by the Lorentz force

$$F = \frac{d\boldsymbol{p}}{dt} = e\,(\boldsymbol{E} + \boldsymbol{v} \times \boldsymbol{B})\;. \tag{6.10}$$

In a uniform magnetic field B of magnitude B_0 the particle with relativistic energy E_0 moves on a circle of radius

$$r_0 = \frac{E_0}{ecB_0}\;,\quad E_0 = \gamma mc^2\;. \tag{6.11}$$

Since we are interested in the dynamics of the muon beam in a ring, we consider a cylindrically symmetric situation. The cylindrical coordinates: $r = \sqrt{x^2 + y^2}$, θ, z are the radial, azimuthal and vertical coordinates of the particle position as shown in Fig. 6.6.

The relativistic equation of motion for the muon in the static cylindrical fields $\boldsymbol{B}(r,z)$ and $\boldsymbol{E}(r,z)$ takes the form

$$\frac{d}{dt}(m\dot{r}) = mr\dot\theta^2 - er\dot\theta B_z + eE_r\;, \tag{6.12}$$

$$\frac{d}{dt}(mr^2\dot\theta) = 0\;, \tag{6.13}$$

$$\frac{d}{dt}(m\dot z) = er\dot\theta B_r + eE_z\;. \tag{6.14}$$

The general form of the electrostatic potential applied is

$$V(r,z) = \frac{V_0}{d^2}\left[r^2 - 2r_0^2\ln\frac{r}{r_0} - r_0^2 - 2z^2\right]$$

where r_0 is the radius of the circle on which $\partial V/\partial r = 0$. This potential is singular along the symmetry axis except in the case $r_0 = 0$. In the latter case

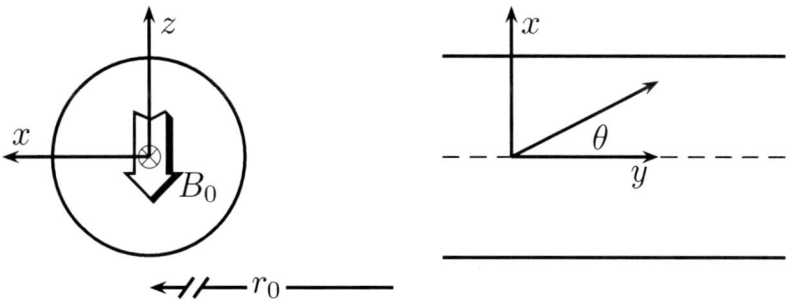

Fig. 6.6. Coordinates for the beam dynamics. View at the beam end (left) $x = r - r_0$ radial, z vertical, with B field in $-z$ direction; $(x,z) = (0,0)$ is the beam position, the negative muon beam points into the plane. View from top (right): y is the direction along the beam

$$V(r,z) = \frac{V_0}{d^2}\left(r^2 - 2z^2\right), \tag{6.15}$$

which is the potential used in an electron trap. Here $(r_0, 0)$ and $(0, z_0)$ are the coordinates of the plates and $d^2 = r_0^2 + 2z_0^2$ (for a symmetric trap $r_0 = \sqrt{2}z_0$).

In the muon $g-2$ experiment $r \to x = r - r_0$ with $|x| \ll r_0$ (see Fig. 6.6) and weak focusing is implemented by a configuration of charged plates as shown in Fig. 6.7.

In order to get a pure quadrupole field one has to use hyperbolic plates with end–caps $z^2 = z_0^2 + x^2/2$ and $z^2 = \frac{1}{2}(x^2 - x_0^2)$ on the ring. While the CERN experiment was using hyperbolic plates, the BLN one uses flat plates which produce 12– and 20–pole harmonics. The length of the electrodes is adjusted to suppress the 12–pole mode leaving a 2% 20–pole admixture. The electric field produces a restoring force in the vertical direction and a repulsive force in the radial direction:

$$\boldsymbol{E} = (E_r, E_\theta, E_z) = (\kappa x, 0, -\kappa z) \tag{6.16}$$

where $x = r - r_0$ and κ a positive constant. In order to keep the beam focused, the restoring force of the vertical magnetic field must be stronger than the repulsive force of the electrical field in the radial direction:

$$0 < \frac{eV_0}{d^2} < \frac{e^2 B^2}{8mc}. \tag{6.17}$$

The radial force is

$$F_r = \frac{\gamma m v^2}{r} - \frac{e}{c} v B_z + e E_r \tag{6.18}$$

and since on the equilibrium orbit $r = r_0$ and $E_r = 0$ we have

$$\frac{\gamma m v^2}{r_0} = \frac{e}{c} v B_z. \tag{6.19}$$

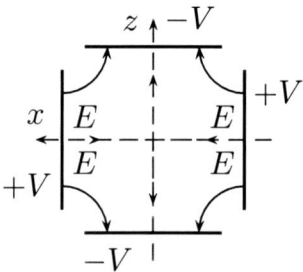

Fig. 6.7. Electric quadrupole field E. The vertical direction is z, the radial x ($x_0 = \sqrt{2}z_0$); $V = V_0/2$ at the plates

As r_0 is large relative to the beam spread, we may expand r about r_0:

$$\frac{1}{r} = \frac{1}{r_0 + x} \simeq \frac{1}{r_0}\left(1 - \frac{x}{r_0}\right).$$

Therefore, using (6.19) we may write

$$F_x = F_r = -e\beta B_z (1-n)\frac{x}{r_0} \Rightarrow \gamma m \ddot{x} = -(1-n)\frac{\gamma m v^2}{r_0^2} x \qquad (6.20)$$

where $\beta = v/c$ and n is the field index

$$n = \frac{\kappa r_0}{\beta B_0}, \quad B_0 = B_z. \qquad (6.21)$$

For the vertical motion we have

$$F_z = -e\kappa z \Rightarrow \gamma m \ddot{z} = -e\kappa z \qquad (6.22)$$

and with $\omega_0 = v/r_0$, using (6.19) and (6.21), the equations of motion take the form

$$\ddot{x} + (1-n)\omega_0^2 x = 0$$
$$\ddot{z} + e\kappa z = 0 \qquad (6.23)$$

with the oscillatory solutions

$$x = A\cos(\sqrt{1-n}\,\omega_0 t),$$
$$z = B\cos(\sqrt{n}\,\omega_0 t). \qquad (6.24)$$

We have used $e\kappa = n\omega_0^2$ following from (6.21). The amplitudes depend on the initial condition of the particle trajectory. This motion is called *betatron oscillation*. The betatron frequencies are $\omega_{y\mathrm{BO}} = \sqrt{n}\,\omega_c$ and $\omega_{x\mathrm{BO}} = \sqrt{1-n}\,\omega_c$ where $\omega_c = \omega_0 = v/r_0$ is the cyclotron frequency. In the experiment a lattice of quadrupoles is distributed along the ring. For the BNL experiment the lattice has a four–fold symmetry and the quadrupoles are covering 43% of the ring. The corresponding dynamics has to be calculated taking into account the geometry of the configuration, but follows the same principle.

The dynamics of an electron in a Penning trap and the principle of electron $g-2$ experiments will be considered briefly in Sect. 6.7 at the end of this part of the book.

6.3 Magnetic Precession for Moving Particles

The precession of spinning particles in magnetic fields is a classic subject investigated long time ago [20]. Our exposition follows closely Bell's lecture. In a magnetic field $\boldsymbol{B}$ the polarization $\boldsymbol{P}$ of a particle changes according to

$$\frac{dP}{dt} = g \frac{e}{2m} P \times B ,$$

the component of P parallel to B remains constant, while the part of P perpendicular to B rotates about B with angular frequency

$$\omega = g \frac{e}{2m} B , \qquad (6.25)$$

the non–relativistic cyclotron frequency. This holds in the rest frame O of the particle. For moving and even fast–moving particles we may get the motion in the laboratory system O' by a Lorentz transformation. In a pure L–transformation $x^{\mu'} = L^\mu_{\ \nu} x^\nu$ $[x^\mu = (ct, \boldsymbol{x})]$ L has the form[2]

$$L = \begin{pmatrix} \gamma & -\gamma \frac{\boldsymbol{v}}{c} \\ -\gamma \frac{\boldsymbol{v}}{c} & \mathbf{1} + (\gamma - 1) \, \boldsymbol{n} \, (\boldsymbol{n} \cdot) \end{pmatrix}$$

where $\boldsymbol{n} = \boldsymbol{v}/v$ and $\gamma = 1/\sqrt{1 - v^2/c^2}$. For accelerated particles, the velocity is changing and in the next moment the velocity is $\boldsymbol{v}' = \boldsymbol{v} + \delta \boldsymbol{v}$. In the laboratory frame we thus have $x^{\mu'} = L^\mu_{\ \nu}(\boldsymbol{v}) x^\nu$ and $x^{\mu''} = L^\mu_{\ \nu}(\boldsymbol{v}') x^\nu$ and expanding to linear order in $\delta \boldsymbol{v}$ one obtains the motion as seen in the laboratory frame as

$$t'' = t' - \delta \boldsymbol{u}' \cdot \boldsymbol{x}'$$
$$\boldsymbol{x}'' = \boldsymbol{x}' + \delta \boldsymbol{\theta}' \times \boldsymbol{x}' - \delta \boldsymbol{u}' t' \qquad (6.26)$$

with

$$\delta \boldsymbol{u}' = \gamma \left(1 + \frac{\gamma - 1}{v^2} \, \boldsymbol{v} \, \boldsymbol{v} \cdot \right) \delta \boldsymbol{v}$$
$$\delta \boldsymbol{\theta}' = \frac{\gamma - 1}{v^2} \, (\delta \boldsymbol{v} \times \boldsymbol{v}) , \qquad (6.27)$$

which tells us that from the two pure boosts we got an infinitesimal transformation which includes both a boost (pure if $\delta \boldsymbol{\theta}' = 0$) and a rotation (pure if $\delta \boldsymbol{u}' = 0$). The transformation (6.26) is the infinitesimal law for transforming vectors in O' to vectors in O''.

The precession equation for accelerated moving particles is then obtained as follows: Let O' be the observer for whom the particle is momentarily at rest. If the particle has no electric dipole moment, what we assume (see end of Sect. 3.3), an electric field does not contribute to the precession and only serves to accelerate the particle

$$\delta \boldsymbol{u}' = \frac{e}{m} \boldsymbol{E}' \, \delta t' , \qquad (6.28)$$

---

[2] L is a matrix operator acting on four–vectors. The · operation at the of the spacial submatrix means forming a scalar product with the spatial part of the vector on which L acts.

6.3 Magnetic Precession for Moving Particles

while the magnetic field provides the precession

$$\delta \boldsymbol{P}' = -g \frac{e}{2m} \boldsymbol{B}' \times \boldsymbol{P}' \, \delta t' \ . \tag{6.29}$$

In the laboratory frame O' the observed polarization is $\boldsymbol{P}' + \delta \boldsymbol{P}'$ where $\boldsymbol{P}' = \boldsymbol{P}$ is the polarization of the particle in its rest frame O. The observer O'' by a boost from O sees a polarization $\boldsymbol{P}'' + \delta \boldsymbol{P}''$ which differs by a rotation $\delta \boldsymbol{\theta}'$ from the previous one: (note that momentarily $\boldsymbol{P}'' = \boldsymbol{P}' = \boldsymbol{P}$)

$$\delta \boldsymbol{P}'' = \delta \boldsymbol{P}' + \delta \boldsymbol{\theta}' \times \boldsymbol{P} \tag{6.30}$$

or

$$\delta \boldsymbol{P}'' = -g \frac{e}{2m} \boldsymbol{B}' \times \boldsymbol{P} \, \delta t' + \frac{(\gamma - 1)}{v^2} (\delta \boldsymbol{v} \times \boldsymbol{v}) \times \boldsymbol{P} \ . \tag{6.31}$$

The precession equation in the laboratory frame may be obtained by applying the L-transformations of coordinates and fields to the lab frame:

$$\delta t' = \gamma \left(\delta t - \frac{\boldsymbol{v} \cdot \delta \boldsymbol{x}}{c^2} \right) = \gamma \, \delta t \left(1 - \frac{v^2}{c^2} \right) = \frac{1}{\gamma} \delta t$$

$$\boldsymbol{B}' = \gamma \left(\boldsymbol{B} - \frac{\boldsymbol{v} \times \boldsymbol{E}}{c^2} \right) + \frac{(1 - \gamma)}{v^2} \boldsymbol{v} \cdot \boldsymbol{B} \, \boldsymbol{v}$$

$$\boldsymbol{E}' = \gamma \left(\boldsymbol{E} + \boldsymbol{v} \times \boldsymbol{B} \right) + \frac{(1 - \gamma)}{v^2} \boldsymbol{v} \cdot \boldsymbol{E} \, \boldsymbol{v} \tag{6.32}$$

and one obtains

$$\frac{\mathrm{d} \boldsymbol{P}}{\mathrm{d} t} = \boldsymbol{\omega}_s \times \boldsymbol{P} \tag{6.33}$$

with

$$\boldsymbol{\omega}_s = \frac{\gamma - 1}{v^2} \frac{\mathrm{d} \boldsymbol{v}}{\mathrm{d} t} \times \boldsymbol{v} - g \frac{e}{2m} \left(\boldsymbol{B} - \frac{\boldsymbol{v} \times \boldsymbol{E}}{c^2} + \frac{1 - \gamma}{\gamma v^2} \boldsymbol{v} \cdot \boldsymbol{B} \, \boldsymbol{v} \right) \ . \tag{6.34}$$

The first term, which explicitly depends on the acceleration, is called *Thomas precession*. The acceleration in the laboratory frame may be obtained in the same way from (6.28) together with (6.27) and (6.32)

$$\frac{\mathrm{d} \boldsymbol{v}}{\mathrm{d} t} = \frac{e}{\gamma m} (\boldsymbol{E} + \boldsymbol{v} \times \boldsymbol{B}) - \frac{e}{\gamma m c^2} \boldsymbol{v} \cdot \boldsymbol{E} \, \boldsymbol{v} \ , \tag{6.35}$$

which is just another form of the usual equation of motion[3] (Lorentz force)

$$\frac{\mathrm{d} \boldsymbol{p}}{\mathrm{d} t} = \frac{\mathrm{d}}{\mathrm{d} t} (\gamma m \boldsymbol{v}) = e (\boldsymbol{E} + \boldsymbol{v} \times \boldsymbol{B}) \ .$$

---
[3] Note that $\mathrm{d} \gamma = \gamma^3 \, \boldsymbol{v} \cdot \mathrm{d} \boldsymbol{v} / c^2$ and the equation of motion implies

$$\boldsymbol{v} \cdot \frac{\mathrm{d} (\gamma m \boldsymbol{v})}{\mathrm{d} t} = m \gamma^3 \boldsymbol{v} \cdot \frac{\mathrm{d} \boldsymbol{v}}{\mathrm{d} t} = e \boldsymbol{v} \cdot \boldsymbol{E}$$

as $\boldsymbol{v} \cdot (\boldsymbol{v} \times \boldsymbol{B}) \equiv 0$. This has been used in obtaining (6.35).

If one uses (6.35) to eliminate the explicit acceleration term from (6.34) together with $(\boldsymbol{v} \times \boldsymbol{B}) \times \boldsymbol{v} = \boldsymbol{B} v^2 - \boldsymbol{v} \cdot \boldsymbol{B} \boldsymbol{v}$ and $\boldsymbol{v} \times \boldsymbol{v} = 0$, one obtains

$$\boldsymbol{\omega}_s = -\frac{e}{\gamma m} \left\{ (1 + \gamma a) \, \boldsymbol{B} + \frac{(1-\gamma)}{v^2} a \, \boldsymbol{v} \cdot \boldsymbol{B} \boldsymbol{v} + \gamma \left(a + \frac{1}{\gamma + 1} \right) \frac{\boldsymbol{E} \times \boldsymbol{v}}{c^2} \right\}, \tag{6.36}$$

where $a = g/2 - 1$ is the anomaly term.

6.3.1 $g - 2$ Experiment and Magic Momentum

In the $g - 2$ experiment one works with purely transversal fields: $\boldsymbol{v} \cdot \boldsymbol{E} = \boldsymbol{v} \cdot \boldsymbol{B} = 0$. Then using $(\boldsymbol{v} \times \boldsymbol{E}) \times \boldsymbol{v} = v^2 \boldsymbol{E}$ (when $\boldsymbol{v} \cdot \boldsymbol{E} = 0$) and $v^2/c^2 = (\gamma^2 - 1)/\gamma^2$ the equation of motion can be written

$$\frac{d\boldsymbol{v}}{dt} = \boldsymbol{\omega}_c \times \boldsymbol{v}, \quad \boldsymbol{\omega}_c = -\frac{e}{\gamma m} \left(\boldsymbol{B} + \frac{\gamma^2}{\gamma^2 - 1} \frac{\boldsymbol{E} \times \boldsymbol{v}}{c^2} \right). \tag{6.37}$$

The velocity $\boldsymbol{v}$ thus rotates, without change of magnitude, with the relativistic cyclotron frequency $\boldsymbol{\omega}_c$. The precession of the polarization $\boldsymbol{P}$, which is to be identified with the muon spin $\boldsymbol{S}_\mu$, for purely transversal fields is then

$$\boldsymbol{\omega}_a = \boldsymbol{\omega}_s - \boldsymbol{\omega}_c = -\frac{e}{m} \left\{ a \boldsymbol{B} + \left(a - \frac{1}{\gamma^2 - 1} \right) \frac{\boldsymbol{E} \times \boldsymbol{v}}{c^2} \right\}. \tag{6.38}$$

This establishes the key formula for measuring a_μ, which we have used and discussed earlier. It was found by Bargmann, Michel and Telegdi in 1959 [20]. Actually, the magnetic transversality condition $\boldsymbol{v} \cdot \boldsymbol{B} = 0$ due to electrostatic focusing is not accurately satisfied (pitch correction) such that the more general formula

$$\boldsymbol{\omega}_a = -\frac{e}{m} \left\{ a \boldsymbol{B} - a \left(\frac{\gamma}{\gamma + 1} \right) \frac{\boldsymbol{v} \cdot \boldsymbol{B} \boldsymbol{v}}{c^2} + \left(a - \frac{1}{\gamma^2 - 1} \right) \frac{\boldsymbol{E} \times \boldsymbol{v}}{c^2} \right\}, \tag{6.39}$$

has to be used.

Since the anomalous magnetic moment for leptons is a very small quantity $a \approx 1.166 \times 10^{-3}$, electrons and muons in a pure magnetic field and initially polarized in the direction of motion ($\boldsymbol{P} \propto \boldsymbol{v}$) only very slowly develop a component of polarization transverse to the direction of motion. The observation of this development provides a sensitive measure of the small but theoretically very interesting anomalous magnetic moment.

In the original muon $g - 2$ experiments only a $\boldsymbol{B}$ field was applied and in order to give some stability to the beam the $\boldsymbol{B}$ was not quite uniform[4], and the particles oscillate about an equilibrium orbit. As a result one of the

---

[4] Magnetic focusing using an inhomogeneous field $B_z = B_0 (r_0/r)^n$, which by Maxwell's equation $\nabla \times \boldsymbol{B} = 0$ implies $B_r \simeq -n/r_0 B_0 z$ for $r \simeq r_0$, leads to identical betatron oscillation equations (6.23) as electrostatic focusing.

main limitations of the precision of those experiments was the difficulty to determine the effective average B to be used in calculating a_μ from the observed oscillation frequencies. To avoid this, in the latest CERN experiment, as later in the BNL experiment, the field B is chosen as uniform as possible and focusing is provided by transverse electric quadrupole fields. To minimize the effect of the electric fields on the precession of P, muons with a special "magic" velocity are used so that the coefficient of the second term in (6.37) is small:

$$a_\mu - \frac{1}{\gamma^2 - 1} \approx 0$$

corresponding to a muon energy of about 3.1 GeV. This elegant method for measuring a_μ was proposed by Bailey, Farley, Jöstlein, Picasso and Wickens and realized as the last CERN muon $g - 2$ experiment and later adopted by the experiment at BNL. The motion of the muons is characterized by the frequencies listed in Table 6.1

Two small, but important, corrections come from the effect of the electric focusing field E on the spin precession ω_a.

The first is the *Radial Electric Field Correction*, the change in ω_a when the momentum p deviates from the magic value $p \neq p_m$ and hence $p = \beta\gamma m = p_m + \Delta p$. In fact, the beam is not monoenergetic and the momentum tune has a small uncertainty of about $\pm 0.5\%$. This effect can be corrected by a change in the effective magnetic field [6] used in extracting a_μ. In cylindrical coordinates Fig. 6.6 using $(v \times E)_z = -v_y E_x = -v E_r$, as $E_y = 0$, we find $a B_z + (a - 1/(\beta^2 \gamma^2)) v E_r / c^2$ or, with $B_0 = -B_z > 0$,

$$B_{0\,\text{eff}} = B_0 \left[1 - \beta \frac{E_r}{B_0} \left(1 - \frac{1}{a_\mu \beta^2 \gamma^2} \right) \right] \equiv C_E B_0 . \quad (6.40)$$

This directly translates into

$$\frac{\Delta \omega_a}{\omega_a} = C_E \simeq -2 \frac{\beta E_r}{B_0} \left(\frac{\Delta p}{p_m} \right) . \quad (6.41)$$

One may apply furthermore the relation $\Delta p / p_m = (1 - n)(x_e / r_0)$, where x_e is the equilibrium position of the particle relative to the center of the aperture

Table 6.1. Frequencies and time periods in the muon $g - 2$ experiment E821. The field index used is $n = 0.137$. It is optimized to avoid unwanted resonances in the muon storage ring

Type	$\nu_i = \omega_i / 2\pi$	Expression	Frequency	Period
Anomalous precession	ν_a	$\frac{e a_\mu B}{2\pi m}$	0.23 MHz	4.37 μs
Cyclotron	ν_c	$\frac{v}{2\pi r_0}$	6.71 MHz	149 ns
Horizontal betatron	ν_x	$\sqrt{1 - n}\,\nu_c$	6.23 MHz	160 ns
Vertical betatron	ν_z	$\sqrt{n}\,\nu_c$	2.48 MHz	402 ns

of the ring. For the BNL experiment typically

$$C_E \simeq 0.5 \text{ ppm} . \tag{6.42}$$

The second effect is the *Vertical Pitch Correction* arising from vertical betatron oscillations [1, 21]. The focusing force due to $\boldsymbol{E}$ changes v_z at the betatron oscillation frequency $\omega_p = \omega_{z\mathrm{BO}}$[5] such that

$$\psi(t) = \psi_0 \sin \omega_p t . \tag{6.43}$$

The muon will follow a spiral path with pitch angle ψ (see Fig. 6.8) given by

$$\frac{v_z}{v} = \sin \psi \simeq \psi \tag{6.44}$$

and ω_a is changed. Now $\boldsymbol{v} \cdot \boldsymbol{B} \neq 0$, which persists as an effect from the focusing field also if running at the magic γ. The corresponding correction follows from (6.39), at $\gamma = \gamma_m$. The motion vertical to the main plane implies

$$\omega_{az} = \frac{e}{m} a B_0 \left[1 - \left(\frac{\gamma}{\gamma+1}\right)\beta_z^2\right]$$
$$= \omega_a \left[1 - \left(\frac{\gamma}{\gamma+1}\right)\beta^2 \frac{v_z^2}{v^2}\right] = \omega_a \left[1 - \left(\frac{\gamma-1}{\gamma}\right)\psi^2\right] \tag{6.45}$$

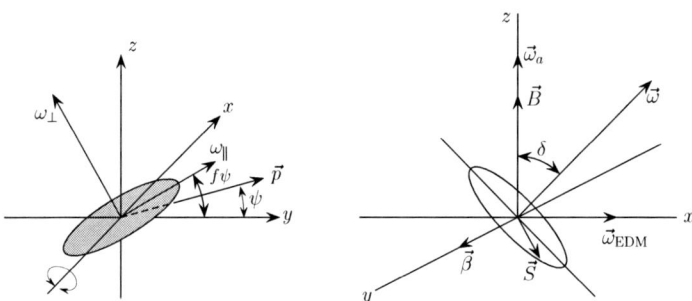

Fig. 6.8. Left: frame for pitch correction. $\boldsymbol{p}$ lies always in the yz-plane. The pitch angle ψ between $\boldsymbol{p}$ and the y-axis (beam direction) oscillates. The spin $\boldsymbol{S}$ then rotates about the x-axis through an angle $f\psi$, where for electric focusing $f = 1+\beta^2\gamma a-\gamma^{-1}$; $f = 1$ at magic γ. Right: frame for EDM correction. As $|\boldsymbol{E}| \ll |\boldsymbol{E}^*| = c|\boldsymbol{\beta} \times \boldsymbol{B}|$, ω_{EDM} points along the x-axis while the unperturbed ω_a points in z-direction. $\delta = \arctan\frac{\eta\beta}{2a} \simeq \frac{\eta}{2a}$

---

[5]The pitch frequency here should not to be confused with the proton precession frequency ω_p appearing in (6.8).

where ω_a is the ideal (unperturbed) precession frequency. Similarly,

$$\omega_{ay} = -\frac{e}{m} a B_0 \left[1 - \left(\frac{\gamma}{\gamma+1}\right) \beta_z \beta_y\right]$$
$$= -\omega_a \left[1 - \left(\frac{\gamma}{\gamma+1}\right) \beta^2 \frac{v_z v_y}{v^2}\right] = -\omega_a \left[1 - \left(\frac{\gamma-1}{\gamma}\right)\psi\right] \quad (6.46)$$

where we used

$$\frac{v_z}{v} = \sin\psi \simeq \psi, \quad \frac{v_y}{v} = \cos\psi \simeq 1.$$

The component of $\boldsymbol{\omega}_a$ parallel to the tilted plane changes sign and in the time average has no effect. The perpendicular component is

$$\omega_\perp = \omega_a = \omega_z \cos\psi - \omega_y \sin\psi \simeq \omega_z - \omega_y \psi \quad (6.47)$$

and hence

$$\omega'_a = \omega_a (1 - C_P) = \omega_a \left(1 - \frac{\psi^2}{2}\right). \quad (6.48)$$

In the time average by (6.43) $\overline{\psi^2} = \frac{1}{2}\psi_0^2$ and thus $C_P = \frac{1}{4}\psi_0^2$. This holds provided $\omega_a \ll \omega_p$ otherwise the correction reads [21]

$$C_P = \frac{1}{4}\psi_0^2 \beta^2 \left(1 - (a\beta\gamma)^2 \frac{\omega_p^2}{(\omega_a^2 - \omega_p^2)}\right), \quad (6.49)$$

with $(a\beta\gamma)^2 = 1/(\beta\gamma)^2$ at magic γ. For the BNL experiment the pitch corrections is of the order

$$C_P \simeq 0.3 \text{ ppm}. \quad (6.50)$$

A third possible correction could be due to an EDM of the muon. If a large enough *electric dipole moment*[6]

$$\boldsymbol{d}_e = \frac{\eta e}{2mc}\boldsymbol{S} \quad (6.51)$$

(see (1.5), p. 32 f. in Sect. 2.1.2 and the discussion at the end of Sect. 3.3) would exist the applied electric field $\boldsymbol{E}$ (which is vanishing at the equilibrium beam position) and the motional electric field induced in the muon rest frame $\boldsymbol{E}^* = \gamma \boldsymbol{\beta} \times \boldsymbol{B}$ would add an extra precession of the spin with a component along $\boldsymbol{E}$ and one about an axis perpendicular to $\boldsymbol{B}$:

$$\boldsymbol{\omega} = \boldsymbol{\omega}_a + \boldsymbol{\omega}_{\text{EDM}} = \boldsymbol{\omega}_a - \frac{\eta e}{2m_\mu}\left(\frac{\boldsymbol{E}}{c} + \boldsymbol{\beta} \times \boldsymbol{B}\right) \quad (6.52)$$

---

[6] Remembering the normalization: the magnetic and electric dipole moments are given by $\mu = \frac{g}{2}\frac{e\hbar}{2mc}$ and $d = \frac{\eta}{2}\frac{e\hbar}{2mc}$, respectively.

or
$$\Delta\omega_a = -2d_\mu\,(\boldsymbol{\beta}\times\boldsymbol{B}) - 2d_\mu\,\boldsymbol{E}$$
which, for $\beta\sim 1$ and $d_\mu\,\boldsymbol{E}\sim 0$, yields
$$\omega_a = B\sqrt{\left(\frac{e}{m_\mu}a_\mu\right)^2 + (2d_\mu)^2}$$
where η is the dimensionless constant, equivalent of magnetic moment g-factors. The result is that the plane of precession in no longer horizontal but tilted at an angle
$$\delta \equiv \arctan\frac{\omega_{\text{EDM}}}{\omega_a} = \arctan\frac{\eta\beta}{2a} \simeq \frac{\eta}{2a} \qquad (6.53)$$
and the precession frequency is increased by a factor
$$\omega_a' = \omega_a\sqrt{1+\delta^2}\,. \qquad (6.54)$$
The tilt gives rise to an oscillating vertical component of the muon polarization, and may be detected by recording separately the electrons which strike the counters above and below the mid–plane of the ring. This measurement has been performed in the last CERN experiment on $g-2$, and a corresponding analysis is in progress at BNL.

6.4 Theory: Production and Decay of Muons

For the $(g-2)_\mu$ experiments one needs polarized muons. Basic symmetries of the weak interaction of the muons make it relatively easy to produce polarized muons. What helps is the maximal parity violation of the charged current weak interactions, mediated by the charged $W^\pm$ gauge bosons, which in its most pronounced form manifests itself in the "non–existence" of right–handed neutrinos ν_R. What it means more precisely is that right handed neutrinos are "sterile" in the sense that they do not interact with any kinds of the gauge bosons, which we know are responsible for electromagnetic (photon), weak (W- and Z-bosons) and strong (gluons) interactions of matter. It means that their production rate in ordinary weak reactions is practically zero which amounts to lepton number conservation for all practical purposes in laboratory experiments[7].

Pion production may be done by shooting protons (accumulated in a proton storage ring) on a target material where pions are the most abundant secondary particles. The most effective pion production mechanism proceeds via

---

[7] Only the recently established phenomenon of neutrino oscillations proves that lepton number in fact is not a perfect quantum number. This requires that neutrinos must have tiny masses and this requires that right–handed neutrinos (ν_R's) must exist. In fact, the smallness of the neutrino masses explains the strong suppression of lepton number violating effects.

decays of resonances. For pions it is dominated by the Δ_{33} isobar ($\Delta_{33} \to N\pi$) [basic processes $p + p \to p + n + \pi^+$ and $p + n \to p + p + \pi^-$]

$$p + (N, Z) \to \Delta^* + X \to "(N+1, Z+1 \mp 1)" + \pi^\pm$$

where the ratio $\sigma(\pi^+)/\sigma(\pi^-) \to 1$ at high Z^8.

We now look more closely to the decay chain

$$\pi \to \mu + \nu_\mu$$
$$\hookrightarrow e + \nu_e + \nu_\mu$$

producing the polarized muons which decay into electrons which carry along in their direction of propagation the knowledge of the muon's polarization (for a detailed discussion see e.g. [22]).

1) Pion decay:
The π^- is a pseudoscalar bound state $\pi^- = (\bar{u}\gamma_5 d)$ of a d quark and a u antiquark $\bar{u}$. The main decay channel is via the diagram:

π-decay

In this two–body decay of the charged spin zero pseudoscalar mesons the lepton energy is fixed (monochromatic) and given by

$$E_\ell = \sqrt{m_\ell^2 + p_\ell^2} = \frac{m_\pi^2 + m_\ell^2}{2m_\pi}, \quad p_\ell = \frac{m_\pi^2 - m_\ell^2}{2m_\pi}.$$

Here the relevant part of the Fermi type effective Lagrangian reads

$$\mathcal{L}_{\text{eff,int}} = -\frac{G_\mu}{\sqrt{2}} V_{ud} \left(\bar{\mu}\gamma^\alpha (1-\gamma_5) \nu_\mu\right) \left(\bar{u}\gamma_\alpha (1-\gamma_5) d\right) + \text{h.c.}$$

where G_μ denotes the Fermi constant and V_{ud} the first entry in the CKM matrix. For our purpose $V_{ud} \sim 1$. The transition matrix–element reads

$$T = {}_\text{out}\langle \mu^-, \bar{\nu}_\mu | \pi^- \rangle_\text{in}$$
$$= -i\frac{G_\mu}{\sqrt{2}} V_{ud} F_\pi \left(\bar{u}_\mu \gamma^\alpha (1-\gamma_5) v_{\nu_\mu}\right) p_\alpha$$

---

[8] At Brookhaven the 24 GeV proton beam extracted from the AGS with 60×10^{12} protons per AGS cycle of 2.5 s impinges on a Nickel target of one interaction length and produces amongst other debris–particles a large number of low energy pions. The pions are momentum selected and then decay in a straight section where about one third of the pions decay into muons. The latter are momentum selected once more before they are injected into the $g-2$ storage ring.

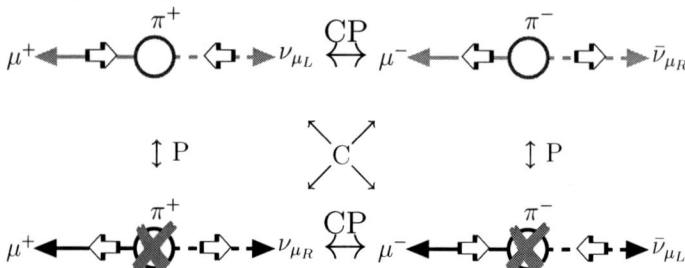

Fig. 6.9. Pion decay is a parity violating weak decay where leptons of definite handedness are produced depending on the given charge. CP is conserved while P and C are violated maximally (unique handedness). μ^- [μ^+] is produced with positive [negative] helicity $h = \boldsymbol{S} \cdot \boldsymbol{p}/|\boldsymbol{p}|$. The existing μ^- and μ^+ decays are related by a CP transformation. The decays obtained by C or P alone are inexistent in nature

where we used the hadronic matrix–element

$$\langle 0| \, \bar{d} \, \gamma_\mu \gamma_5 \, u \, |\pi(p)\rangle \doteq \mathrm{i} F_\pi p_\mu$$

which defines the pion decay constant F_π. As we know the pion is a pseudoscalar such that only the axial part of the weak charged $V - A$ current couples to the pion. By angular momentum conservation, as the π^+ has spin 0 and the emitted neutrino is left–handed $((1-\gamma_5)/2$ projector) the μ^+ must be left–handed as well. Going to the π^- not only particles have to be replaced by antiparticles (C) but also the helicities have to be reversed (P), since a left–handed antineutrino (essentially) does not exist. Note that the decay is possible only due to the non–zero muon mass, which allows for the necessary helicity flip of the muon. The handedness is opposite for the opposite charge. This is illustrated in Fig. 6.9.

The pion decay rate is given by

$$\Gamma_{\pi^- \to \mu^- \bar{\nu}_\mu} = \frac{G_\mu^2}{8\pi} |V_{ud}|^2 F_\pi^2 \, m_\pi \, m_\mu^2 \left(1 - \frac{m_\mu^2}{m_\pi^2}\right)^2 \times (1 + \delta_{\mathrm{QED}}) \;,$$

with CKM matrix–element $V_{ud} \sim 1$ and δ_{QED} the electromagnetic correction.

2) Muon decay:
Muon decay $\mu^- \to e^- \bar{\nu}_e \nu_\mu$ is a three body decay

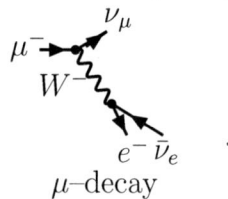

μ–decay

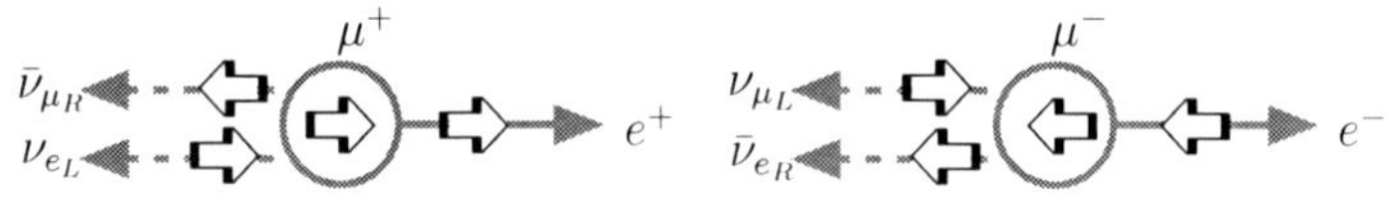

Fig. 6.10. In μ^- [μ^+] decay the produced e^- [e^+] has negative [positive] helicity, respectively

The matrix element can be easily calculated. The relevant part of the effective Lagrangian reads

$$\mathcal{L}_{\text{eff,int}} = -\frac{G_\mu}{\sqrt{2}} \left(\bar{e}\gamma^\alpha (1-\gamma_5) \nu_e\right) \left(\bar{\nu}_\mu \gamma_\alpha (1-\gamma_5) \mu\right) + \text{h.c.}$$

and thus

$$T = {}_{\text{out}}\!<e^-, \bar{\nu}_e \nu_\mu | \mu^- >_{\text{in}}$$
$$= \frac{G_\mu}{\sqrt{2}} \left(\bar{u}_e \gamma^\alpha (1-\gamma_5) v_{\nu_e}\right) \left(\bar{u}_{\nu_\mu} \gamma_\alpha (1-\gamma_5) u_\mu\right)$$

which proves that the μ^- and the e^- have both the same left–handed helicity [the corresponding anti–particles are right–handed] in the massless approximation. This implies the decay scheme Fig. 6.10 for the muon.

The electrons are thus emitted in the direction of the muon spin, i.e. measuring the direction of the electron momentum provides the direction of the muon spin.

After integrating out the two unobservable neutrinos, the differential decay probability to find an $e^\pm$ with reduced energy between x and $x + dx$ emitted at an angle between θ and $\theta + d\theta$ reads (see also (2.46))

$$\frac{d^2 \Gamma^\pm}{dx\, d\cos\theta} = \frac{G_\mu^2 m_\mu^5}{192\pi^3} x^2 \left(3 - 2x \pm P_\mu \cos\theta (2x - 1)\right) \qquad (6.55)$$

and typically is strongly peaked at small angles and may be written in the form (6.3). The reduced $e^\pm$ energy is $x = E_e/W_{\mu e}$ with $W_{\mu e} = \max E_e = (m_\mu^2 + m_e^2)/2m_\mu$, the $e^\pm$ emission angle θ is the angle between the e momentum $\boldsymbol{p}_e$ and the muon polarization vector $\boldsymbol{P}_\mu$. The result above holds in the approximation $x_0 = m_e/W_{e\mu} \sim 9.67 \times 10^{-3} \simeq 0$.

6.5 Muon $g - 2$ Results

First a historical note: before the E821 experiment at Brookhaven the last of a series of measurement of the anomalous g-factor $a_\mu = (g_\mu - 2)/2$ at CERN was published about 30 years ago. At that time a_μ had been measured for muons of both charges in the Muon Storage Ring at CERN. The two results,

$$a_{\mu^-} = 1165937(12) \times 10^{-9}$$
$$a_{\mu^+} = 1165911(11) \times 10^{-9}, \qquad (6.56)$$

are in good agreement with each other, and combine to give a mean

$$a_\mu = 1165924.0(8.5)10^{-9} \ [7\text{ppm}] \,, \tag{6.57}$$

which was very close to the theoretical prediction $1165921.0(8.3)10^{-9}$ at that time. The measurements thus confirmed the remarkable QED calculation plus hadronic contribution, and served as a precise verification of the CPT theorem for muons.

Measured in the experiments is the ratio of the muon precession frequency $\omega_a = \omega_s - \omega_c$ and the proton precession frequency ω_p: $R = \omega_a/\omega_p$ which together with the ratio of the magnetic moment of the muon to the one of the proton $\lambda = \mu_\mu/\mu_p$ determines the anomalous magnetic moment as

$$a_\mu = \frac{R}{\lambda - R} \,. \tag{6.58}$$

The CERN determination of a_μ utilized the value $\lambda = 3.1833437(23)$.

The BNL muon $g-2$ experiment has been able to improve and perfect the method of the last CERN experiments in several respects and was able to achieve an impressive 14-fold improvement in precision. The measurements are $R_{\mu^-} = 0.0037072083(26)$ and $R_{\mu^+} = 0.0037072048(25)$ the difference being $\Delta R = (3.5 \pm 3.4) \times 10^{-9}$. Together with $\lambda = 3.18334539(10)$ [23] one obtains the new values

$$a_{\mu^-} = 11659214(8)(3) \times 10^{-10}$$
$$a_{\mu^+} = 11659204(7)(5) \times 10^{-10} \,. \tag{6.59}$$

Assuming CPT symmetry, as valid in any QFT, and taking into account correlations between systematic errors between the various data sets the new average $R = 0.0037072063(20)$ is obtained. From this result one obtains the new average value

Table 6.2. Summary of CERN and E821 Results

Experiment	Year	Polarity	$a_\mu \times 10^{10}$	Precision [ppm]	Reference
CERN I	1961	μ^+	11 450 000(220000)	4300	[24]
CERN II	1962–1968	μ^+	11 661 600(3100)	270	[25]
CERN III	1974–1976	μ^+	11 659 100(110)	10	[6]
CERN III	1975–1976	μ^-	11 659 360(120)	10	[6]
BNL	1997	μ^+	11 659 251(150)	13	[11]
BNL	1998	μ^+	11 659 191(59)	5	[12]
BNL	1999	μ^+	11 659 202(15)	1.3	[13]
BNL	2000	μ^+	11 659 204(9)	0.73	[14]
BNL	2001	μ^-	11 659 214(9)	0.72	[15]
	Average		11 659 208.0(6.3)	0.54	[15]

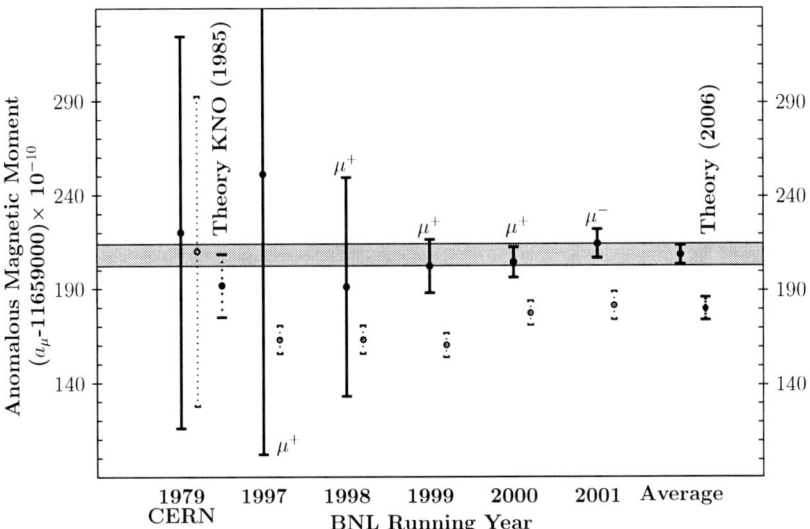

Fig. 6.11. Results for the individual E821 measurements, together with the new world average and the theoretical prediction. The CERN result is shown together with the theoretical prediction by Kinoshita et al. 1985, at about the time when the E821 project was proposed. The dotted vertical bars indicate the theory values quoted by the experiments

$$a_\mu = 11659208.0(5.4)(3.3) \times 10^{-10}, \qquad (6.60)$$

with a relative uncertainty of 0.54 ppm [7]. Where two uncertainties are given the first is statistical and the second systematic, otherwise the total error is given where statistical and systematic errors have been added in quadrature. In Table 6.2 all results from CERN and E821 are collected. The new average is completely dominated by the BNL results. The individual measurements are shown also in Fig. 6.11. The comparison with the theoretical result is devoted to the next section. The achieved improvement and a comparison of the sensitivity to various kinds of physics effects has been shown earlier in Fig. 3.8 at the end of Sect. 3.2.1.

The following two sections are addenda, one on the determination of λ in (6.58) and the other a sketch of the electron $g - 2$ measurement technique.

6.6 Ground State Hyperfine Structure of Muonium

The hyperfine and Zeeman levels of $^2S_{\frac{1}{2}}$ ground state Muonium are shown in Fig. 6.12. The energy levels are described by the Hamiltonian

$$\mathcal{H} = h\,\Delta\nu\,\boldsymbol{I}_\mu \cdot \boldsymbol{J} - \mu_B^\mu\,g'_\mu\,\boldsymbol{I}_\mu \cdot \boldsymbol{B} + \mu_B^e\,g_J\,\boldsymbol{J} \cdot \boldsymbol{B} \qquad (6.61)$$

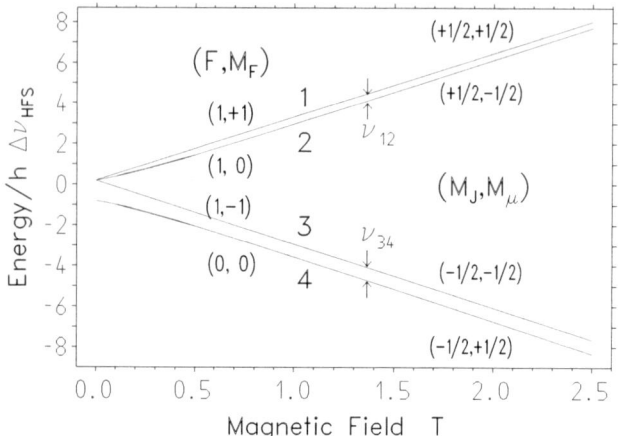

Fig. 6.12. Muonium ground state hyperfine structure Zeeman splitting (Breit-Rabi energy level diagram). At high fields the transitions ν_{12} and ν_{34} are essentially muon spin flip transitions

where $\boldsymbol{I}_\mu$ is the muon spin operator, $\boldsymbol{J}$ is the electron total angular momentum operator and $\boldsymbol{B}$ is the external static magnetic field. The total angular momentum is $\boldsymbol{F} = \boldsymbol{J} + \boldsymbol{I}_\mu$.

Microwave transitions ν_{12} and ν_{34} are measured in a strong magnetic field B of 1.6 Tesla. Also this experiment uses the parity violating correlation of the direction of the muon spin and the positron emission of μ–decay.

The hyperfine splitting (HFS) and the muon magnetic moment are determined from ν_{12} and ν_{34}.

$$\nu_{12} = -\mu_\mu B + \frac{\Delta\nu}{2}\left[1 + x - \sqrt{1+x^2}\right]$$
$$\nu_{34} = +\mu_\mu B + \frac{\Delta\nu}{2}\left[1 - x + \sqrt{1+x^2}\right] \quad (6.62)$$

where $x = (g_J \mu_B^e + g'_\mu \mu_B^\mu) B/(h\Delta\nu)$ is proportional to the magnetic field strength B[9]. The latest experiment at LAMPF at Los Alamos has measured these level splittings very accurately. The Larmor relation, $2\mu_p B = h\nu_p$, and NMR is used to determine B in terms of the free proton precession frequency ν_p and the proton magnetic moment μ_p. Using (6.62) and the measured transition frequencies ν_{12} and ν_{34} both $\Delta\nu$ and μ_μ/μ_p can be determined.

---

[9]The gyromagnetic ratios of the bound electron and muon differ from the free ones by the binding corrections [26]

$$g_J = g_e\left(1 - \frac{\alpha^2}{3} + \frac{\alpha^2}{2}\frac{m_e}{m_\mu} + \frac{\alpha^3}{4\pi}\right) \quad , \quad g'_\mu = g_\mu\left(1 - \frac{\alpha^2}{3} + \frac{\alpha^2}{2}\frac{m_e}{m_\mu}\right) \; .$$

Note that the sum of (6.62) equals to the zero field splitting $\Delta\nu \equiv \Delta\nu_{\mathrm{HFS}}$ independent of the field B, while for high fields the difference measures the magnetic moment μ_μ:

$$\Delta\nu = \nu_{12} + \nu_{34},$$

$$\mu_\mu B = \nu_{34} - \nu_{12} - \Delta\nu\left(\sqrt{1+x^2}-x\right) \approx \nu_{34} - \nu_{12} - \frac{\Delta\nu}{2x}, \quad (x \gg 1).$$

The magnetic moment was measured to be

$$\mu_\mu/\mu_p = 3.183\,345\,24(37)\,(120\text{ ppb}),$$

which translates into a muon–electron mass ratio

$$m_\mu/m_e = \left(\frac{g_\mu}{2}\right)\left(\frac{\mu_p}{\mu_\mu}\right)\left(\frac{\mu_B^e}{\mu_p}\right) = 206.768\,276(24)\,(120\text{ ppb}),$$

when using $g_\mu = 2(1+a_\mu)$ with $a_\mu = 11\,659\,208.0(6.3) \times 10^{-10}$ and $\mu_p/\mu_B^e = 1.521\,032\,206(15) \times 10^{-3}$ [27]. The measured value of the zero field HFS is

$$\Delta\nu^{\mathrm{exp}} = 4\,463\,302\,765(53)\text{ Hz }(12\text{ ppb})$$

in good agreement with the theoretical prediction [28, 18, 29, 30, 19, 27]

$$\Delta\nu^{\mathrm{the}} = \frac{16}{3}cR_\infty \alpha^2 \frac{m_e}{m_\mu}\left(1+\frac{m_e}{m_\mu}\right)^{-3}(1+\delta\mathcal{F}(\alpha, m_e/m_\mu))$$
$$= 4\,463\,302\,905(272)\text{ Hz }(61\text{ ppb}),$$

where the error is mainly due to the uncertainty in m_μ/m_e. The correction $\delta\mathcal{F}(\alpha, m_e/m_\mu)$ depends weakly on α and m_e/m_μ,

$$R_\infty = 10\,973\,731.568\,525(37)\text{ m}^{-1}$$

is the Rydberg constant $\alpha^2 m_e c/2h$ [27]. A combined result was used to determine (6.9) used in the determination of a_μ (see also [31]).

6.7 Single Electron Dynamics and the Electron $g - 2$

The basic principle of a muon $g-2$ experiment is in many respects very similar to the one of electron $g-2$ experiments, although the scale of the experiment is very different and the electron $g-2$ experiment uses atomic spectroscopy type methods to determine the frequencies. The particle dynamics considered in Sect. 6.2 applies to the single electron or single ion Penning trap shown in Fig. 6.13. Electron motion in a hyperboloid Penning trap in the *axial* (vertical) direction is a harmonic oscillation

$$z(t) = A\cos(\omega_z t)$$

370   6 The g − 2 Experiments

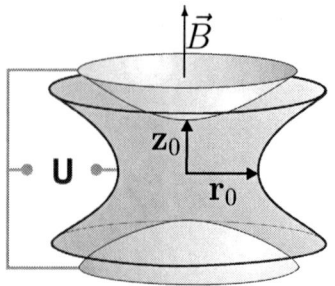

(a) Penning trap: an electrical quadrupole field superimposed on a homogeneous magnetic field ($U = V_0$)

(b) Penning trap device (in units of the size of a 1 € coin)

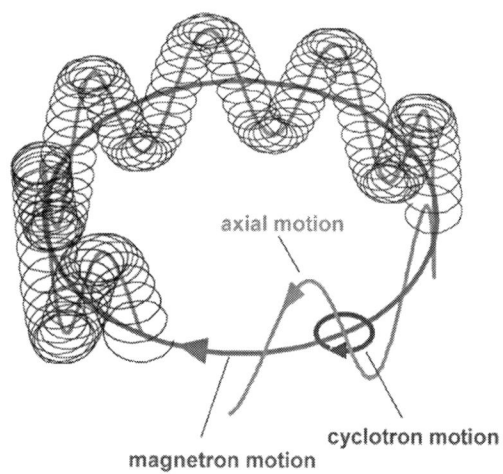

(c) Typical electron trajectory

Fig. 6.13. Electron motion in a hyperbolic Penning trap [Courtesy of G. Werth, Mainz [36]]

with
$$\omega_z = 2\sqrt{eV_0/md^2}$$

(see (6.15)). In the *radial* direction it is an epicycloid motion with

$$x(t) = +\rho_m \cos(\omega_m t) + \rho_c \cos(\omega'_c t),$$
$$y(t) = -\rho_m \sin(\omega_m t) - \rho_c \sin(\omega'_c t).$$

Here
$$\omega'_c = \omega_+ = \frac{1}{2}(\omega_c + \sqrt{\omega_c^2 - 2\omega_z^2}) \simeq \omega_c$$

is the *perturbed cyclotron frequency* and

6.7 Single Electron Dynamics and the Electron $g-2$

$$\omega_m = \omega_- = \frac{1}{2}(\omega_c - \sqrt{\omega_c^2 - 2\omega_z^2}) = \omega_c - \omega_c'$$

the *magnetron frequency*. The frequencies are related by $\omega_c^2 = \omega_+^2 + \omega_-^2 + \omega_z^2$. Typical values for a positron in a magnetic field $B = 3\,\text{T}$, $U = 10\,\text{V}$ and $d = 3.3\,\text{mm}$ are $\nu_c = 48\,\text{GHz}$, $\nu_z = 64\,\text{MHz}$, $\nu_m = 12\,\text{kHz}$ depending on the field strengths determined by B, U and d.

The observation of the splitting of the spin states requires a coupling of the cyclotron and spin motion of the trapped electron to the axial oscillation, which is realized by an extremely weak magnetic bottle modifying the uniform magnetic field by an inhomogeneous component (Dehmelt et al. 1973) (see Fig. 6.14). The latter is imposed by a ferromagnetic ring electrode, such that

$$B = B_0 + B_2\, z^2 + \cdots \tag{6.63}$$

which imposes a force

$$F = m_s\, g_e\, \mu_B\, \text{grad}\, B = m_s\, g_e\, \mu_B\, B_2\, z\;,$$

on the magnetic moment. Because of the cylindric symmetry the force is linear in first order and the motion remains harmonic. The force adds or subtracts a component depending on $m_s = \pm 1/2$ and thus changes the axial frequency by

$$\Delta\omega_z = g_e\, \mu_B\, \frac{B_2}{m_e \omega_z} \tag{6.64}$$

as shown in Fig. 6.14.

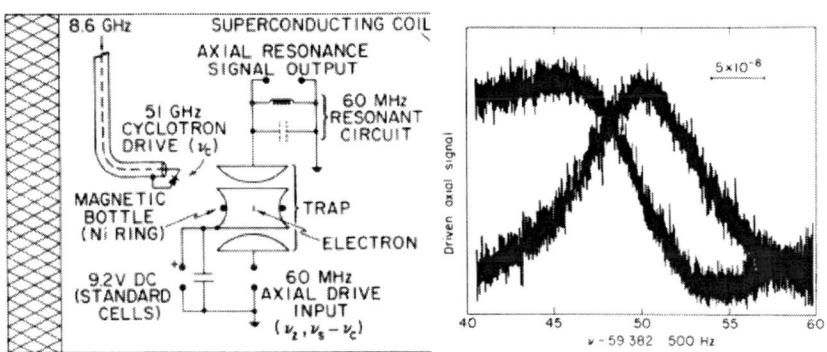

Fig. 6.14. **Left**: schematic of the geonium apparatus (Dehmelt et al. [32]). Hyperbolic endcaps and ring electrodes trap the electron axially while coupling the driven harmonic motion to an external LC circuit tuned to drive the axial frequency. Radial trapping of the electron is produced by the strong magnetic field from a superconducting solenoid. **Right**: frequency shift in the axial resonance signal at $\approx 60\,\text{MHz}$. The signal-to-noise ratio of this $\approx 8\,\text{Hz}$ wide line corresponds to a frequency resolution of 10 ppb. Reprinted with permission from [32]. Copyright (2007) by the American Physical Society

For a trap working at a temperature of $T = 4°\,\text{K}$ the thermic energy is $E = kT = 3.45 \times 10^{-4}\,\text{eV}$. The trapped electron occupies low quantum states, the cyclotron ($n = 0, 1, 2, \cdots$) and spin ($m_s = \pm 1/2$) energy levels,

$$E(n, m_s) = \left(n + \frac{1}{2}\right) \hbar\omega_c' + \frac{g_e}{2} \hbar\omega_c m_s - \frac{\hbar}{2}\delta \left(n + \frac{1}{2} + m_s\right)^2, \quad (6.65)$$

for $\nu_c = 84\,\text{GHz}$ thus $\hbar\omega_c = 3.47 \times 10^{-4}\,\text{eV}$ which implies $n_c = 0, 1$ such that QM is at work (the axial motion corresponds to $n_z \simeq 1000$ and hence is classical). In fact this is not quite true: Gabrielse has shown that in Dehmelt's experiment at $4°\,\text{K}$, because of the spread in the thermic spectrum, still many higher states are populated and, in a field of a few Tesla, only at about $T = 0.1°\,\text{K}$ one reaches the ground state [33]. The third term in (6.65) is the leading relativistic correction of size $\delta/\nu_c \equiv h\nu_c/(mc^2) \approx 10^{-9}$ [34], too small to be important at the present level of accuracy of the experiments. The radiation damping is

$$\frac{dE}{dt} = -\hat{\gamma} E, \quad \hat{\gamma} = \frac{e^2 \omega^2}{6\pi\epsilon_0 mc^3} \quad (6.66)$$

and with $\alpha\hbar c = e^2/(4\pi\epsilon_0) = 1.44\,\text{MeVfm}$ one has $\hat{\gamma}_c = 1.75\,\text{s}^{-1}$. The spontaneous damping by radiation is then $\hat{\gamma}_z \simeq \hat{\gamma}_c/10^6 \simeq 0.15$ per day. The $g - 2$ follows from the spin level splitting Fig. 6.15

$$\Delta E = g_e \mu_B B = \frac{g_e}{2} \hbar\omega_c \equiv \hbar\omega_s \quad (6.67)$$

such that

$$a_e \equiv \frac{g_e - 2}{2} = \frac{\omega_s - \omega_c}{\omega_c} \equiv \frac{\omega_a}{\omega_c}. \quad (6.68)$$

From the spin Larmor precession frequency $\hbar\omega_s = m_e g_e \mu_B B$ (μ_B the Bohr magneton) and the calibration of the magnetic field by the cyclotron frequency of a single ion in the Penning trap $\hbar\omega_c = q_{\text{ion}}/M_{\text{ion}} B$ one obtains

$$g_e = 2\, \frac{\omega_s}{\omega_c} \frac{q_{\text{ion}}}{e} \frac{m_e}{M_{\text{ion}}} \quad (6.69)$$

or if g_e is assumed to be known one may determine the electron mass very precisely. The most precise determination was obtained from g-factor experiments on $^{12}\text{C}^{5+}$ and $^{16}\text{C}^{7+}$ [35] with a cylindrical cryogenic double Penning trap in a magnetic field of 3.8 T [working at frequencies $\nu_c = 25\,\text{MHz}$, $\nu_z = 1\,\text{MHz}$, $\nu_m = 16\,\text{kHz}$].

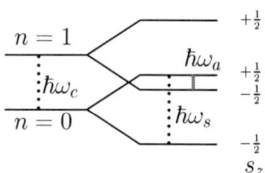

Fig. 6.15. Lowest electron quantum states in a Penning trap

The Harvard electron $g - 2$ experiment performs spectroscopy of a single electron in the lowest cyclotron and spin levels in a cylindrical Penning trap. The problem of a harmonic Penning trap is that it is a cavity and hence allows only certain electromagnetic frequencies. The damping by spontaneous emission affects the cyclotron frequency in a way which is not fully under control. The cylindrical trap which exhibits plenty of higher harmonics solves this problem as it can be operated at well selected frequencies. Working frequencies are $\nu_s \approx \nu_c \approx 149\,\mathrm{GHz}$, $\nu_z \approx 200\,\mathrm{MHz}$, $\nu_m \approx 134\mathrm{kHz}$. For the first time it was possible to work with the lowest quantum states of (6.65) (see Fig. 6.15) in the determination of $g_e - 2$. The result has been discussed in Sect. 3.2.2.

References

1. F. J. M. Farley, E. Picasso, In: *Quantum Electrodynamics*, ed by T. Kinoshita (World Scientific, Singapore 1990) p 479; Adv. Ser. Direct. High Energy Phys. **7** (1990) 479
2. V. W. Hughes, *The anomalous magnetic moment of the muon*. In: Intern. School of Subnuclear Physics: 39th Course: New Fields and Strings in Subnuclear Physics, Erice, Italy, 29 Aug - 7 Sep 2001; Int. J. Mod. Phys. A **18S1** (2003) 215
3. F. J. M. Farley, Y. K. Semertzidis, Prog. Part. Nucl. Phys. **52** (2004) 1
4. D. W. Hertzog, W. M. Morse, Annu. Rev. Nucl. Part. Sci. **54** (2004) 141
5. J. M. Paley, Measurement of the Anomalous Magnetic Moment of the Negative Muon to 0.7 Parts per Million, Boston University Dissertation, 2004, available from the **UMI Thesis Server**
6. J. Bailey et al., Nucl. Phys. B **150** (1979) 1
7. G. W. Bennett et al. [Muon g-2 Collaboration], Phys. Rev. D **73** (2006) 072003
8. G. T. Danby et al., Nucl. Instr. and Methods Phys. Res. **A 457** (2001) 151
9. Y. K. Semertzidis, Nucl. Instr. and Methods Phys. Res. **A503** (2003) 458
10. E. Efstathiadis et al., Nucl. Instr. and Methods Phys. Res. **A496** (2002) 8
11. R. M. Carey et al., Phys. Rev. Lett. **82** (1999) 1632
12. H. N. Brown et al. [Muon (g-2) Collaboration], Phys. Rev. D **62** (2000) 091101
13. H. N. Brown et al. [Muon (g-2) Collaboration], Phys. Rev. Lett. **86** (2001) 2227
14. G. W. Bennett et al. [Muon g-2 Collaboration], Phys. Rev. Lett. **89** (2002) 101804 [Erratum-ibid. **89** (2002) 129903]
15. G. W. Bennett et al. [Muon g-2 Collaboration], Phys. Rev. Lett. **92** (2004) 161802
16. R. Prigl et al., Nucl. Instr. and Methods Phys. Res. **A374** (1996) 118; X. Fei, V. Hughes, R. Prigl, Nucl. Instr. and Methods Phys. Res. **A394** (1997) 349
17. W. Liu et al., Phys. Rev. Lett. **82** (1999) 711
18. T. Kinoshita, M. Nio, Phys. Rev. D **53** (1996) 4909; M. Nio, T. Kinoshita, Phys. Rev. D **55** (1997) 7267; T. Kinoshita, hep-ph/9808351
19. A. Czarnecki, S. I. Eidelman, S. G. Karshenboim, Phys. Rev. D **65** (2002) 053004; S. G. Karshenboim, V. A. Shelyuto, Phys. Lett. B **517** (2001) 32
20. L. H. Thomas, Phil. Mag. **3** (1927) 1; V. Bargmann, L. Michel, V. A. Telegdi, Phys. Rev. Lett. **2** (1959) 435; B. W. Montague, Phys. Rept. **113** (1984) 1; J. S. Bell, CERN-75-11, 38p (1975)

21. F. J. N. Farley, Phys. Lett. B **42** (1972) 66
22. F. Scheck, Leptons, Hadrons and Nuclei, North Holland, Amsterdam (1983)
23. S. Eidelman et al. [Particle Data Group], Phys. Lett. B **592** (2004) 1
24. G. Charpak et al., Phys. Rev. Lett. **6** (1961) 128; G. Charpak et al., Nuovo Cimento **22** (1961) 1043
25. J. Bailey et al., Phys. Lett. B **28** (1968) 287; Nuovo Cimento A **9** (1972) 369
26. H. Grotch, R. A. Hegstrom, Phys. Rev. A **4** (1971) 59; R. Faustov, Phys. Lett. B **33** (1970) 422
27. P. J. Mohr, B. N. Taylor, Rev. Mod. Phys. **72** (2000) 351; **77** (2005) 1
28. K. Pachucki, Phys. Rev. A **54** (1996) 1994; ibid. **56** (1997) 297; S. G. Karshenboim, Z. Phys. D **36** (1996) 11; S. A. Blundell, K. T. Cheng, J. Sapirstein, Phys. Rev. Lett. **78** (1997) 4914; M. I. Eides, H. Grotch, V. A. Shelyuto, Phys. Rev. D **58** (1998) 013008
29. K. Melnikov, A. Yelkhovsky, Phys. Rev. Lett. **86** (2001) 1498
30. R. J. Hill, Phys. Rev. Lett. **86** (2001) 3280
31. K. P. Jungmann, Nucl. Phys. News **12** (2002) 23
32. R. S. Van Dyck, P. B. Schwinberg, H. G. Dehmelt, Phys. Rev. D **34** (1986) 722
33. S. Peil, G. Gabrielse, Phys. Rev. Lett. **83** (1999) 1287
34. L. S. Brown, G. Gabrielse, Rev. Mod. Phys. **58** (1986) 233
35. T. Beier et al., Phys. Rev. Lett. **88** (2002) 011603; Nucl. Instrum. Meth. B **205** (2003) 15
36. F.G. Major, V.N. Gheorghe, G. Werth: Charged Particle Traps (Springer, Berlin, 2005)

7

Comparison Between Theory and Experiment and Future Perspectives

7.1 Experimental Results Confront Standard Theory

The anomalous magnetic moment of the muon provides one of the most precise tests of quantum field theory as a basic framework of elementary particle theory and of QED and the electroweak SM in particular. With what has been reached by the BNL muon $g-2$ experiment (see Table 7.1), namely the reduction of the experimental uncertainty by a factor 14 to $\sim 63 \times 10^{-11}$, a new quality in "diving into the sea of quantum corrections" has been achieved: the 8th order QED [$\sim 381 \times 10^{-11}$] known thanks to the heroic efforts of Kinoshita and Nio, the weak correction up to 2nd order [$\sim 154 \times 10^{-11}$] and the hadronic light–by–light scattering[1] [$\sim 100 \times 10^{-11}$] are now in the focus. The hadronic vacuum polarization effects which played a significant role already for the last CERN experiment now is a huge effect of more than 11 SD's. As a non–perturbative effect it still has to be evaluated largely in terms of experimental data with unavoidable experimental uncertainties which yield the biggest contribution to the uncertainty of theoretical predictions. However, due to substantial progress in the measurement of total hadronic e^+e^-–annihilation cross–sections, the uncertainty from this source has reduced to a remarkable $\sim 56 \times 10^{-11}$ only. This source of error now is only slightly larger than the uncertainty in the theoretical estimates of the hadronic light–by–light scattering contribution [$\sim 40 \times 10^{-11}$]. Nevertheless, we have a solid prediction with a total uncertainty of $\sim 68 \times 10^{-11}$, which is essentially equal to the experimental error of the muon $g-2$ measurement. A graphical representation for the sensitivity and the weight of the various contributions is presented in Fig. 3.8 (see also Table 7.2 and Fig. 7.2 below).

We now have at the same time a new very sensitive test of our current theoretical understanding of the fundamental forces and the particle spectrum and a stringent bound on physics beyond the SM entering at scales below about 1 TeV. But, may be more important is the actual deviation between

---

[1] In this Chapter we adopt a rounded value $(100 \pm 40) \times 10^{-11}$ for the hadronic LbL contribution, in place of $(93 \pm 34) \times 10^{-11}$ estimated in Sect. 5.4.5

Table 7.1. Progress from CERN 1979 to BNL 2006 [*= CPT assumed]

	CERN 1979 [1]	BNL 2006 [2]
a_{μ^+}	$1165911(11) \times 10^{-9}$	$11659204(7)(5) \times 10^{-10}$
a_{μ^-}	$1165937(12) \times 10^{-9}$	$11659214(8)(3) \times 10^{-10}$
a_μ *	$1165924(8.5)10^{-9}$	$11659208(4)(3) \times 10^{-10}$
$(a_{\mu^+} - a_{\mu^-})/a_\mu$	$-(2.2 \pm 2.8) \times 10^{-5}$	$-(8.6 \pm 18.2) \times 10^{-7}$
d_μ (EDM) *	$(3.7 \pm 3.4) \times 10^{-19} \, e \cdot \text{cm}$	$< 2.7 \times 10^{-19} \, e \cdot \text{cm}$
a_μ^{the}	$1165921(8.3)10^{-9}$	$11659179.3(6.8)10^{-10}$
$a_\mu^{\text{the}} - a_\mu^{\text{exp}}$	$(-3.0 \pm 11.9) \times 10^{-9}$	$(-28.7 \pm 9.1) \times 10^{-10}$
$(a_\mu^{\text{the}} - a_\mu^{\text{exp}})/a_\mu^{\text{exp}}$	$-(2.6 \pm 10.2) \times 10^{-6}$	$-(2.5 \pm 0.8) \times 10^{-6}$

theory and experiment at the 3 σ level which is *a clear indication of something missing*. We have to remember that such high precision physics is extremely challenging for both experiment and for theory and it is not excluded that some small effect has been overlooked or underestimated at some place. To our present knowledge, it is hard to imagine that a 3 σ shift could be explained by known physics. Thus New Physics seems a likely interpretation, if it is not an experimental fluctuation (0.27% chance).

It should be noted that among all the solid precision tests, to my knowledge, the muon $g - 2$ shows the largest established deviation between theory and experiment. Actually, the latter has been persisting since the first precise measurement was released at BNL in February 2001 [11], and a press release announced "We are now 99 percent sure that the present Standard Model calculations cannot describe our data". A 2.6 σ deviation was found at that time for a selected choice of the hadronic vacuum polarization and with the wrong sign hadronic LbL scattering contribution[2]. In the meantime

Table 7.2. Standard model theory and experiment comparison

Contribution	Value $\times 10^{10}$	Error $\times 10^{10}$	Reference
QED incl. 4-loops+LO 5-loops	11 658 471.81	0.02	[3]
Hadronic vacuum polarization	692.1	5.6	[4]
Hadronic light–by–light	10.0	4.0	[5, 6, 7, 8]
Hadronic, other 2nd order	−10.0	0.2	[4]
Weak 2-loops	15.4	0.22	[9, 10]
Theory	11 659 179.3	6.8	–
Experiment	11 659 208.0	6.4	[2]
The. - Exp. 3.2 standard deviations	−28.7	9.1	–

---

[2]With the correct sign of the hadronic LbL term the deviation would have been 1.5 σ based on the smallest available hadronic vacuum polarization. With larger values of the latter the difference would have been smaller.

errors went further down experimentally as well as in theory[3], especially the improvement of the experimental e^+e^-–data, indispensable as an input for the "prediction" of the hadronic vacuum polarization, and the remedy of the wrong sign of the π^0 exchange LbL term has brought us forward a big step. The theoretical status, the main theme of this book, has been summarized in Sect. 3.2.3 (see Table 3.5 and Fig. 3.8) the experimental one in Sect. 6.5 (see Table 6.2 and Fig. 6.11). The jump in the precision is best reminded by a look at Table 7.1 which compares the results from the 1979 CERN final report [1] with the one's of the 2006 BNL final report [2].

The CPT test has improved by an order of magnitude. Relativistic QFT in any case guarantees CPT symmetry to hold and we assume CPT throughout in taking averages or estimating new physics effects etc. The world average experimental muon magnetic anomaly, dominated by the very precise BNL result, now is [2]

$$a_\mu^{\text{exp}} = 1.16592080(54)(33) \times 10^{-3}, \tag{7.1}$$

with relative uncertainty 5.4×10^{-7}, which confronts the SM prediction

$$a_\mu^{\text{the}} = 1.16591793(68) \times 10^{-3}, \tag{7.2}$$

and agrees up to the small but non–negligible deviation

$$\delta a_\mu = a_\mu^{\text{exp}} - a_\mu^{\text{the}} = 287 \pm 91 \times 10^{-11}, \tag{7.3}$$

which is a 3.2 σ effect. Errors have been added in quadrature. Some other recent evaluations are collected in Table 7.3. Differences in errors come about

Table 7.3. Some recent evaluations of $a_\mu^{\text{had}(1)}$

$a_\mu \times 10^{10} - 11659000$	$a_\mu^{\text{had}(1)} \times 10^{10}$	data	label	Ref.
181.3[16.]	696.7[12.]	e^+e^-	EJ95	[12]
180.9[8.0]	696.3[7.2]	e^+e^-	DEHZ03	[13]
195.6[6.8]	711.0[5.8]	$e^+e^- + \tau$	DEHZ03	[13]
179.4[9.3]	694.8[8.6]	e^+e^-	GJ03	[14]
169.2[6.4]	684.6[6.4]	e^+e^- TH	SN03	[15]
183.5[6.7]	692.4[6.4]	e^+e^-	HMNT03	[16]
180.6[5.9]	693.5[5.9]	e^+e^-	TY04	[17]
188.9[5.9]	701.8[5.8]	$e^+e^- + \tau$	TY04	[17]
180.5[5.6]	690.8[4.3]	e^+e^- **	DEHZ06	[18]
180.4[5.1]	689.4[4.6]	e^+e^- **	HMNT06	[19]
179.3[6.8]	692.1[5.6]	e^+e^- **	FJ06	[4]

---

[3]To mention the sign error and the issue of the high energy behavior in the LbL contribution or errors in the applied radiative corrections of e^+e^-–data or taking into account or not the isospin rotated τ–data.

mainly by utilizing more "theory–driven" concepts[4]: use of selected data sets only if data are not in satisfactory agreement, extended use of perturbative QCD in place of data [assuming more local duality], sum rule methods, or low energy effective methods [21]. Only the last three (**) results include at least partially the most recent data from KLOE, SND, CMD-2[5], and BaBar[6].

---

[4]The terminology "theory–driven" means that we are not dealing with a solid theory prediction. As in some regions only old data sets are available, some authors prefer to use pQCD in place of the data also in regions where pQCD is not supposed to work reliably. The argument is that even under these circumstances pQCD may be better than the available data. This may be true, but one has to specified what "better" means. In this approach non–perturbative effects are accounted for by referring to local quark–hadron duality in relatively narrow energy intervals. What is problematic is a reliable error estimate. Usually only the pQCD errors are accounted for (essentially only the uncertainty in α_s is taken into account). It is *assumed* that no other uncertainties from non–perturbative effects exist; this is why errors in this approach are systematically lower than in more conservative data oriented approaches. Note that applying pQCD in any case *assumes* quark–hadron duality to hold in large enough intervals, ideally from threshold to ∞ (global duality). My "conservative" evaluation of a_μ^{had} estimates an error of 0.8%, which for the given quality of the data is as progressive as it can be, according to my standards concerning reliability. In spite of big progress in hadronic cross–section measurements the agreement between different measurements is not as satisfactory as one would wish. Also more recent measurements often do not agree within the errors quoted by the experiments. Thus, one may seriously ask the question how such small uncertainties come about. The main point is that results in different energy ranges, as listed in Table 5.2 in Sect. 5.2, are treated as independent and all errors including the systematic ones are added in quadrature. By choosing a finer subdivision, like in the clustering procedure of [16], for example, one may easily end up with smaller errors (down to 0.6%). The subdivision I use was chosen originally in [12] and were more or less naturally associated with the ranges of the different experiments. The problem is that combining systematic errors is not possible on a commonly accepted basis if one goes beyond the plausible procedures advocated by the Particle Data Group.

[5]In the common KLOE energy range (591.6,969.5) MeV individual contributions based on the latest [2004/2006] data are:

	KLOE	CMD-2	SND
a_μ^{had}	391.90(0.52)(5.10)	392.64(1.87)(3.14)	390.47(1.34)(5.08)

in good agreement.

[6]The analysis [19] does not include exclusive data in a range from 1.43 to 2 GeV; therefore also the new BaBar data are not included in that range. It also should be noted that CMD-2 and SND are not fully independent measurements; data are taken at the same machine and with the same radiative correction program. The radiative corrections play a crucial role at the present level of accuracy, and common errors have to be added linearly. In [13, 18] pQCD is used in the extended ranges 1.8–3.7 GeV and above 5.0 GeV; furthermore [18] excludes the KLOE data.

The last entry [4] is based on the evaluation of all data and pQCD is used only where it can be applied safely according to [22, 23] and as discussed in Sect. 5.2.

Note that the experimental uncertainty is still statistics dominated[7]. Thus just running the BNL experiment longer could have substantially improved the result. Originally the E821 goal was $\delta a_\mu^{\mathrm{exp}} \sim 40 \times 10^{-11}$. Fig. 7.1 illustrates the improvement achieved by the BNL experiment. The theoretical predictions mainly differ by the L.O. hadronic effects, which also dominates the theoretical error. Results from different analyses are given in Table 7.3, see also Figs. 5.15, 5.16 in Sect. 5.2.

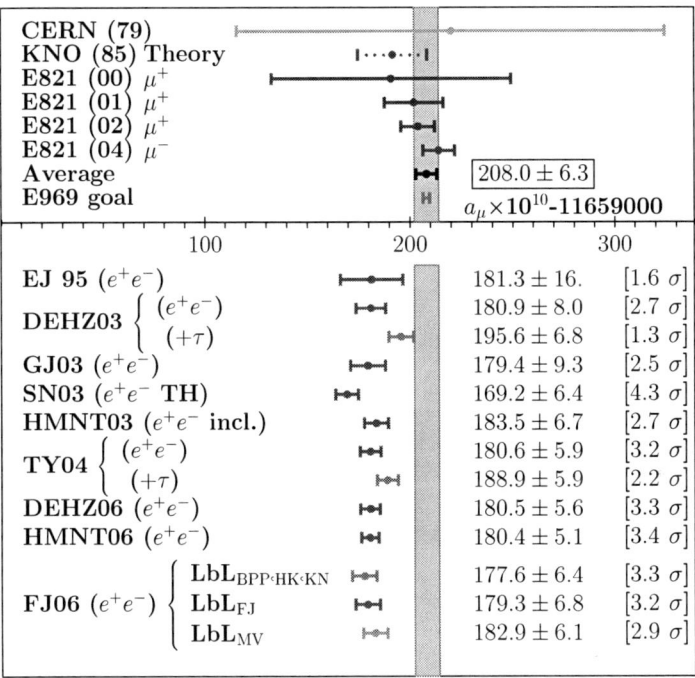

Fig. 7.1. Comparison between theory and experiment. Results differ by different L.O. hadronic vacuum polarizations and variants of the LbL contribution. Some estimates include isospin rotated τ–data ($+\tau$). The last entry FJ06 also illustrates the effect of using different LbL estimations: 1) Bijnens, Pallante, Prades (BPP) [5], Hayakawa, Kinoshita (HK) [6] and Knecht, Nyffeler (KN) [7]; 2) my estimation based on the other evaluations; 3) the Melnikov, Vainshtein (MV) [8] estimate of the LbL contribution. EJ95 vs. FJ06 illustrates the improvement of the e^+e^-–data between 1995 and 2006 (see also Table 7.3 and Fig. 6.11). E969 is a possible follow-up experiment of E821 proposed recently [20]

---

[7] The small spread in the central values does not reflect this fact, however.

380   7 Comparison Between Theory and Experiment and Future Perspectives

As discussed earlier in Sect. 5.2.2, in principle, the $I = 1$ iso–vector part of $e^+e^- \to$ hadrons can be obtained in an alternative way by using the precise vector spectral functions from hadronic τ–decays via an isospin rotation. The unexpectedly large discrepancy, of order 10 to 20% starting just above the ρ peak and increasing with energy, between the appropriately isospin–violations corrected τ–based and the e^+e^- cross–sections makes it difficult to combine the two types of data [13] (see also [24] and references therein for a recent status report). Including the τ–data shifts upwards the evaluation of a_μ by about 2 σ and would lead to a better agreement between theory and experiment, as seen in Fig. 7.1 and Table 7.3.

Possible explanations for the observed difference are so far unaccounted isospin breaking [14] or experimental problems with the data. Because the e^+e^-–data are more directly related to what is needed in the dispersion integral representation of a_μ^{had} and since the dominant $e^+e^- \to \pi^+\pi^-$ channel measured by CMD-2, SND and KLOE (the agreement of the latter is only at the $\pm$ 3-4% level in the distribution, but agrees when integrated) agree much better among each other than with the τ-data (with good agreement between ALEPH and CLEO data, while OPAL data show clear deviations (see e.g. [14])) presently one refrains from taking into account the τ-data as

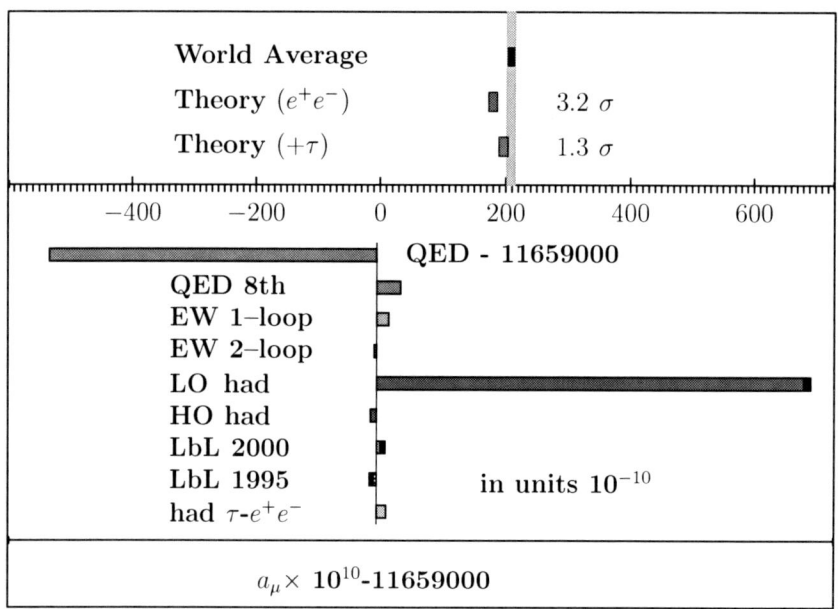

Fig. 7.2. All kinds of physics meet. Shown are the various contributions which add up to the theory prediction relative to the experimental result. The 8th order QED included in the QED part is shown separately. For comparison also the old LbL contribution of wrong sign and the extra contribution obtained by including the isospin rotated hadronic τ–decay data are shown. The black heads on the bars represent the uncertainties

Table 7.4. Progress in a_μ measurements. Theory values as quoted in References (μSR= Muon Storage Ring)

Laboratory	year	Ref.	result (error)$\times 10^3$	precision	theory$\times 10^3$
Columbia	1960	[25]	1.22 (8)		1.16
CERN cyclotron	1961	[26]	μ^+ 1.145 (22)		1.165
CERN cyclotron	1962	[27]	μ^+ 1.162 (05)		1.165
CERN 1st μSR	1966	[28]	μ^- 1.165 (03)		1.165
CERN 1st μSR	1968	[29]	$\mu^\pm$ 1.166 16 (31)		1.1656
CERN 2nd μSR	1977	[30]	$\mu^\pm$ 1.165 924 0 (85)	7 ppm	1.165 921 0 (83)
BNL, 1997 data	1999	[31]	μ^+ 1.165 925 (15)	13 ppm	1.165 916 3 (8)
BNL, 1998 data	2000	[32]	μ^+ 1.165 919 1 (59)	5 ppm	1.165 916 3 (8)
BNL, 1999 data	2001	[11]	μ^+ 1.165 920 2 (15)	1.3 ppm	1.165 916 0 (7)
BNL, 2000 data	2002	[33]	μ^+ 1.165 920 4 (9)	0.73 ppm	1.165 917 7 (7)
BNL, 2001 data	2004	[34]	μ^- 1.165 921 4 (9)	0.72 ppm	1.165 918 1 (8)
World average	2004	[34]	$\mu^\pm$ 1.165 920 80 (63)	0.54 ppm	1.165 917 93 (68)

long as the reason for the discrepancy is not really understood. Figure 7.2 illustrates how different physics contributions add up to the final answer. We note that the theory error is somewhat larger than the experimental one. It is fully dominated by the uncertainty of the hadronic low energy cross–section data, which determine the hadronic vacuum polarization and, partially, by the uncertainty of the hadronic light–by–light scattering contribution. The history of muon $g-2$ measurements together with the theory values with which results were compared are listed once more in Table 7.4.

7.2 New Physics in $g-2$

The question about what is the physics behind the SM was and is the main issue of theoretical particle physics since the emergence of the SM as *the* theory of "fundamental" particle interactions which we know today. Besides the SM's main shortcoming, which is that it lacks to include gravity, it rises many other questions about is structure, and the answers always are attempts of embedding the SM into an extended theory. While the SM is very well established and is able to explain a plenitude of experimental data so well that experimenting starts to be a kind of frustrating, it is well known and as well established that the SM is *not* able to explain a number of facts, like the existence of non–baryonic cold dark matter (at most 10% is normal baryonic matter), the matter–antimatter asymmetry in the universe, which requires baryon–number B and lepton–number L violation, the problem of the cosmological constant and so on. So, new physics is there but where precisely? What can $g-2$ tell us about new physics[8]?

New physics contributions, which, if they exist, are an integral part of the measured numbers, typically are expected to be due to states or interactions which have not been seen by other experiments, either by a lack of sensitivity or, at the high energy frontier, because experimental facilities like accelerators are below the threshold of energy needed for producing the new heavy states or because the signal was still buried in the background. At the high energy frontier LEP and the Tevatron have set limits on many species of possible new particles predicted in a plenitude of models beyond the SM. A partial list of existing bounds is collected in Table 7.5. The simplest possibility is to add a 4th fermion family called sequential fermions, where the neutrino has to have a large mass (> 45 GeV) as additional light (nearly massless) neutrinos have been excluded by LEP.

Another possibility for extending the SM is the Higgs sector where one could add scalar singlets, an additional doublet, a Higgs triplet and so on. Two Higgs doublet models (THDM or 2HDM) are interesting as they predict 4

Table 7.5. Present lower bounds on new physics states. Bounds are 95% C.L. limits from LEP (ALEPH, DELPHI, L3, OPAL) and Tevatron (CDF, D0)

Object	mass bound	comment
Heavy neutrino	$m_{\nu'}^M >$ 39 GeV	Majorana-ν [$\nu \equiv \bar{\nu}$]
Heavy neutrino	$m_{\nu'}^D >$ 45 GeV	Dirac-ν [$\nu \neq \bar{\nu}$]
Heavy lepton	$m_L >$ 100 GeV	
4th family quark b'	$m_{b'} >$ 199 GeV	$p\bar{p}$ NC decays
W'_{SM}	$M_{W'} >$ 800 GeV	SM couplings
W_R	$M_{W_R} >$ 715 GeV	right–handed weak current
Z'_{SM}	$M_{Z'} >$ 825 GeV	SM couplings
Z_{LR} ($g_R = g_L$)	$M_{Z_{LR}} >$ 630 GeV	of $G_{LR} = SU(2)_R \otimes SU(2)_L \otimes U(1)$
Z_χ ($g_\chi = e/\cos\Theta_W$)	$M_{Z_\chi} >$ 595 GeV	of $SO(10) \to SU(5) \otimes U(1)_\chi$
Z_ψ ($g_\psi = e/\cos\Theta_W$)	$M_{Z_\psi} >$ 590 GeV	of $E_6 \to SO(10) \otimes U(1)_\psi$
Z_η ($g_\eta = e/\cos\Theta_W$)	$M_{Z_\eta} >$ 620 GeV	of $E_6 \to G_{LR} \otimes U(1)_\eta$
H Higgs	$m_H >$ 114.4 GeV	SM
$h^0 \equiv H_1^0$ Higgs	$m_{H_1^0} >$ 89.8 GeV	SUSY ($m_{H_1^0} < m_{H_2^0}$)
A^0 pseudoscalar Higgs m_A	$>$ 90.4 GeV	THDM, MSSM
$H^\pm$ charged Higgs	$m_{H^\pm} >$ 79.3 GeV	THDM, MSSM

---

[8] The variety of speculations about new physics is mind–blowing and the number of articles on Physics beyond the SM almost uncountable. This short essay tries to reproduce a few of the main ideas for illustration, since a shift in one number can have many reasons and only in conjunction with other experiments it is possible to find out what is the true cause. My citations may be not very concise and I apologize for the certainly numerous omissions.

additional physical spin 0 bosons one neutral scalar H^0, a neutral pseudoscalar A, as well as the two charged bosons $H^\pm$. Many new real and virtual processes, like $W^\pm H^\mp \gamma$ transitions, are the consequence. Any SUSY extension of the SM requires two Higgs doublets. Similarly, there could exist additional gauge bosons, like from an extra $U(1)'$. This would imply an additional Z boson, a sequential Z' which would mix with the SM Z and the photon. More attractive are extensions which solve some real or thought shortcomings of the SM. This includes Grand Unified Theories (GUT) [35] which attempt to unify the strong, electromagnetic and weak forces, which correspond to three different factors of the local gauge group of the SM, in one big simple local gauge group

$$G_{\mathrm{GUT}} \supset SU(3)_c \otimes SU(2)_L \otimes U(1)_Y \equiv G_{\mathrm{SM}}$$

which is assumed to be spontaneously broken in at least two steps

$$G_{\mathrm{GUT}} \to SU(3)_c \otimes SU(2)_L \otimes U(1)_Y \to SU(3)_c \otimes U(1)_{\mathrm{em}} \ .$$

Coupling unification is governed by the renormalization group evolution of $\alpha_1(\mu)$, $\alpha_2(\mu)$ and $\alpha_3(\mu)$, corresponding to the SM group factors $U(1)_Y$, $SU(2)_L$ and $SU(3)_c$, with the experimentally given low energy values, typically at the Z mass scale, as starting values evolved to very high energies, the GUT scale M_{GUT} where couplings should meet. Within the SM the three couplings do not unify, thus unification requires new physics as predicted by a GUT extension. Also extensions like the left–right (LR) symmetric model are of interest. The simplest possible unifying group is $SU(5)$ which, however, is ruled out by the fact that it predicts protons to decay faster than allowed by observation. GUT models like $SO(10)$ or the exceptional group E_6 not only unify the gauge group, thereby predicting many additional gauge bosons, they also unify quarks and leptons in GUT matter multiplets. Now quarks and leptons directly interact via the *leptoquark* gauge bosons X and Y which carry color, fractional charge ($Q_X = -4/3$, $Q_Y = -1/3$) as well as baryon and lepton number. Thus GUTs are violating B as well as L, yet with $B - L$ still conserved. The proton may now decay via $p \to e^+ \pi^0$ or many other possible channels. The experimental proton lifetime $\tau_{\mathrm{proton}} > 2 \times 10^{29}$ years at 90% C.L. requires the extra gauge bosons to exhibit masses of about $M_{\mathrm{GUT}} > 10^{16}$ GeV and excludes $SU(5)$ as it predicts unification at too low scales. M_{GUT} is the *GUT scale* which is only a factor 1000 below the Planck scale[9]. In general GUTs also have additional normal gauge bosons, extra W's and Z's which mix with the SM gauge bosons.

---

[9]GUT extensions of the SM are not very attractive for the following reasons: the extra symmetry breaking requires an additional heavier Higgs sector which makes the models rather clumsy in general. Also, unlike in the SM, the known matter-fields are *not* in the fundamental representations, while an explanation is missing why the existing lower dimensional representations remain unoccupied. In addition, the three SM couplings (as determined from experiments) allow for unification only with at least one additional symmetry breaking step $G_{\mathrm{GUT}} \to G' \to G_{\mathrm{SM}}$. In non-SUSY GUTs the only possible groups are $G_{\mathrm{GUT}} = E_6$ or $SO(10)$ and $G' = G_{LR} =$

In deriving bounds on New Physics it is important to respect constraints not only from a_μ and the direct bounds of Table 7.5, but also from other precision observables which are sensitive to new physics via radiative corrections. Important examples are the electroweak precision observables [38, 39]:

$$M_W = 80.392(29) \text{ GeV}, \qquad (7.4)$$

$$\sin^2 \Theta^\ell_{\text{eff}} = 0.23153(16), \quad \rho_0 = 1.0002^{+0.0007}_{-0.0004}, \qquad (7.5)$$

which are both precisely measured and precisely predicted by the SM or in extensions of it. The SM predictions use the very precisely known independent input parameters α, G_μ and M_Z, but also the less precisely known top mass

$$m_t = 171.4 \pm 2.1 \text{ GeV}, \qquad (7.6)$$

(the dependence on other fermion masses is usually weak, the one on the unknown Higgs is only logarithmic and already fairly well constrained by experimental data). The effective weak mixing parameter essentially determines $m_H = 114^{+45}_{-33}$ GeV 68% C.L. (not taking into account M_W). The parameter ρ_0 is the tree level (SM radiative corrections subtracted) ratio of the low energy effective weak neutral to charged current couplings: $\rho = G_{\text{NC}}/G_{\text{CC}}$ where $G_{\text{CC}} \equiv G_\mu$. This parameter is rather sensitive to new physics. Equally important are constraints by the B–physics branching fractions [40]

$$\text{BR}(b \to s\gamma) = (3.55 \pm 0.24^{+0.09}_{-0.10} \pm 0.03) \times 10^{-4},$$
$$\text{BR}(B_s \to \mu^+\mu^-) < 1.0 \times 10^{-7} \quad (95\% \text{ C.L.}). \qquad (7.7)$$

Another important object is the electric dipole moment which is a measure of CP–violation and was briefly discussed at the end of Sect. 3.3. Since extensions of the SM in general exhibit additional sources of CP violation, EDMs are very promising probes of new physics. An anomalously large EDM of the muon d_μ would influence on the a_μ extraction from the muon precession data as discussed at the end of Sect. 6.3.1. We may ask whether d_μ could be responsible for the observed deviation in a_μ. In fact (6.54) tells us that a non–negligible d_μ would increase the observed a_μ, and we may estimate

---

$SU(3)_c \otimes SU(2)_R \otimes SU(2)_L \otimes U(1)$ or $G_{PS} = SU(2)_R \otimes SU(2)_L \otimes SU(4)$ [36]. G_{LR} is the left–right symmetric extension of the SM and G_{PS} is the Pati-Salam model, where $SU(3)_c \otimes U(1)_Y$ of the SM is contained in the $SU(4)$ factor. Coupling unification requires the extra intermediate breaking scale to lie very high $M' \sim 10^{10}$ GeV for G_{LR} and $M' \sim 10^{14}$ GeV for G_{PS}. These are the scales of new physics in these extensions, completely beyond of being phenomenologically accessible. The advantage of SUSY GUTs is that they allow for unification of the couplings with the new physics scale being as low as M_Z to 1 TeV [37], and the supersymmetrized $G_{\text{GUT}} = SU(5)$ extension of the SM escapes to be excluded.

$$|d_\mu| = \frac{1}{2}\frac{e}{m_\mu}\sqrt{(a_\mu^{\mathrm{exp}})^2 - (a_\mu^{\mathrm{SM}})^2} = (2.42 \pm 0.41) \times 10^{-19}\, e\cdot\mathrm{cm}\,. \qquad (7.8)$$

This also may be interpreted as an upper limit as given in Table 7.1. Recent advances in experimental techniques will allow to perform much more sensitive experiments for electrons, neutrons and neutral atoms [41]. For new efforts to determine d_μ at much higher precision see [42, 43]. In the following we will assume that d_μ is in fact negligible, and that the observed deviation has other reasons.

As mentioned many times, the general form of contributions from states of mass $M_{\mathrm{NP}} \gg m_\mu$ takes the form

$$a_\mu^{\mathrm{NP}} = \mathcal{C}\,\frac{m_\mu^2}{M_{\mathrm{NP}}^2} \qquad (7.9)$$

where naturally $\mathcal{C} = O(\alpha/\pi)$, like for the weak contributions (4.33), but now from interactions and states not included in the SM. New fermion loops may contribute similarly to a τ–lepton as

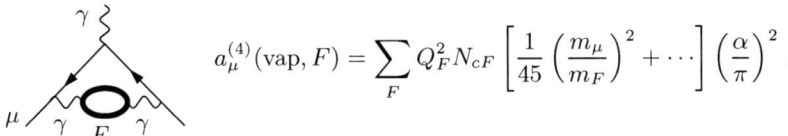

$$a_\mu^{(4)}(\mathrm{vap}, F) = \sum_F Q_F^2 N_{cF}\left[\frac{1}{45}\left(\frac{m_\mu}{m_F}\right)^2 + \cdots\right]\left(\frac{\alpha}{\pi}\right)^2,$$

which means $\mathcal{C} = O((\alpha/\pi)^2)$. Note that the τ contribution to a_μ is 4.2×10^{-10} only, while the 3 σ effect we are looking for is 28.7×10^{-10}. As the direct lower limit for a sequential fermion is about 100 GeV (see Table 7.5) such effects cannot account for the observed deviation. A 100 GeV heavy lepton only yields the tiny contribution[10] 1.34×10^{-13}.

---

[10] It should be noted that heavy sequential fermions are constrained severely be the ρ–parameter (NC/CC effective coupling ratio), if doublet members are not nearly mass degenerate. However, a doublet (ν_L, L) with $m_{\nu_L} = 45$ GeV and $m_L = 100$ GeV only contributes $\Delta\rho \simeq 0.0008$ which is within the limit from LEP electroweak fits (7.5). Not yet included is a similar type contribution from the 4th family (t', b') doublet mass–splitting, which also would add a positive term

$$\Delta\rho = \frac{\sqrt{2}G_\mu}{16\pi^2}\,3\,|m_{t'}^2 - m_{b'}^2| + \cdots$$

In this context it should be mentioned that the so called *custodial symmetry* of the SM which predicts $\rho_0 = 1$ at the tree level (independent of any parameter of the theory, which implies that it is not subject to subtractions due to parameter renormalization) is one of the severe constraints to extensions of the SM. The virtual top effect contributing to the radiative corrections of ρ allowed a determination of the top mass prior to the discovery of the top by direct production at Fermilab in 1995. The LEP precision determination of $\Delta\rho = \frac{\sqrt{2}G_\mu}{16\pi^2}\,3\,|m_t^2 - m_b^2|$ (up to subleading terms) from precision measurements of Z resonance parameters yields $m_t = 172.3^{+10.2}_{-7.6}$ GeV in excellent agreement with the direct determination $m_t = 171.4(2.1)$ GeV at the

Table 7.6. Typical New Physics scales required to satisfy $\Delta a_\mu^{\mathrm{NP}} = \delta a_\mu$ (7.3)

$\mathcal{C}$	1	α/π	$(\alpha/\pi)^2$
M_{NP}	$2.0^{+0.4}_{-0.3}$ TeV	100^{+21}_{-13} GeV	5^{+1}_{-1} GeV

A rough estimate of the scale M_{NP} required to account for the observed deviation is given in Table 7.6. An effective tree level contribution would extend the sensibility to the very interesting 2 TeV range, however, no compelling scenario I know of exists for this case.

......

For a different point of view see [45]. The argument is that the same interactions and heavy states which could contribute to a_μ^{NP} according to Fig. 7.3 would contribute to the muon self energy according to Fig. 7.4. By imposing chiral symmetry to the SM, i.e. setting the SM Yukawa couplings to zero, lepton masses could be radiatively induced by flavor changing $f\bar{\psi}_\mu \psi_F S +$ h.c. and $f\bar{\psi}_\mu i\gamma_5 \psi_F P +$ h.c. interactions (F a heavy fermion, S a scalar and P a pseudoscalar) in a hierarchy $m_\mu \ll M_F \ll M_S, M_P$. Then with $m_\mu \propto f^2 M_F$ and $a_\mu \propto f^2 m_\mu M_F/M_{S,P}^2$ one obtains $a_\mu = \mathcal{C}\, m_\mu^2/M_{S,P}^2$ with $\mathcal{C} = O(1)$, and the interaction strength f has dropped from the ratio. The problem is that a convincing approach of generating the lepton/fermion spectrum by radiative effects is not easy to accommodate. Of course it is a very attractive idea to replace the Yukawa term, put in by hand in the SM, by a mechanism which allows us to understand or even calculate the known fermion mass-spectrum, exhibiting a tremendous hierarchy of about 13 orders of magnitude of vastly different couplings/masses [from m_{ν_e} to m_t]. The radiatively induced values must reproduce this pattern and one has to explain why the same effects which make up the muon mass do not contribute to the electron mass. Again the needed hierarchy of fermion masses is only obtained by putting it in by hand in some way. In the scenario of radiatively induced lepton masses one has to require the family hierarchy like $f_e^2 M_{F_e}/f_\mu^2 M_{F_\mu} \simeq m_e/m_\mu$, $f_P \equiv f_S$ in order to get a finite cut-off independent answer, and $M_0 \to M_S \neq M_P$, such that $m_\mu = \frac{f_\mu^2 M_{F_\mu}}{16\pi^2} \ln \frac{M_S^2}{M_P^2}$ which is positive provided $M_S > M_P$.

......

Common to many of the extensions of the SM are predictions of new states: scalars S, pseudoscalars P, vectors V or axialvectors A, neutral or charged. They contribute via one–loop lowest order type diagrams shown in Fig. 7.3. Here, we explicitly assume all fermions to be Dirac fermions. Besides the SM fermions, μ in particular, new heavy fermions F of mass M may be involved,

---

Tevatron. In extensions of the SM in which ρ depends on physical parameters on the classical level, like in GUT models or models with Higgs triplets etc. one largely looses this prediction and thus one has a fine tuning problem [44]. But, also "extensions" which respect custodial symmetry like simply adding a 4th family of fermions should not give a substantial contribution to $\Delta\rho$, otherwise also this would spoil the indirect top mass prediction.

but fermion number is assumed to be conserved, like in $\Delta \mathcal{L}_S = f\bar{\psi}_\mu \psi_F S + \text{h.c.}$, which will be different in SUSY extensions discussed below, where fermion number violating Majorana fermions necessarily must be there.

We explicitly discuss contributions from diagram a) and c), the others give similar results. Exotic neutral bosons of mass M_0 coupling to muons $(m = m_\mu)$ with coupling strength f would contribute [46]

$$\Delta a_\mu^{\text{NP}} = \frac{f^2}{4\pi^2} \frac{m_\mu^2}{M_0^2} L, \; L = \frac{1}{2} \int_0^1 dx \, \frac{Q(x)}{(1-x)(1-\lambda^2 x) + (\epsilon\lambda)^2 x} \quad (7.10)$$

where $Q(x)$ is a polynom in x which is depending on the type of coupling:

Scalar   : $Q_S = x^2 (1 + \epsilon - x)$
Pseudoscalar : $Q_P = x^2 (1 - \epsilon - x)$
Vector   : $Q_V = 2x(1-x)(x - 2(1-\epsilon)) + x^2(1+\epsilon-x)\lambda^2(1-\epsilon)^2$
Pseudovector : $Q_A = 2x(1-x)(x - 2(1+\epsilon)) + x^2(1-\epsilon-x)\lambda^2(1+\epsilon)^2$

with $\epsilon = M/m$ and $\lambda = m/M_0$. As an illustration we only consider one regime explicitly, since the others yields qualitatively similar results. For a heavy boson of mass M_0 and $m, M \ll M_0$ one gets

$$\begin{aligned} L_S &= \frac{M}{m}\left(\ln \frac{M_0}{M} - \frac{3}{4}\right) + \frac{1}{6} \stackrel{M=m}{=} \ln \frac{M_0}{m} - \frac{7}{12} \\ L_P &= -\frac{M}{m}\left(\ln \frac{M_0}{M} - \frac{3}{4}\right) + \frac{1}{6} \stackrel{M=m}{=} -\ln \frac{M_0}{m} + \frac{11}{12} \\ L_V &= \frac{M}{m} - \frac{2}{3} \stackrel{M=m}{=} \frac{1}{3} \\ L_A &= -\frac{M}{m} - \frac{2}{3} \stackrel{M=m}{=} -\frac{5}{3} \end{aligned} \quad (7.11)$$

where it is more realistic to assume a flavor conserving neutral current $M = m = m_\mu$ as used in the second form[11]. Typical contributions are shown in

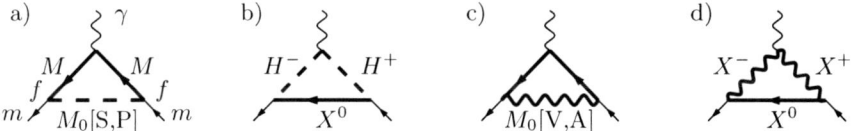

Fig. 7.3. Possible New Physics contributions: neutral boson exchange a) scalar or pseudoscalar and b) scalars or pseudoscalars, c) vector or axialvector, flavor changing or not, new charged bosons d) vector or axialvector

[11] As we will see later, in SUSY extensions the leading contributions actually come from the regime $m \ll M, M_0, M \sim M_0$, which is of enhanced FCNC type, and thus differs from the case just presented in (7.11). For the combinations of fixed chirality up to terms of order $O(m/M)$ one gets

$$L_S + L_P = +\frac{1}{6} \frac{1}{(1-x)^4} \left[2 + 3x - 6x^2 + x^3 + 6x \ln x\right] = \frac{1}{12} F_1^C(x)$$

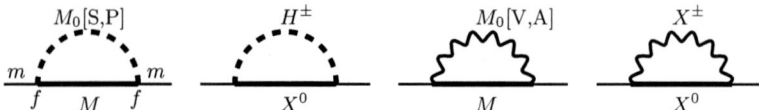

Fig. 7.4. Lepton self–energy contributions induced by the new interactions appearing in Fig. 7.3 may generate m_μ as a radiative correction effect

Fig. 7.5. Taking the coupling small enough such that a perturbative expansion in f makes sense, we take $f/(2\pi) = 0.1$, only the scalar exchange could account for the observed deviation with a scalar mass 480 GeV $< M_0 <$ 690 GeV. Pseudoscalar and pseudovector yield the wrong sign. The vector exchange is too small.

However, after neutrino oscillations and herewith right–handed singlet neutrinos and neutrino masses have been established, also lepton number violating transitions like $\mu^\pm \to e^\pm \gamma$ Fig. 7.6 are in the focus of further searches. The corresponding contributions here read

$$L_S^\mu \simeq \tfrac{1}{6} \quad , \quad L_S^e \simeq \tfrac{m_\mu}{m_e}\left(\ln \tfrac{M_0}{m_\mu} - \tfrac{3}{4}\right)$$
$$L_P^\mu \simeq \tfrac{1}{6} \quad , \quad L_P^e \simeq -\tfrac{m_\mu}{m_e}\left(\ln \tfrac{M_0}{m_\mu} - \tfrac{3}{4}\right)$$
$$L_V^\mu \simeq \tfrac{2}{3} \quad , \quad L_V^e \simeq \tfrac{m_\mu}{m_e}$$
$$L_A^\mu \simeq -\tfrac{2}{3} \quad , \quad L_A^e \simeq -\tfrac{m_\mu}{m_e} \; .$$

The latter flavor changing transitions are strongly constrained, first by direct rare decay search experiments which were performed at the Paul Scherrer Institute (PSI) and second, with the advent of the much more precise measurement of a_e.

For example, for a scalar exchange mediating $e \to \mu \to e$ with $f^2/(4\pi^2) \simeq 0.01$, $M_0 \simeq 100$ GeV we obtain

$$\Delta a_e^{NP} \simeq 33 \times 10^{-11}$$

which is ruled out by $a_e^{\exp} - a_e^{\text{the}} \sim 1 \times 10^{-11}$ (see p. 165 in Sect. 3.2.2). Either M_0 must be heavier or the coupling smaller: $f^2/(4\pi^2) < 0.0003$. The present

$$L_S - L_P = -\frac{M}{2m}\frac{1}{(1-x)^3}\left[3 - 4x + x^2 + 2\ln x\right] = \frac{M}{3m} F_2^C(x)$$

$$L_V + L_A = -\frac{1}{6}\frac{1}{(1-x)^4}\left[8 - 38x + 39x^2 - 14x^3 + 5x^4 - 18x^2 \ln x\right] = -\frac{12}{13} F_3^C(x)$$

$$L_V - L_A = +\frac{M}{2m}\frac{1}{(1-x)^3}\left[4 - 3x - x^3 + 6x \ln x\right] = \frac{M}{m} F_4^C(x) \qquad (7.12)$$

where $x = (M/M_0)^2 = O(1)$ and the functions F_i^C normalized to $F_i^C(1) = 1$. The possible huge enhancement factors M/m_μ, in some combination of the amplitudes, typical for flavor changing transitions, may be compensated due to radiative contributions to the muon mass (as discussed above) or by a corresponding Yukawa coupling $f \propto y_\mu = m_\mu/v$, as it happens in SUSY extensions of the SM (see below).

7.2 New Physics in g − 2

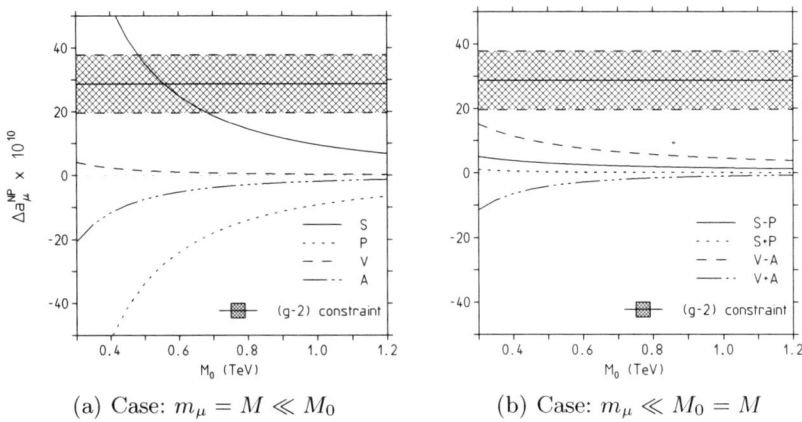

(a) Case: $m_\mu = M \ll M_0$ (b) Case: $m_\mu \ll M_0 = M$

Fig. 7.5. Single particle one–loop induced NP effects for $f^2/(4\pi^2) = 0.01$ (note, a typical EW SM coupling would be $e^2/(4\pi^2 \cos^2\Theta_W) = 0.003$). S, P, V, A denote scalar, pseudoscalar, vector and axialvector exchange. Panel (**a**) shows (7.11) for $M = m = m_\mu$, panel (**b**) the chiral combinations (7.12) for $m = m_\mu$ and $M = M_0$, with the large combinations $L_S - L_P$ and $L_V - L_A$ rescaled by the muon Yukawa coupling m_μ/v in order to compensate for the huge prefactor M/m_μ (see text)

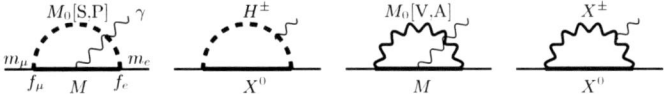

Fig. 7.6. $\mu \to e\gamma$ transitions by new interactions (overall flavor changing version of Fig. 7.3)

limit for the branching fraction $Br(\mu \to e\gamma)$ is 1.2×10^{-11}, which will be improved to 10^{-13} at PSI by a new experiment [47]. Note that

$$\Gamma(\mu \to e\gamma) = \frac{e^2 f_\mu^2 f_e^2}{16\pi^2} m_\mu^5 \left(|F_M^L|^2 + |F_M^R|^2\right), \qquad (7.13)$$

where $F_M^{L,R}$ are the left– and right–handed zero–momentum transfer magnetic $\mu e\gamma$ form factors. In the SM

$$Br(\mu \to e\gamma) \propto \frac{\alpha^3}{G_\mu^2} \frac{(\Delta m_\nu^2)_{\mu e}^2}{M_W^8}, \qquad (7.14)$$

is extremely tiny. Ony new physics can give rates in experimentally interesting ranges. In the quark sector CKM flavor mixing via the charged current is comparably huge and the $b \to s\gamma$ transitions is an established effect. This process also acquires enhanced SUSY contributions which makes it an excellent monitor for new physics [48], as we will see below.

Another simple illustration of the one–loop sensitivity to new physics are heavier gauge bosons with SM couplings. From direct searches we know that they must be at least as heavy as 800 GeV. Contributions then follow from the weak one-loop contributions by rescaling with $(M_W/M_{W'_{SM}})^2 \sim 0.01$ and hence 1% of 19.5×10^{-10} only, an effect much too small to be of relevance.

At $O((\alpha/\pi)^2)$ new physics may enter via vacuum polarization and we may write corresponding contributions as a dispersion integral (3.141):

$$\Delta a_\mu^{NP} = \frac{\alpha}{\pi} \int_0^\infty \frac{ds}{s} \frac{1}{\pi} \operatorname{Im} \Delta \Pi_\gamma^{NP}(s) \, K(s) \,.$$

Since, we are looking for contributions from heavy yet unknown states of mass $M \gg m_\mu$, and $\operatorname{Im} \Delta\Pi_\gamma^{NP}(s) \neq 0$ for $s \geq 4M^2$ only, we may safely approximate

$$K(s) \simeq \frac{1}{3} \frac{m_\mu^2}{s} \quad \text{for} \quad s \gg m_\mu^2$$

such that

$$\Delta a_\mu^{NP} = \frac{1}{3} \frac{\alpha}{\pi} \left(\frac{m_\mu}{M}\right)^2 L$$

where due to the optical theorem $\frac{1}{\pi} \operatorname{Im} \Delta\Pi_\gamma^{NP}(s) = \frac{\alpha(s)}{\pi} R^{NP}(s)$ above threshold is positive (see Sect. 3.7.1)

$$\frac{L}{M^2} = \frac{\alpha}{3\pi} \int_0^\infty \frac{ds}{s^2} R^{NP}(s) \,.$$

An explicit example was given above for the case of a heavy lepton. A heavy narrow vector meson resonance of mass M_V and electronic width $\Gamma(V \to e^+e^-)$ (which is $O(\alpha^2)$) contributes $R_V(s) = \frac{9\pi}{\alpha^2} M_V \, \Gamma(V \to e^+e^-) \, \delta(s - M_V^2)$ such that $L = \frac{3\Gamma(V \to e^+e^-)}{\alpha M_V}$ and hence

$$\Delta a_\mu^{NP} = \frac{m_\mu^2 \, \Gamma(V \to e^+e^-)}{\pi M_V^3} = \frac{4\alpha^2 \gamma_V^2 \, m_\mu^2}{3 M_V^2} \,. \tag{7.15}$$

Here we applied the Van Royen-Weisskopf formula [49], which for a $J^{PC} = 1^{--}$ vector state predicts

$$\Gamma(V \to e^+e^-) = 16\pi\alpha^2 Q_q^2 \frac{|\psi_V(0)|^2}{M_V^2} = \frac{4}{3}\pi\alpha^2 \gamma_V^2 M_V$$

where $\psi_V(0)$ is the meson wave function at the origin (dim 3) and γ_V is the dimensionless effective photon vector-meson coupling defined by $j_{em}^\mu(x) = \gamma_V M_V^2 V^\mu(x)$ with $V^\mu(x)$ the interpolating vector-meson field. γ_V characterizes the strong interaction properties of the $\gamma - V$ coupling and typically has values 0.2 for the ρ to 0.02 for the Υ. For $\gamma_V = 0.1$ and $M_V = 200$ GeV we get

$\Delta a_\mu \sim 2 \times 10^{-13}$. The hadronic contribution of a 4th family quark doublet assuming $m_{b'} = m_{t'} = 200\,\text{GeV}$ would yield $\Delta a_\mu \sim 5.6 \times 10^{-14}$ only. Unless there exists a new type of strong interactions like Technicolor[12] [50], new strong interaction resonances are not expected, because new heavy sequential quarks would be too shortlived to be able to form resonances. As we know, due to the large mass and the large mass difference $m_t \gg m_b$, the top quark is the first quark which decays, via $t \to Wb$, as a bare quark before it has time to form hadronic resonances. This is not so surprising as the top Yukawa coupling responsible for the weak decay is stronger than the strong interaction constant.

New physics effects here may be easily buried in the uncertainties of the hadronic vacuum polarization. In any case, we expect $O((\alpha/\pi)^2)$ terms from heavy states not yet seen to be too small to play a role here.

In general the effects related to single diagrams, discussed in this paragraph, are larger than what one expects in a viable extension of the SM, usually required to be a renormalizable QFT[13] and to exhibit gauge interactions which typically cause large cancellations between different contributions. But even if one ignores possible cancellations, all the examples considered so far show how difficult it actually is to reconcile the observed deviation with NP effects not ruled out already by LEP or Tevatron new physics searches.

---

[12] Searches for Technicolor states like color–octet techni–ρ were negative up to 260 to 480 GeV depending on the decay mode.

[13] Of course, there are more non-renormalizable extensions of the SM than renormalizable ones. For the construction of the electroweak SM itself renormalizability was the key guiding principle which required the existence of neutral currents, of the weak gauge bosons, the quark-lepton family structure and last but not least the existence of the Higgs, which we are still hunting for. However, considered as a low energy effective theory one expects all kinds of higher dimension transition operators coming into play at higher energies. Specific scenarios are anomalous gauge couplings, a Higgsless SM, little Higgs models, models with extra space–dimensions à la Kaluza-Klein, or infrared free extensions of the SM like the ones proposed in [51]. In view of the fact that non-renormalizable interactions primarily change the high energy behavior of the theory, we expect corresponding effects to show up primarily at the high energy frontier. The example of anomalous $W^+W^-\gamma$ couplings, considered in the following subsection, confirms such an expectation. Also in non-renormalizable scenarios effects are of the generic form (7.9) possibly with M_NP replaced by a cut-off Λ_NP. On a fundamental level we expect the Planck scale to provide the cut-off, which would imply that effective interactions of non-renormalizable character show up at the 1 ppm level at about 10^{16} GeV. It is conceivable that at the Planck scale a sort of cut-off theory which is modelling an "ether" is more fundamental than its long distance tail showing up as a renormalizable QFT. Physics-wise such an effective theory, which we usually interpret to tell us the fundamental laws of nature, is different in character from what we know from QCD where chiral perturbation theory or the resonance Lagrangian type models are non-renormalizable low energy tails of a known renormalizable theory, as is Fermi's non-renormalizable low energy effective current–current type tail within the SM.

Apparently a more sophisticated extension of the SM is needed which is able to produce substantial radiative corrections in the low energy observable a_μ while the new particles have escaped detection at accelerator facilities so far and only produce small higher order effects in other electroweak precision observables. In fact supersymmetric extensions of the SM precisely allow for such a scenario, as we will discuss below.

7.2.1 Anomalous Couplings

Besides new states with new interactions also possible anomalous couplings of SM particles are very interesting. In particular the non–Abelian gauge boson self–interactions have to be checked for possible deviations. In the SM these couplings are dictated by the local gauge principle of Yang-Mills, once the interaction between the gauge bosons and the matter–fields (4.30) is given. For $g-2$ in particular the anomalous W–boson couplings are of interest, which occur in the 1st of the weak one–loop diagrams in Fig. 4.10. Possible is an anomalous magnetic dipole moment (see [52] and references therein)

$$\mu_W = \frac{e}{2m_W}(1 + \kappa + \lambda) \tag{7.16}$$

and an anomalous electric quadrupole moment

$$Q_W = -\frac{e}{2m_W}(\kappa - \lambda). \tag{7.17}$$

In the SM local gauge symmetry, which is mandatory for renormalizability of the SM, requires $\kappa = 1$ and $\lambda = 0$. The contribution to a_μ due to the deviation from the SM may be calculated and as a result one finds [53]

$$a_\mu(\kappa, \lambda) \simeq \frac{G_\mu m_\mu^2}{4\sqrt{2}\pi^2} \left[(\kappa - 1) \ln \frac{\Lambda^2}{m_W^2} - \frac{1}{3}\lambda \right]. \tag{7.18}$$

Actually, the modification spoils renormalizability and one has to work with a cut–off Λ in order to get a finite answer and the result has to be understood as a low energy effective answer. For $\Lambda \simeq 1$ TeV the BNL constraint (7.3) would yield

$$\kappa - 1 = 0.24 \pm 0.08, \quad \lambda = -3.58 \pm 1.17 \quad \text{(BNL 04)}, \tag{7.19}$$

on the axes of the $(\Delta\kappa, \lambda)$–plane. Of course from one experimental number one cannot fix two or more parameters. In fact arbitrary large deviations from the SM are still possible described by the band Fig. 7.7: $\lambda = 3 \ln \frac{\Lambda^2}{m_W^2} \Delta\kappa - \tilde{a}_\mu$ with $\tilde{a}_\mu = \frac{12\sqrt{2}\pi^2 \, \delta a_\mu}{G_\mu m_\mu^2} \simeq 3.58 \pm 1.17$, as an interval on the λ-axis and a slope of about 15. This possibility again is already ruled out by $e^+e^- \to W^+W^-$ data from LEP [54, 55] $\kappa - 1 = -0.027 \pm 0.045$, $\lambda = -0.028 \pm 0.021$. Applying the LEP bounds we can get not more than $a_\mu(\kappa, 0) \simeq (-3.3 \pm 5.3) \times 10^{-10}$,

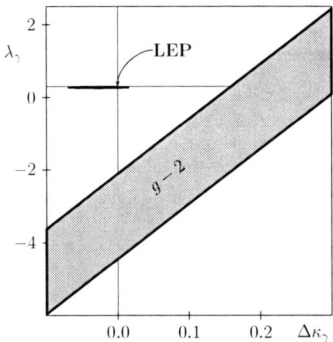

Fig. 7.7. Bounds on triple gauge couplings in $WW\gamma$

$a_\mu(1,\lambda) \simeq (0.2 \pm 1.6) \times 10^{-10}$, and thus the observed deviation cannot be due to anomalous $WW\gamma$ couplings. The constraint on those couplings from $g-2$ is at least an order of magnitude weaker than the one from LEP. Much more promising is the next example, the supersymmetrized SM.

7.2.2 Supersymmetry

The most promising theoretical scenarios are supersymmetric (SUSY) extensions of the SM, in particular the Minimal Supersymmetric Standard Model (MSSM). Supersymmetry implements a symmetry mapping

$$\text{boson} \stackrel{Q}{\leftrightarrow} \text{fermion}$$

between bosons and fermions, by changing the spin by $\pm 1/2$ units [56]. The SUSY algebra [graded Lie algebra]

$$\{Q_\alpha, \overline{Q}_\beta\} = -2\,(\gamma^\mu)_{\alpha\beta}\,P_\mu\,;\ P_\mu = (H, \boldsymbol{P})$$

P_μ the generators of space–time translations, Q_α four component Majorana (neutral) spinors and $\overline{Q}_\alpha = (Q^+\gamma^0)_\alpha$ the Pauli adjoint, is the only possible non–trivial unification of internal and space–time symmetry in a quantum field theory. The Dirac matrices in the Majorana representation play the role of the structure constants. The SUSY extension of the SM associates with each SM state X a supersymmetric "sstate" $\tilde{X}$ where sfermions are bosons and sbosons are fermions as shown in Table 7.7.

SUSY is a global symmetry imposed on the SM particle spectrum, the SM gauge group remains untouched and there are no new gauge bosons. Also the matter fields remain the same. SUSY and gauge invariance are compatible only if a second Higgs doublet field is introduced where H_1 induces the masses of all down–type fermions and H_2 the masses of all up–type fermions. A second complex Higgs doublet is also required for the anomaly cancellation of the fermionic sboson sector. This means 4 additional scalars ($H^0, A^0, H^\pm$) and

Table 7.7. The particle spectrum of a MSSM

SM particles ($R_p = +1$)	SUSY partners ($R_p = -1$)	
$\begin{pmatrix}\nu_e\\e^-\end{pmatrix}_L, \begin{pmatrix}\nu_\mu\\\mu^-\end{pmatrix}_L, \begin{pmatrix}\nu_\tau\\\tau^-\end{pmatrix}_L$	$\begin{pmatrix}\tilde\nu_e\\\tilde e^-\end{pmatrix}_L, \begin{pmatrix}\tilde\nu_\mu\\\tilde\mu^-\end{pmatrix}_L, \begin{pmatrix}\tilde\nu_\tau\\\tilde\tau^-\end{pmatrix}_L$	sneutrinos, sleptons
$\nu_{eR}, e_R^-, \nu_{\mu R}, \mu_R^-, \nu_{\tau R}, \tau_R^-$	$\tilde\nu_{eR}, \tilde e_R^-, \tilde\nu_{\mu R}, \tilde\mu_R^-, \tilde\nu_{\tau R}, \tilde\tau_R^-$	
$\begin{pmatrix}u\\d\end{pmatrix}_L, \begin{pmatrix}c\\s\end{pmatrix}_L, \begin{pmatrix}t\\b\end{pmatrix}_L$	$\begin{pmatrix}\tilde u\\\tilde d\end{pmatrix}_L, \begin{pmatrix}\tilde c\\\tilde s\end{pmatrix}_L, \begin{pmatrix}\tilde t\\\tilde b\end{pmatrix}_L$	squarks (stop, ...)
$u_R, d_R, c_R, s_R, t_R, b_R$	$\tilde u_R, \tilde d_R, \tilde c_R, \tilde s_R, \tilde t_R, \tilde b_R$	
$W^\pm, H^\pm$	$\tilde W^\pm, \tilde H^\pm \to \tilde\chi^\pm_{1,2}$	charginos
γ, Z, h^0, H^0, A^0	$\tilde\gamma, \tilde Z, \tilde h^0, \tilde H^0, \tilde A^0 \to \tilde\chi^0_{1,2,3,4}$	neuralinos
g, G	$\tilde g, \tilde G$	gluino, gravitino

their SUSY partners. The lighter neutral scalar denoted by h^0 corresponds to the SM Higgs H. Both Higgs fields exhibit a neutral scalar which acquire vacuum expectation values v_1 and v_2. The parameter $\tan\beta = v_2/v_1$ is one of the very important basic parameters as we will see. As $m_t \propto v_2$ and $m_b \propto v_1$ in such a scenario the large mass splitting $m_t/m_b \sim 40$ could be "explained" by a large ratio v_2/v_1, which means a large $\tan\beta$, i.e., values $\tan\beta \sim 40$ GeV look natural.

Digression on Supergravity and SUSY Breaking

A very interesting question is what happens if one attempts to promote global SUSY to local SUSY. As SUSY entangles internal with space–time symmetries of special relativity local SUSY implies supergravity (SUGRA) as one has to go from global Poincaré transformations to local ones, which means general coordinate invariance which in turn implies gravity (general relativity). SUGRA must include the spin 2 graviton and its superpartner, the spin 3/2 gravitino. Such a QFT is necessarily non–renormalizable [57]. Nevertheless is is attractive to consider the MSSM as a low energy effective theory of a non–renormalizable SUGRA scenario with $M_{\text{Planck}} \to \infty$ [58]. SUSY is spontaneously broken in the hidden sector by fields with no $SU(3)_c \otimes SU(2)_L \otimes U(1)_Y$ quantum numbers and which couple to the observable sector only gravitationally. M_{SUSY} denotes the SUSY breaking scale and the gravitino acquires a mass

$$m_{3/2} \sim \frac{M_{\text{SUSY}}^2}{M_{\text{Planck}}}$$

with M_{Planck} the inherent scale of gravity. SUSY is not realized as a perfect symmetry in nature. SUSY partners of the known SM particles have not yet been observed because sparticles in general are heavier than the known particles. Like the SM G_{SM} symmetry is broken by the Higgs mechanism, SUGRA is broken at some higher scale M_{SUSY} by a super–Higgs mechanism. The Lagrangian takes the form

$$\mathcal{L}^{MSSM} = \mathcal{L}^{SUSY}_{global} + \mathcal{L}^{SUSY}_{breaking}$$

with

$$\mathcal{L}^{SUSY}_{global} = \mathcal{L}^{SUSY}(SU(3)_c \otimes SU(2)_L \otimes U(1)_Y; W)$$

with W the following gauge invariant and B and L conserving superpotential[14]

$$W = W_Y - \mu H_1 H_2 \,;\, W_Y = \sum_F (h_U \tilde{Q}_L \tilde{U}^c_L H_2 + h_D \tilde{Q}_L \tilde{D}^c_L H_1 + h_L \tilde{L} \tilde{E}^c_L H_1)$$

(Y=Yukawa; F=families) where[15] $\tilde{Q}_L$ and $\tilde{L}$ denote the $SU(2)_L$ doublets $(\tilde{U}_L, \tilde{D}_L)$, $(\tilde{N}_L, \tilde{E}_L)$ and $\tilde{U}^c_L$, $\tilde{D}^c_L$, $\tilde{E}^c_L$ are the scalar partners of the right–handed quarks and leptons, written as left–handed fields of the antiparticle (^{c}=charge conjugation). $SU(2)_L$ and $SU(3)_c$ indices are summed over. h_U, h_D and h_L are the Yukawa couplings, the complex 3×3 matrices in family space of the SM. In the minimal SUGRA (mSUGRA) scheme, also called "Constrained MSSM" (CMSSM), one assumes universality of all soft parameters. In this case the SUSY breaking term has the form

$$\mathcal{L}^{SUSY}_{breaking} = -m^2 \sum_i |\varphi_i|^2 - M \sum_a \lambda_a \lambda_a + (A\, m\, W_Y - B\, m\, \mu H_1 H_2 + h.c.) \,.$$

The essential new parameters are

- μ the supersymmetric higgsino mass
- m is the universal mass term for all scalars φ_i
- M is the universal mass term to all gauginos λ_a
- A, B are the breaking terms in the superpotential W.

Thus in addition to the SM parameters we have 5 new parameters

$$\mu, m, M, A \text{ and } B.$$

The SUSY breaking lifts the degeneracy between particles and sparticle and essentially makes all sparticles to be heavier than all particles, as illustrated in the figure.

---

[14] One could add other gauge invariant couplings like

$$(\tilde{U}^c_L \tilde{D}^c_L \tilde{D}^c_L)\,,\, (\tilde{Q}_L \tilde{L} \tilde{D}^c_L)\,,\, m(\tilde{L} H_2)\,,\, (\tilde{L} \tilde{L} \tilde{E}^c_L)$$

which violate either B or L, however. In the minimal model they are absent.

[15] We label $U = (u, c, t)$, $D = (d, s, b)$, $N = (\nu_e, \nu_\mu, \nu_\tau)$ and $E = (e, \mu, \tau)$.

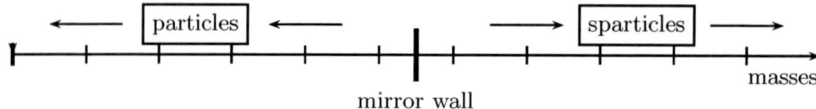

This scenario leads to universal masses for all SUSY partners:

- s–matter: $m_{\tilde{q}} = m_{\tilde{\ell}} = m_{\tilde{H}} = m_{1/2} \sim m_{3/2}$
- gauginos: $M_3 = M_2 = M_1 = m_0 \sim m_{3/2}$

where M_3, M_2 and M_1 are the mass scales of the spartners of the gauge bosons in $SU(3)_c$, $SU(2)_L$ and $U(1)_Y$, respectively. The non–observation of any sparticles so far requires a mass bound of about $m_{3/2} \sim 100 \div 1000$ GeV, which is of the order of the weak scale 246 GeV.

In general one expects different masses for the different types of gauginos:

- M' the $U(1)_Y$ gaugino mass
- M the $SU(2)_L$ gaugino mass
- $m_{\tilde{g}}$ the $SU(3)_c$ gluino mass.

However, the grand unification assumption

$$M' = \frac{5}{3} \tan^2 \Theta_W M = \frac{5}{3} \frac{\alpha}{\cos^2 \Theta_W \alpha_s} m_{\tilde{g}} .$$

leads back to the mSUGRA scenario. A very attractive feature of this scenario is the fact that the known SM Yukawa couplings now may be understood by evolving couplings from the GUT scale down to low energy by the corresponding RG equations. This also implies the form of the muon Yukawa coupling $y_\mu \propto \tan \beta$, as

$$y_\mu = \frac{m_\mu}{v_1} = \frac{m_\mu g_2}{\sqrt{2} M_W \cos \beta} \qquad (7.20)$$

where $g_2 = e/\sin \Theta_W$ and $1/\cos \beta \approx \tan \beta$. This enhanced coupling is central for the discussion of the SUSY contributions to a_μ. In spite of the fact that SUSY and GUT extensions of the SM have completely different motivations and in a way are complementary, supersymmetrizing a GUT is very popular as it allows coupling constant unification together with a low GUT breaking scale which promises nearby new physics. Actually, supersymmetric $SU(5)$ circumvents the problems of the normal $SU(5)$ GUT and provides a viable phenomenological framework. The extra GUT symmetry requirement is attractive also because it reduces the number of independent parameters.

End of the Digression

While supersymmetrizing the SM fixes all gauge and Yukawa couplings of the sparticles (see Fig. 7.8), there are a lot of free parameters to fix the SUSY breaking and masses, such that mixing of the sparticles remain quite arbitrary: the mass eigenstates of the gaugino–Higgsino sector are obtained by unitary transformations which mix states with the same conserved quantum numbers (in particular the charge)

$$\chi_i^+ = V_{ij}\psi_j^+ \,,\ \chi_i^- = U_{ij}\psi_j^- \,,\ \chi_i^0 = N_{ij}\psi_j^0 \qquad (7.21)$$

where ψ_j^a denote the spin 1/2 sparticles of the SM gauge bosons and the two Higgs doublets. In fact, a SUSY extension of the SM in general exhibits more than 100 parameters, while the SM has 28 (including neutrino masses and mixings). Also, in general SUSY extensions of the SM lead to Flavor Changing Neutral Currents (FCNC) and unsuppressed CP–violation, which are absent or small, respectively, in the SM and known to be suppressed in nature. Actually, just a SUSY extension of the SM, while solving the naturalness problem of the SM Higgs sector, creates its own naturalness problem as it leads to proton decay and the evaporation of baryonic matter in general. An elegant way to get rid of the latter problem is to impose the so called R–parity, which assigns $R_p = +1$ to all normal particles and $R_p = -1$ to all sparticles. If R–parity is conserved sparticles can only be produced in pairs and there must exist a stable Lightest Supersymmetric Particle (LSP), the lightest neutralino. Thus all sparticles at the end decay into the LSP plus normal matter. The LSP is a Cold Dark Matter (CDM) candidate [59] if it is neutral and colorless. From the precision mapping of the anisotropies in the cosmic microwave background, the Wilkinson Microwave Anisotropy Probe (WMAP) collaboration has determined the relict density of cold dark matter to [60]

$$\Omega_{\rm CDM}h^2 = 0.1126 \pm 0.0081 \,. \qquad (7.22)$$

This sets severe constraints on the SUSY parameter space [61, 62]. Note that SUSY is providing a new source for CP–violation which could help in understanding the matter–antimatter asymmetry $n_B = (n_b - n_{\bar b})/n_\gamma \simeq 6 \times 10^{-10}$ present in our cosmos.

However, what should cause R–parity to be conserved is another question. It just means that certain couplings one usually would assume to be there

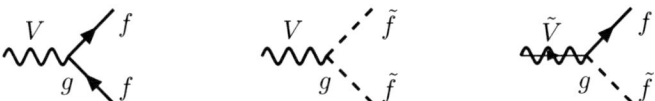

Fig. 7.8. Yukawa coupling=gauge coupling in the MSSM

are excluded. If R is not conserved sparticles may be produced singly and the LSP is not stable and would not provide a possible explanation of CDM.

The main theoretical motivation for a supersymmetric extension of the SM is the **hierarchy** or **naturalness** problem[16] of the latter: chiral symmetry requires fermions to be massless, local gauge symmetries require the gauge bosons to be massless, so the only SM particle which is not required to be massless before the spontaneous symmetry breaking by the Higgs mechanism is the scalar Higgs boson, together with the mass–degenerate later Higgs-ghosts (all fields in the Higgs doublet). As a consequence one would expect the Higgs boson to be much heavier than all other SM particles which acquire a mass proportional to the Higgs vacuum expectation value $v = 1/(\sqrt{2}G_\mu) = 246.221(1)$ GeV. Indirect Higgs mass bounds from LEP require the Higgs to be relatively light $m_H < 200$ GeV, i.e. not heavier than the other SM particles, including the heaviest ones. Therefore we think a symmetry should protect the Higgs from being much heavier than other SM states[17]. The only known symmetry which requires scalar particles to be massless in the symmetry limit is supersymmetry. Simply because a scalar is now always a supersymmetric partner of a fermion which is required to be massless be chiral symmetry. Thus only in a supersymmetric theory it is natural to have a "light" Higgs, in fact in a SUSY extension of the SM the lightest scalar h^0, which corresponds to the SM Higgs, is bounded to have mass $m_{h^0} \leq M_Z$ at tree level. This bound receives large radiative corrections from the $t/\tilde{t}$ sector, which changes the upper bound to [63]

---

[16]Stating that a small parameter (like a small mass) is unnatural unless the symmetry is increased by setting it to zero.

[17]Within the electroweak SM the Higgs mass is a free input parameter fixed by a renormalization condition to whatever input value will be determined by experiment. True, in the unbroken phase the Higgs doublet mass counter term represents the only quadratic divergence in the SM, which carries over to the broken phase. Since in a renormalized QFT counter terms are never observable there is nothing wrong with a counter–term getting very large and this should not be confused with the *hierarchy problem*. Renormalization is always a fine tuning. A completely different situation we have if the SM is considered as a low energy effective theory, what likely everybody does. Then the cut–off has a physical meaning as a new physics scale which very likely is the *Planck scale* $M_{\mathrm{Planck}} \sim 10^{19}$ GeV, where all other particle forces are expected to unify with gravity. Then the relation between the bare cut–off theory and the renormalized low energy effective theory is physics and in principle it becomes observable. So far nobody has measured a counter term, however. This is in contrast to condensed matter systems, where microscopic and macroscopic properties are obviously related and both are experimentally accessible, although, in most cases not under quantitative control of theory. In the SM the Higgs mass term is the only dimension 2 operator and scales like $m_H^2(E) \sim (\Lambda/E)^2 m_H^2(\Lambda)$ for $E \ll \Lambda$. We thus would expect m_H to be extremely large at a low scale E, unless for some reason (symmetry) it is extremely small at the high scale Λ.

Table 7.8. Present lower bounds (95% C.L.) on SUSY states. Bounds from LEP (ALEPH, DELPHI, L3, OPAL) and Tevatron (CDF, D0)

Object	mass bound	comment
sleptons	$m_{\tilde{e},\tilde{\mu},\tilde{\tau}} > 73, 94, 82$ GeV	$m_{\tilde{\mu}\tilde{\tau}} - m_{\tilde{\chi}_1^0} > 10, 15$ GeV
sbottom, stop	$m_{\tilde{b},\tilde{t}} > 89, 96$ GeV	for $m_{\tilde{b},\tilde{t}} - m_{\tilde{\chi}_1^0} = 8, 10$ GeV
squarks$\neq \tilde{t}, \tilde{b}$	$m_{\tilde{q}} > 250$ GeV	
chargino	$m_{\tilde{\chi}_1^\pm} > 104$ GeV	for $m_{\tilde{\nu}} > 300$ GeV
gluino	$m_{\tilde{g}} > 195[300]$ GeV	any $m_{\tilde{q}}[m_{\tilde{g}} = m_{\tilde{q}}]$

$$m_{h^0} \leq \left(1 + \frac{\sqrt{2}G_\mu}{2\pi^2 \sin^2\beta} 3m_t^4 \ln\left(\frac{m_{\tilde{t}_1} m_{\tilde{t}_2}}{m_t^2}\right) + \cdots\right) M_Z \quad (7.23)$$

which in any case is well below 200 GeV. For an improved bound obtained by including the 2–loop corrections I refer to [64]. In Table 7.8 some important direct search bounds on sparticle masses are listed.

It is worthwhile to mention that in an exactly supersymmetric theory the anomalous magnetic moment must vanish, as observed by Ferrara and Remiddi in 1974 [65]:

$$a_\mu^{\text{tot}} = a_\mu^{\text{SM}} + a_\mu^{\text{SUSY}} = 0 \,.$$

Thus, since $a_\mu^{\text{SM}} > 0$, in the SUSY limit, in the unbroken theory, we must have

$$a_\mu^{\text{SUSY}} < 0 \,.$$

However, we know that SUSY must be drastically broken, not a single super-symmetric partner has been observed so far. All super–partners of existing particles seem to be too heavy to be produced up to now. If SUSY is broken a_μ may have either sign. In fact, the 3 standard deviation $(g_\mu - 2)$–discrepancy requires $a_\mu^{\text{SUSY}} > 0$, of the same sign as the SM contribution and of at least the size of the weak contribution $[\sim 200 \times 10^{-11}]$ (see Fig. 3.8).

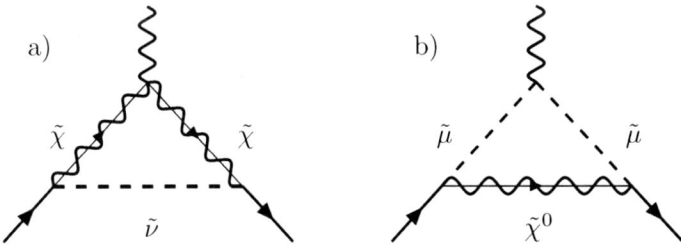

Fig. 7.9. Physics beyond the SM: leading SUSY contributions to $g-2$ in super-symmetric extension of the SM. Diagrams **a)** and **b)** correspond to diagrams **a)** and **b)** of Fig. 7.3, respectively

The leading SUSY contributions, like the weak SM contributions, are due to one–loop diagrams. Most interesting are the ones which get enhanced for large $\tan\beta$. Such supersymmetric contributions to a_μ stem from sneutrino–chargino and smuon–neutralino loops Fig. 7.9 and yield [66, 67, 68]:

$$a_\mu^{\text{SUSY (1)}} = a_\mu^{\chi^\pm} + a_\mu^{\chi^0} \tag{7.24}$$

with

$$a_\mu^{\chi^\pm} = \frac{m_\mu}{16\pi^2} \sum_k \left\{ \frac{m_\mu}{12 m_{\tilde\nu_\mu}^2} (|c_k^L|^2 + |c_k^R|^2) F_1^C(x_k) + \frac{m_{\chi_k^\pm}}{3 m_{\tilde\nu_\mu}^2} \operatorname{Re}[c_k^L c_k^R] F_2^C(x_k) \right\}$$

$$a_\mu^{\chi^0} = \frac{m_\mu}{16\pi^2} \sum_{i,m} \left\{ -\frac{m_\mu}{12 m_{\tilde\mu_m}^2} (|n_{im}^L|^2 + |n_{im}^R|^2) F_1^N(x_{im}) + \frac{m_{\chi_i^0}}{3 m_{\tilde\mu_m}^2} \operatorname{Re}[n_{im}^L n_{im}^R] F_2^N(x_{im}) \right\}$$

and $k = 1, 3$ and $i = 1, ..., 4$ denote the chargino and neutralino indices, $m = 1, 2$ is the smuon index, and the couplings are given by

$$c_k^L = -g_2 V_{k1} ,$$
$$c_k^R = y_\mu U_{k2} ,$$
$$n_{im}^L = \frac{1}{\sqrt{2}} (g_1 N_{i1} + g_2 N_{i2}) U_{m1}^{\tilde\mu\,*} - y_\mu N_{i3} U_{m2}^{\tilde\mu\,*} ,$$
$$n_{im}^R = \sqrt{2} g_1 N_{i1} U_{m2}^{\tilde\mu} + y_\mu N_{i3} U_{m1}^{\tilde\mu} .$$

The kinematical variables are the mass ratios $x_k = m_{\chi_k^\pm}^2/m_{\tilde\nu_\mu}^2$, $x_{im} = m_{\chi_i^0}^2/m_{\tilde\mu_m}^2$, and the one–loop vertex functions read

$$F_1^C(x) = \frac{2}{(1-x)^4} [2 + 3x - 6x^2 + x^3 + 6x \ln x] ,$$

$$F_2^C(x) = \frac{3}{2(1-x)^3} [-3 + 4x - x^2 - 2 \ln x] ,$$

$$F_1^N(x) = \frac{2}{(1-x)^4} [1 - 6x + 3x^2 + 2x^3 - 6x^2 \ln x] ,$$

$$F_2^N(x) = \frac{3}{(1-x)^3} [1 - x^2 + 2x \ln x] ,$$

and are normalized to $F_i^J(1) = 1$. The functions $F_i^C(x)$ are the ones calculated in (7.12). The couplings g_i denote the $U(1)$ and $SU(2)$ gauge couplings $g_1 = e/\cos\Theta_W$ and $g_2 = e/\sin\Theta_W$, respectively, and y_μ is the muon's Yukawa coupling (7.20). The interesting aspect of the SUSY contribution to a_μ is that they are enhanced for large $\tan\beta$ in contrast to SUSY contributions to electroweak precision observables, which mainly affect $\Delta\rho$ which determines the ρ–parameter and contributes to M_W. The anomalous magnetic moment thus may be used to constrain the SUSY parameter space and an expansion in $1/\tan\beta$ and because SUSY partners of SM particles are heavier (as mentioned above SUSY is broken in such a way that the SUSY partners essentially all are heavier than the SM particles) one usually also expands in M_W/M_{SUSY} leading to the handy approximation

$$a_\mu^{\chi^\pm} = \frac{g_2^2}{32\pi^2} \frac{m_\mu^2}{M_{\text{SUSY}}^2} \, \text{sign}(\mu M_2) \, \tan\beta \left[1 + O(\frac{1}{\tan\beta}, \frac{M_W}{M_{\text{SUSY}}})\right],$$

$$a_\mu^{\chi^0} = \frac{g_1^2 - g_2^2}{192\pi^2} \frac{m_\mu^2}{M_{\text{SUSY}}^2} \, \text{sign}(\mu M_2) \, \tan\beta \left[1 + O(\frac{1}{\tan\beta}, \frac{M_W}{M_{\text{SUSY}}})\right],$$

where parameters have been taken to be real and M_1 and M_2 of the same sign. One thus obtains

$$a_\mu^{\text{SUSY}} \simeq \text{sign}(\mu) \frac{\alpha(M_Z)}{8\pi \sin^2 \Theta_W} \frac{(5 + \tan^2 \Theta_W)}{6} \frac{m_\mu^2}{\tilde{m}^2} \tan\beta \left(1 - \frac{4\alpha}{\pi} \ln \frac{\tilde{m}}{m_\mu}\right) \quad (7.25)$$

$\tilde{m} = M_{\text{SUSY}}$ a typical SUSY loop mass and μ is the Higgsino mass term. Here we also included the leading 2–loop QED logarithm as an RG improvement factor [69]. In Fig. 7.10 contributions are shown for various values of $\tan\beta$. Above $\tan\beta \sim 5$ and $\mu > 0$ the SUSY contributions from the diagrams Fig. 7.9 easily could explain the observed deviation (7.3) with SUSY states of masses in the interesting range 100 to 500 GeV.

In the large $\tan\beta$ regime we have

$$|a_\mu^{\text{SUSY}}| \simeq 123 \times 10^{-11} \left(\frac{100 \text{ GeV}}{\tilde{m}}\right)^2 \tan\beta. \quad (7.26)$$

a_μ^{SUSY} generally has the same sign as the μ–parameter. The deviation (7.3) requires positive $\text{sign}(\mu)$ and if identified as a SUSY contribution

$$\tilde{m} \simeq (65.5 \text{ GeV}) \sqrt{\tan\beta}. \quad (7.27)$$

Negative μ models give the opposite sign contribution to a_μ and are strongly disfavored. For $\tan\beta$ in the range $2 \sim 40$ one obtains

$$\tilde{m} \simeq 93 - 414 \text{ GeV}, \quad (7.28)$$

precisely the range where SUSY particles are often expected. For a more elaborate discussion and further references I refer to [45]. The effects in a_μ from two doublet Higgs models (which include the Higgs sector of SUSY extensions of the SM) are discussed in [70].

A remarkable 2–loop calculation within the MSSM has been performed by Heinemeyer, Stöckinger and Weiglein [71]. They evaluated the exact 2–loop correction of the SM 1–loop contributions Figs. 4.1 and 4.10. These are all diagrams where the μ–lepton number is carried only by μ and/or ν_μ. In other words SM diagrams with an additional insertion of a closed sfermion– or charginos/neutralino–loop. Thus the full 2–loop result from the class of diagrams with closed sparticle loops is known. This class of SUSY contributions is interesting because it has a parameter dependence completely different from the one of the leading SUSY contribution and can be large in regions of parameter space where the 1–loop contribution is small. The second class of corrections are the 2–loop corrections to the SUSY 1–loop diagrams Fig. 7.9, where the μ–lepton number is carried also by $\tilde{\mu}$ and/or $\tilde{\nu}_\mu$. This class of corrections is expected to have the same parameter dependence as the leading SUSY 1–loop ones and only the leading 2–loop QED corrections are known [69] as already included in (7.25).

The prediction of a_μ as a function of the mass of the Lightest Observable SUSY Particle $M_{\mathrm{LOSP}} = \min(m_{\tilde{\chi}_1^\pm}, m_{\tilde{\chi}_2^0}, m_{\tilde{f}_i})$, from a MSSM parameter scan with $\tan\beta = 50$, including the 2–loop effects is shown in Fig. 7.11. Plotted is the maximum value of a_μ obtained by a scan of that part of SUSY parameter space which is allowed by the other observables like m_h, M_W and the b–decays. The 2–loop corrections in general are moderate (few %). However, not so for lighter M_{LOSP} in case of heavy smuons and sneutrinos when corrections become large (see also [72]). The remaining uncertainty of the calculation has been estimated to be below 3×10^{-10}, which is satisfactory in the present situation. This may however depend on details of the SUSY scenario and of the parameter range considered. A comprehensive review on supersymmetry, the different symmetry breaking scenarios and the muon magnetic moment has been presented recently by Stöckinger [68]. Low energy precision test of supersymmetry and present experimental constraints also are reviewed and discussed, in [73].

In comparison to $g_\mu - 2$, the SM prediction of M_W [75], as well as of other electroweak observables, as a function of m_t for given α, G_μ and M_Z, is in much better agreement with the experimental result (1 σ), although the

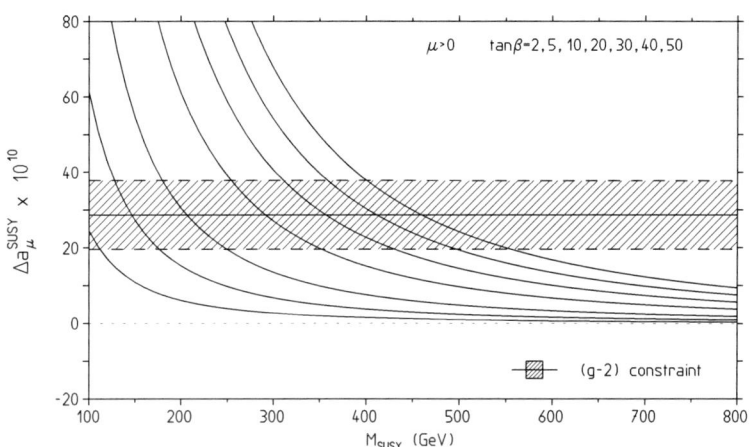

Fig. 7.10. Constraint on large $\tan\beta$ SUSY contributions as a function of M_{SUSY}

MSSM prediction for suitably chosen MSSM parameters is slightly favored by the data, as shown in Fig. 7.12. Thus large extra corrections to the ones of the SM are not tolerated. The radiative shift of M_W is represented by (4.32) and the leading SUSY contributions mainly come in via $\Delta\rho$. As we know, $\Delta\rho$ is most sensitive to weak isospin splitting and in the SM is dominated by the contribution from the (t,b)–doublet. In the SUSY extension of the SM these effects are enhanced by the contributions from the four SUSY partners $\tilde{t}_{L,R}, \tilde{b}_{L,R}$ of t,b, which can be as large as the SM contribution itself for $m_{1/2} \ll m_t$ [light SUSY], and tends to zero for $m_{1/2} \gg m_t$ [heavy SUSY]. It is important to note that these contributions are not enhanced by $\tan\beta$. Thus, provided $\tan\beta$ enhancement is at work, it is quite natural to get a larger SUSY contribution to $g_\mu - 2$ than to M_W, otherwise some tension between the two constraints would be there as M_W prefers the heavy SUSY domain.

Assuming the CMSSM scenario, besides the direct limits from LEP and Tevatron, presently, the most important constraints come from $(g-2)_\mu$, $b \to s\gamma$ and from the dark matter relic density (cosmological bound on CDM) given in (7.22) [61, 62]. Due to the precise value of Ω_{CDM} the lightest SUSY fermion (sboson) of mass m_0 is given as a function of the lightest SUSY boson (sfermion) with mass $m_{1/2}$ within a narrow band. This is illustrated in Fig. 7.13 together with the constraints from $g_\mu - 2$ (7.3) and $b \to s\gamma$ (7.7). Since m_h for given $\tan\beta$ is fixed by $m_{1/2}$ via (7.23) with $\min(m_{\tilde{t}_i}; i = 1, 2) \sim m_{1/2}$,

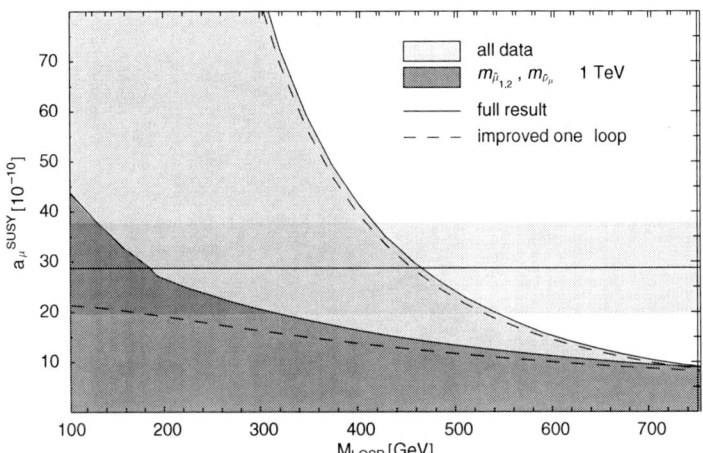

Fig. 7.11. Allowed values of MSSM contributions to a_μ as a function of M_{LOSP}, from an MSSM parameter scan with $\tan\beta = 50$. The 3σ region corresponding to the deviation (7.3) is indicated as a horizontal band. The light gray region corresponds to all input parameter points that satisfy the experimental constraints from b–decays, m_h and $\Delta\rho$. In the middle gray region, smuons and sneutrinos are heavier than 1 TeV. The dashed lines correspond to the contours that arise from ignoring the 2-loop corrections from chargino/neutralino– and sfermion–loop diagrams. Courtesy of D. Stöckinger [68]

the allowed region is to the right of the (almost vertical) line $m_h = 114$ GeV which is the direct LEP bound. Again there is an interesting tension between the SM like lightest SUSY Higgs mass m_h which in case the Higgs mass goes up from the present limit to higher values requires heavier sfermion masses and/or lower $\tan\beta$, while a_μ prefers light sfermions and large $\tan\beta$. Another lower bound from LEP is the line characterizing $m_{\chi^\pm} > 104$ GeV. The CDM bound gives a narrow hyperbola like shaped band. The cosmology bound is harder to see in the $\tan\beta = 40$ plot, but it is the strip up the $\chi - \tilde{\tau}$ degeneracy line, the border of the excluded region (dark) which would correspond to a charged LSP which is not allowed. The small shaded region in the upper left is excluded due to no electroweak symmetry breaking (EWSB) there. The latter

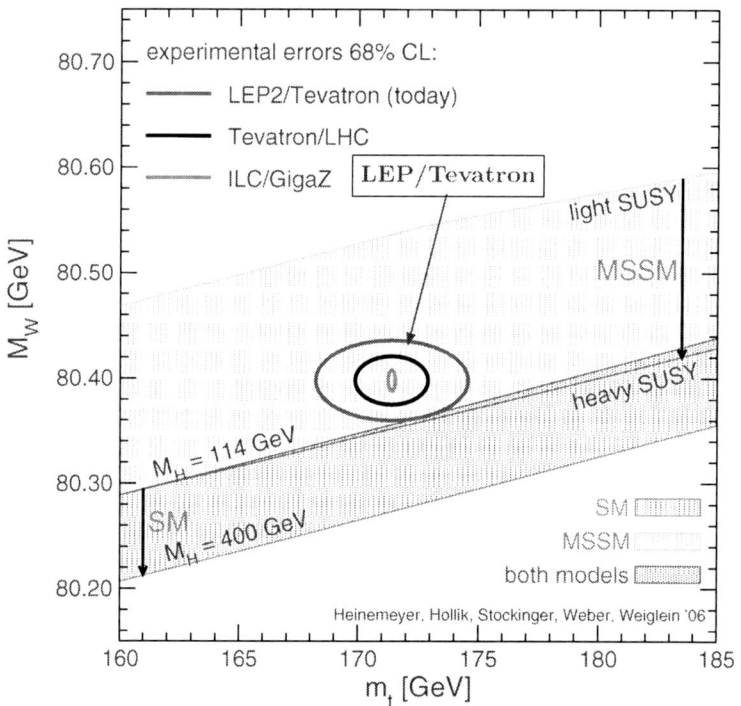

Fig. 7.12. Prediction for M_W in the MSSM and the SM as a function of m_t in comparison with the present experimental results for M_W and m_t and the prospective accuracies (using the current central values) at the Tevatron/LHC and at the ILC. The allowed region in the MSSM, corresponding to the light-shaded and dark-shaded bands, results from varying the SUSY parameters independently of each other in a random parameter scan. The allowed region in the SM, corresponding to the medium-shaded and dark-shaded bands, results from varying the mass of the SM Higgs boson from $m_H = 114$ GeV to $m_H = 400$ GeV. Values in the very light shaded region can only be obtained in the MSSM if at least one of the ratios $m_{\tilde{t}_2}/m_{\tilde{t}_1}$ or $m_{\tilde{b}_2}/m_{\tilde{b}_1}$ exceeds 2.5. Courtesy of S. Heinemeyer et al. [74]

must be tuned to reproduce the correct value for M_Z. The $\tan\beta = 40$ case is much more favorable, since $g_\mu - 2$ selects the part of the WMAP strip which has a Higgs above the LEP bound. Within the CMSSM the discovery of the Higgs and the determination of its mass would essentially fix m_0 and $m_{1/2}$.

Since the SM prediction [76] for the $b \to s\gamma$ rate BR($b \to s\gamma$) = $(3.15 \pm 0.23) \times 10^{-4}$ is in good agreement with the experimental value (7.7), only small extra radiative corrections are allowed (1.5 σ). In SUSY extensions of the SM [77], this excludes light $m_{1/2}$ and m_0 from light to larger values depending on $\tan\beta$. Reference [76] also illustrates the updated $b \to s\gamma$ bounds on M_{H^+} (> 295 GeV for $2 \leq \tan\beta$) in the TDHM (Type II) [78].

It is truly remarkable that in spite of the different highly non–trivial dependencies on the MSSM parameters, with $g - 2$ favoring definitely $\mu > 0$, $\tan\beta$ large and/or light SUSY states, there is a common allowed range, although a quite narrow one, depending strongly on $\tan\beta$.

The case sketched above is the constrained MSSM motivated by a minimal SUGRA breaking scheme. Already now cold dark matter constraints in conjunction with R–party conserving mSUGRA (CMSSM) scenarios constrain the SUSY parameter space dramatically. For a discussion of other SUSY models based on different SUSY breaking mechanisms I refer to the review [68] and the references therein (see in particular [79]). One should be aware that even when a SUSY extension of the SM should be realized in nature, many

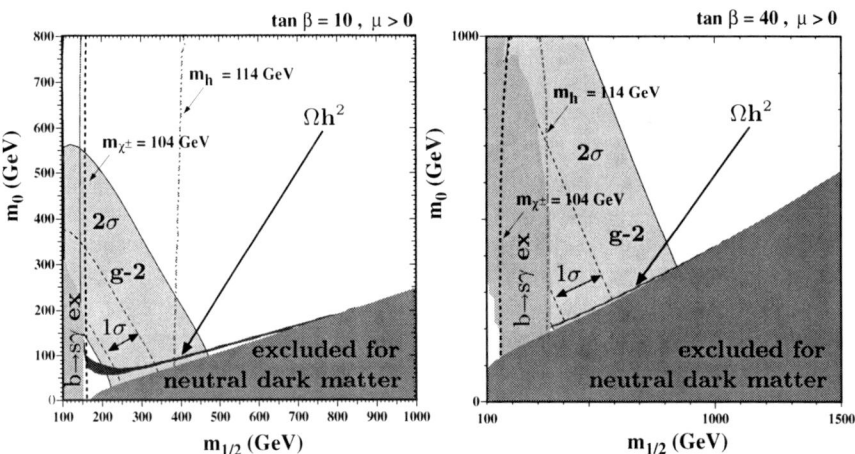

Fig. 7.13. The $(m_0, m_{1/2})$ plane for $\mu > 0$ for (a) $\tan\beta = 10$ and (b) $\tan\beta = 40$ in the constrained MSSM (mSUGRA) scenario. The allowed region by the cosmological neutral dark matter constraint (7.22) is shown by the black parabolic shaped region. The disallowed region where $m_{\tilde{\tau}_1} < m_\chi$ has dark shading. The regions excluded by $b \to s\gamma$ have medium shading (left). The $g_\mu - 2$ favored region at the 2 σ $[(287 \pm 182) \times 10^{-11}]$ (between dashed lines the 1 σ $[(287 \pm 91) \times 10^{-11}]$ band) level has medium shading. The LEP constraint on $m_{\chi^\pm} = 104$ GeV and $m_h = 114$ GeV are shown as near vertical lines. Plot courtesy of K. Olive updated from [61]

specific assumptions made, leading to the MSSM and specific versions of it, may not be realized in nature and other variants may be needed. In particular, one should keep in mind that $b \to s\gamma$ branching fractions are rather assumption dependent ("SUSY flavor" and "SUSY CP" problems). In contrast, the possible SUSY contribution to a_μ is quite universal, provided $\tan\beta$ – enhancement is at work. As mentioned, the fact that there is a non–empty region in SUSY parameter space, is remarkable and an interesting input for SUSY phenomenology at the colliders LHC and ILC. Thereby the a_μ constraint is crucial as it definitely requires new particles "around the corner", if we assume that the deviation is real.

7.3 Perspectives for the Future

The electron's spin and magnetic moment were evidenced from the deflection of atoms in an inhomogeneous magnetic field and the observation of fine structure by optical spectroscopy [80, 81]. Ever since, magnetic moments and g–values of particles in general and the $g - 2$ experiments with the electron and the muon in particular together with high precision atomic spectroscopy have played a central role in establishing the modern theoretical framework for particle physics: relativistic quantum field theory in general and quantum electrodynamics in particular the prototype theory which developed further into the SM of electromagnetic, weak and strong interactions based on a local gauge principle and spontaneous symmetry breaking[18]. Not only particle physics, also precision atomic physics and nuclear theory are based on relativistic QFT methods[19].

New milestones have been achieved now with the BNL muon $g - 2$ experiment together with the Harvard electron $g - 2$ experiment. Both experiments exploited all ingenuity to reach the next level of precisions and together with theory efforts maybe the next level of understanding of how it works. On the theory side, what we learn from the BNL experiment and from possible succeeding experiments will depend on how well we can solidify the theoretical prediction. There is certainly common agreement that the hadronic light–by–light scattering contribution is the most problematic one, since no theoretically established method exists to calculate this contribution in a model independent way so far.

A big hope for the long term future are the non–perturbative calculations of electromagnetic current correlators by means of lattice QCD [82]. This

---

[18]With local gauge group $SU(3)_c \otimes SU(2)_L \otimes U(1)_Y$ spontaneously broken to $SU(3)_c \otimes U(1)_{em}$.

[19]Not to forget the role of QFT for other systems of infinite (large) numbers of degrees of freedom: condensed matter physics and critical phenomena in phase transitions. The Higgs mechanism as a variant of the Ginzburg-Landau theory of superconductivity and the role QFT and the renormalization group play in the theory of phase transitions are good examples for synergies between elementary particle physics and condensed matter physics.

has to go in steps from two–point amplitudes (vacuum polarization and/or Adler function) to three–point form factors (NP effects in VVA correlator) and the four–point function linked to light–by–light scattering. Very important is to watch for more experimental information for better modeling by effective theories. An example is the $\pi^0\gamma^*\gamma^*$ form factor for both photons off–shell or direct light–by–light scattering in $e^+e^- \to e^+e^-\gamma^*\gamma^* \to e^+e^-\gamma\gamma$ or $e^+e^-\gamma^*\gamma$ with the virtual final state photon converting to a pair.

As a worst case scenario, the a_μ measurement may be used one day to fix the value of the hadronic light–by–light contribution after physics at colliders will have established the relevant part of the spectrum of physics beyond the SM. Thus confronting $a_\mu = a_\mu(\text{SM; predictable}) + a_\mu(\text{SM; unpredictable}) + a_\mu(\text{NP})$, making the reasonable assumption that the relevant NP part is also calculable in a then more or less established extension of the SM, we may determine $a_\mu(\text{SM; unpredictable})$ by comparing this "prediction" with the experimental result a_μ^{exp}. An alternative strategy, proposed by Remiddi some time ago, would be to increase the precision of independent measurements of α and a_e by a factor 20 each together with improved QED and SM calculations and determine $a_\mu(\text{unpredicted})$ utilizing the fact that this is proportional to m_l^2: thus from

$$a_\mu = a_\mu(\text{predicted}) + a_\mu(\text{unpredicted})$$
$$a_e = a_e(\text{predicted}) + (m_e/m_\mu)^2 \times a_\mu(\text{unpredicted})$$

we could determine $a_\mu(\text{unpredicted})$. This would include such unaccounted new physics which also scales proportional to m_l^2.

The hadronic vacuum–polarization in principle may be substantially improved by continuing $e^+e^- \to$ hadrons cross–section measurements with higher precision. Existing deviations at the few % level between the KLOE result on the one hand and the CMD-2 and SND results[20] on the other hand and the deviations at the 10% level between e^+e^-–data and the appropriately corrected hadronic τ–decay spectral functions, are waiting for being clarified before we can fully trust that we understand what we are doing. Fortunately, new efforts are in progress at Beijing with BES-III [83], Cornell CLEO-c [84] and Novosibirsk with VEPP-2000 [85] as well as ongoing measurements at the B–factories (radiative return) at SLAC and KEK and

---

[20] The KLOE measurement is a radiative return measurement witch is a next to leading order approach. On the theory side one expects that the handling of the photon radiation requires one order in α more than the scan method for obtaining the same accuracy. Presently a possible deficit is on the theory side. What is urgently needed are full $O(\alpha^2)$ QED calculations, for Bhabha luminosity monitoring, μ–pair production as a reference and test process, and π–pair production in sQED as a first step and direct measurements of the final state radiation from hadrons. The CMD-2 and SND measurements take data at the same accelerator (same luminosity/normalization uncertainties) and use identical radiative corrections, such that for that part they are strongly correlated and this should be taken into account appropriately in combining the data.

Frascati is discussing plans to set up a new scan experiment going up to 2.5 GeV with DANAE/KLOE-2 [86]. The need for a more precise determination of the hadronic vacuum polarization effects for the future precision physics projects including a new high energy e^+e^-–annihilation facility, the International Linear Collider ILC, has been elaborated recently in [4].

There is no doubt that performing doable improvements on both the theory and the experimental side allows to substantially sharpen (or diminish) the apparent gap between theory and experiment. Yet, even the present situation gives ample reason for speculations. No other experimental result has as many problems to be understood in terms of SM physics. One point should be noted in this context, however. An experiment at that level of accuracy, going one order of magnitude beyond any previous experiment, is a real difficult enterprise and only one such experiment has been performed so far. There is also a certain possibility to overlook some new problem which only shows up at higher precision and escaped the list of explicitly addressed problems by the experiment. It is for instance not 100% clear that what is measured in the experiment is precisely what theoreticians calculate. For example, it is believed that, because radiative corrections in $g-2$ are infrared finite to all orders, real photon radiation can be completely ignored, in spite of the fact that we know that due to the electric interaction via charges a naive S–matrix in QED does not exist. Muons, like any charged particles, produce and absorb continuously photon radiation and therefore are dressed by a photon cloud which is thought not to affect the $g-2$ measurement. The question has been addressed to leading order by Steinmann [87]. Possible effects at higher orders have not been estimated to my knowledge. Also multiple interactions with the external field usually are not accounted for, beyond the classical level. One also should keep in mind that the muon is unstable and the on-shell projection technique (see Sect. 3.5) usually applied in calculating a_μ in principle has its limitation. As $\Gamma_\mu \simeq 3 \times 10^{-16}$ MeV $\ll m_\mu \simeq 105.658$ MeV, it is unlikely that treating the muon to be stable could cause any problem.

Another question one may ask is whether the measurement of the magnetic field strength could not change the magnitude of the field by a tiny but non–negligible amount[21]. On the theory side one should be aware that the important 4–loop contribution has not been cross–checked by a completely independent calculation. What about the hadronic vacuum polarization, in the unlikely case that the τ–data are the right answer and not the e^+e^-–data a substantial part of the difference would disappear, not to forget the problems with the light–by–light calculation. Nonetheless, according to the best of our knowledge, the present status of both theory and experiment is as reflected by the systematic errors which have been estimated. Therefore most probably, the difference must be considered as a real indication of a missing piece on the theory side.

---

[21] Of course such questions have been carefully investigated, and a sophisticated magnetic probe system has been developed by the E821 collaboration.

The anomalous magnetic moment of the muon is a beautiful example of "the closer we look the more we see"[22], however, the efforts to dig even deeper into the structure of matter remains a big adventure also in future.

The $g - 2$ measurement is like a peek through the keyhole, you see at the same time an overlay of all things to a certain depth in one projection, but to make sure that what you see is there, you have to open the door and go to check. This will be a matter of accelerator physics, and an ILC would be the preferred and ideal facility to clarify the details. Of course and luckily, the LHC will tell us much sooner the gross direction new physics will go and it is expected to reach the physics at much higher scales. But, it is not the physics at the highest scales you see first in $g - 2$ as we learned by the above considerations.

The Muon Storage Ring experiment on $g_\mu - 2$ and similarly the Penning Trap experiment on $g_e - 2$ are like microscopes which allow us to look into the subatomic world and the scales which we have reached with a_μ is about 100 GeV, i.e., the scale of the weak gauge bosons W and Z which is the LEP energy scale (as a_μ is effectively by a factor 52 more sensitive to new heavy physics the mass scale which is tested by a_e is about $100/\sqrt{52} \sim 14$ GeV only, an energy region which we know as it has been explored by other means). So $g_e - 2$, if it would not be used to determine α, would be the ideal observable to test the deeper quantum levels of a known theory which is QED as well as the rest of the SM in this case.

Remember that at LEP-I by electron–positron annihilation predominantly "heavy light" particles Z or at LEP-II predominantly W^+W^-–pairs have been produced, states which were produced in nature mostly in the very early universe[23].

Particle accelerators and storage rings are microscopes which allow us to investigates the nature in the subatomic range at distance $< 10^{-15}$ m and at the same time have the aim to directly produce new forms of matter, by pair creation, for example. The size of such machines is essentially determined by two parameters: the energy which determines the resolution $\lambda = hc/E_{c.m.} \simeq 1.2 \text{GeV}/E_{c.m.}(\text{GeV}) \times 10^{-15}$ m and the collision rate $\Delta N/\Delta t = L \times \sigma \simeq 10^{32}\, \sigma(\text{cm}^2)/\text{cm}^2$ sec (luminosity L as for LEP). Usually projectiles must be stable particles or antiparticles like electrons, positrons and protons and antiprotons. The Muon Storage Ring experiments work with the rather unstable muons which are boosted to highly relativistic quasi–stable

---

[22] which is not always true, for example if we read a newspaper or if you read this book.

[23] This time is given by (see (19.43) in [88]) $t = 2.4/\sqrt{N(T)}\, (1\text{ MeV}/kT)^2$ sec. Here $N(T) = \sum_{\text{bosons } B} g_B(T) + \frac{7}{8} \sum_{\text{fermions } F} g_F(T)$ is the effective number of degrees of freedom excited at temperature T, where $g_{B/F}(T)$ is the number of bosonic/fermionic degrees of freedom in the massless limit. For $m_b \ll kT \simeq M_W$ all SM particles except $W^\pm, Z, H$ and the top quark t are contributing. Counting spin, color and charge appropriately gives $N(T) = 345/4$, which yields the time $t \sim 0.3 \times 10^{-10}$ sec after the Big Bang for $T \sim 100$ GeV.

muons well selected in energy and polarization before they are injected into the storage ring which more acts as a detector rather than an accelerator as it usually does in the case of typical high energy machines. This allows to study the motion of the muons at incredible precision with very little background.

A very very different type of experiment will start soon at the LHC, which is kind of the other extreme in conditions. At LHC one will produce enormous amounts of events, billions per second, of which the overwhelming part of events are too complex to be understood and the interesting "gold platted" events which will tell us about new physics have to be digged out like "searching for a needle in a haystack". Nevertheless, the physics there hopefully tells us what we see in $g-2$. At LEP a big machine was able to measure about 20 different observables associated with different final states at the level of 1ppm. The strength of the LHC is that it will enable us to go far beyond what we have reached so far in the energy scale.

So we hope we may soon add more experimentally established terms to the SM Lagrangian and extent our predictions to include the yet unknown. Thats how it worked in the past with minimal extensions on theoretical grounds. Why this works so successfully nobody really knows. One observes particles, one associates with it a field, interactions are the simplest non–trivial products of fields (triple and quartic) at a spacetime point, one specifies the interaction strength, puts everything into a renormalizable relativistic QFT and predicts what should happen and it "really" happened essentially without exception. Maybe the muon $g-2$ is the first exception!

This book tried to shed light on the physics encoded in a single real number. Such a single number in principle encodes an infinity of information, as each new significant digit (each improvement should be at least by a factor ten in order to establish the next significant digit) is a new piece of information. It is interesting to ask, what would we know if we would know this number to infinite precision. Of course one cannot encode all we know in that single number. Each observable is a new view to reality with individual sensitivity to the deep structure of matter. All these observables are cornerstones of *one reality* unified self–consistently to our present knowledge by the knowledge of the Lagrangian of a renormalizable quantum field theory. Theory and experiments of the anomalous magnetic moment are one impressive example what it means to understand physics at a fundamental level. The muon $g-2$ reveals the major ingredients of the SM and as we know now maybe even more.

On the theory part the fascinating thing is the technical complexity of higher order SM (or beyond) calculations of in the meantime thousands of diagrams which can only be managed by the most powerful computers in analytical as well as in the numerical part of such calculations. This book only gives little real insight into the technicalities of such calculation. Performing higher order Feynman diagram calculation could look like formal nonsense but at the end results in a number which experimenters indeed measure. Much of theoretical physics today takes place beyond the Galilean rules, namely that sensible predictions must be testable. With the anomalous magnetic moment at least we still follow the successful tradition set up by Galileo, we definitely can check it, including all the speculations about it.

A next major step in this field of research would be establishing experimentally the electric dipole moment. This seems to be within reach thanks to a breakthrough in the experimental techniques. The electric dipole moments are an extremely fine monitor for CP violation beyond the SM which could play a key role for understanding the origin of the baryon matter–antimatter asymmetry in the universe.

But first, we are waiting for the LHC data to tell us where we go!

References

1. J. Bailey et al., Nucl. Phys. B **150** (1979) 1
2. G. W. Bennett et al. [Muon g-2 Collaboration], Phys. Rev. D **73** (2006) 072003
3. T. Kinoshita, M. Nio, Phys. Rev. D **73** (2006) 053007
4. F. Jegerlehner, Nucl. Phys. Proc. Suppl. **162** (2006) 22 [hep-ph/0608329]
5. J. Bijnens, E. Pallante, J. Prades, Phys. Rev. Lett. **75** (1995) 1447 [Erratum-ibid. **75** (1995) 3781]; Nucl. Phys. B **474** (1996) 379 [Erratum-ibid. **626** (2002) 410]
6. M. Hayakawa, T. Kinoshita, Phys. Rev. D **57** (1998) 465 [Erratum-ibid. D **66** (2002) 019902]
7. M. Knecht, A. Nyffeler, Phys. Rev. D **65** (2002) 073034
8. K. Melnikov, A. Vainshtein, Phys. Rev. D **70** (2004) 113006
9. A. Czarnecki, W. J. Marciano, A. Vainshtein, Phys. Rev. D **67** (2003) 073006; M. Knecht, S. Peris, M. Perrottet, E. de Rafael, JHEP **0211** (2002) 003; E. de Rafael, *The muon g–2 revisited*, hep-ph/0208251
10. S. Heinemeyer, D. Stöckinger, G. Weiglein, Nucl. Phys. B **699** (2004) 103; T. Gribouk, A. Czarnecki, Phys. Rev. D **72** (2005) 053016
11. H. N. Brown et al. [Muon (g-2) Collaboration], Phys. Rev. Lett. **86** (2001) 2227
12. S. Eidelman, F. Jegerlehner, Z. Phys. C **67** (1995) 585; F. Jegerlehner, In: *Radiative Corrections*, ed by J. Solà (World Scientific, Singapore 1999) p. 75
13. M. Davier, S. Eidelman, A. Höcker, Z. Zhang, Eur. Phys. J. C **27** (2003) 497; Eur. Phys. J. C **31** (2003) 503
14. S. Ghozzi, F. Jegerlehner, Phys. Lett. B **583** (2004) 222
15. S. Narison, Phys. Lett. B **568** (2003) 231
16. K. Hagiwara, A. D. Martin, D. Nomura, T. Teubner, Phys. Lett. B **557** (2003) 69; Phys. Rev. D **69** (2004) 093003
17. J. F. De Troconiz, F. J. Yndurain, Phys. Rev. D **65** (2002) 093001; Phys. Rev. D **71** (2005) 073008
18. S. Eidelman, Proceedings of the XXXIII International Conference on High Energy Physics, July 27 – August 2, 2006, Moscow (Russia), World Scientific, to appear; M. Davier, Nucl. Phys. Proc. Suppl. **169** (2007) 288
19. K. Hagiwara, A. D. Martin, D. Nomura, T. Teubner, Phys. Lett. B **649** (2007) 173
20. B. L. Roberts Nucl. Phys. B (Proc. Suppl.) **131** (2004) 157; R. M. Carey et al., Proposal of the BNL Experiment E969, 2004; J-PARC Letter of Intent L17

21. H. Leutwyler, *Electromagnetic form factor of the pion*, hep-ph/0212324; G. Colangelo, Nucl. Phys. Proc. Suppl. **131** (2004) 185; ibid. **162** (2006) 256
22. K. G. Chetyrkin, J. H. Kühn, Phys. Lett. B **342** (1995) 356; K. G. Chetyrkin, J. H. Kühn, A. Kwiatkowski, Phys. Rept. **277** (1996) 189; K. G. Chetyrkin, J. H. Kühn, M. Steinhauser, Phys. Lett. B **371**, 93 (1996); Nucl. Phys. B **482**, 213 (1996); **505**, 40 (1997); K.G. Chetyrkin, R. Harlander, J.H. Kühn, M. Steinhauser, Nucl. Phys. B **503**, 339 (1997) K. G. Chetyrkin, R. V. Harlander, J. H. Kühn, Nucl. Phys. B **586** (2000) 56 [Erratum-ibid. B **634** (2002) 413]
23. R. V. Harlander, M. Steinhauser, Comput. Phys. Commun. **153** (2003) 244
24. M. Davier, A. Höcker, Z. Zhang, Nucl. Phys. Proc. Suppl. **169** (2007) 22
25. R. L. Garwin, D. P. Hutchinson, S. Penman, G. Shapiro, Phys. Rev. **118** (1960) 271
26. G. Charpak, F. J. M. Farley, R. L. Garwin, T. Muller, J. C. Sens, V. L. Telegdi, A. Zichichi, Phys. Rev. Lett. **6** (1961) 128; G. Charpak, F. J. M. Farley, R. L. Garwin, T. Muller, J. C. Sens, A. Zichichi, Nuovo Cimento **22** (1961) 1043
27. G. Charpak, F. J. M. Farley, R. L. Garwin, T. Muller, J. C. Sens, A. Zichichi, Phys. Lett. 1B (1962) 16; Nuovo Cimento **37** (1965) 1241
28. F. J. M. Farley, J. Bailey, R. C. A. Brown, M. Giesch, H. Jöstlein, S. van der Meer, E. Picasso, M. Tannenbaum, Nuovo Cimento **45** (1966) 281
29. J. Bailey et al., Phys. Lett. B **28** (1968) 287; Nuovo Cimento A **9** (1972) 369
30. J. Bailey et al. [CERN Muon Storage Ring Collaboration], Phys. Lett. B **55** (1975) 420; Phys. Lett. B **67** (1977) 225 [Phys. Lett. B **68** (1977) 191]; J. Bailey et al. [CERN-Mainz-Daresbury Collaboration], Nucl. Phys. B **150** (1979) 1
31. R. M. Carey et al. [Muon (g-2) Collaboration], Phys. Rev. Lett. **82** (1999) 1632
32. H. N. Brown et al. [Muon (g-2) Collaboration], Phys. Rev. D **62** (2000) 091101
33. G. W. Bennett et al. [Muon (g-2) Collaboration], Phys. Rev. Lett. **89** (2002) 101804 [Erratum-ibid. **89** (2002) 129903]
34. G. W. Bennett et al. [Muon (g-2) Collaboration], Phys. Rev. Lett. **92** (2004) 161802
35. J. C. Pati, A. Salam, Phys. Rev. Lett. **31** (1973) 661; Phys. Rev. D **8** (1973) 1240; H. Georgi, S. L. Glashow, Phys. Rev. Lett. **32** (1974) 438
36. A. Galli, Nuovo Cim. A **106** (1993) 1309
37. J. R. Ellis, S. Kelley, D. V. Nanopoulos, Phys. Lett. B **249** (1990) 441; ibid **260** (1991) 131; U. Amaldi, W. de Boer, H. Fürstenau, Phys. Lett. B **260** (1991) 447; P. Langacker, M. x. Luo, Phys. Rev. D **44** (1991) 817
38. LEP Electroweak Working Group (LEP EWWG), http://lepewwg.web.cern.ch/LEPEWWG/plots/summer2006
[ALEPH, DELPHI, L3, OPAL, SLD Collaborations], *Precision electroweak measurements on the Z resonance*, Phys. Rept. **427** (2006) 257 [hep-ex/0509008]; http://lepewwg.web.cern.ch/LEPEWWG/Welcome.html
39. J. Erler, P. Langacker, *Electroweak model and constraints on new physics* in W. M. Yao et al. [Particle Data Group], J. Phys. G **33** (2006) 1
40. Heavy Flavor Averaging Group (HFAG), http://www.slac.stanford.edu/xorg/hfag/ http://www-cdf.fnal.gov/physics/new/bottom/bottom.html
41. I. B. Khriplovich, S. K. Lamoreaux, CP Violation Without Strangeness: Electric Dipole Moments Of Particles, Atoms, And Molecules, (Springer, Berlin, 1997)
42. F. J. M. Farley et al., Phys. Rev. Lett. **93** (2004) 052001; M. Aoki et al. [J-PARC Letter of Intent]: *Search for a Permanent Muon Electric Dipole Moment at the* $\times 10^{-24}$ $e \cdot cm$ *Level*, http://www-ps.kek.jp/jhf-np/LOIlist/pdf/L22.pdf

43. A. Adelmann, K. Kirch, hep-ex/0606034
44. M. Czakon, J. Gluza, F. Jegerlehner, M. Zrałek, Eur. Phys. J. C **13** (2000) 275
45. A. Czarnecki, W. J. Marciano, Phys. Rev. D **64** (2001) 013014
46. B. E. Lautrup, A. Peterman, E. de Rafael, Phys. Reports 3C (1972) 193
47. S. Ritt [MEG Collaboration], Nucl. Phys. Proc. Suppl. **162** (2006) 279
48. R. Barbieri, L. J. Hall, Phys. Lett. B **338** (1994) 212; R. Barbieri, L. J. Hall, A. Strumia, Nucl. Phys. B **445** (1995) 219; J. Hisano, D. Nomura, Phys. Rev. D **59** (1999) 116005
49. R. Van Royen, V. F. Weisskopf, Nuovo Cim. A **50** (1967) 617 [Erratum-ibid. A **51** (1967) 583]
50. C. T. Hill, E. H. Simmons, Phys. Rept. **381** (2003) 235 [Erratum-ibid. **390** (2004) 553]
51. F. Jegerlehner, Helv. Phys. Acta **51** (1978) 783; F. Jegerlehner, The 'etherworld' and elementary particles. In: *Theory of Elementary Particles*, ed by H. Dorn, D. Lüst, G. Weight, (WILEY-VCH, Berlin, 1998) p. 386, hep-th/9803021
52. F. Jegerlehner, Nucl. Phys. B (Proc. Suppl.) **37** (1994) 129
53. P. Mery, S. E. Moubarik, M. Perrottet, F. M. Renard, Z. Phys. C **46** (1990) 229
54. W. M. Yao et al. [Particle Data Group], J. Phys. G **33** (2006) 1
55. LEP Electroweak Working Group (LEP EWWG), http://lepewwg.web.cern.ch/LEPEWWG/lepww/tgc/
56. J. Wess, B. Zumino, Nucl. Phys. B **70** (1974) 39; R. Haag, J. T. Lopuszanski, M. Sohnius, Nucl. Phys. B **88** (1975) 257.
57. D. Z. Freedman, P. van Nieuwenhuizen, S. Ferrara, Phys. Rev. D **13** (1976) 3214; S. Deser, B. Zumino, Phys. Lett. B **62** (1976) 335
58. H. P. Nilles, Phys. Rep. **110** (1984) 1; H. E. Haber, G. L. Kane, Phys. Rep. **117** (1985) 75; L. Ibáñez, Beyond the Standard Model. In: CERN Yellow Report, CERN 92-06 (1992) 131–237
59. J. R. Ellis, J. S. Hagelin, D. V. Nanopoulos, K. A. Olive, M. Srednicki, Nucl. Phys. B **238** (1984) 453
60. C. L. Bennett et al. [WMAP Collaboration], Astrophys. J. Suppl. **148** (2003) 1; D. N. Spergel et al. [WMAP Collaboration], Astrophys. J. Suppl. **148** (2003) 175
61. J. R. Ellis, K. A. Olive, Y. Santoso, V. C. Spanos, Phys. Lett. B **565** (2003) 176; Phys. Rev. D **71** (2005) 095007
62. H. Baer, A. Belyaev, T. Krupovnickas, A. Mustafayev, JHEP **0406** (2004) 044; J. Ellis, S. Heinemeyer, K. A. Olive, G. Weiglein, *Indications of the CMSSM mass scale from precision electroweak data*, hep-ph/0604180
63. H. E. Haber, R. Hempfling, Phys. Rev. Lett. **66** (1991) 1815
64. R. Hempfling, A. H. Hoang, Phys. Lett. B **331** (1994) 99; S. Heinemeyer, W. Hollik, G. Weiglein, Phys. Lett. B **455** (1999) 179; Phys. Rept. **425** (2006) 265
65. S. Ferrara, E. Remiddi, Phys. Lett. B **53** (1974) 347
66. J. L. Lopez, D. V. Nanopoulos, X. Wang, Phys. Rev. D **49** (1994) 366; U. Chattopadhyay, P. Nath, Phys. Rev. D **53** (1996) 1648; T. Moroi, Phys. Rev. D **53** (1996) 6565 [Erratum-ibid. D **56** (1997) 4424]
67. S. P. Martin, J. D. Wells, Phys. Rev. D **64** (2001) 035003
68. D. Stöckinger, J. Phys. G: Nucl. Part. Phys. 34 (2007) 45 [hep-ph/0609168]
69. G. Degrassi, G. F. Giudice, Phys. Rev. **58D** (1998) 053007
70. M. Krawczyk, PoS **HEP2005** (2006) 335 [hep-ph/0512371]

71. S. Heinemeyer, D. Stöckinger, G. Weiglein, Nucl. Phys. B **690** (2004) 62; ibid **699** (2004) 103
72. T. F. Feng, X. Q. Li, L. Lin, J. Maalampi, H. S. Song, Phys. Rev. D **73** (2006) 116001
73. M. J. Ramsey-Musolf, S. Su, *Low energy precision test of supersymmetry*, hep-ph/0612057
74. S. Heinemeyer, W. Hollik, D. Stöckinger, A. M. Weber, G. Weiglein, JHEP **0608** (2006) 052 [hep-ph/0604147]
75. M. Awramik, M. Czakon, A. Freitas, G. Weiglein, Phys. Rev. D **69** (2004) 053006
76. M. Misiak et al., Phys. Rev. Lett. **98** (2007) 022002
77. R. Barbieri, G. F. Giudice, Phys. Lett. B **309** (1993) 86; M. Carena, D. Garcia, U. Nierste, C. E. M. Wagner, Phys. Lett. B **499** (2001) 141
78. L. F. Abbott, P. Sikivie, M. B. Wise, Phys. Rev. D **21** (1980) 1393; M. Ciuchini, G. Degrassi, P. Gambino, G. F. Giudice, Nucl. Phys. B **527** (1998) 21
79. S. P. Martin, J. D. Wells, Phys. Rev. D **67** (2003) 015002
80. W. Gerlach, O. Stern, Zeits. Physik **8** (1924) 110
81. G. E. Uhlenbeck, S. Goudsmit, Naturwissenschaften **13** (1925) 953; Nature **117** (1926) 264
82. C. Aubin, T. Blum, Nucl. Phys. Proc. Suppl. **162** (2006) 251
83. F. A. Harris, Nucl. Phys. Proc. Suppl. **162** (2006) 345
84. S. A. Dytman [CLEO Collaboration], Nucl. Phys. Proc. Suppl. **131** (2004) 32
85. S. Eidelman, Nucl. Phys. Proc. Suppl. **162** (2006) 323
86. F. Ambrosino et al., Eur. Phys. J. C **50** (2007) 729; G. Venanzoni, Nucl. Phys. Proc. Suppl. **162** (2006) 339
87. O. Steinmann, Commun. Math. Phys. **237** (2003) 181
88. K. A. Olive, J. A. Peacock, Big-Bang cosmology, in S. Eidelman et al. [Particle Data Group], Phys. Lett. B **592** (2004) pp. 191–201

List of Acronyms*

ABJ	Adler-Bell-Jackiw (anomaly)
AF	Asymptotic Freedom
AGS	Alternating Gradient Synchrotron
BNL	Brookhaven National Laboratory
BO	Betatron Oscillations
BPP	Bijnens-Pallante-Prades
BW	Breit-Wigner (resonance)
C	Charge-conjugation
CC	Charged Current
CDM	Cold Dark Matter
CERN	European Organization for Nuclear Research
CHPT	Chiral Perturbation Theory
CKM	Cabibbo-Kobayashi-Maskawa

---

[23]*KLOE, CMD, SND, MD, BaBar, Belle, BES, E821, NA7, CLEO, CELLO, TASSO are names of detectors, experiments or collaborations see Tab. 5.1. ALEPH, DELPHI, L3 and OPAL are LEP detector/collaborations, CDF and D0 are TEVATRON detectors/collaborations.

List of Acronyms

C.L. .... Confidence Level

CM or c.m. .... Center of Mass

CP .... parity × charge-conjugation (symmetry)

CPT .... time-reversal × parity × charge-conjugation (symmetry)

CQM .... Constituent Quark Model

CS .... Callan-Symanzik

CVC .... Conserved Vector Current

DESY .... Deutsches Elektronen-Synchrotron

DIS .... Deep Inelastic Scattering

DR .... Dispersion Relation/Dimensional Regularization

ED .... Extra Dimension ($D - 4 \geq 1$)

EDM .... Electric Dipole Moment

EFT .... Effective Field Theory

em .... electromagnetic

ENJL .... Extended Nambu-Jona-Lasinio (model)

EW .... Electro Weak

EWSB .... Electro Weak Symmetry Breaking

exp (suffix/index) .... experimental

FCNC .... Flavor Changing Neutral Currents

FNAL .... Fermi National Accelerator Laboratory (Batavia, Illinois, USA)

FP .... Faddeev-Popov (Lagrangian)

F.P. .... Finite Part (integral)

List of Acronyms

FSR . . . . Final State Radiation

GF . . . . Gauge Fixing (Lagrangian)

GOR . . . . Gell-Mann, Oakes and Renner

GS . . . . Gounaris-Sakurai (parametrization)

h.c. . . . . hermitian conjugate

HFS . . . . Hyper Fine Structure

HK . . . . Hayakawa-Kinoshita

HKS . . . . Hayakawa-Kinoshita-Sanda

HLS . . . . Hidden Local Symmetry

H.O. or HO . . . . Higher Order

ILC . . . . International Linear Collider (future e^+e^- collider)

IR . . . . InfraRed

ISR . . . . Initial State Radiation

J-PARC . . . . Japan Proton Accelerator Research Complex

KEK . . . . High Energy Accelerator Research Organization, KEK, Japan

KLN . . . . Kinoshita-Lee-Nauenberg

KN . . . . Knecht-Nyffeler

KNO . . . . Kinoshita-Nizic-Okamoto

LAMPF . . . . Los Alamos Meson Physics Facility

LbL . . . . Light-by-Light

L.D. or LD . . . . Long Distance

LEP . . . . Large Electron Positron (collider)

List of Acronyms

LHC .... Large Hadron Collider

LL .... Leading Logarithm

LMD .... Leading Meson Dominance

LNC .... Large N_c

L.O. or LO .... Lowest Order (Leading Order)

LOSP .... Lightest Observable SUSY Particle

LSP .... Lightest Supersymmetric Particle

LSZ .... Lehmann, Symanzik, Zimmermann

MS .... Minimal Subtraction

μSR .... Muon Storage Ring

MV .... Melnikov-Vainshtein

NC .... Neutral Current

NJL .... Nambu-Jona-Lasinio (model)

NLL .... Next to Leading Logarithm

NMR .... Nuclear Magnetic Resonance

NP .... New Physics/Non-Perturbative

1PI .... One Particle Irreducible

OPE .... Operator Product Expansion

P .... Parity (Space-reflection)

PCAC .... Partially Conserved Axialvector Current

PMT .... Photo Multiplier Tube

pQCD .... perturbative QCD

List of Acronyms

PSI	Paul Scherrer Institut
QCD	Quantum Chromodynamics
QED	Quantum Electrodynamics
QFT	Quantum Field Theory
QM	Quantum Mechanics
QPM	Quark Parton Model
RG	Renormalization Group
RLA	Resonance Lagrangian Approach
S.D. or SD	Short Distance
SD	Standard Deviation (1 SD = 1 σ)
SLAC	Stanford Linear Accelerator Center
SM	Standard Model
sQED	scalar QED
SSB	Spontaneous Symmetry Breaking
SUGRA	Supergravity
SUSY	Supersymmetry
SVZ	Shifman-Vainshtein-Zakharov
T	Time-reversal
the (suffix/index)	theoretical
TEVATRON	TeV Proton-Antiproton Collider at FNAL
TDHM	Two Doublet Higgs Model
UV	UltraViolet

VEV	Vacuum Expectation Value
VMD	Vector Meson Dominance
VP	Vacuum Polarization
VVA	Vector-Vector-Axialvector (amplitude)
WMAP	Wilkinson Microwave Anisotropy Probe
WT	Ward-Takahashi (identity)
WZW	Wess-Zumino-Witten (Lagrangian)
YM	Yang-Mills

Index

Adler–function, 198, 267, 287, 290
Adler-Bell-Jackiw anomaly, 129, 158, 161, 230, 231, 233
a_e
  experiment, 369–373
  experimental value, 144, 163, 165
  lowest order result, 99
  QED prediction, 164
  SM prediction, 165
  theory, 162–166
a_μ, 11
  experiment, 143, 347–352
  experimental value, 366, 377
  hadronic contribution
    leading, 155, 156, 294, 379
    subleading, 157, 312
  hadronic light–by–light scattering, 153, 158, 316–341
  lowest order result, 99
  QED prediction, 166, 207–224
  SM prediction, 377
  theory, 166–168
  weak bosonic corrections, 256
  weak contribution, 14, 160, 162, 259
  weak fermionic corrections, 230
analyticity, 67, 69–71
anapole moment, 170
annihilation operator, 25, 46, 52
anomalous dimension, 108
anomalous precession, 359
anomaly
  cancellation, 161, 230, 233, 331, 393
anti–screening, 127

anti–unitarity, 30, 31, 304
anticommutation relations, 25
asymptotic condition, 52
asymptotic freedom, 8, 108, 109, 127, 284

baryon number
  conservation, 129
  violation, 382
betatron oscillations, 359
Bhabha scattering, 272
Bloch-Nordsieck prescription, 112, 282
Bohr magneton, 5, 6, 141
boost, 24, 38
Bose condensate, 34
bosons, 26
bremsstrahlung, 112, 293
  collinear, 119
  cuts, 270
  hard, 118, 282
  soft, 113, 282
    exponentiation, 119

C , see charge conjugation 8
Cabibbo-Kobayashi-Maskawa matrix , see CKM matrix 33
canonical scaling, 108
Casimir operator, 39
causality, 67, 69, 154
  Einstein, 29
charge conjugation, 8, 30
chiral
  currents, 239, 240
  fields, 27

perturbation theory, 129, 158, 239–242, 266, 300, 314
 symmetry, 129
 symmetry breaking, 239
chronological products , see time ordered products 46
CKM matrix, 33, 171, 226, 299, 363
Cold Dark Matter, 382, 397–403
color, 43, 125
 factor, 84, 226
commutation relations, 26
computer algebra, 13
confinement, 128
conformal invariance, 290
conformal mapping, 85, 305
constituent quarks, 239, 244, 301–320, 330
 masses, 231, 320
 model, 231
counter terms, 56
covariant derivative, 44, 121, 125
CP
 symmetry, 8
 violation, 171, 226, 384
CPT
 symmetry, 8, 9
 theorem, 9
creation operator, 25, 46, 52
cross–section, 53, 55
 bremsstrahlung, 113
 data, 273
 differential, 53
 dressed, 269, 284
 exclusive, 117, 272
 inclusive, 114, 117, 271
 total, 54, 270
 undressed, 269, 284, 293
crossing
 particle–antiparticle, 51
current
 conserved, 92, 239
 dilatation, 103
 electromagnetic, 30, 291
 partially conserved, 239
current quarks, 239
 masses, 146, 231
custodial symmetry, 385
Cutkosky rules, 190, 237
CVC, 239, 299

cyclotron frequency, 144, 359
cyclotron motion, 347

d' Alembert equation, 26
decay–rate, 53, 54
decay law, 54
decoupling, 208
decoupling theorem, 150
deep inelastic scattering, 109
detector acceptance, 270
detector efficiency, 270
dilatation current, 290
dipole moment, 5
 non–relativistic limit, 172
Dirac
 algebra, 27
  helicity representation, 43
  standard representation, 27
 equation, 27, 135
 field, 26, 29
 matrices, 27
 spinor, 27, 28
  adjoint, 28
dispersion integral, 155
dispersion relation, 181, 182
dispersive approach, 123
duality
 quark–hadron, 275
 quark–lepton, 230, 250
Dyson
 series, 78
 summation, 78, 183

electric dipole moment, 5, 9, 10, 33, 142, 170, 171, 226, 361, 385
electromagnetic
 current, 30, 45, 182
  hadronic, 291
 vertex, 92
electron
 charge, 45
 EDM, 172
 mass, 145
electron–positron annihilation, 155
e^+e^- cross–section, 273
 in pQCD, 284–286
e^+e^-–data, 155, 266–279
equation of motion, 135–140
error

correlations, 277
propagation, 277
Euclidean field theory, 67
exclusion principle, 29
exponentiation
　Coulomb singularity, 286
　soft photon, 116

factorization, 117, 315
Faddeev-Popov
　ghosts, 126, 228
　term, 126
fermion
　loops, 49
　strings, 49
fermions, 25
Feynman propagator, 47, 67
Feynman rules, 46
　EFT, 322
　QCD, 127
　QED, 47, 57
　resonance Lagrangian, 322
　sQED, 121
field
　left–handed, 27
　right–handed, 27
field strength tensor
　Abelian, 26
　dual, 234, 249
　electromagnetic, 26
　non–Abelian, 125, 235
Final State Radiation, 117, 293, 313
fine structure constant, 55, 145
　effective, 183, 185, 284, 297–298
flavor
　conservation, 226
　mixing, 226, 389
　violation, 388
Foldy-Wouthuysen transformation, 136
four–momentum, 27
　conservation, 51
four–spinor, 27
Fourier transformation, 27

g–factor, 6, 141, 355, 368, 372
gauge
　coupling, 44, 125, 226, 228, 392, 397
　Feynman, 78, 89, 90, 95, 232
　fixing, 126

　group, 44, 125
　invariance, 29, 44, 160, 171
　Landau, 104, 105
　parameter, 45, 126
　symmetry, 171
　unitary, 34, 160, 171, 228, 232
gauge theory
　Abelian, 8
　non–Abelian, 8
gauge boson
　masses, 145, 228
gauge transformation
　Abelian, 29
Gell-Mann Low formula, 46
gluons, 43, 125, 153, 238, 248, 251, 285, 319
　jet, 288
Goldstone bosons, 240, 241, 300
Gordon identity, 97, 169
GOR relation, 242, 253
Grand Unified Theory, 383, 396
　scale, 383, 396
Green function, 46, 52
　time ordered, 70

hadronic
　light–by–light scattering, 306
hadronic contribution, 197
hadronic effects, 13
hadronization, 287
handedness , see helicity 11
helicity, 11, 34, 235, 364, 365
Hermitian transposition, 30
hierarchy problem, 398
Higgs, 145, 161, 227, 228
　boson, 160, 225
　contribution, 229, 257, 259
　ghosts, 171, 228
　mass, 145, 227, 258
　mechanism, 34, 129, 171
　phase, 171
　two doublet model, 394
　vacuum expectation value, 227, 394
Hilbert space, 24

imaginary time, 67, 68
infrared behavior, 109
infrared problem, 44, 51, 88, 97, 99
infrared save, 120

424   Index

Initial State Radiation, 117, 269, 281, 282, 293
integral
　contour, 67, 181
　form factor, 73
　self–energy, 73
　tadpole, 73
interaction
　electromagnetic, 44
　final state, 305
　hadronic, 153, 265
　strong, 125, 153
　weak, 160, 225
invariance
　C, P, T, 31
　dilatation, 103
　gauge, 44
　relativistic, 24
　scale, 290
isospin
　symmetry, 299
　symmetry breaking, 300
isospin violation, 157

Jarlskog invariant, 171
jets, 288
　gluon jet, 288
　Sterman-Weinberg formula, 120

Kinoshita-Lee-Nauenberg, 117, 293
Klein-Gordon equation, 26, 71

Landau pole, 110, 112, 195
Larmor precession, 143
lattice QCD, 70, 129, 131, 267, 314, 406
lepton–quark family, 225, 231, 233
leptons, 161
　masses, 145
　quantum numbers, 225
lepton number
　violation, 382
lifetime, 54
light–by–light scattering, 151
　hadronic, 157, 312–341
　pion–pole dominance, 330
logarithm
　leading, 103
　Sudakov, 116
long distance behavior, 109
Lorentz

boost, 28, 41, 136
contraction, 24
factor, 349
force, 353, 357
invariant distance, 24
transformation, 24, 356
LSZ reduction formula, 52, 236
luminosity, 54

magnetic moment, 5, 139
　anomalous, 99, 141
meson
　exchange, 253
minimal
　coupling, 44
　substitution, 121
minimal subtraction, 63, 64
　scheme, 73
momenta
　non-exceptional, 101
$\mu^\pm$–decay, 364
muon
　EDM, 172, 384
　lifetime, 4
　magic momentum, 358
　magnetic precession, 355–358
　mass, 4, 145
　orbital motion, 352–355
　storage ring, 13, 143
muonium, 144
　hyperfine splitting, 368

Nambu-Jona-Lasinio, 319
naturalness problem, 398
non–perturbative effects, 248, 274, 288
non–relativistic limit, 135–140, 172–173
nuclear magnetic resonance, 144, 350

Omnès representation, 303
1PI , see one–particle irreducible 50
one–particle irreducible, 50, 77
one loop integrals, 387
　scalar, 72, 73
OPE, 241, 244–253, 274, 288, 333
operator product expansion , see OPE 241
optical theorem, 155
orderparameter, 251
oscillations

betatron, 355
magnetron, 371

parity, 8, 24, 30
   violation, 226, 350
partons, 287
Pauli
   equation, 135, 139
   matrices, 27
   term, 10, 60, 144
PCAC, 239, 332
Penning trap, 144, 369
   cylindrical, 373
perturbation expansion, 46
$\pi^{\pm}$–decay, 363
pion
   decay constant, 240
   form–factor, 276, 280, 282, 293, 300–306
      Brodsky-Lepage, 328
      data, 273
   mass, 145
   scattering, 304
      lengths, 305
      phase shift, 304
pitch correction, 361
Planck scale, 394, 398
Poincaré group, 24
   ray representation, 25
polarization, 28, 364, 365
polarization vector
   photon, 29
pole mass, 87
polylogarithms, 146
power-counting theorem
   Dyson, 101
   Weinberg, 101
precession frequency, 144

QCD
   asymptotic freedom, 155, 272
   perturbative, 155, 272
   renormalization group, 129
   running coupling, 154
QED in external field, 135
QPM , see quark parton model 154
Quantum Chromodynamics, 8, 48, 125–131, 155
Quantum Electrodynamics, 4, 44–46

quantum mechanics
   time evolution, 25
   transition probability, 24
quantum field theory, 3, 23–44
quantum mechanics
   state space, 24
quark
   condensates, 240, 242, 252, 274, 288
quark parton model, 154, 231, 285
quarks, 43, 125, 153, 161, 248, 285
   quantum numbers, 225

R–parity, 397
radial electric field correction, 360
radiation
   final state, 283, 293, 310
   initial state, 117, 269, 281, 282, 293
radiative corrections, 76, 116, 140, 207
radiative return, 281, 296
regularization, 47, 55
renormalizability, 60, 125, 144, 161, 225, 391
renormalization, 47, 55
   charge, 92
   coupling constant, 56
   group, 102
   mass, 56
   $\overline{MS}$ scheme, 110
   on–shell scheme, 110
   scale, 74
   theorem, 56
   wave function, 56, 184
renormalization group, 232
representation
   finite dimensional, 38
   fundamental, 37, 226, 383
   non–unitary, 38
   unitary, 38
resonance, 155, 287
   Breit-Wigner, 123
   narrow width, 123
ρ–meson, 155
$\rho - \omega$ mixing, 274, 302
ρ–parameter, 227, 400, 385
rotation, 24, 38
running α_s, 131
running charge, 184

S–matrix, 45, 51, 304

scaling, 108
s-channel, 272, 279
self-energy
    lepton, 86
    photon, 76
short distance behavior, 107
space-like, 185, 198, 267, 287
space-reflection , see parity 8
special Lorentz transformation , see boost 24
spectral condition, 26
spectral function, 187
spin, 6, 34–44
    operator, 6, 139
spinor
    representation, 29
spontaneous symmetry breaking, 171, 251, 300
Standard Model, 4, 14, 48, 224, 233
supersymmetry, 393–405

T , see time-reversal 8
T-matrix, 51
    element, 52
tadpole, 80
τ-data, 157, 298–300, 380
t-channel, 272
tensor
    antisymmetric, 24
    decomposition, 63
    energy momentum, 290
    integral, 74
    metric, 23
    permutation, 27
    vacuum polarization, 77
theorem
    Adler-Bardeen non–renormalization, 233, 238, 332
    Cauchy's, 181
    CPT, 31
    decoupling, 150, 208
    Furry's, 49
    Kinoshita-Lee-Nauenberg, 117, 293
    Noether's, 77
    optical, 122, 155, 189, 192
    Osterwalder-Schrader, 69
    renormalization, 56
    spin–statistics, 29, 40
    Watson's, 304
Thomas precession, 357

Thomson limit, 58, 184
threshold, 83, 287
time-like, 83
time-reversal, 8, 24, 30
time dilatation, 5, 143, 349
time ordered products, 46
translation, 24

ultraviolet behavior, 107
ultraviolet problem, 55, 58
unitarity, 46, 154, 304

vacuum, 25
vacuum expectation value, 34, 227
vacuum polarization, 76, 84, 148, 182
    hadronic, 154
Van Royen-Weisskopf formula, 390
vector–meson, 253, 293
    dominance, 125, 158, 312
vertex
    dressed, 50
    electromagnetic, 50
vertex functions, 100
VMD model, 312, 314, 320, 336

Ward-Takahashi identity, 59, 92
wave function, 27
weak
    gauge bosons, 160
    hadronic effects, 230, 233
    hypercharge, 225
    interaction, 225
    isospin, 225
Wess-Zumino-Witten Lagrangian, 158, 242
Wick
    ordering, 45, 47
    rotation, 67
Wigner state, 25

Yang-Mills
    structure, 160
    theory, 8, 225, 228
Yennie-Frautschi-Suura, 116
Yukawa
    coupling, 34, 396, 397
    interaction, 34, 395

Zeeman effect, 6
    anomalous, 6